Design of Structural Elements

Second Edition

*Concrete, steelwork, masonry and timber design to
British Standards and Eurocodes*

Design of Structural Elements

Second Edition

Concrete, steelwork, masonry and timber design to British Standards and Eurocodes

Chanakya Arya

Department of Civil and Environmental
Engineering, University College London, UK

Spon Press
Taylor & Francis Group

LONDON AND NEW YORK

First published 1994
by E & FN Spon

Reprinted 1994, 1995, 1997, 1998

Reprinted 2001, 2004
by Spon Press
11 New Fetter Lane, London EC4P 4EE

Simultaneously published in the USA and Canada
by Spon Press
29 West 35th Street, New York, NY 10001

Second edition first published 2003

Spon Press is an imprint of the Taylor & Francis Group

Typeset in Plantin by Graphicraft Limited, Hong Kong
Printed and bound in Great Britain by Alden Press, Oxford

British Library Cataloguing in Publication Data
A catalogue record for this book is available from the British Library

Library of Congress Cataloging in Publication Data
A catalogue record for this book has been requested

ISBN 0-415-26844-3 (Hbk)
ISBN 0-415-26845-1 (Pbk)

Contents

Contents

Contents

List of Worked Examples

Preface

Structural design is a key element of all degree and diploma courses in civil and structural engineering. It involves the study of principles and procedures contained in the latest codes of practice for structural design for a range of materials, including concrete, steel, masonry and timber.

Most textbooks on structural design consider only one construction material and, therefore, the student may end up buying several books on the subject. This is undesirable both from the viewpoint of cost but also because it makes it difficult for the student to unify principles of structural design, because of differing presentation approaches adopted by the authors.

There are a number of combined textbooks which include sections on several materials. However, these tend to concentrate on application of the codes and give little explanation of the structural principles involved or, indeed, an awareness of material properties and their design implications. Moreover, none of the books refer to the new Eurocodes for structural design, some of which were scheduled to replace the equivalent British Standards around 1998.

The purpose of this book, then, is to describe the background to the principles and procedures contained in the latest British Standards and Eurocodes on the structural use of concrete, steelwork, masonry and timber. It is primarily aimed at students on civil and structural engineering degree and diploma courses. Allied professionals such as architects, builders and surveyors will also find it appropriate. In so far as it includes five chapters on the structural Eurocodes it will be of considerable interest to practising engineers too.

The subject matter is divided into 11 chapters and 3 parts:

Part One   contains two chapters and explains the principles and philosophy of structural design, focusing on the limit state approach. It also explains how the overall loading on a structure and individual elements can be assessed, thereby enabling the designer to size the element.

Part Two   contains four chapters covering the design and detailing of a number of structural elements, e.g. floors, beams, walls, columns, connections and foundations to the latest British codes of practice for concrete, steelwork, masonry and timber design.

Part Three   contains five chapters on the Eurocodes for these materials. The first of these describes the purpose, scope and problems associated with drafting the Eurocodes. The remaining chapters describe the layout and contents of EC2, EC3, EC5 and EC6 for design in concrete, steelwork, timber and masonry respectively.

At the end of Chapters 1–6 a number of design problems have been included for the student to attempt, typical answers for which will subsequently be given in a companion to this book.

Although most of the tables and figures from the British Standards referred to in the text have been reproduced, it is expected that the reader will have either the full Standard or the publication *Extracts from British Standards for Students of Structural Design*, obtainable from BSI Sales, Linford Wood, Milton Keynes, Bucks, in order to gain the most from this book.

Chapters 1, 4 and 9 were co-authored by my colleague Mr Peter Wright to whom I would like to express my gratitude.

C. Arya
London
UK

Preface to the second edition

The main motivation for preparing this new edition was to update the text in *Chapters 4* and *6* on steel and timber design to conform with the latest editions of respectively BS 5950: Part 1 and BS 5268: Part 2. The opportunity has also been taken to add new material to *Chapters 3* and *4*. Thus, *Chapter 3* on concrete design now includes a new section and several new worked examples on the analysis and design of continuous beams and slabs. Examples illustrating the analysis and design of two-way spanning slabs and columns subject to axial load and bending have also been added. The section on concrete slabs has been updated. A discussion on flooring systems for steel framed structures is featured in *Chapter 4* together with a section and several worked examples on composite floor design.

Work on converting Parts 1.1 of the Eurocodes for concrete, steel, timber and masonry structures to full EN status is still ongoing. Until such time that these documents are approved the design rules in pre-standard form, designated by ENV, remain valid. The material in Chapters 8, 9 and 11 to the ENV versions of EC2, EC3 and EC5 are still current. The first part of Eurocode 6 on masonry design was published in pre-standard form in 1996, some three years after publication of the first edition of this book. The material in *Chapter 10* has therefore been revised, so it now conforms with the guidance given in the ENV.

An online solutions manual is available online at http://www.sponpress.com/civeng/support.htm

Acknowledgements

I am once again indebted to Tony Threlfall, formerly of the British Cement Association and currently an independent training consultant, who very kindly reviewed the new material in *Chapter 3*, Fred Lambert, formerly of South Bank University and currently an independent consulting engineer, who reviewed the material in *Chapter 4* on composite design and Peter Watt of the Brick Development Association who reviewed *Chapter 10*. Thanks are also due to Charles Goodchild of the British Cement Association for his comments on section 3.10 on slabs. My sincere gratitude goes to Dr Colin Bailey of the Building Research Establishment for preparing *Appendix C*.

I am grateful to Corus plc, the Construction Industry Research and Information Association and the British Cement Association for permission to use extracts from their publications. Extracts from British Standards are reproduced with the permission of BSI under licence number 2001SK/0439. Complete standards can be obtained from BSI Customer Services, 389 Chiswick High Road, London, W4 4AL.

In the memory of Biji

PART ONE

INTRODUCTION TO STRUCTURAL DESIGN

The primary aim of all structural design is to ensure that the structure will perform satisfactorily during its design life. Specifically, the designer must check that the structure is capable of carrying the loads safely and that it will not deform excessively due to the applied loads. This requires the designer to make realistic estimates of the strengths of the materials composing the structure and the loading to which it may be subject during its design life. Furthermore, the designer will need a basic understanding of structural behaviour.

The work that follows has two objectives:

1. to describe the philosophy of structural design;
2. to introduce various aspects of structural and material behaviour.

Towards the first objective, *Chapter 1* discusses the three main philosophies of structural design, emphasizing the limit state philosophy which forms the bases of design in many of the modern codes of practice. *Chapter 2* then outlines a method of assessing the design loading acting on individual elements of a structure and how this information can be used, together with the material properties, to size elements.

Chapter 1

Philosophy of design

This chapter is concerned with the philosophy of structural design. The chapter describes the overall aims of design and the many inputs into the design process. The primary aim of design is seen as the need to ensure that at no point in the structure do the design loads exceed the design strengths of the materials. This can be achieved by using the permissible stress or load factor philosophies of design. However, both suffer from drawbacks and it is more common to design according to limit state principles which involve considering all the mechanisms by which a structure could become unfit for its intended purpose during its design life.

1.1 Introduction

The task of the structural engineer is to design a structure which satisfies the needs of the client and the user. Specifically the structure should be safe, economical to build and maintain, and aesthetically pleasing. But what does the design process involve?

Design is a word that means different things to different people. In dictionaries the word is described as a mental plan, preliminary sketch, pattern, construction, plot or invention. Even among those closely involved with the built environment there are considerable differences in interpretation. Architects, for example, may interpret design as being the production of drawings and models to show what a new building will actually look like. To civil and structural engineers, however, design is taken to mean the entire planning process for a new building structure, bridge, tunnel, road, etc., from outline concepts and feasibility studies through mathematical calculations to working drawings which could show every last nut and bolt in the project. Together with the drawings there will be bills of quantities, a specification and a contract, which will form the necessary legal and organizational framework within which a contractor, under

the supervision of engineers and architects, can construct the scheme.

There are many inputs into the engineering design process as illustrated by *Fig. 1.1* including:

1. client brief
2. experience
3. imagination
4. a site investigation
5. model and laboratory tests
6. economic factors
7. environmental factors.

The starting-point for the designer is normally a conceptual brief from the client, who may be a private developer or perhaps a government body. The conceptual brief may simply consist of some sketches prepared by the client or perhaps a detailed set of architect's drawings. Experience is crucially important, and a client will always demand that the firm he is employing to do the design has previous experience designing similar structures.

Although imagination is thought by some to be entirely the domain of the architect, this is not so. For engineers and technicians an imagination of how elements of structure interrelate in three dimensions is essential, as is an appreciation of the loadings to which structures might be subject in certain circumstances. In addition, imaginative solutions to engineering problems are often required to save money, time, or to improve safety or quality.

A site investigation is essential to determine the strength and other characteristics of the ground on which the structure will be founded. If the structure is unusual in any way, or subject to abnormal loadings, model or laboratory tests may also be used to help determine how the structure will behave.

In today's economic climate a structural designer must be constantly aware of the cost implications of his or her design. On the one hand design should aim to achieve economy of materials in the structure, but over-refinement can lead to an excessive

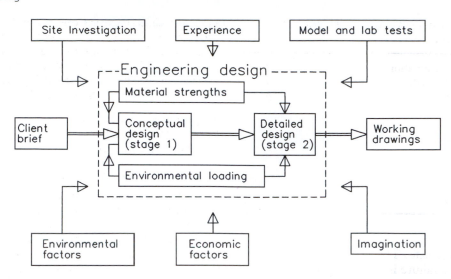

Fig. 1.1 *Inputs into the design process.*

number of different sizes and components in the structure, and labour costs will rise. In addition the actual cost of the designer's time should not be excessive, or this will undermine the employer's competitiveness. The idea is to produce a workable design achieving reasonable economy of materials, while keeping manufacturing and construction costs down, and avoiding unnecessary design and research expenditure. Attention to detailing and buildability of structures cannot be overemphasized in design. Most failures are as a result of poor detailing rather than incorrect analysis.

Designers must also understand how the structure will fit into the environment for which it is designed. Today many proposals for engineering structures stand or fall on this basis, so it is part of the designer's job to try to anticipate and reconcile the environmental priorities of the public and government.

The engineering design process can often be divided into two stages: (1) a feasibility study involving a comparison of the alternative forms of structure and selection of the most suitable type and (2) a detailed design of the chosen structure. The success of stage 1, the conceptual design, relies to a large extent on engineering judgement and instinct, both of which are the outcome of many years' experience of designing structures. Stage 2, the detailed design, also requires these attributes but is usually more dependent upon a thorough understanding of the codes of practice for structural design, e.g. BS 8110 and BS 5950. These documents are based on the amassed experience of

many generations of engineers, and the results of research. They help to ensure safety and economy of construction, and that mistakes are not repeated. For instance, after the infamous disaster at the Ronan Point block of flats in Newham, London, when a gas explosion caused a serious partial collapse, research work was carried out, and codes of practice were amended so that such structures could survive a gas explosion, with damage being confined to one level.

The aim of this book is to look at the procedures associated with the detailed design of structural elements such as beams, columns and slabs. *Chapter 2* will help the reader to revise some basic theories of structural behaviour. *Chapters 3–6* deal with design to British Standard (BS) codes of practice for the structural use of concrete (BS 8110), structural steelwork (BS 5950), masonry (BS 5628) and timber (BS 5268). *Chapter 7* introduces the new Eurocodes (EC) for structural design and *Chapters 8–11* then describe the layout and design principles in EC2, EC3, EC6 and EC5 for concrete, steelwork, masonry and timber respectively.

1.2 Basis of design

Table 1.1 illustrates some risk factors that are associated with activities in which people engage. It can be seen that some degree of risk is associated with air and road travel. However, people normally accept that the benefits of mobility outweigh the risks. Staying in buildings, however, has always been

Table 1.1 Comparative death risk per 10^8 persons exposed

Mountaineering (international)	2700
Air travel (international)	120
Deep water trawling	59
Car travel	56
Coal mining	21
Construction sites	8
Manufacturing	2
Accidents at home	2
Fire at home	0.1
Structural failures	0.002

regarded as fairly safe. The risk of death or injury due to structural failure is extremely low, but as we spend most of our life in buildings this is perhaps just as well.

As far as the design of structures for safety is concerned, it is seen as the process of ensuring that stresses due to loading at all critical points in a structure have a very low chance of exceeding the strength of materials used at these critical points. *Figure 1.2* illustrates this in statistical terms.

In design there exist within the structure a number of critical points (e.g. beam mid-spans) where the design process is concentrated. The normal distribution curve on the left of *Fig. 1.2* represents the actual maximum material stresses at these critical points due to the loading. Because loading varies according to occupancy and environmental conditions, and because design is an imperfect process, the material stresses will vary about a modal value – the peak of the curve. Similarly the normal distribution curve on the right represents material strengths at these critical points, which are also not constant due to the variability of manufacturing conditions.

The overlap between the two curves represents a possibility that failure may take place at one of the critical points, as stress due to loading exceeds the strength of the material. In order for the structure to be safe the overlapping area must be kept to a minimum. The degree of overlap between the two curves can be minimized by using one of three distinct design philosophies, namely:

1. permissible stress design
2. load factor method
3. limit state design.

1.2.1 PERMISSIBLE STRESS DESIGN

In permissible stress design, sometimes referred to as modular ratio or elastic design, the stresses in the structure at working loads are not allowed to exceed a certain proportion of the yield stress of the construction material, i.e. the stress levels are limited to the elastic range. By assuming that the stress–strain relationship over this range is linear, it is possible to calculate the actual stresses in the material concerned. Such an approach formed the basis of the design methods used in CP 114 (the forerunner of BS 8110) and BS 449 (the forerunner of BS 5950).

However, although it modelled real building performance under actual conditions, this philosophy had two major drawbacks. Firstly, permissible design methods sometimes tended to overcomplicate the design process and also led to conservative solutions. Secondly, as the quality of materials increased and the safety margins decreased, the assumption that stress and strain are directly proportional became unjustifiable for materials such as concrete, making it impossible to estimate the true factors of safety.

1.2.2 LOAD FACTOR DESIGN

Load factor or plastic design was developed to take account of the behaviour of the structure once the yield point of the construction material had been reached. This approach involved calculating the collapse load of the structure. The working load was derived by dividing the collapse load by a load factor. This approach simplified methods of analysis and allowed actual factors of safety to be calculated. It was in fact permitted in CP 114 and BS 449 but was slow in gaining acceptance and was eventually superseded by the more comprehensive limit state approach.

The reader is referred to *Appendix A* for an example illustrating the differences between the permissible stress and load factor approaches to design.

1.2.3 LIMIT STATE DESIGN

Originally formulated in the former Soviet Union in the 1930s and developed in Europe in the 1960s,

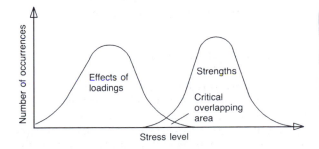

Fig. 1.2 *Relationship between stress and strength.*

limit state design can perhaps be seen as a compromise between the permissible and load factor methods. It is in fact a more comprehensive approach which takes into account both methods in appropriate ways. Most modern structural codes of practice are now based on the limit state approach. BS 8110 for concrete, BS 5950 for structural steelwork, BS 5400 for bridges and BS 5628 for masonry are all limit state codes. The principal exceptions are the code of practice for design in timber, BS 5268, and the old (but still current) structural steelwork code, BS 449, both of which are permissible stress codes. It should be noted, however, that the Eurocode for timber (EC5), which is expected to replace BS 5268 around 2008, is based on limit state principles.

As limit state philosophy forms the basis of the design methods in most modern codes of practice for structural design, it is essential that the design methodology is fully understood. This then is the purpose of the following subsections.

1.2.3.1 Ultimate and serviceability limit states

The aim of limit state design is to achieve acceptable probabilities that a structure will not become unfit for its intended use during its design life, that is, the structure will not reach a limit state. There are many ways in which a structure could become unfit for use, including excessive conditions of bending, shear, compression, deflection and cracking (*Fig. 1.3*). Each of these mechanisms is a limit state whose effect on the structure must be individually assessed.

Some of the above limit states, e.g. deflection and cracking, principally affect the appearance of the structure. Others, e.g. bending, shear and compression, may lead to partial or complete collapse of the structure. Those limit states which can cause failure of the structure are termed ultimate limit states. The others are categorized as serviceability limit states. The ultimate limit states enable the designer to calculate the strength of the structure. Serviceability limit states model the behaviour of the structure at working loads. In addition, there may be other limit states which may adversely affect the performance of the structure, e.g. durability and fire resistance, and which must therefore also be considered in design.

It is a matter of experience to be able to judge which limit states should be considered in the design of particular structures. Nevertheless, once this has been done, it is normal practice to base the design on the most critical limit state and then check for the remaining limit states. For example, for reinforced concrete beams the ultimate limit states of bending and shear are used to size the beam. The design is then checked for the remaining limit states, e.g. deflection and cracking. On the other hand, the serviceability limit state of deflection is normally critical in the design of concrete slabs. Again, once the designer has determined a suitable depth of slab, he/she must then make sure that the design satisfies the limit states of bending, shear and cracking.

In assessing the effect of a particular limit state on the structure, the designer will need to assume certain values for the loading on the structure and the strength of the materials composing the structure. This requires an understanding of the concepts of characteristic and design values which are discussed below.

1.2.3.2 Characteristic and design values

As stated at the outset, when checking whether a particular member is safe, the designer cannot be certain about either the strength of the material composing the member or, indeed, the load which the member must carry. The material strength may be less than intended (a) because of its variable composition, and (b) because of the variability of manufacturing conditions during construction, and other effects such as corrosion. Similarly the load in the member may be greater than anticipated (a) because of the variability of the occupancy or environmental loading, and (b) because of unforeseen circumstances which may lead to an increase in the general level of loading, errors in the analysis, errors during construction, etc.

In each case, item (a) is allowed for by using a **characteristic** value. The characteristic strength is the value **below** which the strength lies in only a small number of cases. Similarly the characteristic load is the value **above** which the load lies in only a small percentage of cases. In the case of strength the characteristic value is determined from test results using statistical principles, and is normally defined as the value below which not more than 5% of the test results fall. However, at this stage there are insufficient data available to apply statistical principles to loads. Therefore the characteristic loads are normally taken to be the design loads from other codes of practice, e.g. BS 648 and BS 6399.

The overall effect of items under (b) is allowed for using a partial safety factor: γ_m for strength

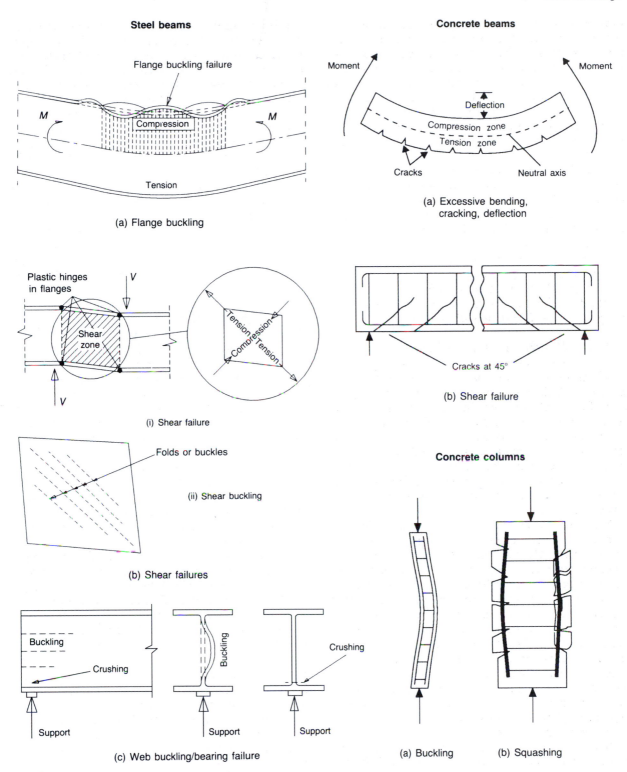

Fig. 1.3 *Typical modes of failure for beams and columns.*

and γ_f for load. The design strength is obtained by dividing the characteristic strength by the partial safety factor for strength:

$$\text{Design strength} = \frac{\text{characteristic strength}}{\gamma_m} \quad (1.1)$$

The design load is obtained by multiplying the characteristic load by the partial safety factor for the load:

$$\text{Design load} = \text{characteristic load} \times \gamma_f \quad (1.2)$$

The value of γ_m will depend upon the properties of the actual construction material being used. Values for γ_f depend on other factors which will be discussed more fully in *Chapter 2*.

In general, once a preliminary assessment of the design loads has been made it is then possible to calculate the maximum bending moments, shear forces and deflections in the structure (*Chapter 2*). The construction material must be capable of withstanding these forces otherwise failure of the structure may occur, i.e.

$$\text{Design strength} \geqslant \text{design load} \quad (1.3)$$

Simplified procedures for calculating the moment, shear and axial load capacities of structural elements together with acceptable deflection limits are described in the appropriate codes of practice.

These allow the designer to rapidly assess the suitability of the proposed design. However, before discussing these procedures in detail, *Chapter 2* describes in general terms how the design loads acting on the structure are estimated and used to size individual elements of the structure.

1.3 Summary

This chapter has examined the bases of three philosophies of structural design: permissible stress, load factor and limit state. The chapter has concentrated on limit state design since it forms the basis of the design methods given in the codes of practice for concrete (BS 8110), structural steelwork (BS 5950) and masonry (BS 5628). The aim of limit state design is to ensure that a structure will not become unfit for its intended use, that is, it will not reach a limit state during its design life. Two categories of limit states are examined in design: ultimate and serviceability. The former is concerned with overall stability and determining the collapse load of the structure; the latter examines its behaviour under working loads. Structural design principally involves ensuring that the loads acting on the structure do not exceed its strength and the first step in the design process then is to estimate the loads acting on the structure.

Questions

1. Explain the difference between conceptual design and detailed design.
2. What is a code of practice and what is its purpose in structural design?
3. List the principal sources of uncertainty in structural design and discuss how these uncertainties are rationally allowed for in design.
4. The characteristic strengths and design strengths are related via the partial safety factor for materials. The partial safety factor for concrete is higher than for steel reinforcement. Discuss why this should be so.
5. Describe in general terms the ways in which a beam and column could become unfit for use.

Chapter 2

Basic structural concepts and material properties

This chapter is concerned with general methods of sizing beams and columns in structures. The chapter describes how the characteristic and design loads acting on structures and on the individual elements are determined. Methods of calculating the bending moments, shear forces and deflections in beams are outlined. Finally, the chapter describes general approaches to sizing beams according to elastic and plastic criteria and sizing columns subject to axial loading.

2.1 Introduction

All structures are composed of a number of inter-connected elements such as slabs, beams, columns, walls and foundations. Collectively, they enable the internal and external loads acting on the structure to be safely transmitted down to the ground. The actual way that this is achieved is difficult to model and many simplifying, but conservative, assumptions have to be made. For example, the degree of fixity at column and beam ends is usually uncertain but, nevertheless, must be estimated as it significantly affects the internal forces in the element. Furthermore, it is usually assumed that the reaction from one element is a load on the next and that the sequence of load transfer between elements occurs in the order: ceiling/floor loads to beams to columns to foundations to ground (*Fig. 2.1*).

At the outset, the designer must make an assessment of the future likely level of loading, including self-weight, to which the structure may be subject during its design life. Using computer methods or hand calculations the design loads acting on individual elements can then be evaluated. The design loads are used to calculate the bending moments, shear forces and deflections at critical points along the elements. Finally, suitable dimensions for the element can be determined. This aspect requires an understanding of the elementary theory of bending and the behaviour of elements subject to

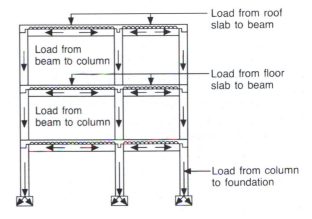

Fig. 2.1 *Sequence of load transfer between elements of a structure.*

compressive loading. These steps are summarized in *Fig. 2.2* and the following sections describe the procedures associated with each step.

2.2 Design loads acting on structures

The loads acting on a structure are divided into three basic types: dead, imposed and wind. For each type of loading there will be characteristic and design values, as discussed in *Chapter 1*, which must be estimated. In addition, the designer will have to determine the particular combination of loading which is likely to produce the most adverse effect on the structure in terms of bending moments, shear forces and deflections.

2.2.1 DEAD LOADS, G_k, g_k

Dead loads are all the permanent loads acting on the structure including self-weight, finishes, fixtures and partitions. The characteristic dead loads can be

9

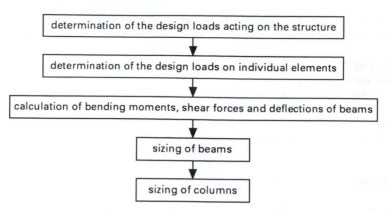

Fig. 2.2 *Design process.*

Example 2.1 Self-weight of a reinforced concrete beam

Calculate the self-weight of a reinforced concrete beam of breadth 300 mm, depth 600 mm and length 6000 mm.

From *Table 2.1*, unit mass of reinforced concrete is 2400 kg m^{-3}. Assuming that the gravitational constant is 10 m s^{-2} (strictly 9.807 m s^{-2}), the unit weight of reinforced concrete, ρ, is

$$\rho = 2400 \times 10 = 24\,000 \text{ N m}^{-3} = 24 \text{ kN m}^{-3}$$

Hence, the self-weight of beam, SW, is

$$SW = \text{volume} \times \text{unit weight}$$
$$= (0.3 \times 0.6 \times 6)24 = 25.92 \text{ kN}$$

---

estimated using the schedule of weights of building materials given in BS 648 (*Table 2.1*) or from manufacturers' literature. The symbols G_k and g_k are normally used to denote the total and uniformly distributed characteristic dead loads respectively.

Estimation of the self-weight of an element tends to be a cyclic process since its value can only be assessed once the element has been designed which requires prior knowledge of the self-weight of the element. Generally, the self-weight of the element is likely to be small in comparison with other dead and live loads and any error in estimation will tend to have a minimal effect on the overall design (*Example 2.1*).

2.2.2 IMPOSED LOADS Q_k, q_k
Imposed load, sometimes also referred to as live load, represents the load due to the proposed occupancy and includes the weights of the occupants, furniture and roof loads including snow. Since imposed loads tend to be much more variable than dead loads they are more difficult to predict.

BS 6399: Part 1: 1984: *Code of Practice for Dead and Imposed Loads* gives typical characteristic imposed floor loads for different classes of structure, e.g. residential dwellings, educational institutions, hospitals, and parts of the same structure, e.g. balconies, corridors and toilet rooms (*Table 2.2*).

2.2.3 WIND LOADS
Wind pressure can either add to the other gravitational forces acting on the structure or, equally well, exert suction or negative pressures on the structure. Under particular situations, the latter may well lead to critical conditions and must be considered in design. The characteristic wind loads acting on a structure can be assessed in accordance with the recommendations given in CP 3: Chapter V: Part 2: 1972 *Wind Loads* or Part 2 of BS 6399: *Code of Practice for Wind Loads*.

Wind loading is important in the design of masonry panel walls (*Chapter 5*). However beyond that, wind loading is not considered further since the emphasis in this book is on the design of elements rather

Table 2.1 Schedule of unit masses of building materials (based on BS 648)

Asphalt			**Plaster**	
Roofing 2 layers, 19 mm thick	42 kg m^{-2}		Two coats gypsum, 13 mm thick	22 kg m^{-2}
Damp-proofing, 19 mm thick	41 kg m^{-2}			
Roads and footpaths, 19 mm thick	44 kg m^{-2}		**Plastics sheeting (corrugated)**	4.5 kg m^{-2}
Bitumen roofing felts			**Plywood**	
Mineral surfaced bitumen	3.5 kg m^{-2}		per mm thick	0.7 kg m^{-2}
Blockwork			**Reinforced concrete**	2400 kg m^{-3}
Solid per 25 mm thick, stone aggregate	55 kg m^{-2}		**Rendering**	
Aerated per 25 mm thick	15 kg m^{-2}		Cement: sand (1:3), 13 mm thick	30 kg m^{-2}
Board			**Screeding**	
Blockboard per 25 mm thick	12.5 kg m^{-2}		Cement: sand (1:3), 13 mm thick	30 kg m^{-2}
Brickwork			**Slate tiles**	
Clay, solid per 25 mm thick medium density	55 kg m^{-2}		(depending upon thickness and source)	$24\text{–}78 \text{ kg m}^{-3}$
Concrete, solid per 25 mm thick	59 kg m^{-2}		**Steel**	
Cast stone	2250 kg m^{-3}		Solid (mild)	7850 kg m^{-3}
			Corrugated roofing sheets, per mm thick	10 kg m^{-2}
Concrete				
Natural aggregates	2400 kg m^{-3}		**Tarmacadam**	
Lightweight aggregates (structural)	$1760 + 240/$ -160 kg m^{-3}		25 mm thick	60 kg m^{-2}
Flagstones			**Terrazzo**	
Concrete, 50 mm thick	120 kg m^{-2}		25 mm thick	54 kg m^{-2}
Glass fibre			**Tiling, roof**	
Slab, per 25 mm thick	$2.0\text{–}5.0 \text{ kg m}^{-2}$		Clay	70 kg m^{-2}
Gypsum panels and partitions			**Timber**	
Building panels 75 mm thick	44 kg m^{-2}		Softwood	590 kg m^{-3}
			Hardwood	1250 kg m^{-3}
Lead				
Sheet, 2.5 mm thick	30 kg m^{-2}		**Water**	1000 kg m^{-3}
Linoleum			**Woodwool**	
3 mm thick	6 kg m^{-2}		Slabs, 25 mm thick	15 kg m^{-2}

than structures, which generally involves investigating the effects of dead and imposed loads only.

2.2.4 LOAD COMBINATIONS AND DESIGN LOADS

The design loads are obtained by multiplying the characteristic loads by the partial safety factor for loads, γ_f (*Chapter 1*). The value for γ_f depends on several factors including the limit state under consideration, i.e. ultimate or serviceability, the accuracy of predicting the load and the particular combination of loading which will produce the worst possible effect on the structure in terms of bending moments, shear forces and deflections.

Table 2.2 Imposed loads for residential occupancy class

Floor area usage	Intensity of distributed load kN m^{-2}	Concentrated load kN
Type 1. Self-contained dwelling units		
All	1.5	1.4
Type 2. Apartment houses, boarding houses, lodging houses, guest houses, hostels, residential clubs and communal areas in blocks of flats		
Boiler rooms, motor rooms, fan rooms and the like including the weight of machinery	7.5	4.5
Communal kitchens, laundries	3.0	4.5
Dining rooms, lounges, billiard rooms	2.0	2.7
Toilet rooms	2.0	–
Bedrooms, dormitories	1.5	1.8
Corridors, hallways, stairs, landings, footbridges, etc.	3.0	4.5
Balconies	Same as rooms to which they give access but with a minimum of 3.0	1.5 per metre run concentrated at the outer edge
Cat walks	–	1.0 at 1 m centres
Type 3. Hotels and motels		
Boiler rooms, motor rooms, fan rooms and the like, including the weight of machinery	7.5	4.5
Assembly areas without fixed seating, dance halls	5.0	3.6
Bars	5.0	–
Assembly areas with fixed seating[a]	4.0	–
Corridors, hallways, stairs, landings, footbridges, etc.	4.0	4.5
Kitchens, laundries	3.0	4.5
Dining rooms, lounges, billiard rooms	2.0	2.7
Bedrooms	2.0	1.8
Toilet rooms	2.0	–
Balconies	Same as rooms to which they give access but with a minimum of 4.0	1.5 per metre run concentrated at the outer edge
Cat walks	–	1.0 at 1 m centres

Note. [a] Fixed seating is seating where its removal and the use of the space for other purposes are improbable.

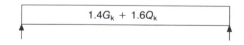

Fig. 2.3

In most of the simple structures which will be considered in this book, the worst possible combination will arise due to the maximum dead and maximum imposed loads acting on the structure together. In such cases, the partial safety factors for dead and imposed loads are 1.4 and 1.6 respectively (*Fig. 2.3*) and hence the design load is given by

Design load = $1.4G_k + 1.6Q_k$

However, it should be appreciated that theoretically the design dead loads can vary between the characteristic and ultimate values, i.e. $1.0G_k$ and $1.4G_k$. Similarly, the design imposed loads can vary between zero and the ultimate value, i.e. $0.0Q_k$ and $1.6Q_k$. Thus for a simply supported beam with an overhang (*Fig. 2.4(a)*) the load cases shown in *Figs 2.4(b)–(d)* will need to be considered in order to determine the design bending moments and shear forces in the beam.

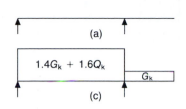

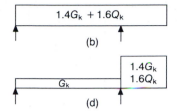

Fig. 2.4

2.3 Design loads acting on elements

Once the design loads acting on the structure have been estimated it is then possible to calculate the design loads acting on individual elements. As was pointed out at the beginning of this chapter, this usually requires the designer to make assumptions regarding the support conditions and how the loads will eventually be transmitted down to the ground. *Figures 2.5(a)* and *(b)* illustrate some of the more commonly assumed support conditions at the ends of beams and columns respectively.

In design it is common to assume that all the joints in the structure are pinned and that the sequence of load transfer occurs in the order: ceiling/floor loads to beams to columns to foundations to ground. These assumptions will considerably simplify calculations and lead to conservative estimates of the design loads acting on individual elements of the structure. The actual calculations to determine the forces acting on the elements are best illustrated by a number of worked examples as follows.

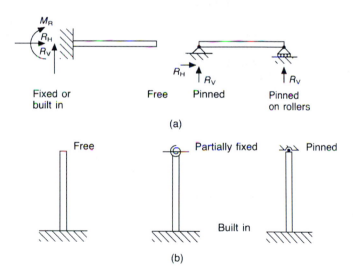

Fig. 2.5 *Typical beams and column support conditions.*

Example 2.2 Design loads on a floor beam

A composite floor consisting of a 150 mm thick reinforced concrete slab supported on steel beams spanning 5 m and spaced at 3 m centres is to be designed to carry an imposed load of 3.5 kN m^{-2}. Assuming that the unit mass of the steel beams is 50 kg m^{-1} run, calculate the design loads on a typical internal beam.

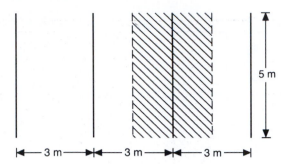

UNIT WEIGHTS OF MATERIALS

Reinforced concrete

From *Table 2.1*, unit mass of reinforced concrete is 2400 kg m^{-3}. Assuming the gravitational constant is 10 m s^{-2}, the unit weight of reinforced concrete is

$$2400 \times 10 = 24\,000 \text{ N m}^{-3} = 24 \text{ kN m}^{-3}$$

Steel beams

Unit mass of beam $= 50$ kg m^{-1} run
Unit weight of beam $= 50 \times 10 = 500$ N m^{-1} run $= 0.5$ kN m^{-1} run

LOADING

Slab

Slab dead load (g_k) $=$ self-weight $= 0.15 \times 24 = 3.6$ kN m^{-2}
Slab imposed load (q_k) $= 3.5$ kN m^{-2}
Slab ultimate load $= 1.4g_k + 1.6q_k = 1.4 \times 3.6 + 1.6 \times 3.5$
                          $= 10.64$ kN m^{-2}

Beam

Beam dead load (g_k) $=$ self-weight $= 0.5$ kN m^{-1} run
Beam ultimate load $= 1.4g_k$ $= 1.4 \times 0.5 = 0.7$ kN m^{-1} run

DESIGN LOAD

Each internal beam supports a uniformly distributed load from a 3 m width of slab (hatched \\\\\\\\\\) plus self-weight. Hence

$$\text{Design load on beam} = \text{slab load} + \text{self-weight of beam}$$
$$= 10.64 \times 5 \times 3 + 0.7 \times 5$$
$$= 159.6 + 3.5 = 163.1 \text{ kN}$$

Example 2.3 Design loads on floor beams and columns

The floor shown below with an overall depth of 225 mm is to be designed to carry an imposed load of 3 kN m^{-2} plus floor finishes and ceiling loads of 1 kN m^{-2}. Calculate the design loads acting on beams B1–C1, B2–C2 and B1–B3 and columns B1 and C1. Assume that all the column heights are 3 m and that the beam and column weights are 70 and 60 kg m^{-1} run respectively.

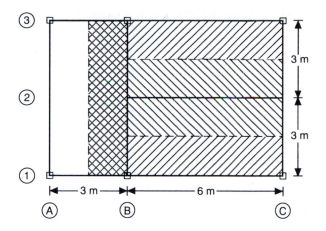

UNIT WEIGHTS OF MATERIALS

Reinforced concrete

From *Table 2.1*, unit mass of reinforced concrete is 2400 kg m^{-3}. Assuming the gravitational constant is 10 m s^{-2}, the unit weight of reinforced concrete is

$$2400 \times 10 = 24\,000 \text{ N m}^{-3} = 24 \text{ kN m}^{-3}$$

Steel beams

Unit mass of beam   = 70 kg m^{-1} run
Unit weight of beam = 70 × 10 = 700 N m^{-1} run = 0.7 kN m^{-1} run

Steel columns

Unit mass of column   = 60 kg m^{-1} run
Unit weight of column = 60 × 10 = 600 N m^{-1} run = 0.6 kN m^{-1} run

LOADING

Slab

Slab dead load (g_k)   = self-weight + finishes
                        = 0.225 × 24 + 1 = 6.4 kN m^{-2}
Slab imposed load (q_k) = 3 kN m^{-2}
Slab ultimate load   = 1.4g_k + 1.6q_k
                        = 1.4 × 6.4 + 1.6 × 3 = 13.76 kN m^{-2}

Beam

Beam dead load (g_k) = self-weight = 0.7 kN m^{-1} run
Beam ultimate load   = 1.4g_k = 1.4 × 0.7 = 0.98 kN m^{-1} run

Column

Column dead load (g_k) = 0.6 kN m^{-1} run
Column ultimate load   = 1.4g_k = 1.4 × 0.6 = 0.84 kN m^{-1} run

DESIGN LOADS

Beam B1–C1

Assuming that the slab is simply supported, beam B1–C1 supports a uniformly distributed load from a 1.5 m width of slab (hatched //////) plus self-weight of beam. Hence

Design load on beam B1–C1 = slab load + self-weight of beam

$$= 13.76 \times 6 \times 1.5 + 0.98 \times 6$$

$$= 123.84 + 5.88 = 129.72 \text{ kN}$$

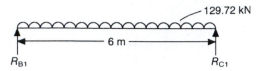

Since the beam is symmetrically loaded,

$$R_{B1} = R_{C1} = 129.72/2 = 64.86 \text{ kN}$$

Beam B2–C2

Assuming that the slab is simply supported, beam B2–C2 supports a uniformly distributed load from a 3 m width of slab (hatched \\\\\\) plus its self-weight. Hence

Design load on beam B2–C2 = slab load + self-weight of beam

$$= 13.76 \times 6 \times 3 + 0.98 \times 6$$

$$= 247.68 + 5.88 = 253.56 \text{ kN}$$

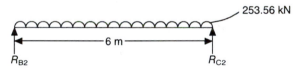

Since the beam is symmetrically loaded, R_{B2} and R_{C2} are the same and equal to $253.56/2 = 126.78$ kN.

Beam B1–B3

Assuming that the slab is simply supported, beam B1–B3 supports a uniformly distributed load from a 1.5 m width of slab (shown cross-hatched) plus the self-weight of the beam and the reaction transmitted from beam B2–C2 which acts as a point load at mid-span. Hence

Design load on beam B1–B3 = uniformly distributed load from slab plus self-weight of beam
+ point load from reaction R_{B2}

$$= (13.76 \times 1.5 \times 6 + 0.98 \times 6) + 126.78$$

$$= 129.72 + 126.78 = 256.5 \text{ kN}$$

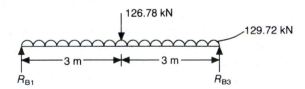

Since the beam is symmetrically loaded,

$$R_{B1} = R_{B3} = 256.5/2 = 128.25 \text{ kN}$$

Column B1

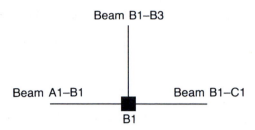

Column B1 supports the reactions from beams A1–B1, B1–C1 and B1–B3 and its self-weight. From the above, the reaction at B1 due to beam B1–C1 is 64.86 kN and from beam B1–B3 is 128.25 kN. Beam A1–B1 supports only its self-weight = $0.98 \times 3 = 2.94$ kN. Hence reaction at B1 due to A1–B1 is $2.94/2 = 1.47$ kN. Since the column height is 3 m, self-weight of column = $0.84 \times 3 = 2.52$ kN. Hence

$$\text{Design load on column B1} = 64.86 + 128.25 + 1.47 + 2.52$$

$$= 197.1 \text{ kN}$$

Column C1

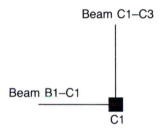

Column C1 supports the reactions from beams B1–C1 and C1–C3 and its self-weight. From the above, the reaction at C1 due to beam B1–C1 is 64.86 kN. Beam C1–C3 supports the reactions from B2–C2 (= 126.78 kN) and its self-weight (= 0.98×6 = 5.88 kN). Hence the reaction at C1 is $(126.78 + 5.88)/2 = 66.33$ kN. Since the column height is 3 m, self-weight = $0.84 \times 3 = 2.52$ kN. Hence

$$\text{Design load on column C1} = 64.86 + 66.33 + 2.52 = 133.71 \text{ kN}$$

2.4 Structural analysis

The design axial loads can be used directly to size columns. Column design will be discussed more fully in *section 2.5*. However, before flexural members such as beams can be sized, the design bending moments and shear forces must be evaluated. Such calculations can be performed by a variety of methods as noted below, depending upon the complexity of the loading and support conditions:

1. equilibrium equations
2. formulae
3. computer methods.

Hand calculations are suitable for analysing statically determinate structures such as simply supported beams and slabs (*section 2.4.1*). For various standard load cases, formulae for calculating the maximum bending moments, shear forces and deflections are available which can be used to rapidly analyse beams, as will be discussed in *section 2.4.2*. Alternatively, the designer may resort to using various commercially available computer packages, e.g. SAND. Their use is not considered in this book.

2.4.1 EQUILIBRIUM EQUATIONS

It can be demonstrated that if a body is in equilibrium under the action of a system of external forces,

17

all parts of the body must also be in equilibrium. This principle can be used to determine the bending moments and shear forces along a beam. The actual procedure simply involves making fictitious 'cuts' at intervals along the beam and applying the equilibrium equations given below to the cut portions of the beam.

$$\Sigma \text{ moments } (M) = 0 \qquad (2.1)$$

$$\Sigma \text{ vertical forces } (V) = 0 \qquad (2.2)$$

Example 2.4 Design moments and shear forces in beams using equilibrium equations

Calculate the design bending moments and shear forces in beams B2–C2 and B1–B3 of *Example 2.3*.

BEAM B2–C2

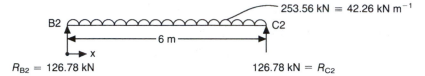

Let the longitudinal centroidal axis of the beam be the x axis and $x = 0$ at support B2.

x = 0
By inspection,

$$\text{Moment at } x = 0 \ (M_{x=0}) = 0$$

$$\text{Shear force at } x = 0 \ (V_{x=0}) = R_{B2} = 126.78 \text{ kN}$$

x = 1
Assuming that the beam is cut 1 m from support B2, i.e. $x = 1$ m, the moments and shear forces acting on the cut portion of the beam will be those shown in the free body diagram below:

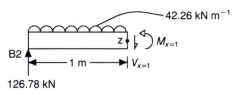

From equation 2.1, taking moments about Z gives

$$126.78 \times 1 - 42.26 \times 1 \times 0.5 - M_{x=1} = 0$$

Hence

$$M_{x=1} = 105.65 \text{ kN m}$$

From equation 2.2, summing the vertical forces gives

$$126.78 - 42.26 \times 1 - V_{x=1} = 0$$

Hence

$$V_{x=1} = 84.52 \text{ kN}$$

x = 2

The free body diagram for the beam, assuming that it has been cut 2 m from support B2, is shown below:

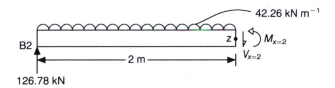

From equation 2.1, taking moments about Z gives

$$126.78 \times 2 - 42.26 \times 2 \times 1 - M_{x=2} = 0$$

Hence

$$M_{x=2} = 169.04 \text{ kN m}$$

From equation 2.2, summing the vertical forces gives

$$126.78 - 42.26 \times 2 - V_{x=2} = 0$$

Hence

$$V_{x=2} = 42.26 \text{ kN}$$

If this process is repeated for values of x equal to 3, 4, 5 and 6 m, the following values of the moments and shear forces in the beam will result:

x(m)	0	1	2	3	4	5	6
M(kN m)	0	105.65	169.04	190.17	169.04	105.65	0
V(kN)	126.78	84.52	42.26	0	−42.26	−84.52	−126.78

This information is better presented diagrammatically as shown below:

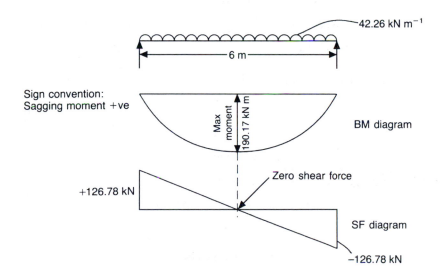

Hence, the design moment (M) is 190.17 kN m. Note that this occurs at mid-span and coincides with the point of zero shear force. The design shear force (V) is 126.78 kN and occurs at the supports.

BEAM B1–B3

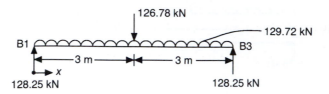

Again, let the longitudinal centroidal axis of the beam be the x axis of the beam and set $x = 0$ at support B1. The steps outlined earlier can be used to determine the bending moments and shear forces at $x = 0$, 1 and 2 m and since the beam is symmetrically loaded and supported, these values of bending moment and shear forces will apply at $x = 4$, 5 and 6 m respectively.

The bending moment at $x = 3$ m can be calculated by considering all the loading immediately to the left of the point load as shown below:

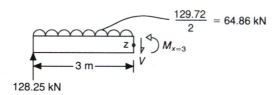

From equation 2.1, taking moments about Z gives

$$128.25 \times 3 - 64.86 \times 3/2 - M_{x=3} = 0$$

Hence

$$M_{x=3} = 287.46 \text{ kN m}$$

In order to determine the shear force at $x = 3$ the following two load cases need to be considered:

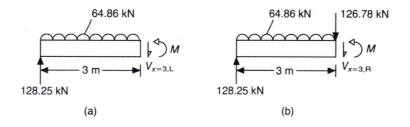

In (a) it is assumed that the beam is cut immediately to the left of the point load. In (b) the beam is cut immediately to the right of the point load. In (a), from equation 2.2, the shear force to the left of the cut, $V_{x=3,\text{L}}$, is given by

$$128.25 - 64.86 - V_{x=3,\text{L}} = 0$$

Hence

$$V_{x=3,\text{L}} = 63.39 \text{ kN}$$

In (b), from equation 2.2, the shear force to the right of the cut, $V_{x=3,R}$, is given by

$$128.25 - 64.86 - 126.78 - V_{x=3,R} = 0$$

Hence

$$V_{x=3,R} = -63.39 \text{ kN}$$

Summarizing the results in tabular and graphical form gives

x(m)	0	1	2	3	4	5	6
M(kN m)	0	117.44	213.26	287.46	213.26	117.44	0
V(kN)	128.25	106.63	85.01	63.39 | −63.39	−85.01	−106.63	−128.25

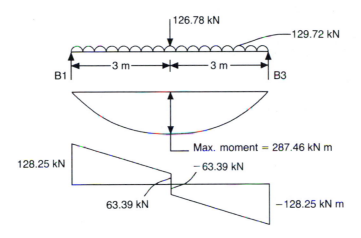

Hence, the design moment for beam B1–B3 is 287.46 kN m and occurs at mid-span and the design shear force is 128.25 kN and occurs at the supports.

2.4.2 FORMULAE

An alternative method of determining the design bending moments and shear forces in beams involves using the formulae quoted in *Table 2.3*. The table also includes formulae for calculating the maximum deflections for each load case. The formulae can be derived using a variety of methods for analysing statically indeterminate structures, e.g. slope deflection and virtual work, and the interested reader is referred to any standard work on this subject for background information. There will be many instances in practical design where the use of standard formulae is not convenient and equilibrium methods are preferable.

Table 2.3 Bending moments, shear forces and deflections for various standard load cases

Loading	Maximum bending moment	Maximum shearing force	Maximum deflection
	$\dfrac{WL}{4}$	$\dfrac{W}{2}$	$\dfrac{WL^3}{48EI}$
	$\dfrac{WL}{6}$	$\dfrac{W}{2}$	$\dfrac{23WL^3}{1296EI}$
	$\dfrac{WL}{8}$	$\dfrac{W}{2}$	$\dfrac{11WL^3}{768EI}$
	$\dfrac{WL}{8}$	$\dfrac{W}{2}$	$\dfrac{5WL^3}{384EI}$
	$\dfrac{WL}{6}$	$\dfrac{W}{2}$	$\dfrac{WL^3}{60EI}$
	$\dfrac{WL}{8}$ (at supports and at midspan)	$\dfrac{W}{2}$	$\dfrac{WL^3}{192EI}$
	$\dfrac{WL}{12}$ at supports $\dfrac{WL}{24}$ at midspan	$\dfrac{W}{2}$	$\dfrac{WL^3}{384EI}$
	WL	W	$\dfrac{WL^3}{3EI}$
	$\dfrac{WL}{2}$	W	$\dfrac{WL^3}{8EI}$

Example 2.5 Design moments and shear forces in beams using formulae

Repeat *Example 2.4* using the formulae given in *Table 2.3*.

BEAM B2–C2

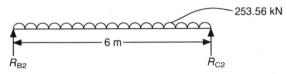

By inspection, maximum moment and maximum shear force occur at beam mid-span and supports respectively. From *Table 2.3*, design moment, M, is given by

$$M = \frac{Wl}{8} = \frac{253.56 \times 6}{8} = 190.17 \text{ kN m}$$

and design shear force, V, is given by

$$V = \frac{W}{2} = \frac{253.56}{2} = 126.78 \text{ kN}$$

BEAM B1–B3

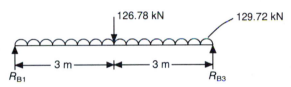

This load case can be solved using the principle of superposition which can be stated in general terms as follows: 'The effect of several actions taking place simultaneously can be reproduced exactly by adding the effects of each case separately.' Thus, the loading on beam B1–B3 can be considered to be the sum of a uniformly distributed load (W_{udl}) of 129.72 kN and a point load (W_{pl}) at mid-span of 126.78 kN.

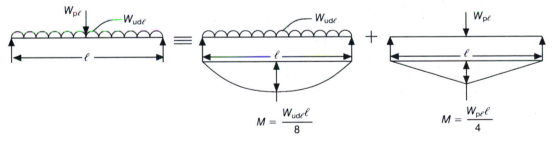

By inspection, the maximum bending moment and shear force for both load cases occur at beam mid-span and supports respectively. Thus, the design moment, M, is given by

$$M = \frac{W_{udl}l}{8} + \frac{W_{pl}l}{4} = \frac{129.72 \times 6}{8} + \frac{126.78 \times 6}{4} = 287.46 \text{ kN m}$$

and the design shear force, V, is given by

$$V = \frac{W_{udl}}{2} + \frac{W_{pl}}{2} = \frac{129.72}{2} + \frac{126.78}{2} = 128.25 \text{ kN}$$

2.5 Beam design

Having calculated the design bending moment and shear force, all that now remains to be done is to assess the size and strength of beam required. Generally, the ultimate limit state of bending will be critical for medium-span beams which are moderately loaded and shear for short-span beams which are heavily loaded. For long-span beams the serviceability limit state of deflection may well be critical. Irrespective of the actual critical limit state, once a preliminary assessment of the size and strength of beam needed has been made, it must be checked for the remaining limit states that may influence its long-term integrity.

The processes involved in such a selection will depend on whether the construction material behaves (i) elastically or (ii) plastically. If the material is elastic, it obeys Hooke's law, that is, the stress in the material due to the applied load is directly proportional to its strain (*Fig. 2.6*) where

$$\text{Stress } (\sigma) = \frac{\text{force}}{\text{area}}$$

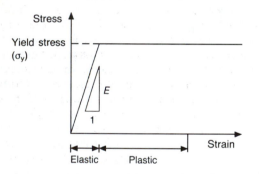

Fig. 2.6 *Stress–strain plot for steel.*

and $\text{Strain } (\varepsilon) = \dfrac{\text{change in length}}{\text{original length}}$

The slope of the graph of stress vs. strain (*Fig. 2.6*) is therefore constant, and this gradient is normally referred to as the elastic or Young's modulus and is denoted by the letter *E*. It is given by

$$\text{Young's modulus } (E) = \frac{\text{stress}}{\text{strain}}$$

Note that strain is dimensionless but that both stress and Young's modulus are usually expressed in N mm^{-2}.

A material is said to be plastic if it strains without a change in stress. Plasticine and clay are plastic materials but so is steel beyond its yield point (*Fig. 2.6*). As will be seen in *Chapter 3*, reinforced concrete design also assumes that the material behaves plastically.

The structural implications of elastic and plastic behaviour are best illustrated by considering how bending is resisted by the simplest of beams – a rectangular section *b* wide and *d* deep.

2.5.1 ELASTIC CRITERIA

When any beam is subject to load it bends as shown in *Fig. 2.7(a)*. The top half of the beam is put into compression and the bottom half into tension. In the middle, there is neither tension nor compression. This axis is normally termed the neutral axis.

If the beam is elastic and stress and strain are directly proportional for the material, the variation in strain and stress from the top to middle to bottom is linear (*Figs 2.7(b)* and *(c)*). The maximum stress in compression and tension is σ_y. The average stress in compression and tension is $\sigma_y/2$. Hence the compressive force, F_c, and tensile force, F_t, acting on the section are equal and are given by

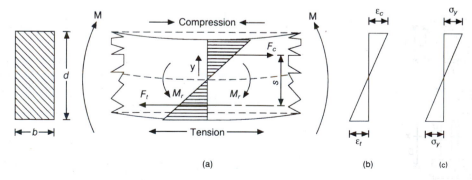

Fig. 2.7 *Strain and stress in an elastic beam.*

$$F_c = F_t = F = \text{stress} \times \text{area}$$

$$= \frac{\sigma_y}{2}\frac{bd}{2} = \frac{bd\sigma_y}{4}$$

The tensile and compression forces are separated by a distance s whose value is equal to $2d/3$ (*Fig. 2.7(a)*). Together they make up a couple, or moment, which acts in the opposite sense to the design moment. The value of this moment of resistance, M_r, is given by

$$M_r = Fs = \frac{bd\sigma_y}{4}\frac{2d}{3} = \frac{bd^2\sigma_y}{6} \qquad (2.3)$$

At equilibrium, the design moment in the beam will equal the moment of resistance, i.e.

$$M = M_r \qquad (2.4)$$

Provided that the yield strength of the material, i.e. σ_y is known, equations 2.3 and 2.4 can be used to calculate suitable dimensions for the beam needed to resist a particular design moment. Altern-

atively if b and d are known, the required material strength can be evaluated.

Equation 2.3 is more usually written as

$$M = \sigma Z \qquad (2.5)$$

or

$$M = \frac{\sigma I}{y} \qquad (2.6)$$

where Z is the elastic section modulus and is equal to $bd^2/6$ for a rectangular beam, I is the moment of inertia or, more correctly, the second moment of area of the section and y the distance from the neutral axis (*Fig. 2.7(a)*). The elastic section modulus can be regarded as an index of the strength of the beam in bending. The second moment of area about an axis x–x in the plane of the section is defined by

$$I_{xx} = \int y^2 \mathrm{d}A \qquad (2.7)$$

Second moments of area of some common shapes are given in *Table 2.4*.

2.5.2 PLASTIC CRITERIA

While the above approach would be suitable for design involving the use of materials which have a linear elastic behaviour, materials such as reinforced concrete and steel have a substantial plastic performance. In practice this means that on reaching an elastic yield point the material continues to deform but with little or no change in maximum stress.

Figure 2.8 shows what this means in terms of stresses in the beam. As the loading on the beam increases extreme fibre stresses reach the yield point, σ_y, and remain constant as the beam continues to bend. A zone of plastic yielding begins to penetrate into the interior of the beam, until a point is reached immediately prior to complete failure, when practically all the cross-section has yielded plastically. The average stress at failure is σ_y, rather than $\sigma_y/2$ as was

Table 2.4 Second moments of area

Shape		Second moment of area
Rectangle		$I_{xx} = bd^3/12 \quad I_{aa} = \dfrac{bd^3}{3}$ $I_{yy} = db^3/12$
Triangle		$I_{xx} = bh^3/36$ $I_{aa} = bh^3/12$
Disk		$I_{xx} = \pi R^4/4$ $I_{polar} = 2I_{xx}$
Hollow rectangle		$I_{xx} = \dfrac{BD^3 - bd^3}{12}$ $I_{yy} = \dfrac{DB^3 - db^3}{12}$
I-section		$I_{xx} = \dfrac{BD^3 - bd^3}{12}$ $I_{yy} = \dfrac{2TB^3}{12} + \dfrac{dt^3}{12}$

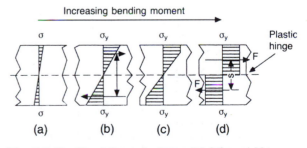

Fig. 2.8 Bending failure of a beam (a) below yield (elastic); (b) at yield point (elastic); (c) beyond yield (partially plastic); (d) beyond yield (fully plastic).

found to be the case when the material was assumed to have a linear-elastic behaviour. The moment of resistance assuming plastic behaviour is given by

$$M_r = Fs = \frac{bd\sigma_y}{2}\frac{d}{2} = \frac{bd^2\sigma_y}{4} = S\sigma_y \qquad (2.8)$$

where S is the plastic section modulus and is equal to $bd^2/4$ for rectangular beams. By setting the design moment equal to the moment of resistance of the beam its size and strength can be calculated according to plastic criteria.

2.6 Column design

Generally, column design is relatively straightforward. The design process simply involves making sure that the design load does not exceed the load capacity of the column, i.e.

$$\text{Load capacity} \geq \text{design axial load} \qquad (2.9)$$

Where the column is required to resist a predominantly axial load, its load capacity is given by

$$\begin{aligned}\text{Load capacity} = \text{design stress} \times \text{area of} \\ \text{column cross-section} \qquad (2.10)\end{aligned}$$

The design stress is related to the crushing strength of the material(s) concerned. However, not all columns fail in this mode. Some fail due to a combination of buckling followed by crushing of the material(s) (*Fig. 1.3*). These columns will tend to have lower load-carrying capacities, a fact which is taken into account by reducing the design stresses in the column. The design stresses are related to the slenderness ratio of the column, which is found to be a function of the following factors:

1. geometric properties of the column cross-section, e.g. lateral dimension of column, radius of gyration;
2. length of column;
3. support conditions (*Fig. 2.5*).

The radius of gyration, r, is a sectional property which provides a measure of the column's ability to resist buckling. It is given by

$$r = (I/A)^{1/2} \qquad (2.11)$$

Generally, the higher the slenderness ratio, the greater the tendency for buckling and hence the lower the load capacity of the column. Most practical reinforced concrete columns are designed to fail by crushing and the design equations for this medium are based on equation 2.10 (*Chapter 3*). Steel columns, on the other hand, are designed to fail by a combination of buckling and crushing. Empirical relations have been derived to predict the design stress in steel columns in terms of the slenderness ratio (*Chapter 4*).

Example 2.6 Elastic and plastic moments of resistance of a beam section

Calculate the moment of resistance of a beam 50 mm wide by 100 mm deep with $\sigma_y = 20$ N mm^{-2} according to (i) elastic criteria and (ii) plastic criteria.

ELASTIC CRITERIA
From equation 2.3,

$$M_{r,el} = \frac{bd^2}{6}\sigma_y = \frac{50 \times 100^2 \times 20}{6} = 1.67 \times 10^6 \text{ N mm}$$

PLASTIC CRITERIA
From equation 2.8,

$$M_{r,pl} = \frac{bd^2\sigma_y}{4} = \frac{50 \times 100^2 \times 20}{4} = 2.5 \times 10^6 \text{ N mm}$$

Hence it can be seen that the plastic moment of resistance of the section is greater than the maximum elastic moment of resistance. This will always be the case but the actual difference between the two moments will depend upon the shape of the section.

In most real situations, however, the loads will be applied eccentric to the axes of the column. Columns will therefore be required to resist axial loads and bending moments about one or both axes. In some cases, the bending moments may be small and can be accounted for in design simply by reducing the allowable design stresses in the material(s). Where the moments are large, for instance columns along the outside edges of buildings, the analysis procedures become more complex and the designer may have to resort to the use of design charts. However, as different codes of practice treat this in different ways, the details will be discussed separately in the chapters which follow.

2.7 Summary

This chapter has presented a simple but conservative method for assessing the design loads acting on individual members of building structures. This assumes that all the joints in the structure are pinended and that the sequence of load transfer occurs in the order: ceiling/floor loads to beams to columns to foundations to ground. The design loads are used to calculate the design bending moments, shear forces, axial loads and deflections experienced by members. In combination with design strengths and other properties of the construction medium, the sizes of the structural members are determined.

Example 2.7 Analysis of column section

Determine whether the reinforced concrete column cross-section shown below would be suitable to resist an axial load of 1500 kN. Assume that the design compressive strengths of the concrete and steel reinforcement are 14 and 345 N mm^{-2} respectively.

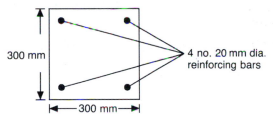

Area of steel bars = $4 \times (\pi 20^2/4)$ = 1256 mm^2

Net area of concrete = $300 \times 300 - 1256$ = 88 744 mm^2

Load capacity of column = force in concrete + force in steel

$$= 14 \times 88\,744 + 345 \times 1256$$

$$= 1\,675\,736 \text{ N} = 1675.7 \text{ kN}$$

Hence the column cross-section would be suitable to resist the design load of 1500 kN.

Questions

Basic structural concepts and material properties

Questions

1. For the beams shown, calculate and sketch the bending moment and shear force diagrams.

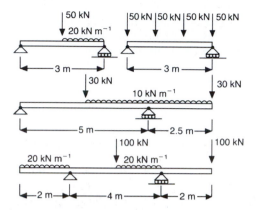

2. For the three load cases shown in *Fig. 2.4*, sketch the bending moment and shear force diagrams and hence determine the design bending moments and shear forces for the beam. Assume the main span is 6 m and the overhang is 2 m. The characteristic dead and imposed loads are respectively 20 kN/m and 10 kN/m.

3. (a) Calculate the area and the major axis moment of inertia, elastic modulus, plastic modulus and the radius of gyration of the steel I-section shown below.

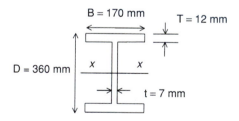

(b) Assuming the design strength of steel, σ_y, is 275 N/mm², calculate the moment of resistance of the I-section

according to (i) elastic and (ii) plastic criteria. Use your results to determine the working and collapse loads of a beam with this cross-section, 6 m long and simply supported at its ends. Assume the loading on the beam is uniformly distributed.

4. What are the most common ways in which columns can fail? List and discuss the factors that influence the load carrying capacity of columns.

5. The water tank shown in *Fig. 2.9* is subjected to the following characteristic dead (G_k), imposed (Q_k) and wind loads (W_k) respectively
   (i) 200 kN, 100 kN and 50 kN (Load case 1)
   (ii) 200 kN, 100 kN and 75 kN (Load case 2).

Assuming the support legs are pinned at the base, determine the design axial forces in both legs by considering the following load combinations:
(a) dead plus imposed
(b) dead plus wind
(c) dead plus imposed plus wind.
Refer to Table 3.4 for relevant load factors.

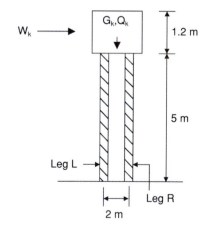

Fig. 2.9

28

PART TWO

STRUCTURAL DESIGN TO BRITISH STANDARDS

Part One has discussed the principles of limit state design and outlined general approaches towards assessing the sizes of beams and columns in building structures. Since limit state design forms the basis of the design methods in most modern codes of practice on structural design, there is considerable overlap in the design procedures presented in these codes.

The aim of this part of the book is to give detailed guidance on the design of a number of structural elements in the four media: concrete, steel, masonry and timber to the current British Standards. The work has been divided into four chapters as follows:

1. *Chapter 3* discusses the design procedures in BS 8110: *Code of Practice for the Structural Use of Concrete* relating to beams, slabs, pad foundations, retaining walls and columns.
2. *Chapter 4* discusses the design procedures in BS 5950: *Code of Practice for the Structural Use of Steelwork in Buildings* relating to beams, columns, floors and connections.
3. *Chapter 5* discusses the design procedures in BS 5628: *Code of Practice for Use of Masonry* relating to unreinforced loadbearing and panel walls.
4. *Chapter 6* discusses the design procedures in BS 5268: *Code of Practice for Structural Use of Timber* relating to beams, columns and stud walling.

Chapter 3

Design of reinforced concrete elements to BS 8110

This chapter is concerned with the detailed design of reinforced concrete elements to British Standard 8110. A general discussion of the different types of commonly occurring beams, slabs, walls, foundations and columns is given together with a number of fully worked examples covering the design of the following elements: singly and doubly reinforced beams, continuous beams, one-way and two-way spanning solid slabs, pad foundation, cantilever retaining wall and short braced columns supporting axial loads and uni-axial or bi-axial bending. The section which deals with singly reinforced beams is, perhaps, the most important since it introduces the design procedures and equations which are common to the design of the other elements mentioned above, with the exception of columns.

3.1 Introduction

Reinforced concrete is one of the principal materials used in structural design. It is a composite material, consisting of steel reinforcing bars embedded in concrete. These two materials have complimentary properties. Concrete, on the one hand, has high compressive strength but low tensile strength. Steel bars, on the other, can resist high tensile stresses but will buckle when subjected to comparatively low compressive stresses. Steel is much more expensive than concrete. By providing steel bars predominantly in those zones within a concrete member which will be subjected to tensile stresses, an economical structural material can be produced which is both strong in compression and strong in tension. In addition, the concrete provides corrosion protection and fire resistance to the more vulnerable embedded steel reinforcing bars.

Reinforced concrete is used in many civil engineering applications such as the construction of structural frames, foundations, retaining walls, water retaining structures, highways and bridges. They are normally designed in accordance with the recommendations given in various documents including BS 5400: *Code of Practice for the Design of Steel, Concrete and Composite Bridges*, BS 8007: *Code of Practice for the Design of Concrete Structures for Retaining Aqueous Liquids* and BS 8110: *Code of Practice for the Structural Use of Concrete in Buildings and Structures*. Since the primary aim of this book is to give guidance on the design of structural elements, this is best illustrated by considering the contents of BS 8110.

BS 8110 is divided into the following three parts:

Part 1: *Code of Practice for Design and Construction.*
Part 2: *Code of Practice for Special Circumstances.*
Part 3: *Design Charts for Singly Reinforced Beams, Doubly Reinforced Beams and Rectangular Columns.*

Part 1 covers most of the material required for everyday design. Given that most of this chapter is concerned with the contents of Part 1, it should be assumed that all references to BS 8110 refer to Part 1 exclusively. Part 2 covers subjects such as torsional resistance, calculation of deflections and estimation of elastic deformations. These aspects of design are beyond the scope of this book and Part 2, therefore, is not discussed here. Part 3 of BS 8110 contains charts for use in the design of singly reinforced beams, doubly reinforced beams and rectangular columns. A number of design examples illustrating the use of these charts are included in the relevant sections of this chapter.

3.2 Objectives and scope

All reinforced concrete building structures are composed of various categories of elements including slabs, beams, columns, walls and foundations (*Fig. 3.1*). Within each category is a range of element types. The aim of this chapter is to describe the element types and, for selected elements, to give guidance on their design.

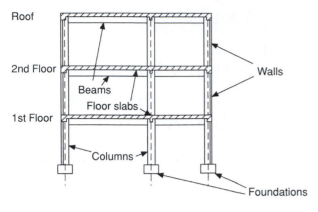

Fig. 3.1 Some elements of a structure.

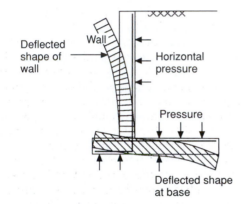

Fig. 3.2 Cantilever retaining wall.

A great deal of emphasis has been placed in the text to highlight the similarities in structural behaviour and, hence, design of the various categories of elements. Thus, certain slabs can be regarded for design purposes as a series of transversely connected beams. Columns may support slabs and beams but columns may also be supported by slabs and beams, in which case the latter are more commonly referred to as foundations. Cantilever retaining walls are usually designed as if they consist of three cantilever beams as shown in *Fig. 3.2*. Columns are different in that they are primarily compression members rather than beams and slabs which predominantly resist bending. Therefore columns are dealt with separately at the end of the chapter.

Irrespective of the element being designed, the designer will need a basic understanding of the following aspects which are discussed next

1. symbols
2. basis of design
3. material properties
4. loading
5. stress – strain relationships
6. durability and fire resistance.

The detailed design of beams, slabs, foundations, retaining walls and columns will be discussed in sections 3.9, 3.10, 3.11, 3.12 and 3.13 respectively.

3.3 Symbols

For the purpose of this book, the following symbols have been used. These have largely been taken from BS 8110. Note that in one or two cases the same symbol is differently defined. Where this occurs the reader should use the definition most appropriate to the element being designed.

GEOMETRIC PROPERTIES

b	width of section
d	effective depth of the tension reinforcement
h	overall depth of section
x	depth to neutral axis
z	lever arm
d'	depth to the compression reinforcement
ℓ	effective span
c	nominal cover to reinforcement

BENDING

F_k	characteristic load
g_k, G_k	characteristic dead load
q_k, Q_k	characteristic imposed load
w_k, W_k	characteristic wind load
f_k	characteristic strength
f_{cu}	characteristic compressive cube strength of concrete
f_y	characteristic tensile strength of reinforcement
γ_f	partial safety factor for load
γ_m	partial safety factor for material strengths
K	coefficient given by $M/f_{cu}bd^2$
K'	0.156 when redistribution does not exceed 10%
M	design ultimate moment
M_u	design ultimate moment of resistance
A_s	area of tension reinforcement
A_s'	area of compression reinforcement
Φ	diameter of main steel
Φ'	diameter of links

SHEAR

f_{yv}	characteristic strength of links
s_v	spacing of links along the member

V design shear force due to ultimate loads

v design shear stress

v_c design concrete shear stress

A_{sv} total cross-sectional area of shear reinforcement

COMPRESSION

b width of column

h depth of column

ℓ_o clear height between end restraints

ℓ_e effective height

ℓ_{ex} effective height in respect of x–x axis

ℓ_{ey} effective height in respect of y–y axis

N design ultimate axial load

A_c net cross-sectional area of concrete in a column

A_{sc} area of vertical reinforcement

3.4 Basis of design

The design of reinforced concrete elements to BS 8110 is based on limit state philosophy. As discussed in Chapter 1, the two principal categories of limit states normally considered in design are:

(i) ultimate limit state
(ii) serviceability limit state.

The ultimate limit state models the behaviour of the element at failure due to a variety of mechanisms including excessive bending, shear and compression or tension. The serviceability limit state models the behaviour of the member at working loads and in the context of reinforced concrete design is principally concerned with the limit states of deflection and cracking.

Having identified the relevant limit states, the design process simply involves basing the design on the most critical one and then checking for the remaining limit states. This requires an understanding of

1. material properties
2. loadings.

3.5 Material properties

The two materials whose properties must be known are concrete and steel reinforcement. In the case of concrete, the property with which the designer is primarily concerned is its compressive strength. For steel, however, it is its tensile strength capacity which is important.

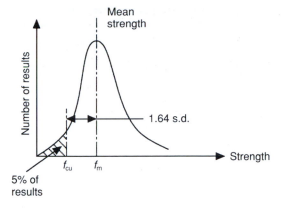

Fig. 3.3 *Normal frequency distribution of strengths.*

3.5.1 CHARACTERISTIC COMPRESSIVE STRENGTH OF CONCRETE, f_{cu}

Concrete is a mixture of water, coarse and fine aggregate and a cementitious binder (normally Portland cement) which hardens to a stone like mass. As can be appreciated, it is difficult to produce a homogeneous material from these components. Furthermore, its strength and other properties may vary considerably due to operations such as transportation, compaction and curing.

The compressive strength of concrete is usually determined by carrying out compression tests on 28 day old cubes which have been prepared using a standard procedure laid down in BS 1881. If a large number of compression tests were carried out on concrete cubes made from the same mix, it would be found that a plot of crushing strength against frequency of occurrence would approximate to a normal distribution (*Fig. 3.3*).

For design purposes it is necessary to assume a unique value for the strength of the mix. However, choosing too high a value will result in a high probability that most of the structure will be constructed with concrete having a strength below this value. Conversely, too low a value will result in inefficient use of the material. As a compromise between economy and safety, BS 8110 refers to the characteristic strength (f_{cu}) which is defined as the value below which not more than 5 per cent of the test results fall.

The characteristic and mean strength (f_m) of a sample are related by the expression:

$$f_{cu} = f_m - 1.64 \text{ s.d.}$$

where s.d. is the standard deviation. Thus assuming that the mean strength is 35 N/mm² and standard deviation is 3 N/mm², the characteristic strength of the mix is $35 - 1.64 \times 3 = 30$ N/mm².

Table 3.1 Concrete compressive strengths

Concrete grade	Characteristic strength, f_{cu} (N/mm²)
C25	25
C30	30
C35	35
C40	40
C45	45
C50	50

Concrete of a given strength can be identified by its 'grade'. For example a grade 30 concrete (C30) has a characteristic strength of 30 N/mm². For reinforced concrete made with normal aggregates, BS 8110 recommends that the lowest grade of concrete should be C25 although, in practice, a C30 mix is invariably necessary because of durability considerations (*section 3.8*). *Table 3.1* shows the characteristic strengths of various grades of concrete normally specified in reinforced concrete design.

3.5.2 CHARACTERISTIC STRENGTH OF REINFORCEMENT, f_y

Concrete is strong in compression but weak in tension. Because of this it is normal practice to provide steel reinforcement in those areas where tensile stresses in the concrete are most likely to develop. Consequently, it is the tensile strength of the reinforcement which most concerns the designer.

The tensile strength of reinforcement can be determined using the procedure laid down in BS EN 10002: Part 1. The tensile strength will also vary 'normally' with specimens of the same composition. Using the same reasoning as above, BS 8110 recommends that design should be based on the characteristic strength of the reinforcement (f_y) and gives typical values for mild steel and high yield steel reinforcement, the two types used in UK practice, of 250 N/mm² and 460 N/mm² respectively (*Table 3.2*).

Table 3.2 Strength of reinforcement (Table 3.1, BS 8110)

Reinforcement type	Characteristic strength, f_y (N/mm²)
Hot rolled mild steel	250
High yield steel (hot rolled or cold worked)	460

Table 3.3 Values of γ_m for the ultimate limit state (Table 2.2, BS 8110)

Material/stress type	Partial safety factor, γ_m
Reinforcement	1.05
Concrete in flexure or axial load	1.50
Concrete shear strength without shear reinforcement	1.25
Concrete bond strength	1.40
Concrete, others (e.g. bearing stress)	≥ 1.50

3.5.3 DESIGN STRENGTH

Tests to determine the characteristic strengths of concrete and steel reinforcement are carried out on near perfect specimens, which have been prepared under laboratory conditions. Such conditions will seldom exist in practice. Therefore it is undesirable to use characteristic strengths to size members.

To take account of differences between actual and laboratory values, local weaknesses and inaccuracies in assessment of the resistances of sections, the characteristic strengths (f_k) are divided by appropriate partial safety factor for strengths (γ_m), obtained from *Table 3.3*. The resulting values are termed design strengths and it is the design strengths which are used to size members.

$$\text{Design strength} = \frac{f_k}{\gamma_m} \qquad (3.1)$$

It should be noted that the partial safety factor for reinforcement (γ_{ms}) is always 1.05 but for concrete (γ_{mc}) assumes different values depending upon the stress type under consideration. Furthermore, the partial safety factors for concrete are all greater than that for reinforcement since concrete quality is less controllable.

3.6 Loading

In addition to material properties, the designer needs to know the type and magnitude of the loading to which the structure may be subject during its design life.

The loads acting on a structure are divided into three basic types: dead, imposed and wind (*section 2.2*). Associated with each type of loading there are characteristic and design values, which must be assessed before the individual elements of the structure can be designed. These aspects are discussed next.

Table 3.4 Values of γ_f for various load combinations (based on Table 2.1, BS 8110)

Load combination	Dead, G_k		Imposed, Q_k		Wind, W_k
	Adverse	*Beneficial*	*Adverse*	*Beneficial*	
1. Dead and imposed	1.4	1.0	1.6	0	–
2. Dead and wind	1.4	1.0	–	–	1.4
3. Dead and wind and imposed	1.2	1.2	1.2	1.2	1.2

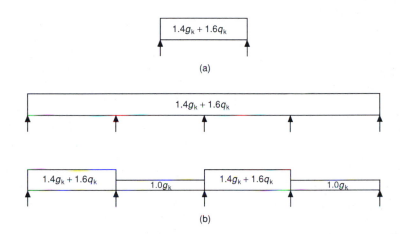

Fig. 3.4 *Ultimate design loads: (a) single span beam; (b) continuous beam.*

3.6.1 CHARACTERISTIC LOAD

As noted in Chapter 2, it is not possible to apply statistical principles to determine characteristic dead (G_k), imposed (Q_k) and wind (W_k) loads simply because there are insufficient data. Therefore, the characteristic loads are taken to be those given in the following documents:

1. BS 648: *Schedule of Weights for Building Materials.*
2. BS 6399: *Design Loadings for Buildings*, Part 1: *Code of Practice for Dead and Imposed Loads*; Part 2: *Code of Practice for Wind Loads.*

3.6.2 DESIGN LOAD

Variations in the characteristic loads may arise due to a number of reasons such as errors in the analysis and design of the structure, constructional inaccuracies and possible unusual load increases. In order to take account of these effects, the characteristic loads (F_k) are multiplied by the appropriate partial safety factor for loads (γ_f), taken from Table 3.4, to give the design loads acting on the structure:

$$\text{Design load} = \gamma_f F_k \qquad (3.2)$$

Generally, the 'adverse' factors will be used to derive the design loads acting on the structure. For example, for simple beams subject to only dead and imposed loads the appropriate values of γ_f are generally 1.4 and 1.6 respectively (*Fig. 3.4(a)*). However, for continuous beams, load cases must be analysed which may include maximum and minimum design loads on alternate spans (*Fig. 3.4(b)*).

The design loads are used to calculate the distribution of bending moments and shear forces in the structure usually using elastic analysis methods as discussed in *Chapter 2*. At no point should they exceed the corresponding design strengths of the member, otherwise failure of the structure may arise.

The design strength is a function of the distribution of stresses in the member. Thus, for the simple case of a steel bar in direct tension the design

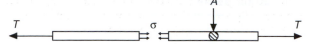

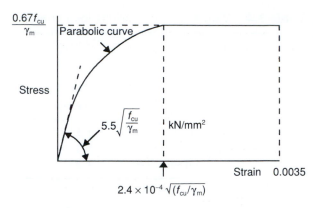

Fig. 3.7 Design stress-strain curve for concrete in compression (*Fig. 2.1, BS 8110*).

Fig. 3.5 Design strength of the bar. Design strength, $T, = \sigma.A$, where σ is the average stress at failure and A the cross-sectional area of the bar.

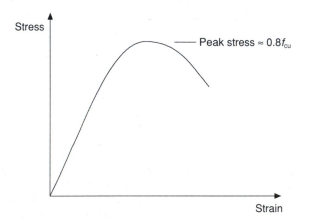

Fig. 3.6 Actual stress-strain curve for concrete in compression.

strength is equal to the cross-sectional area of the bar multiplied by the average stress at failure (*Fig. 3.5*). The distribution of stresses in reinforced concrete members is usually more complicated, but can be estimated once the stress-strain behaviour of the concrete and steel reinforcement is known. This aspect is discussed next.

3.7 Stress–strain curves

3.7.1 STRESS-STRAIN CURVE FOR CONCRETE

Figure 3.6 shows a typical stress-strain curve for a concrete cylinder under uniaxial compression. Note that the stress-strain behaviour is never truly linear and that the maximum compressive stress at failure is approximately $0.8 \times$ characteristic strength (i.e. $0.8f_{cu}$).

However, the actual behaviour is rather complicated to model mathematically and, therefore, BS 8110 uses the modified stress-strain curve shown in *Fig. 3.7* for design. This assumes that the peak stress is only 0.67 (rather than 0.8) times the characteristic strength and hence the design stress for concrete is given by

$$\begin{array}{l} \text{Design compressive} \\ \text{stress for concrete} \end{array} = \frac{0.67 f_{cu}}{\gamma_{mc}} \approx 0.45 f_{cu} \quad (3.3)$$

In other words, the failure stress assumed in design is approximately $0.45/0.8 = 56\%$ of the actual stress at failure when near perfect specimens are tested.

3.7.2 STRESS-STRAIN CURVE FOR STEEL REINFORCEMENT

A typical tensile stress-strain curve for steel reinforcement is shown in *Fig. 3.8*. It can be divided into two regions: (i) an elastic region where strain is proportional to stress and (ii) a plastic region where small increases in stress produce large increases in strain. The change from elastic to plastic behaviour occurs at the yield stress and is significant since it defines the characteristic strength of reinforcement (f_y).

Once again, the actual material behaviour is rather complicated to model mathematically and therefore BS 8110 modifies it to the form shown in *Fig. 3.9* which also includes the idealised stress-strain relationship for reinforcement in compression. The maximum design stress for reinforcement in tension and compression is given by

$$\text{Design stress for reinforcement} = \frac{f_y}{\gamma_{ms}} \quad (3.4)$$

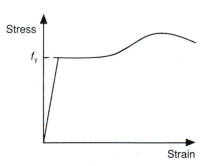

Fig. 3.8 Actual stress-strain curve for reinforcement.

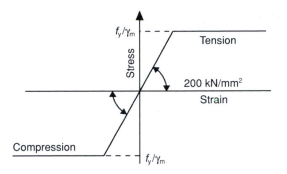

Fig. 3.9 *Design stress-strain curve for reinforcement (Fig. 2.2, BS 8110).*

From the foregoing it is possible to determine the distribution of stresses at a section and hence calculate the design strength of the member. The latter is normally carried out using the equations given in BS 8110. However, before considering these in detail, it is useful to pause for a moment in order to introduce BS 8110's requirements in respect of durability and fire resistance since these requirements are common to several of the elements which will be subsequently discussed.

3.8 Durability and fire resistance

Apart from the need to ensure that the design is structurally sound, the designer must also verify the proper performance of the structure in service. Principally this will involve the two limit states of (i) durability and (ii) fire resistance.

3.8.1 DURABILITY

Many concrete structures are showing signs of severe deterioration after only a few years of service. Repair of these structures is both difficult and very costly. Therefore, over recent years, much effort has been directed towards improving the durability requirements, particularly with respect to the protection of embedded metal from corrosion. This has been largely achieved by specifying limits for:

1. minimum depth of cover to the reinforcement;
2. minimum cement content;
3. maximum water/cement ratios;
4. maximum allowable crack widths.

Table 3.5 gives the nominal (i.e. minimum plus 5 mm) depths of concrete cover to all reinforcement for specified exposure conditions, minimum cement contents and maximum water/cement ratios. The limits on water/cement ratio and cement content will automatically be assured by specifying the minimum grades of concrete indicated in the table.

BS 8110 further recommends that the maximum crack width should not exceed 0.3 mm in order to avoid corrosion of the reinforcing bars. This

Table 3.5 Exposure conditions and nominal covers to all reinforcement (including links) to meet durability requirements (based on Tables 3.2 and 3.3, BS 8110).

Conditions of exposure	Nominal cover (mm)				
Mild: for example, protected against weather or aggressive condition	25	20	20	20	20
Moderate: for example, sheltered from severe rain and freezing while wet; subject to condensation or continuously under water; in contact with non-aggressive soil	–	35	30	25	20
Severe: for example, exposed to severe rain; alternate wetting and drying; occasional freezing or severe condensation	–	–	40	30	25
Very Severe: for example, exposed to sea water spray, de-icing salts, corrosive fumes or severe wet freezing	–	–	50	40	30
Most Severe: for example, frequent exposure to sea water spray or de-icing salts	–	–	–	–	50
Abrasive: for example, exposed to abrasive action (water carrying solids, machinery or metal tyred vehicles)	–	–	–	See Note 1	
Maximum free water/cement ratio	0.65	0.60	0.55	0.50	0.45
Minimum cement content (kg/m³)	275	300	325	350	400
Lowest concrete grade	C30	C35	C40	C45	C50

Note 1. Cover should not be less than the nominal value corresponding to the relevant environmental category plus any allowance for loss of cover due to abrasion.

Table 3.6 Nominal cover to all reinforcement to meet specified periods of fire resistance (based on Table 3.4, BS 8110)

Fire resistance (hours)	Nominal cover (mm)				
	Beams		Floors		Columns
	Simply supported	Continuous	Simply supported	Continuous	
0.5	20	20	20	20	20
1.0	20	20	20	20	20
1.5	20	20	25	20	20
2.0	40	30	35	25	25
3.0	60	40	45	35	25
4.0	70	50	55	45	25

requirement will generally be satisfied by observing the detailing rules given in BS 8110 with regards to:

1. minimum reinforcement areas;
2. maximum clear spacings between reinforcing bars.

These requirements will be discussed individually for beams, slabs and columns in *sections 3.9.1.6, 3.10.2.4* and *3.13.6* respectively.

3.8.2 FIRE PROTECTION

Fire protection of reinforced concrete members is largely achieved by specifying limits for:

1. minimum thickness of cover to the reinforcement;
2. minimum dimensions of members.

Table 3.6 gives the actual values of the nominal depths of concrete covers to all reinforcement for specified periods of fire resistance and member types. The covers in the *Table 3.6* may need to be increased because of durability considerations. The minimum dimensions of members for fire resistance are shown in *Fig. 3.10*.

Having discussed these more general aspects relating to structural design, the detailed design of beams is considered next.

3.9 Beams

Beams in reinforced concrete structures can be defined according to:

1. cross-section
2. position of reinforcement
3. support conditions.

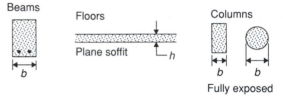

Fire resistance (hours)	Minimum dimension (mm)		
	Beam width (b)	Floor thickness (h)	Exposed column width (b)
0.5	200	75	150
1.0	200	95	200
1.5	200	110	250
2.0	200	125	300
3.0	240	150	400
4.0	280	170	450

Fig. 3.10 *Minimum dimensions of reinforced concrete members for fire resistance (based on Fig. 3.2, BS 8110).*

Some common beam sections are shown in *Fig. 3.11*. Beams reinforced with tension steel only are referred to as singly reinforced. Beams reinforced with tension and compression steel are termed doubly reinforced. Inclusion of compression steel will increase the moment capacity of the beam and hence allow more slender sections to be used. Thus, doubly reinforced beams are used in preference to singly reinforced beams when there is some restriction on the construction depth of the section.

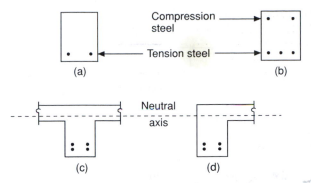

Fig. 3.11 *Beam sections: (a) singly reinforced; (b) doubly reinforced; (c) T-section; (d) L-section.*

Fig. 3.12 *Support conditions: (a) simply supported; (b) continuous.*

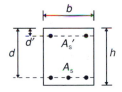

Fig. 3.13 *Notation.*

Under certain conditions, T and L beams are more economical than rectangular beams since some of the concrete below the dotted line (neutral axis), which serves only to contain the tension steel, is removed resulting in a reduced unit weight of beam. Furthermore, beams may be simply supported at their ends or continuous, as illustrated in *Fig. 3.12*.

Figure 3.13 illustrates some of the notation used in beam design. Here b is the width of the beam, h the overall depth of section, d the effective depth of tension reinforcement, d' the depth of compression reinforcement, A_s the area of tension reinforcement and A_s' is the area of compression reinforcement.

The following sub-sections consider the design of:

1. singly reinforced beams
2. doubly reinforced beams
3. continuous, L and T beams.

3.9.1 SINGLY REINFORCED BEAM DESIGN

All beams may fail due to excessive bending or shear. In addition, excessive deflection of beams must be avoided otherwise the efficiency or appearance of the structure may become impaired. As discussed in *section 3.4*, bending and shear are ultimate states while deflection is a serviceability state. Generally, structural design of concrete beams primarily involves consideration of the following aspects which are discussed next.

1. bending
2. shear
3. deflection.

3.9.1.1 Bending (clause 3.4.4.4, BS 8110)

Consider the case of a simply supported, singly reinforced, rectangular beam subject to a uniformly distributed load ω as shown in *Figs 3.14* and *3.15*.

The load causes the beam to deflect downwards, putting the top portion of the beam into compression and the bottom portion into tension. At some distance x below the compression face, the section is neither in compression or tension and therefore the strain at this level is zero. This axis is normally referred to as the neutral axis.

Assuming that plane sections remain plane, the strain distribution will be triangular (*Fig. 3.15(b)*). The stress distribution in the concrete above the neutral axis is initially triangular (*Fig. 3.15(c)*), for low values of strain, because stress and strain are directly proportional (*Fig. 3.7*). The stress in the concrete below the neutral axis is zero, however, since it is assumed that the concrete is cracked, being unable to resist any tensile stress. All the tensile stresses in the member are assumed to be resisted by the steel reinforcement and this is

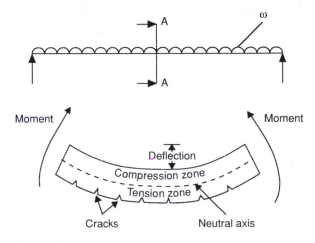

Fig. 3.14

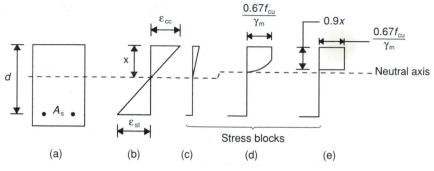

Fig. 3.15 *Stress and strain distributions at section A-A: (a) section; (b) strain; (c) triangular (low strain); (d) rectangular parabolic (large strain); (e) equivalent rectangular.*

reflected in a peak in the tensile stress at the level of the reinforcement.

As the intensity of loading on the beam increases, the mid-span moment also increases and the distribution of stresses change from that shown in *Fig. 3.15(c)* to *3.15(d)*. The stress in the reinforcement increases linearly with strain up to the yield point. Thereafter it remains at a constant value (*Fig. 3.9*). As the strain in the concrete increases, the stress distribution is assumed to follow the parabolic form of the stress-strain relationship for concrete under compression (*Fig. 3.7*).

The final stress distribution at a given section and the mode of failure of the beam will depend upon whether the section is (1) under-reinforced or (2) over-reinforced. If the section is over-reinforced the steel does not yield and failure occurs due to crushing of the concrete as a result of the concrete's compressive capacity being exceeded. Steel is expensive and, therefore, over-reinforcing will lead to uneconomical design. Furthermore, with this type of failure there may be no external warning signs; just sudden, catastrophic collapse.

If the section is under-reinforced, the steel yields and failure will again occur due to crushing of the concrete. However the beam experiences considerable deflection, which results in severe cracking and spalling from the tension face. Ample warning signs of failure are therefore provided. Moreover, this form of design is more economical since a greater proportion of the steel strength is utilised. For these reasons, it is normal practice to design sections which are under-reinforced rather than over-reinforced.

In an under-reinforced section, since the reinforcement will have yielded, the tensile force in the steel (F_{st}) at the ultimate limit state can be readily calculated using the following:

$$F_{st} = \text{design stress} \times \text{area}$$

$$= \frac{f_y A_s}{\gamma_{ms}} \quad \text{(using equation 3.4)} \quad (3.5)$$

where
f_y = yield stress
A_s = area of reinforcement
γ_{ms} = factor of safety for reinforcement (= 1.05)

However, it is not an easy matter to calculate the compressive force in the concrete because of the complicated pattern of stresses in the concrete. To simplify the situation, BS 8110 replaces the rectangular-parabolic stress distribution with an equivalent rectangular stress distribution (*Fig. 3.15(e)*). And it is the rectangular stress distribution which is used in order to develop the design formulae for rectangular beams given in clause 3.4.4.4 of BS 8110. Specifically, the code gives formulae for the following design parameters which are derived below.

1. ultimate moment of resistance
2. area of tension reinforcement
3. lever arm.

(i) Ultimate moment of resistance, M_u. Consider the singly reinforced beam shown in *Fig. 3.16*. The loading on the beam gives rise to an ultimate design moment (M) at mid-span. The resulting curvature of the beam produces a compression force in the concrete (F_{cc}) and a tensile force in the reinforcement (F_{st}). Since there is no resultant axial force on the beam, the force in the concrete must equal the force in the reinforcement:

$$F_{cc} = F_{st} \quad (3.6)$$

These two forces are separated by a distance z, the moment of which forms a couple (M_u) which

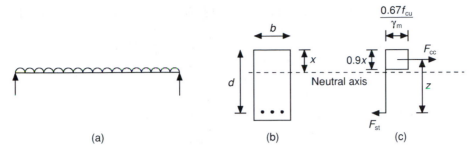

Fig. 3.16 *Ultimate moment of resistance for singly reinforced section.*

opposes the design moment. For structural stability $M_u \geq M$ where

$$M_u = F_{cc}z = F_{st}z \qquad (3.7)$$

From the stress block shown in *Fig. 3.16(c)*

$$F_{cc} = \text{stress} \times \text{area} = \frac{0.67f_{cu}}{\gamma_{mc}}0.9xb \qquad (3.8)$$

and

$$z = d - 0.9x/2 \qquad (3.9)$$

In order to ensure that the section is under-reinforced, BS 8110 limits the depth of the neutral axis (x) to a maximum of $0.5d$, where d is the effective depth. Hence

$$x \leq 0.5d \qquad (3.10)$$

By combining equations 3.7–3.10 and putting $\gamma_{mc} = 1.5$ (*Table 3.3*) it can be shown that the ultimate moment of resistance is given by:

$$M_u = 0.156f_{cu}bd^2 \qquad (3.11)$$

Note that M_u depends only on the properties of the concrete and not the steel reinforcement. Provided that the design moment does not exceed M_u (i.e. $M \leq M_u$), a beam whose section is singly reinforced will be sufficient to resist the design moment. The following section derives the equation necessary to calculate the area of reinforcement needed for such a case.

(ii) Area of tension reinforcement, A_s. At the limiting condition $M_u = M$, equation 3.7 becomes

$$M = F_{st}\cdot z = \frac{f_yA_s}{\gamma_{ms}}z \quad \text{(from equation 3.5)}$$

Rearranging and putting $\gamma_{ms} = 1.05$ (*Table 3.3*) gives

$$A_s = \frac{M}{0.95f_yz} \qquad (3.12)$$

Solution of this equation requires an expression for z which can either be obtained graphically (*Fig. 3.17*) or by calculation as discussed below.

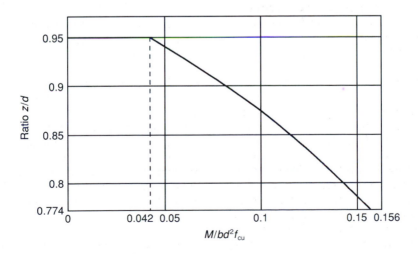

Fig. 3.17 *Lever-arm curve.*

(iii) Lever arm, z. At the limiting condition $M_u = M$, equation 3.7 becomes

$$M = F_{cc}z = \frac{0.67f_{cu}}{\gamma_{mc}}0.9bxz \quad \text{(from equation 3.8)}$$

$$= 0.4f_{cu}bzx \quad \text{(putting } \gamma_{mc} = 1.5\text{)}$$

$$= 0.4f_{cu}bz2\frac{(d-z)}{0.9} \quad \text{(from equation 3.9)}$$

$$= \frac{8}{9}f_{cu}bz(d-z)$$

Dividing both sides by $f_{cu}bd^2$ gives

$$\frac{M}{f_{cu}bd^2} = \frac{8}{9}(z/d)(1 - z/d)$$

Substituting $K = \dfrac{M}{f_{cu}bd^2}$ and putting $z_o = z/d$ gives

$$0 = z_o^2 - z_o + 9K/8$$

This is a quadratic and can be solved to give

$$z_o = z/d = 0.5 + \sqrt{(0.25 - K/0.9)}$$

This equation is used to draw the lever arm curve shown in *Fig. 3.17* and is usually expressed in the following form

$$z = d[0.5 + \sqrt{(0.25 - K/0.9)}] \quad (3.13)$$

Once z has been determined, the area of tension reinforcement, A_s, can be calculated using equation 3.12. In clause 3.4.4.1 of BS 8110 it is noted that z should not exceed $0.95d$ in order to give a reasonable concrete area in compression. Moreover, it should be remembered that equation 3.12 can only be used to determine A_s provided that $M \leqslant M_u$ or $K \leqslant K'$ where

$$K = \frac{M}{f_{cu}bd^2} \quad \text{and} \quad K' = \frac{M_u}{f_{cu}bd^2}$$

To summarize, design for bending requires the calculation of the maximum design moment (M) and corresponding ultimate moment of resistance of the section (M_u). Provided $M \leqslant M_u$, only tension reinforcement is needed and the area of steel can be calculated using equation 3.12 via equation 3.13. Where $M > M_u$ the designer has the option to either increase the section sizes (i.e. then $M \leqslant M_u$) or design as a doubly reinforced section. The latter option is discussed more fully in *section 3.9.2*.

Example 3.1 Design of bending reinforcement for a singly reinforced beam (BS 8110)

A simply supported rectangular beam of 7 m span carries characteristic dead (including self weight of beam), g_k, and imposed, q_k, loads of 12 kN/m and 8 kN/m respectively (*Fig. 3.18*). The beam dimensions are breadth, b, 275 mm and effective depth, d, 450 mm. Assuming the following material strengths, calculate the area of reinforcement required.

$$f_{cu} = 30 \text{ kN/m}^2$$

$$f_y = 460 \text{ kN/m}^2$$

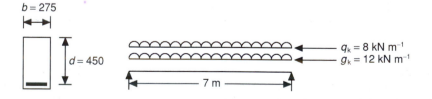

Fig. 3.18

Ultimate load (w) = $1.4g_k + 1.6q_k$

$$= 1.4 \times 12 + 1.6 \times 8 = 29.6 \text{ kN/m}$$

Design moment $(M) = \dfrac{w\ell^2}{8} = \dfrac{29.6 \times 7^2}{8} = 181.3$ kN m

Ultimate moment of resistance $(M_u) = 0.156 f_{cu} bd^2$

$$= 0.156 \times 30 \times 275 \times 450^2 \times 10^{-6} = 260.6 \text{ kN m}$$

Since $M_u > M$ design as a singly reinforced beam.

$$K = \frac{M}{f_{cu}bd^2} = \frac{181.3 \times 10^6}{30 \times 275 \times 450^2} = 0.1085$$

$$z = d[0.5 + \sqrt{(0.25 - K/0.9)}]$$

$$= 450[0.5 + \sqrt{(0.25 - 0.1085/0.9)}]$$

$$= 386.8 \text{ mm} \leqslant 0.95d \ (= 427.5 \text{ mm}) \quad \text{OK}$$

$$A_s = \frac{M}{0.95 f_y z} = \frac{181.3 \times 10^6}{0.95 \times 460 \times 386.8} = 1073 \text{ mm}^2$$

For detailing purposes this area of steel has to be transposed into a certain number of bars of a given diameter. This is usually achieved using steel area tables similar to that shown in *Table 3.7*. Thus it can be seen that four 20 mm diameter bars have a total cross-sectional area of 1260 mm² and would therefore be suitable. Hence provide 4T20. (N.B. T refers to high yield steel bars ($f_y = 460$ N/mm²); R refers to mild steel bars ($f_y = 250$ N/mm²).

Table 3.7   Cross-sectional areas of groups of bars (mm²)

Bar size (mm)	Number of bars									
	1	2	3	4	5	6	7	8	9	10
6	28.3	56.6	84.9	113	142	170	198	226	255	283
8	50.3	101	151	201	252	302	352	402	453	503
10	78.5	157	236	314	393	471	550	628	707	785
12	113	226	339	452	566	679	792	905	1 020	1 130
16	201	402	603	804	1 010	1 210	1 410	1 610	1 810	2 010
20	314	628	943	1 260	1 570	1 890	2 200	2 510	2 830	3 140
25	491	982	1 470	1 960	2 450	2 950	3 440	3 930	4 420	4 910
32	804	1 610	2 410	3 220	4 020	4 830	5 630	6 430	7 240	8 040
40	1 260	2 510	3 770	5 030	6 280	7 540	8 800	10 100	11 300	12 600

3.9.1.2 Design charts

An alternative method of determining the area of tensile steel required in singly reinforced rectangular beams is by using the design charts given in Part 3 of BS 8110. BSI issued these charts when the partial safety factor for steel reinforcement was 1.15 and not 1.05. Use of these charts will therefore produce conservative estimates of tensile steel areas. A modified version of chart 2 which incorporates the amended partial safety factor is reproduced here as *Fig. 3.19* and is appropriate for use with grade 460 reinforcement.

The design charts in BS 8110 are based on the rectangular parabolic stress distribution shown in *Fig. 3.15(d)* rather than the simplified rectangular distribution in *Fig. 3.15(e)*, used to develop the design formulae in the previous section. Therefore,

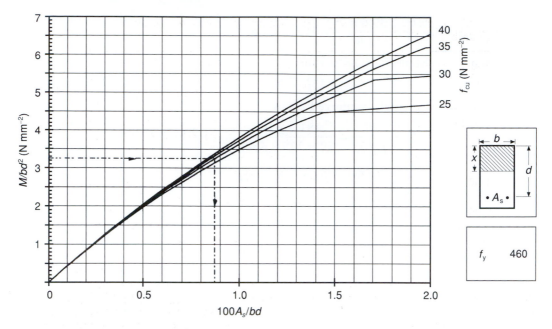

Fig. 3.19 *Design chart for singly reinforced beam (based on chart No. 2, BS 8110: Part 3).*

design charts may tend to produce slightly more economical estimates of the area of tensile steel than those obtained by using the design formulae.

The procedure involves the following steps:

1. Check $M \leqslant M_u$.
2. Select appropriate chart from Part 3 of BS 8110 based on the grade of tensile reinforcement.
3. Calculate M/bd^2.
4. Plot M/bd^2 ratio on chart and read off corresponding $100A_s/bd$ value using curve appropriate to grade of concrete selected for design.
5. Calculate A_s.

Using the figures given in Example 3.1,

$$M = 181.3 \text{ kN m} \leqslant M_u = 260.6 \text{ kN m}$$

$$\frac{M}{bd^2} = \frac{181.3 \times 10^6}{275 \times 450^2} = 3.26$$

From *Fig. 3.19*, using the $f_{cu} = 30$ N/mm² curve

$$\frac{100A_s}{bd} = 0.87$$

Hence, $A_s = 1076$ mm²

Therefore provide **4T20** ($A_s = 1260$ mm²)

3.9.1.3 Shear (clause 3.4.5, BS 8110)

Another way in which failure of a beam may arise is due to its shear capacity being exceeded.

Shear failure may arise in several ways, but the two principal failure mechanisms are shown in *Fig. 3.20*. With reference to *Fig. 3.20(a)*, as the loading increases, an inclined crack rapidly develops between the edge of the support and the load point, resulting in splitting of the beam into two pieces. This is normally termed diagonal tension failure and can be prevented by providing shear reinforcement.

The second failure mode, termed diagonal compression failure (*Fig. 3.20(b)*), occurs under the action of large shear forces acting near the support, resulting in crushing of the concrete. This type of failure is avoided by limiting the maximum shear stress to 5 N/mm² or $0.8\sqrt{f_{cu}}$, whichever is the lesser.

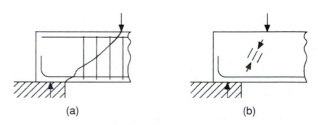

(a)   (b)

Fig. 3.20 *Types of shear failure: (a) diagonal tension; (b) diagonal compression.*

Table 3.8 Values of design concrete shear stress, v_c (N/mm²)
(Table 3.8, BS 8110)

$\dfrac{100A_s}{bd}$	Effective depth (d) mm							
	125	150	175	200	225	250	300	≥ 400
≤ 0.15	0.45	0.43	0.41	0.40	0.39	0.38	0.36	0.34
0.25	0.53	0.51	0.49	0.47	0.46	0.45	0.43	0.40
0.50	0.67	0.64	0.62	0.60	0.58	0.56	0.54	0.50
0.75	0.77	0.73	0.71	0.68	0.66	0.65	0.62	0.57
1.00	0.84	0.81	0.78	0.75	0.73	0.71	0.68	0.63
1.50	0.97	0.92	0.89	0.86	0.83	0.81	0.78	0.72
2.00	1.06	1.02	0.98	0.95	0.92	0.89	0.86	0.80
≥ 3.00	1.22	1.16	1.12	1.08	1.05	1.02	0.98	0.91

The design shear stress, v, at any cross-section can be calculated from:

$$v = V/bd \qquad (3.14)$$

where
V design shear force due to ultimate loads
b breadth of section
d effective depth of section

In order to determine whether shear reinforcement is required, it is necessary to calculate the shear resistance, or using BS 8110 terminology the design concrete shear stress, at critical sections along the beam. The design concrete shear stress, v_c, is found to be composed of three major components, namely:

1. concrete in the compression zone;
2. aggregate interlock across the crack zone;
3. dowel action of the tension reinforcement.

The design concrete shear stress can be determined using *Table 3.8*. The values are in terms of the percentage area of longitudinal tension reinforcement ($100A_s/bd$) and effective depth of the section (d). The table assumes that the grade of concrete is 25. For other grades of concrete, up to a maximum of 40, the design shear stresses can be determined by multiplying the values in the table by the factor $(f_{cu}/25)^{1/3}$.

Generally, where the design shear stress exceeds the design concrete shear stress, shear reinforcement will be needed. This is normally done by providing

1. vertical shear reinforcement commonly referred to as 'links' and/or
2. inclined (or bent-up) bars as shown below.

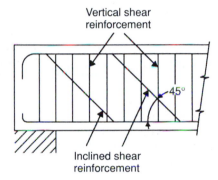

Vertical shear reinforcement

45°

Inclined shear reinforcement

Beam with vertical and inclined shear reinforcement.

The former is the most widely used method and will therefore be the only one discussed here. The following section derives the design equations for calculating the area and spacing of links.

(i) Shear resistance of links. Consider a reinforced concrete beam with links uniformly spaced at a distance s_v, under the action of a shear force V. The resulting failure plane is assumed to be inclined approximately 45° to the horizontal as shown in *Fig. 3.21*.

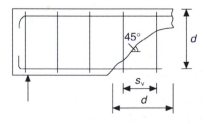

Fig. 3.21 *Shear resistance of links.*

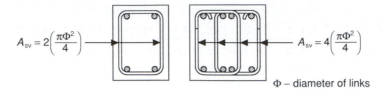

Fig. 3.22 A_{sv} *for varying shear reinforcement arrangements.*

Table 3.9 Form and area of links in beams (Table 3.7, BS 8110)

Values of v (N/mm^2)	Area of shear reinforcement to be provided
$v < 0.5v_c$ throughout the beam	No links required but normal practice to provide nominal links in members of structural importance
$0.5v_c < v < (v_c + 0.4)$	Nominal (or minimum) links for whole length of beam $A_{sv} \geqslant \dfrac{0.4bs_v}{0.95f_{yv}}$
$(v_c + 0.4) < v < 0.8\sqrt{f_{cu}}$ or 5 N/mm^2	Design links $A_{sv} \geqslant \dfrac{bs_v(v - v_c)}{0.95f_{yv}}$

The number of links intersecting the potential crack is equal to d/s_v and it follows therefore that the shear resistance of these links, V_{link}, is given by

$$V_{link} = \text{number of links} \times \text{total cross-sectional area of links } (Fig.\ 3.22) \times \text{design stress}$$

$$= (d/s_v) \times A_{sv} \times 0.95f_{yv}$$

The shear resistance of concrete, V_{conc}, can be calculated from:

$$V_{conc} = v_c bd \quad \text{(using equation 3.14)}$$

The design shear force due to ultimate loads, V, must be less than the sum of the shear resistance of the concrete (V_{conc}) plus the shear resistance of the links (V_{link}), otherwise failure of the beam may arise. Hence

$$V \leqslant V_{conc} + V_{link}$$

$$\leqslant v_c bd + (d/s_v)A_{sv}0.95f_{yv}$$

$$V/bd \leqslant v_c + (1/bs_v)A_{sv}0.95f_{yv}$$
$$\text{(dividing both sides by } bd)$$

$$v \leqslant v_c + (1/bs_v)A_{sv}0.95f_{yv}$$
$$\text{(from equation 3.14)}$$

rearranging gives $\dfrac{A_{sv}}{s_v} = \dfrac{b(v - v_c)}{0.95f_{yv}}$ (3.15)

Where $(v - v_c)$ is less than 0.4 N/mm^2 then links should be provided according to

$$\frac{A_{sv}}{s_v} = \frac{0.4b}{0.95f_{yv}} \quad (3.16)$$

Equations 3.15 and 3.16 provide a basis for calculating the minimum area and spacing of links. The details are discussed next.

(ii) Form, area and spacing of links. Shear reinforcement should be provided in beams according to the criteria given in *Table 3.9.*

Thus where the design shear stress is less than half the design concrete shear stress (i.e. $v < 0.5v_c$), no shear reinforcement will be necessary although, in practice, it is normal to provide nominal links in all beams of structural importance. Where $0.5v_c < v < (v_c + 0.4)$ nominal links based on equation 3.16 should be provided. Where $v > v_c + 0.4$, design links based on equation 3.15 should be provided.

BS 8110 further recommends that the spacing of links in the direction of the span should not exceed $0.75d$. This will ensure that at least one link crosses the potential crack.

Example 3.2 Design of shear reinforcement for a beam (BS 8110)

Design the shear reinforcement for the beam shown in *Fig. 3.23* using mild steel ($f_y = 250$ N/mm^2) links for the following load cases:

(i) $q_k = 0$
(ii) $q_k = 10$ kN/m
(iii) $q_k = 29$ kN/m
(iv) $q_k = 45$ kN/m

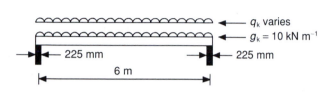

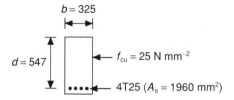

Fig. 3.23

DESIGN CONCRETE SHEAR STRESS, v_c

$$\frac{100A_s}{bd} = \frac{100 \times 1960}{325 \times 547} = 1.10$$

From *Table 3.8*, $v_c = 0.65$ N/mm^2 (see below)

$\dfrac{100A_s}{bd}$	*Effective depth (mm)*	
	300	*≥ 400*
	N/mm^2	N/mm^2
1.00	0.68	0.63
1.10		0.65
1.50	0.78	0.72

$q_k = 0$
Design shear stress (v)

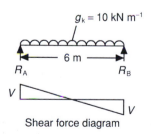

Total ultimate load, W, is

$$W = (1.4g_k + 1.6q_k)\text{span} = (1.4 \times 10 + 1.6 \times 0)6 = 84 \text{ kN}$$

Since beam is symmetrically loaded

$$R_A = R_B = W/2 = 42 \text{ kN}$$

Ultimate shear force (V) = 42 kN and design shear stress, v, is

$$v = \frac{V}{bd} = \frac{42 \times 10^3}{325 \times 547} = 0.24 \text{ N/mm}^2$$

Diameter and spacing of links
By inspection

$$v < v_c/2$$

i.e. 0.24 N/mm^2 < 0.32 N/mm^2. Hence from *Table 3.9*, shear reinforcement may not be necessary.

$q_k = 10$ kN/m

Design shear stress (v)

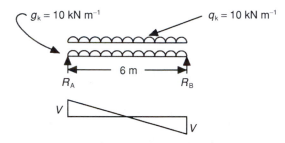

Total ultimate load, W, is

$$W = (1.4g_k + 1.6q_k)\text{span} = (1.4 \times 10 + 1.6 \times 10)6 = 180 \text{ kN}$$

Since beam is symmetrically loaded

$$R_A = R_B = W/2 = 90 \text{ kN}$$

Ultimate shear force (V) = 90 kN and design shear stress, v, is

$$v = \frac{V}{bd} = \frac{90 \times 10^3}{325 \times 547} = 0.51 \text{ N/mm}^2$$

Diameter and spacing of links
By inspection

$$v_c/2 < v < (v_c + 0.4)$$

i.e. 0.32 < 0.51 < 1.05. Hence from *Table 3.9*, provide nominal links for whole length of beam according to

$$\frac{A_{sv}}{s_v} = \frac{0.4b}{0.95f_{yv}} = \frac{0.4 \times 325}{0.95 \times 250} = 0.55$$

This value has to be translated into a certain bar size and spacing of links and is usually achieved using shear reinforcement tables similar to *Table 3.10*. The spacing of links should not exceed $0.75d = 0.75 \times 547 = 410$ mm. From *Table 3.10* it can be seen that 10 mm diameter links spaced at 275 mm centres provide a A_{sv}/s_v ratio of 0.571 and would therefore be suitable. Hence provide R10 at 275 mm centres for whole length of beam.

Table 3.10 Values of A_{sv}/s_v

Diameter (mm)	Spacing of links (mm)										
	85	90	100	125	150	175	200	225	250	275	300
8	1.183	1.118	1.006	0.805	0.671	0.575	0.503	0.447	0.402	0.336	0.335
10	1.847	1.744	1.570	1.256	1.047	0.897	0.785	0.698	0.628	0.571	0.523
12	2.659	2.511	2.260	1.808	1.507	1.291	1.130	1.004	0.904	0.822	0.753
16	4.729	4.467	4.020	3.216	2.680	2.297	2.010	1.787	1.608	1.462	1.340

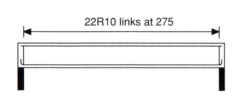

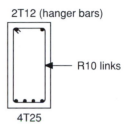

q_k = 29 kN/m

Design shear stress (*v*)

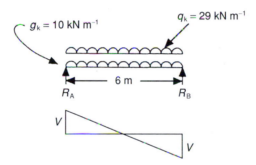

Total ultimate load, W, is
$$W = (1.4g_k + 1.6q_k)\text{span} = (1.4 \times 10 + 1.6 \times 29)6 = 362.4 \text{ kN}$$
Since beam is symmetrically loaded
$$R_A = R_B = W/2 = 181.2 \text{ kN}$$
Ultimate shear force (V) = 181.2 kN and design shear stress, v, is
$$v = \frac{V}{bd} = \frac{181.2 \times 10^3}{325 \times 547} = 1.02 \text{ N/mm}^2$$

Diameter and spacing of links
By inspection
$$v_c/2 < v < (v_c + 0.4)$$
i.e. $0.32 < 1.02 < 1.05$. Hence from *Table 3.9*, provide nominal links for whole length of beam according to
$$\frac{A_{sv}}{s_v} = \frac{0.4b}{0.95f_{yv}} = \frac{0.4 \times 325}{0.95 \times 250} = 0.55$$
Therefore as in case (ii) (q_k = 10 kN/m), provide R10 at 275 mm centres.

$q_k = 45$ kN/m

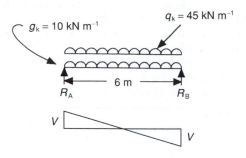

Design shear stress (v)

Total ultimate load, W, is

$$W = (1.4g_k + 1.6q_k)\text{span} = (1.4 \times 10 + 1.6 \times 45)6 = 516 \text{ kN}$$

Since beam is symmetrically loaded

$$R_A = R_B = W/2 = 258 \text{ kN}$$

Ultimate shear force $(V) = 258$ kN and design shear stress, v, is

$$v = \frac{V}{bd} = \frac{258 \times 10^3}{325 \times 547} = 1.45 \text{ N/mm}^2 < \text{maximum} = 0.8\sqrt{25} = 4 \text{ N/mm}^2$$

Diameter and spacing of links

Where $v < (v_c + 0.4) = 0.65 + 0.4 = 1.05$ N/mm^2, nominal links are required according to

$$\frac{A_{sv}}{s_v} = \frac{0.4b}{0.95f_{yv}} = \frac{0.4 \times 325}{0.95 \times 250} = 0.55$$

Hence from *Table 3.10*, provide R10 links at 275 mm centres where $v < 1.05$ N/mm^2 i.e. 2.172 m either side of the mid-span of beam.

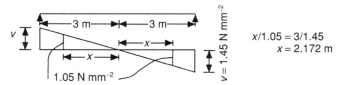

Where $v > (v_c + 0.4) = 1.05$ N/mm^2 design links required according to

$$\frac{A_{sv}}{s_v} = \frac{b(v - v_c)}{0.95f_{yv}} = \frac{325(1.45 - 0.65)}{0.95 \times 250} = 1.09$$

Hence from *Table 3.12*, provide R10 at 125 mm centres $(A_{sv}/s_v = 1.26)$ where $v > 1.05$ N/mm^2 i.e. 0.828 m in from both supports.

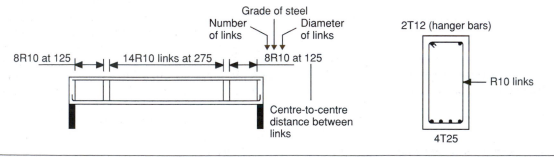

3.9.1.4 Deflection (clause 3.4.6, BS 8110)

In addition to checking that failure of the member does not arise due to the ultimate limit states of bending and shear, the designer must ensure that the deflections under working loads do not adversely affect either the efficiency or appearance of the structure. BS 8110 describes the following criteria for ensuring the proper performance of rectangular beams:

1. Final deflection should not exceed span/250.
2. Deflection after construction of finishes and partitions should not exceed span/500 or 20 mm, whichever is the lesser, for spans up to 10 m.

However, it is rather difficult to make accurate predictions of the deflections that may arise in concrete members principally because the member may be cracked under working loads and the degree of restraint at the supports is uncertain. Therefore, BS 8110 uses an approximate method based on permissible ratios of the span/effective depth. Before discussing this method in detail it is worth clarifying what is meant by the effective span of a beam.

(i) Effective span (clause 3.4.1.2, BS 8110). All calculations relating to beam design should be based on the effective span of the beam. For a simply supported beam this should be taken as the lesser of (1) the distance between centres of bearings, A, or (2) the clear distance between supports, D, plus the effective depth, d, of the beam (*Fig. 3.24*). For a continuous beam the effective span should normally be taken as the distance between the centres of supports.

(ii) Span/effective depth ratio. Generally, the deflection criteria in (1) and (2) above will be satisfied provided that the span/effective depth ratio of the beam does not exceed the appropriate limiting values given in *Table 3.11*. The reader is referred to the Handbook to BS 8110 which outlines the basis of this approach.

The span/effective depth ratio given in the table apply to spans up to 10 m long. Where the span exceeds 10 m, these ratios should be multiplied by 10/span (except for cantilevers). The basic ratios

Fig. 3.24 *Effective span of simply supported beam.*

Table 3.11 Basic span/effective depth ratio for rectangular or flanged beams (Table 3.9, BS 8110)

Support conditions	Rectangular sections	Flanged beams with $\dfrac{\text{width of beam}}{\text{width of flange}} \leqslant 0.3$
Cantilever	7	5.6
Simply supported	20	16.0
Continuous	26	20.8

Table 3.12 Modification factors for compression reinforcement (Table 3.11, BS 8110)

$\dfrac{100 A'_{s,\text{prov}}}{bd}$	Factor
0.00	1.0
0.15	1.05
0.25	1.08
0.35	1.10
0.5	1.14
0.75	1.20
1	1.25
1.5	1.33
2.0	1.40
2.5	1.45
$\geqslant 3.0$	1.5

may be further modified by factors taken from *Tables 3.12* and *3.13*, depending upon the amount of compression and tension reinforcement respectively. Deflection is usually critical in the design of slabs rather than beams and, therefore, modifications factors will be discussed more fully in the context of slab design (*section 3.10*).

3.9.1.5 Member sizing

The dual concepts of span/effective depth ratios and maximum design concrete shear stress can be used not only to assess the performance of members with respect to deflection and shear but also for preliminary sizing of members. *Table 3.14* gives modified span/effective depth ratios for estimating the effective depth of a concrete beam provided that its span is known. The width of the beam can then be determined by limiting the maximum design concrete shear stress to around (say) 1.2 N/mm^2.

Table 3.13 Modification factors for tension reinforcement (based on Table 3.10, BS 8110)

Service stress	M/bd^2								
	0.50	0.75	1.00	1.50	2.00	3.00	4.00	5.00	6.00
100	2.00	2.00	2.00	1.86	1.63	1.36	1.19	1.08	1.01
150	2.00	2.00	1.98	1.69	1.49	1.25	1.11	1.01	0.94
($f_y = 250$) 167	2.00	2.00	1.91	1.63	1.44	1.21	1.08	0.99	0.92
200	2.00	1.95	1.76	1.51	1.35	1.14	1.02	0.94	0.88
250	1.90	1.70	1.55	1.34	1.20	1.04	0.94	0.87	0.82
300	1.60	1.44	1.33	1.16	1.06	0.93	0.85	0.80	0.76
($f_y = 460$) 307	1.56	1.41	1.30	1.14	1.04	0.91	0.84	0.79	0.76

Note 1. The values in the table derive from the equation:

$$\text{Modification factor} = 0.55 + \frac{(477 - f_s)}{120\left(0.9 + \dfrac{M}{bd^2}\right)} \leqslant 2.0 \quad \text{(equation 7)}$$

where
f_s is the design service stress in the tension reinforcement
M is the design ultimate moment at the centre of the span or, for a cantilever, at the support.
Note 2. The design service stress in the tension reinforcement may be estimated from the equation:

$$f_s = \frac{2f_y A_{s,\text{req}}}{3A_{s,\text{prov}}} \times \frac{1}{\beta_b} \quad \text{(equation 8)}$$

where β_b is the percentage of moment redistribution, equal to 1 for simply supported beams.

Example 3.3 Sizing a concrete beam (BS 8110)

A simply supported beam has an effective span of 8 m and supports characteristic dead (g_k) and live (q_k) loads of 15 kN/m and 10 kN/m respectively. Determine suitable dimensions for the effective depth and width of the beam.

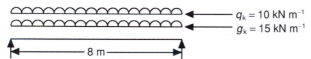

From *Table 3.14*, span/effective depth ratio for a simply supported beam is 12. Hence effective depth, d, is

$$d = \frac{\text{span}}{12} = \frac{8000}{12} \approx 670 \text{ mm}$$

Total ultimate load = $(1.4g_k + 1.6q_k)$span = $(1.4 \times 15 + 1.6 \times 10)8 = 296$ kN
Design shear force $(V) = 296/2 = 148$ kN and design shear force, v, is

$$v = \frac{V}{bd} = \frac{148 \times 10^3}{670b}$$

Assuming v is equal to (say) 1.2 N/mm², gives width of beam, b, of

$$b = \frac{V}{dv} = \frac{148 \times 10^3}{670 \times 1.2} \approx 185 \text{ mm}$$

Hence a beam of width 185 mm and effective depth 670 mm would be suitable to support the given design loads.

Table 3.14 Span/effective depth ratios for initial design

Support condition	Span/effective depth
Cantilever	6
Simply supported	12
Continuous	15

3.9.1.6 Reinforcement details (clause 3.12, BS 8110)

The previous sections have covered much of the theory required to design singly reinforced concrete beams. However, there are a number of code provisions with regards to:

1. maximum and minimum reinforcement areas
2. spacing of reinforcement
3. curtailment and anchorage of reinforcement
4. lapping of reinforcement.

These need to be taken into account since they may affect the final design.

1. Reinforcement areas (clause 3.12.5.3 and 3.12.6.1, BS 8110). As pointed out in *section 3.8*, there is a need to control cracking of the concrete because of durability and aesthetic considerations. This is usually achieved by providing minimum areas of reinforcement in the member. However, too large an area of reinforcement should also be avoided since it will hinder proper placing and adequate compaction of the concrete around the reinforcement.

For rectangular beams with overall dimensions b and h, the area of tension reinforcement, A_s, should lie within the following limits:

$$0.24\%bh \leq A_s \leq 4\%bh \quad \text{when } f_y = 250 \text{ N/mm}^2$$
$$0.13\%bh \leq A_s \leq 4\%bh \quad \text{when } f_y = 460 \text{ N/mm}^2$$

2. Spacing of reinforcement (clause 3.12.11.1, BS 8110). BS 8110 specifies minimum and maximum distances between tension reinforcement. The actual limits vary, depending upon the grade of reinforcement. The minimum distance is based on the need to achieve good compaction of the concrete around the reinforcement. The limits on the maximum distance between bars arises from the need to ensure that the maximum crack width does not exceed 0.3 mm in order to prevent corrosion of embedded bars (*section 3.8*).

For singly reinforced simply supported beams the clear horizontal distance between tension bars, s_b, should lie within the following limits:

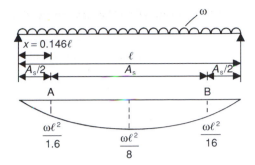

Fig. 3.25

$$h_{agg} + 5 \text{ mm or bar size} \leq s_b \leq 300 \text{ mm}$$
$$\text{when } f_y = 250 \text{ N/mm}^2$$

$$h_{agg} + 5 \text{ mm or bar size} \leq s_b \leq 160 \text{ mm}$$
$$\text{when } f_y = 460 \text{ N/mm}^2$$

where h_{agg} is the maximum size of the coarse aggregate.

3. Curtailment and anchorage of bars (clause 3.12.9, BS 8110). The design process for simply supported beams, in particular the calculations relating to the design moment and area of bending reinforcement, is concentrated at mid-span. However, the bending moment decreases either side of the mid-span and it follows, therefore, that it should be possible to reduce the corresponding area of bending reinforcement by curtailing bars. For the beam shown in *Fig. 3.25*, theoretically 50% of the main steel can be curtailed at points *A* and *B*. In order to develop the design stress in the reinforcement (i.e. $0.95f_y$), these bars must be fully anchored into the concrete. Except at end supports, this is normally achieved by extending the bars beyond the point at which they are theoretically no longer required, by a distance equal to the greater of (i) the effective depth of the member and (ii) 12 times the bar size.

Where a bar is stopped off in the tension zone, e.g. beam shown in *Fig. 3.25*, this distance should be increased to the full anchorage bond length in accordance with the values given in *Table 3.15*. Simplified rules for the curtailment of bars are also given in clause 3.12.10.2 of BS 8110. These are shown diagrammatically in *Fig. 3.26* for simply supported and continuous beams.

The code also gives rules for the anchorage of bars at supports. Thus, at a simply supported end each tension bar will be properly anchored provided the bar extends a length equal to one of the following: (a) 12 times the bar size beyond the centre

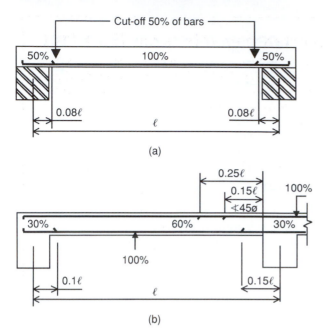

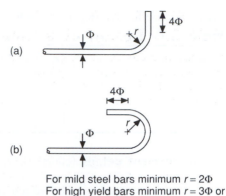

For mild steel bars minimum $r = 2\Phi$
For high yield bars minimum $r = 3\Phi$ or
4Φ for sizes 25 mm and above

Fig. 3.28 *Anchorage lengths for hooks and bends (a) anchorage length for 90° bend = 4r but not greater than 12φ; (b) anchorage length for hook = 8r but not greater than 24φ.*

Fig. 3.26 *Simplified rules for curtailment of bars in beams: (a) simply supported ends; (b) continuous beam.*

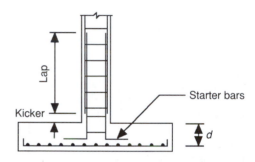

Fig. 3.29 *Lap lengths.*

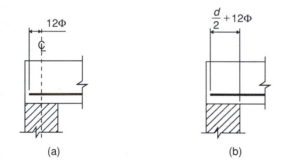

Fig. 3.27 *Anchorage requirements at simple supports.*

line of the support, or (b) 12 times the bar size plus $d/2$ from the face of the support (*Fig. 3.27*).

Sometimes it is not possible to use straight bars due to limitations of space and, in this case, anchorage must be provided by using hooks or bends in the reinforcement. The anchorage values of hooks and bends are shown in *Fig. 3.28*. Where hooks or bends are provided, BS 8110 states that they should not begin before the centre of the support for rule (a) or before $d/2$ from the face of the support for rule (b).

4. Laps in reinforcement (clause 3.12.8, BS 8110). It is not possible nor, indeed, practicable to construct the reinforcement cage for an indi-

vidual element or structure without joining some of the bars. This is normally achieved by lapping bars (*Fig. 3.29*). Bars which have been joined in this way must act as a single length of bar. This means that the lap length should be sufficiently long in order that stresses in one bar can be transferred to the other. The minimum lap length should not be less than 15 times the bar diameter or 300 mm. For **tension laps** it should normally be equal to the tension anchorage length but will often need to be increased as outlined in clause 3.12.8.13 of BS 8110. The anchorage length (L) is calculated using

$$L = L_A \times \Phi \qquad (3.17)$$

where
Φ is the diameter of the (smaller) bar
L_A is obtained from *Table 3.15* and depends upon the stress type, grade of concrete and reinforcement type.

For **compression laps** the lap length should be at least 1.25 times the compression anchorage length.

Example 3.4 Design of a simply supported concrete beam (BS 8110)

A reinforced concrete beam which is 300 mm wide and 600 mm deep is required to span 6.0 m between the centres of supporting brick piers 300 mm wide (*Fig. 3.30*). The beam carries dead and imposed loads of 25 kN/m and 19 kN/m respectively. Assuming $f_{cu} = 30$ N/mm², $f_y = 460$ N/mm², $f_{yv} = 250$ N/mm² and the exposure conditions are mild, design the beam.

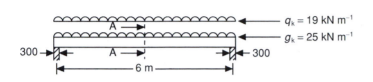

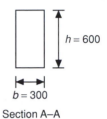

Section A–A

Fig. 3.30

(I) DESIGN MOMENT, M

Loading

Dead

Self weight of beam = $0.6 \times 0.3 \times 24 = 4.32$ kN/m

Total dead load (g_k) = 25 + 4.32 = 29.32 kN/m

Imposed

Total imposed load (q_k) = 19 kN/m

Ultimate load

Total ultimate load $(W) = (1.4g_k + 1.6q_k)\text{span}$

$$= (1.4 \times 29.32 + 1.6 \times 19)6$$

$$= 428.7 \text{ kN}$$

Design moment

Maximum design moment $(M) = \dfrac{W\ell}{8} = \dfrac{428.7 \times 6}{8} = 321.5$ kN m

(II) ULTIMATE MOMENT OF RESISTANCE, M_u

Effective depth, d

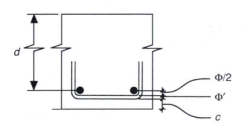

Assume diameter of main bars (Φ) = 25 mm
Assume diameter of links (Φ') = 10 mm
From *Table 3.5*, cover for mild conditions of exposure (c) = 25 mm.

$$d = h - c - \Phi' - \Phi/2$$

$$= 600 - 25 - 10 - 25/2 = 552 \text{ mm}$$

Ultimate moment

$$M_u = 0.156 f_{cu} b d^2 = 0.156 \times 30 \times 300 \times 552^2$$

$$= 427.8 \times 10^6 \text{ N mm} = 427.8 \text{ kN m} > M$$

Since $M_u > M$ no compression reinforcement required.

(III) MAIN STEEL, A_s

$$K = \frac{M}{f_{cu} b d^2} = \frac{321.5 \times 10^6}{30 \times 300 \times 552^2} = 0.117$$

$$z = d[0.5 + \sqrt{(0.25 - K/0.9)}]$$

$$= 552[0.5 + \sqrt{(0.25 - 0.117/0.9)}] = 467 \text{ mm}$$

$$A_s = \frac{M}{0.95 f_y z} = \frac{321.5 \times 10^6}{0.95 \times 460 \times 467} = 1576 \text{ mm}^2$$

Hence from *Table 3.7*, provide 4T25 $(A_s = 1960 \text{ mm}^2)$.

(IV) SHEAR REINFORCEMENT

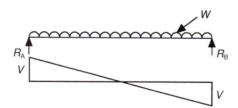

Ultimate design load, $W = 428.7$ kN

Shear stress, v
Since beam is symmetrically loaded

$$R_A = R_B = W/2 = 214.4 \text{ kN}$$

Ultimate shear force $(V) = 214.4$ kN and design shear stress, v, is

$$v = \frac{V}{bd} = \frac{214.4 \times 10^3}{300 \times 552} = 1.29 \text{ N/mm}^2 < \text{maximum} = 0.8\sqrt{30} = 4.38 \text{ N/mm}^2$$

Design concrete shear stress, v_c

$$\frac{100 A_s}{bd} = \frac{100 \times 1960}{300 \times 552} = 1.18$$

From *Table 3.8*,

$$v_c = (30/25)^{1/3} \times 0.66 = 0.70 \text{ N/mm}^2$$

Diameter and spacing of links

Where $v < (v_c + 0.4) = 0.7 + 0.4 = 1.1$ N/mm^2, nominal links are required according to

$$\frac{A_{sv}}{s_v} = \frac{0.4b}{0.95f_{yv}} = \frac{0.4 \times 300}{0.95 \times 250} = 0.505$$

Hence from *Table 3.10*, provide R10 links at 300 mm centres where $v < 1.10$ N/mm^2, i.e. 2.558 m either side of the mid-span of beam.

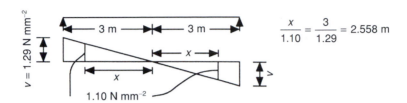

$$\frac{x}{1.10} = \frac{3}{1.29} = 2.558 \text{ m}$$

Where $v > (v_c + 0.4) = 1.10$ N/mm^2 design links required according to

$$\frac{A_{sv}}{s_v} = \frac{b(v - v_c)}{0.95f_{yv}} = \frac{300(1.29 - 0.70)}{0.95 \times 250} = 0.75$$

Maximum spacing of links is $0.75d = 0.75 \times 552 = 414$ mm. Hence from *Table 3.12*, provide 10 mm links at 200 mm centres ($A_{sv}/s_v = 0.787$) where $v > 1.10$ N/mm^2, i.e. 0.442 m in from both supports.

(V) EFFECTIVE SPAN

The above calculations were based on an effective span of 6 m but this needs to be confirmed. As stated in section 3.9.1.4 the effective span is the lesser of (1) centre-to-centre distance between support, i.e. 6 m, and (2) clear distance between supports plus the effective depth, i.e. $5700 + 552 = 6252$ mm. Therefore assumed span length of 6 m is correct.

(VI) DEFLECTION

Actual span/effective depth ratio = 6000/552 = 10.8

$$\frac{M}{bd^2} = \frac{321.5 \times 10^6}{300 \times 552^2} = 3.5$$

and from equation 8 (*Table 3.13*)

$$f_s = \frac{2}{3} \times f_y \times \frac{A_{s,req}}{A_{s,prov}} = \frac{2}{3} \times 460 \times \frac{1576}{1960} = 247 \text{ N/mm}^2$$

From *Table 3.11*, basic span/effective depth ratio for a simply supported beam is 20 and from *Table 3.13*, modification factor ≈ 0.98. Hence permissible span/effective depth ratio = $20 \times 0.98 = 19.7 >$ actual ($= 10.8$) and the beam therefore satisfies the deflection criteria in BS 8110.

(VII) REINFORCEMENT DETAILS

The sketch below shows the main reinforcement requirements for the beam. For reasons of buildability, the actual reinforcement details may well be slightly different and the reader is referred to the following publications for further information on this point:

1. *Design and Detailing* (BS 8110: 1997), Higgins, J. B. and Rogers, B. R., British Cement Association, 1989.

2. *Standard Methods of Detailing Structural Concrete*, the Concrete Society and the Institution of Structural Engineers, London, 1989.

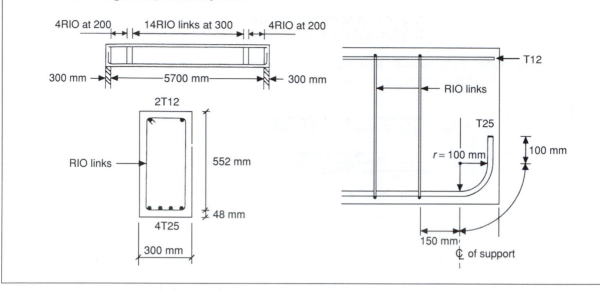

Example 3.5 Analysis of a singly reinforced concrete beam (BS 8110)

A singly reinforced concrete beam in which $f_{cu} = 30$ N/mm^2 and $f_y = 460$ N/mm^2 contains 1960 mm^2 of tension reinforcement (*Fig. 3.31*). If the effective span is 7 m and the density of reinforced concrete is 24 kN/m^3, calculate the maximum imposed load that the beam can carry assuming that the load is (a) uniformly distributed and (b) occurs as a point load at mid-span.

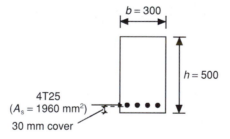

Fig. 3.31

(A) MAXIMUM UNIFORMLY DISTRIBUTED IMPOSED LOAD, q_k

(i) Moment capacity of section, M
Effective depth, d, is

$$d = h - \text{cover} - \phi/2 = 500 - 30 - 25/2 = 457 \text{ mm}$$

$$\frac{100 A_s}{bd} = \frac{100 \times 1960}{300 \times 457} = 1.43$$

From *Fig. 3.19*

$$\frac{M}{bd^2} = 4.8 \Rightarrow M = 4.8 \times 300 \times 457^2 \times 10^{-6} = 300.7 \text{ kN m}$$

(ii) Maximum uniformly distributed imposed load, q_k

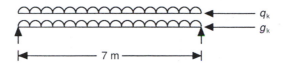

Dead load
Self weight of beam $(g_k) = 0.5 \times 0.3 \times 24 = 3.6$ kN/m

Ultimate load
Total ultimate load $(W) = (1.4g_k + 1.6q_k)$span

$$= (1.4 \times 3.6 + 1.6q_k)7$$

Imposed load

Maximum design moment $(M) = \dfrac{W\ell}{8} = \dfrac{(5.04 + 1.6q_k)7^2}{8} = 300.7$ kN m   (from above)

Hence the maximum uniformly distributed imposed load the beam can support is

Loading $q_k = \dfrac{(300.7 \times 10^6 \times 8)/7^2 - 5.04}{1.6} = 27.5$ kN/m

(B) MAXIMUM POINT LOAD AT MID-SPAN, Q_k

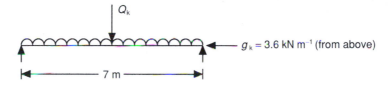

Ultimate load
Ultimate dead load (W_D) $= 1.4g_k \times$ span $= 1.4 \times 3.6 \times 7 = 35.3$ kN

Ultimate imposed load $(W_I) = 1.6Q_k$

Imposed load
Maximum design moment, M, is

$$M = \frac{W_D\ell}{8} + \frac{W_I\ell}{4} \quad \text{(Example 2.5, beam B1–B3)}$$

$$= \frac{35.3 \times 7}{8} + \frac{1.6Q_k \times 7}{4} = 300.7 \text{ kN m} \quad \text{(from above)}$$

Hence the maximum point load which the beam can support at mid-span is

$$Q_k = \frac{(300.7 - 35.3 \times 7/8)4}{1.6 \times 7} = 96.3 \text{ kN}$$

Table 3.15 Anchorage lengths as multiples of bar size (based on Table 3.27, BS 8110)

	L_A			
$f_{cu} = 25$	30	35	40 or more
Plain (250)				
Tension	43	39	36	34
Compression	34	32	29	27
Deformed Type 1 (460)				
Tension	55	50	47	44
Compression	44	40	38	35
Deformed Type 2 (460)				
Tension	44	40	38	35
Compression	35	32	30	28

3.9.2 DESIGN OF DOUBLY REINFORCED BEAMS

If the design moment is greater than the ultimate moment of resistance, i.e. $M > M_u$, or $K > K'$ where $K = M/f_{cu}bd^2$ and $K' = M_u/f_{cu}bd^2$ the concrete will have insufficient strength in compression to generate this moment and maintain an under-reinforced mode of failure.

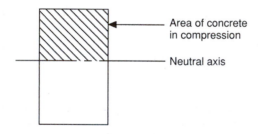

Area of concrete in compression

Neutral axis

The required compressive strength can be achieved by increasing the proportions of the beam, particularly its overall depth. However, this may not always be possible due to limitations on the headroom in the structure, and in such cases it will be necessary to provide reinforcement in the compression face. The compression reinforcement will be designed to resist the moment in excess of M_u. This will ensure that the compressive stress in the concrete does not exceed the permissible value and ensure an under-reinforced failure mode.

Beams which contain tension and compression reinforcement are termed doubly reinforced. They are generally designed in the same way as singly reinforced beams except in respect of the calculations needed to determine the areas of tension and compression reinforcement. This aspect is discussed below.

3.9.2.1 Compression and tensile steel areas (clause 3.4.4.4, BS 8110)

The area of compression steel (A_s') is calculated from

$$A_s' = \frac{M - M_u}{0.95 f_y (d - d')} \qquad (3.18)$$

where d' is the depth of the compression steel from the compression face (*Fig. 3.32*).

The area of tension reinforcement is calculated from

$$A_s = \frac{M_u}{0.95 f_y z} + A_s' \qquad (3.19)$$

where $z = d[0.5 + \sqrt{(0.25 - K'/0.9)}]$ and $K' = 0.156$.

Equations 3.18 and 3.19 can be derived using the stress block shown in *Fig. 3.32*. This is basically the same stress block used in the analysis of a singly reinforced section (*Fig. 3.16*) except for the additional compression force (F_{sc}) in the steel.

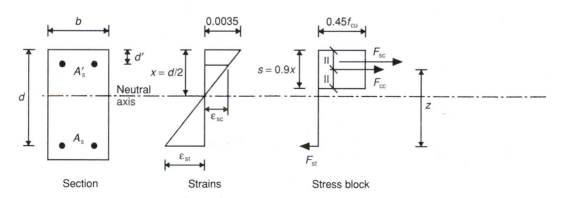

Fig. 3.32 Section with compression reinforcement.

In the derivation of equations 3.18 and 3.19 it is assumed that the compression steel has yielded (i.e. design stress = $0.95f_y$) and this condition will be met only if

$$\frac{d'}{x} \leqslant 0.37 \quad \text{or} \quad \frac{d'}{d} \leqslant 0.19 \quad \text{where } x = \frac{d - z}{0.45}$$

If $d'/x > 0.37$, the compression steel will not have yielded and, therefore, the compressive stress will be less than $0.95f_y$. In such cases, the design stress can be obtained using *Fig. 3.9*.

Example 3.6 Design of bending reinforcement for a doubly reinforced beam (BS 8110)

The reinforced concrete beam shown in *Fig. 3.33* has an effective span of 9 m and carries uniformly distributed dead (including self weight of beam) and imposed loads of 4 and 5 kN/m respectively. Design the bending reinforcement assuming the following:

$$f_{cu} = 30 \text{ N/mm}^2$$

$$f_y = 460 \text{ N/mm}^2$$

$$\text{cover to main steel} = 40 \text{ mm}$$

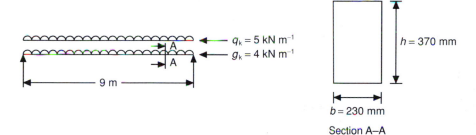

Fig. 3.33

(I) DESIGN MOMENT, M

Loading

Ultimate load
Total ultimate load $(W) = (1.4g_k + 1.6q_k)\text{span}$

$$= (1.4 \times 4 + 1.6 \times 5)9 = 122.4 \text{ kN}$$

Design moment

Maximum design moment $(M) = \dfrac{W\ell}{8} = \dfrac{122.4 \times 9}{8} = 137.7 \text{ kN m}$

(II) ULTIMATE MOMENT OF RESISTANCE, M_u

Effective depth, d

Assume diameter of tension bars $(\Phi) = 25$ mm:

$$d = h - \Phi/2 - \text{cover}$$

$$= 370 - 25/2 - 40 = 317 \text{ mm}$$

Ultimate moment

$$M_u = 0.156 f_{cu} b d^2$$
$$= 0.156 \times 30 \times 230 \times 317^2$$
$$= 108.2 \times 10^6 \text{ N mm} = 108.2 \text{ kN m}$$

Since $M > M_u$ compression reinforcement is required.

(III) COMPRESSION REINFORCEMENT

Assume diameter of compression bars (ϕ) = 16 mm. Hence

$$d' = \text{cover} + \phi/2 = 40 + 16/2 = 48 \text{ mm}$$

$$z = d[0.5 + \sqrt{(0.25 - K'/0.9)}] = 317[0.5 + \sqrt{(0.25 - 0.156/0.9)}] = 246 \text{ mm}$$

$$x = \frac{d - z}{0.45} = \frac{317 - 246}{0.45} = 158 \text{ mm}$$

$$\frac{d'}{x} = \frac{48}{158} = 0.3 < 0.37, \text{ i.e. compression steel has yielded.}$$

$$A'_s = \frac{M - M_u}{0.95 f_y (d - d')} = \frac{(137.7 - 108.2)10^6}{0.95 \times 460(317 - 48)} = 251 \text{ mm}^2$$

Hence from *Table 3.7*, provide 2T16 ($A'_s = 402 \text{ mm}^2$)

(IV) TENSION REINFORCEMENT

$$A_s = \frac{M_u}{0.95 f_y z} + A'_s = \frac{108.2 \times 10^6}{0.95 \times 460 \times 246} + 251 = 1258 \text{ mm}^2$$

Hence provide 3T25 ($A_s = 1470 \text{ mm}^2$).

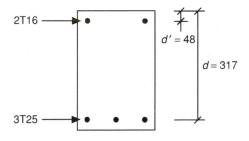

3.9.2.2 Design charts

Rather than solving equations 3.18 and 3.19 it is possible to determine the area of tension and compression reinforcement simply by using the design charts for doubly reinforced beams given in Part 3 of BS 8110. Such charts are available for design involving the use of reinforcement grade 460, concrete grades 25, 30, 35, 40, 45 and 50 and d'/d ratios of 0.1, 0.15 and 0.2. As previously mentioned, BSI issued these charts when the partial safety factor for steel reinforcement was 1.15 and not 1.05 and,

therefore, use of these charts will produce conservative estimates of A_s and A'_s. *Fig. 3.34* presents a modified version of chart 7 which incorporates the amended partial safety factor for reinforcement.

The procedure involves the following steps:

1. Check $M_u < M$.
2. Calculate d'/d.
3. Select appropriate chart from Part 3 of BS 8110 based on grade of concrete and d'/d ratio.
4. Calculate M/bd^2.

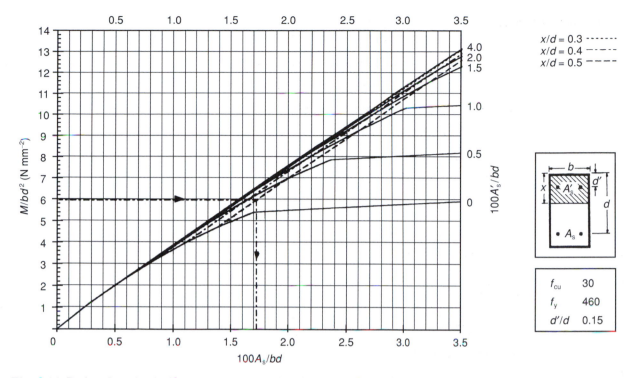

Fig. 3.34 *Design chart for doubly reinforced beams (based on chart 7, BS 8110: Part 3).*

5. Plot M/bd^2 ratio on chart and read off corresponding $100A'_s/bd$ and $100A_s/bd$ values (*Fig. 3.34*).
6. Calculate A'_s and A_s.

Using the figures given in Example 3.6, $M_u = 108.2$ kN m $< M = 137.7$ kN m.

Since d'/d (= 48/317) = 0.15 and $f_{cu} = 30$ N/mm^2, chart 7 is appropriate. Furthermore, since the beam is simply supported, no redistribution of moments is possible, therefore, use $x/d = 0.5$ construction line in order to determine areas of reinforcement.

$$\frac{M}{bd^2} = \frac{137.7 \times 10^6}{230 \times 317^2} = 5.95$$

$$100A'_s/bd = 0.33 \Rightarrow A'_s = 243 \text{ mm}^2$$

$$100A_s/bd = 1.72 \Rightarrow A_s = 1254 \text{ mm}^2$$

Hence from *Table 3.7*, provide 2T16 compression steel and 3T25 tension steel.

3.9.3 CONTINUOUS, *L* AND *T* BEAMS

In most real situations, the beams in buildings are seldom single span but continuous over the supports, e.g. beams 1, 2, 3 and 4 in *Fig. 3.35(a)*. The design process for continuous beams is similar to

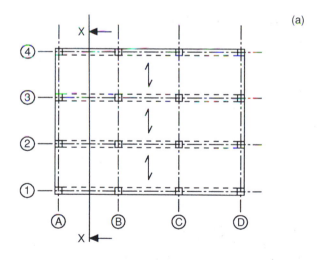

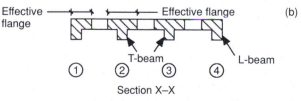

Fig. 3.35 *Floor slab: (a) plan (b) cross-section.*

Table 3.16 Design ultimate moments and shear forces for continuous beams (Table 3.5, BS 8110)

	End support	End span	Penultimate support	Interior span	Interior support
Moment	0	$0.09F\ell$	$-0.11F\ell$	$0.07F\ell$	$-0.08F\ell$
Shear	0.45F	–	0.6F	–	0.55F

$F = 1.4G_k + 1.6Q_k$; ℓ = effective span

that outlined above for single span beams. The main difference is the fact that the designer will need to consider the various loading arrangements discussed in section 3.6.2 in order to determine the design moments and shear forces along the beam. Continuous beams can be analysed by moment distribution as discussed in *section 3.9.3.1* or, provided the conditions in clause 3.4.3 of BS 8110 are satisfied (Example 3.9), by using the coefficients given in *Table 3.5* of BS 8110, reproduced as *Table 3.16*. Once this has been done, the beam can be sized and the area of bending reinforcement calculated as discussed in *section 3.9.1* or *3.9.2*. At the internal supports, the bending moment is reversed and it should be remembered that the tensile reinforcement would occur in the top half of the beam and compression reinforcement in the bottom half of the beam.

Generally, beams and slabs are cast monolithically, that is, they are structurally tied. At mid-span, it is more economical in such cases to design the beam as an *L* or *T* section by including the adjacent areas of the slab (*Fig. 3.35(b)*). The actual width of slab that acts together with the beam is normally termed the effective flange. According to clause 3.4.1.5 of BS 8110, the effective flange width should be taken as the lesser of (a) the actual flange width and (b) the web width plus $\ell_z/5$ (for *T*-beams) or $\ell_z/10$ (for *L*-beams), where ℓ_z is the distance between points of zero moments which for a continuous beam may be taken as 0.7 times the distance between the centres of supports.

The depth of the neutral axis in relation to the depth of flange will influence the design process and must therefore be determined. The depth of the neutral axis, *x*, can be calculated using equation 3.9 derived in *section 3.9.1*, i.e.

$$x = \frac{d - z}{0.45}$$

Where the neutral axis lies within the flange, which will normally be the case in practice, the beam can be designed as being singly reinforced taking the breadth of the beam, *b*, equal to the

effective flange width. At the supports of a continuous member, e.g. at columns B2, B3, C2 and C3, due to the moment reversal, *b* should be taken as the actual width of the beam.

3.9.3.1 Analysis of continuous beams

Continuous beams (and continuous slabs that span in one direction) are not statically determinate and more advanced analytical techniques must be used to obtain the bending moments and shear forces acting along the member. A straightforward method of calculating the moments at the supports of continuous members and hence the bending moments and shear forces in the span is by moment distribution. Essentially the moment-distribution method involves the following steps:

1. Calculate fixed end moments (FEM) in each span using the formulae given in *Table 3.17* and elsewhere. Note that clockwise moments are conventionally positive and anticlockwise moments negative.
2. Determine the stiffness factor for each span. The stiffness factor is the moment required to produce unit rotation at the end of the member. A uniform member (i.e. constant *EI*) of length *L* that is pinned at one end and fixed at the other (*Fig. 3.36(a)*) has a stiffness factor of *4EI/L*. If the member is pinned at both ends its stiffness factor reduces to (3/4)*4EI/L* (*Fig. 3.36(b)*).
3. Evaluate distribution factors for each member meeting at a joint. The factors indicate what proportion of the moment applied to a joint is distributed to each member attached to it in order to maintain continuity of slope. Distribution factors are simply ratios of the stiffnesses of individual members and the sum of the stiffnesses of all the members meeting at a joint. As such, the distribution factors at any joint should sum to unity.
4. Release each joint in turn and distribute the out-of-balance moments between the members meeting at the joint in proportion to their distribution factors. The out-of-balance moment is equal in magnitude but opposite in sense to the sum of the moments in the members meeting at a joint.

Table 3.17 Fixed end moments for uniform beams

	M_{AB}	M_{BA}
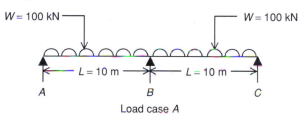*W* at center, L/2 + L/2	$-\dfrac{WL}{8}$	$\dfrac{WL}{8}$
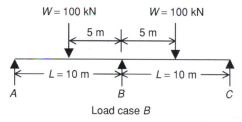ω per unit length, L/2	$-\dfrac{\omega L^2}{12}$	$\dfrac{\omega L^2}{12}$

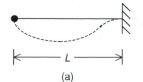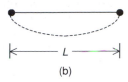

Fig. 3.36 *Stiffness factors for uniform beams: (a) pinned-fixed beam = 4EI/L; (b) pinned-pinned beam = 3EI/L.*

5. Determine the moment developed at the far end of each member via the carry-over factor. If the far end of the member is fixed, the carry over factor is half and a moment of one-half of the applied moment will develop at the fixed end. If the far end is pinned, the carry over factor is zero and no moment is developed at the far end.
6. Repeat steps (4) and (5) until all the out of balance moments are negligible.
7. Determine the end moments for each span by summing the moments at each joint.

Example 3.7 Analysis of a two–span continuous beam using moment distribution

Evaluate the critical moments and shear forces in the beams shown below assuming that they are of constant section and the supports provide no restraint to rotation.

$W = 100$ kN $\qquad$ $W = 100$ kN

$L = 10$ m $\qquad$ $L = 10$ m

A $\qquad$ B $\qquad$ C

Load case *A*

$W = 100$ kN $\qquad$ $W = 100$ kN

5 m $\qquad$ 5 m

$L = 10$ m $\qquad$ $L = 10$ m

A $\qquad$ B $\qquad$ C

Load case *B*

LOAD CASE *A*

(i) Fixed end moments (FEM)
From *Table 3.17*

$$M_{AB} = M_{BC} = \frac{-WL}{12} = \frac{-100 \times 10}{12} = -83.33 \text{ kN m}$$

$$M_{BA} = M_{CB} = \frac{WL}{12} = \frac{100 \times 10}{12} = 83.33 \text{ kN m}$$

(ii) Stiffness factors

Since both spans are effectively pinned at both ends, the stiffness factors for members AB and BC (K_{AB} and K_{BC} respectively), are $(3/4)4EI/10$.

(iii) Distribution factors

$$\text{Distribution factor at end } BA = \frac{\text{Stiffness factor for member } AB}{\text{Stiffness factor for member } AB + \text{Stiffness factor for member } BC}$$

$$= \frac{K_{AB}}{K_{AB} + K_{BC}} = \frac{(3/4)4EI/10}{(3/4)4EI/10 + (3/4)4EI/10} = 0.5$$

Similarly the distribution factor at end $BC = 0.5$

(iv) End moments

Joint	A	B		C
End	AB	BA	BC	CB
Distribution factors		0.5	0.5	
FEM (kN m)	−83.33	83.33	−83.33	83.33
Release A & C^1	+83.33			−83.33
Carry over ^{2}		41.66	−41.66	
Release B		0		
Sum^3 (kN m)	0	125	−125	0

^{1} Since ends A and C are pinned, the moments here must be zero. Applying moments that are equal in magnitude but opposite in sense to the fixed end moments, i.e. +83.33 kN m and −83.33 kN m, satisfies this condition.
^{2} Since joint B is effectively fixed, the carry-over factors for members AB and BC are both 0.5 and a moment of one-half of the applied moment will be induced at the fixed end.
^{3} Summing the values in each column obtains the support moments.

(v) Support reactions and mid-span moments

The support reactions and mid-span moments are obtained using statics.

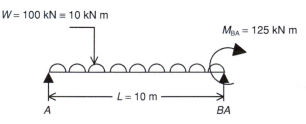

Taking moments about end BA obtains the reaction at end A, R_A, as follows

$$10R_A = WL/2 - M_{BA} = 100 \times (10/2) - 125 = 375 \text{ kN m} \Rightarrow R_A = 37.5 \text{ kN}$$

Reaction at end BA, $R_{BA} = W - R_{AB} = 100 - 37.5 = 62.5$ kN. Since the beam is symmetrically loaded, the reaction at end BC, $R_{BC} = R_{BA}$. Hence, reaction at support B, $R_B = R_{BA} + R_{BC} = 62.5 + 62.5 = 125$ kN.

The span moments, M_x, are obtained from

$$M_x = 37.5x - 10x^2/2$$

Maximum moment occurs when $\partial M/\partial x = 0$, i.e. $x = 3.75$ m $\Rightarrow M = 70.7$ kN m. Hence the bending moment and shear force diagrams are as follows

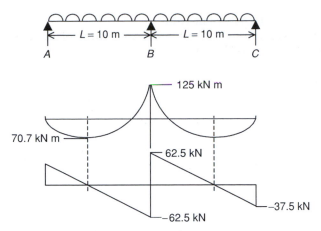

The results can be used to obtain moment and reaction coefficients by expressing in terms of W and L, where W is the load on one span only, i.e. 100 kN and L is the length of one span, i.e. 10 m, as shown in *Fig. 3.37*. The coefficients enable the bending moments and shear forces of any two equal span continuous beams, subjected to uniformly distributed loading, to be rapidly assessed.

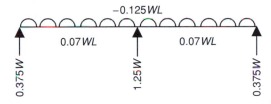

Fig. 3.37 *Bending moment and reaction coefficients for two equal span continuous beams subjected to a uniform load of W on each span.*

LOAD CASE B

(i) Fixed end moments

$$M_{AB} = M_{BC} = \frac{-WL}{8} = \frac{-100 \times 10}{8} = -125 \text{ kN m}$$

$$M_{BA} = M_{CB} = \frac{WL}{8} = \frac{100 \times 10}{8} = 125 \text{ kN m}$$

(ii) Stiffness and distribution factors
The stiffness and distribution factors are unchanged from the values calculated above.

(iii) End moments

Joint	A	B		C
End	AB	BA	BC	CB
Distribution factors		0.5	0.5	
FEM (kN m)	−125	125	−125	125
Release A & C	+125			−125
Carry over		62.5	−62.5	
Release B		0		
Sum	0	187.5	−187.5	0

(iv) Support reactions and mid–span moments
The support reactions and mid-span moments are again obtained using statics.

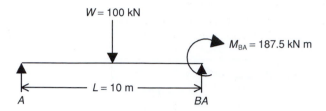

By taking moments about end BA, the reaction at end A, R_A, is

$$10R_A = WL/2 - M_{BA} = 100 \times (10/2) - 187.5$$

$$= 312.5 \text{ kN m} \Rightarrow R_A = 31.25 \text{ kN}$$

The reaction at end BA, $R_{BA} = W - R_A = 100 - 31.25 = 68.75 \text{ kN} = R_{BC}$. Hence the total reaction at support B, $R_B = R_{BA} + R_{BC} = 68.75 + 68.75 = 137.5 \text{ kN}$.

By inspection, the maximum sagging moment occurs at the point load, i.e. $x = 5$ m, and is given by

$$M_{x=5} = 31.25x = 31.25 \times 5 = 156.25 \text{ kN m}$$

The bending moment and shear force distributions along the beam are shown in *Fig. 3.38. Fig. 3.39* records the moment and reaction coefficients for the beam.

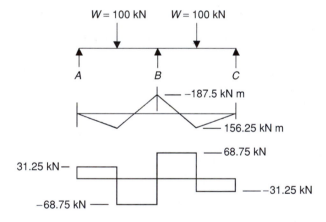

Fig. 3.38

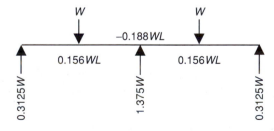

Fig. 3.39 *Bending moment and reaction coefficients for a two equal span continuous beam subjected to concentrated loads of W at each mid-span.*

Example 3.8 Analysis of a three span continuous beam using moment distribution

A three span continuous beam of constant section on simple supports is subjected to the uniformly distributed loads shown below. Evaluate the critical bending moments and shear forces in the beam using moment distribution.

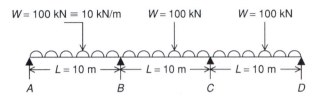

(i) Fixed end moments (FEM)

$$M_{AB} = M_{BC} = M_{CD} = \frac{-WL}{12} = \frac{-100 \times 10}{12} = -83.33 \text{ kN m}$$

$$M_{BA} = M_{CB} = M_{DC} = \frac{WL}{12} = \frac{100 \times 10}{12} = 83.33 \text{ kN m}$$

(ii) Stiffness factors

The outer spans are effectively pinned at both ends. Therefore, the stiffness factors for members AB and CD (K_{AB} and K_{CD} respectively), are $(3/4)4EI/10$.

During analysis, span BC is effectively pinned at one end and fixed at the other and its stiffness factor, K_{BC}, is therefore $4EI/10$.

(iii) Distribution factors

Distribution factor at end BA

$$= \frac{\text{Stiffness factor for member } AB}{\text{Stiffness factor for member } AB + \text{Stiffness factor for member } BC}$$

$$= \frac{K_{AB}}{K_{AB} + K_{BC}} = \frac{(3/4)4EI/10}{(3/4)4EI/10 + 4EI/10} = \frac{3}{7}$$

Distribution factor at end BC

$$= \frac{\text{Stiffness factor for member } BC}{\text{Stiffness factor for member } AB + \text{Stiffness factor for member } BC}$$

$$= \frac{K_{AB}}{K_{AB} + K_{BC}} = \frac{4EI/10}{(3/4)4EI/10 + 4EI/10} = \frac{4}{7}$$

Similarly the distribution factors for ends CB and CD are, respectively, 4/7 and 3/7.

(iv) End moments

Joint	A	B		C		D
End	AB	BA	BC	CB	CD	DC
Distribution factors		3/7	4/7	4/7	3/7	
FEM (kN m)	−83.33	83.33	−83.33	83.33	−83.33	83.33
Release A & D	+83.33					−83.33
Carry over		41.66			−41.66	
Release B & C		−17.86	−23.8	23.8	17.86	
Carry over			11.9	−11.9		
Release B & C		−5.1	−6.8	6.8	5.1	
Carry over			3.4	−3.4		
Release B & C		−1.46	−1.94	1.94	1.46	
Carry over			0.97	−0.97		
Release B & C		−0.42	−0.55	0.55	0.42	
Carry over			0.28	−0.28		
Release B & C		−0.12	−0.16	0.16	0.12	
Sum	0	100.04	−100.04	100.04	−100.04	0

(v) Support reactions and mid-span moments

Span AB

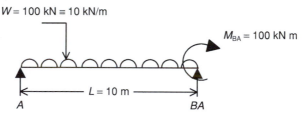

Taking moments about end BA obtains the reaction at end A, R_A, as follows

$$10R_A = WL/2 - M_{BA} = 100 \times (10/2) - 100 = 400 \text{ kN m} \Rightarrow R_A = 40 \text{ kN}$$

Reaction at end BA, $R_{BA} = W - R_{AB} = 100 - 40 = 60$ kN
The span moments, M_x, are obtained from

$$M_x = 40x - 10x^2/2$$

Maximum moment occurs when $\partial M/\partial x = 0$, i.e. $x = 4.0$ m $\Rightarrow M = 80$ kN m.

Span BC

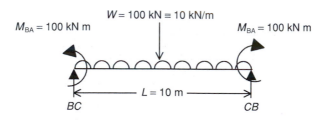

By inspection, reaction at end BC, R_{BC}, = reaction at end CB, R_{CB} = 50 kN

Therefore, total reaction at support B, $R_B = R_{BA} + R_{BC} = 60 + 50 = 110$ kN

By inspection, maximum moment occurs at mid-span of beam, i.e. $x = 5$ m. Hence, maximum moment is given by

$$M_{x=5} = 50x - 10x^2/2 - 100 = 50 \times 5 - 10 \times 5^2/2 - 100 = 25 \text{ kN m}$$

The bending moment and shear force diagrams plus the moment and reaction coefficients for the beam are shown below.

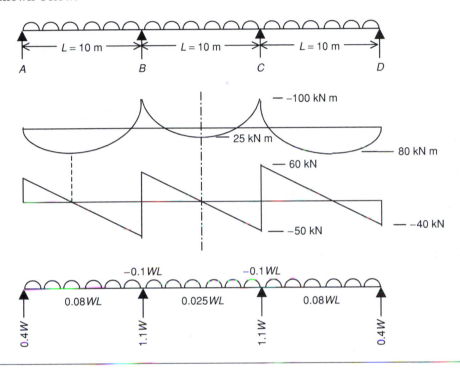

Example 3.9 Continuous beam design (BS 8110)

A typical floor plan of a small building structure is shown in *Fig. 3.40*. Design continuous beams 3A/D and $B1/5$ assuming the slab supports an imposed load of 4 kN/m^2 and finishes of 1.5 kN/m^2. The overall sizes of the beams and slab are indicated on the drawing. The columns are 400 × 400 mm. The characteristic strength of the concrete is 35 N/mm^2 and of the steel reinforcement is 460 N/mm^2. The cover to all reinforcement may be assumed to be 30 mm.

GRID LINE 3

Loading

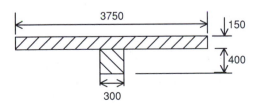

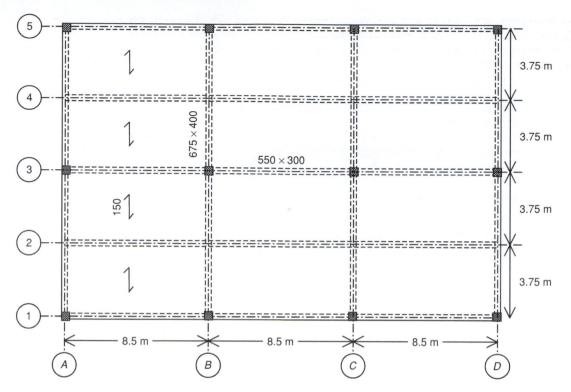

Fig. 3.40

Dead load, g_k, is the sum of

weight of slab = 0.15 × 3.75 × 24　　= 13.5
weight of downstand = 0.3 × 0.4 × 24 =　2.88
finishes = 1.5 × 3.75　　　　　　　= 　5.625
　　　　　　　　　　　　　　　　22.0 kN/m

Imposed load, q_k = 4 × 3.75 = 15 kN/m
Design uniformly distributed load, $\omega = (1.4g_k + 1.6q_k) = (1.4 × 22 + 1.6 × 15) = 54.8$ kN/m
Design load per span, $F = \omega × \text{span} = 54.8 × 8.5 = 465.8$ kN

Design moments and shear forces

From clause 3.4.3 of BS 8110, as $g_k > q_k$, the loading on the beam is substantially uniformly distributed and the spans are of equal length, the coefficients in *Table 3.16* can be used to calculate the design ultimate moments and shear forces. The results are shown in the table below. It should be noted however that these values are conservative estimates of the true design moments and shear forces along the beam since the coefficients in *Table 3.16* are based on simple supports at the ends of the beam. In reality, beam 3A/D is part of a monolithic frame and significant restraint moments will occur at the end supports.

Steel reinforcement

Middle of 3A/B (and middle of 3C/D). Assume diameter of main steel, ϕ = 25 mm, diameter of links, ϕ' = 10 mm and nominal cover, c = 30 mm. Hence

$$\text{Effective depth, } d = h - \frac{\phi}{2} - \phi' - c = 550 - \frac{25}{2} - 10 - 30 = 498 \text{ mm}$$

Position	Bending moment	Shear force
Support 3A	0	$0.45 \times 465.8 = 209.6$ kN
Near middle of 3A/B	$0.09 \times 465.8 \times 8.5 = 356.3$ kN m	0
		$0.6 \times 465.8 = 279.5$ kN (support 3B/A*)
Support 3B	$-0.11 \times 465.8 \times 8.5 = -435.5$ kN m	
		$0.55 \times 465.8 = 256.2$ kN (support 3B/C**)
Middle of 3B/C	$0.07 \times 465.8 \times 8.5 = 277.2$ kN m	0

* shear force at support 3B towards A   ** shear force at support 3B towards C

The effective width of beam is the lesser of

(a) actual flange width = 3750 mm
(b) web width + $\ell_z/5$, where ℓ_z is the distance between points of zero moments which for a continuous beam may be taken as 0.7 times the distance between centres of supports. Hence

$$\ell_z = 0.7 \times 8500 = 5950 \text{ mm} \quad \text{and} \quad b = 300 + 5950/5 = 1490 \text{ mm} \quad \text{(critical)}$$

$$K = \frac{M}{f_{cu}bd^2} = \frac{356.3 \times 10^6}{35 \times 1490 \times 498^2} = 0.0276$$

$$z = d\left(0.5 + \sqrt{(0.25 - K/0.9)}\right) \leqslant 0.95d = 0.95 \times 498 = 473 \text{ mm} \quad \text{(critical)}$$

$$= 498\left(0.5 + \sqrt{(0.25 - 0.0276/0.9)}\right) = 498 \times 0.968 = 482 \text{ mm}$$

$x = (d - z)/0.45 = (498 - 473)/0.45 = 55$ mm $<$ flange thickness (= 150 mm). Hence

Area of steel reinforcement, $A_s = \dfrac{M}{0.95 f_y z} = \dfrac{356.3 \times 10^6}{0.95 \times 460(0.95 \times 498)} = 1723$ mm^2

Provide 4T25 ($A_s = 1960$ mm^2).

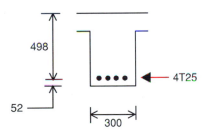

At support 3B (and 3C). Assume the main steel consists of two layers of 25 mm diameter bars, diameter of links, $\phi' = 10$ mm and nominal cover, $c = 30$ mm. Hence

$$\text{Effective depth, } d = h - \phi - \phi' - c = 550 - 25 - 10 - 30 = 485 \text{ mm}$$

Since the beam is in hogging, $b = 300$ mm

$$M_u = 0.156 f_{cu}bd^2 = 0.156 \times 35 \times 300 \times 485^2 \times 10^{-6} = 385.3 \text{ kN m}$$

Since $M_u < M$ (= 435.5 kN m), compression reinforcement is required.
Assume diameter of compression steel, $\Phi = 25$ mm, diameter of links, $\phi' = 10$ mm, and cover to reinforcement, c, is 30 mm. Hence effective depth of compression steel d' is

$$d' = c + \phi' + \Phi/2 = 30 + 10 + 25/2 = 53 \text{ mm}$$

Lever arm, $z = d\left(0.5 + \sqrt{(0.25 - K'/0.9)}\right) = 485\left(0.5 + \sqrt{(0.25 - 0.156/0.9)}\right) = 376.8$ mm

Depth to neutral axis, $x = (d - z)/0.45 = (485 - 376.8)/0.45 = 240$ mm
$d'/x = 53/240 = 0.22 < 0.37$. Therefore, the compression steel has yielded, i.e. $f'_s = 0.95f_y$ and

Area of compression steel, $A'_s = \dfrac{M - M_u}{0.95f_y(d - d')} = \dfrac{(435.5 - 385.3)10^6}{0.95 \times 460(494 - 53)} = 260$ mm^2

Provide 2T25 ($A'_s = 982$ mm^2)

Area of tension steel, $A_s = \dfrac{M_u}{0.95f_y z} - A'_s = \dfrac{385.3 \times 10^6}{0.95 \times 460 \times 376.8} + 260 = 2600$ mm^2

Provide 6T25 as shown ($A_s = 2950$ mm^2)

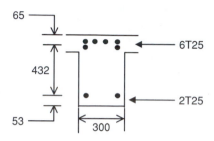

Middle of 3B/C. From above, effective depth, $d = 498$ mm and effective width of beam, $b = 1490$ mm.
Hence, A_s is

$$A_s = \frac{M}{0.95f_y z} = \frac{277.2 \times 10^6}{0.95 \times 460(0.95 \times 498)} = 1341 \text{ mm}^2$$

Provide 3T25 ($A_s = 1470$ mm^2).
Figure 3.41 shows a sketch of the bending reinforcement for spans 3A/B and 3B/C. The curtailment lengths indicated on the sketch are in accordance with the simplified rules for beams given in clause 3.12.10.2 of BS 8110 (*Fig. 3.26*).

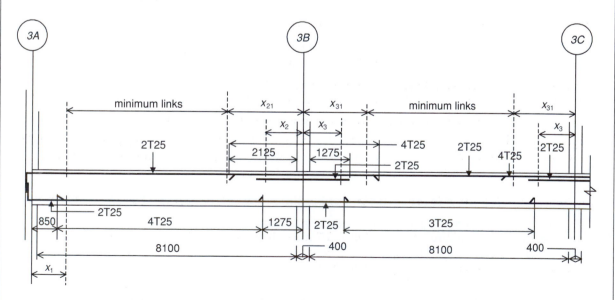

Fig. 3.41

Shear

As discussed in *section 3.9.1.3*, design of the shear reinforcement requires knowledge of the area of tensile steel reinforcement present at critical sections along the beam. Near to supports and at mid-spans it is relatively easy to see what this area of steel should be. At intervening positions along the beam, however, this task becomes more difficult because the points of zero bending moment are unknown. It is normal practice therefore to use conservative estimates of A_s to design the shear reinforcement without obtaining detailed knowledge of the bending moment distribution as illustrated below.

(a) Span 3A/B (3B/C and 3C/D). The minimum tension steel at any point in the span is 2T25. Hence $A_s = 980$ mm^2 and

$$\frac{100A_s}{bd} = \frac{100 \times 980}{300 \times 498} = 0.656$$

From *Table 3.8*, $v_c = \sqrt[3]{\frac{35}{25}} \times 0.54 = 0.6$ N/mm^2

Provide minimum links where $V \leqslant (v_c + 0.4)bd = (0.6 + 0.4)300 \times 498 \times 10^{-3} = 149.4$ kN (*Fig. 3.41*), according to

$$\frac{A_{sv}}{s_v} = \frac{0.4b}{0.95f_{yv}} = \frac{0.4 \times 300}{0.95 \times 250} = 0.505 \text{ with } s_v \leqslant 0.75d = 373 \text{ mm}$$

From *Table 3.10* select R10 at 300 centres ($A_{sv}/s_v = 0.523$).

(b) Support 3A (and 3D). According to clause 3.4.5.10 of BS 8110 for beams carrying generally uniform load the critical section for shear may be taken at distance d beyond the face of the support, i.e. $0.2 + 0.498 = 0.7$ m from the column centreline. Here the design shear force, V_D, is

$$V_D = V_{3A} - 0.7\omega = 209.6 - 0.7 \times 54.8 = 171.3 \text{ kN}$$

Shear stress, $v = \dfrac{V_D}{bd} = \dfrac{171.3 \times 10^3}{498 \times 300} = 1.15$ N/mm^2

From *Fig. 3.41* it can be seen that 50% of main steel is curtailed at support A. The effective area of tension steel is 2T25, hence $A_s = 980$ mm^2, $\dfrac{100A_s}{bd} = 0.656$ and $v_c = 0.6$ N/mm^2

Since $v > (v_c + 0.4)$ provide design links according to

$$\frac{A_{sv}}{s_v} \geqslant \frac{b(v - v_c)}{0.95f_{yv}} = \frac{300(1.15 - 0.6)}{0.95 \times 250} = 0.70.$$

From *Table 3.10* select R10 links at 200 mm centres ($A_{sv}/s_v = 0.785$).
Note that the shear resistance obtained with minimum links, V_1, is

$$V_1 = (v_c + 0.4)bd = (0.6 + 0.4)300 \times 498 \times 10^{-3} = 149.4 \text{ kN}$$

This shear force occurs at $x_1 = (V_{3A} - V_1)/\omega = (209.6 - 149.4)/54.8 = 1.1$ m (*Fig. 3.41*). Assuming the first link is fixed 75 mm from the front face of column $3A$ (i.e. 275 mm from the centerline of the column), then six R10 links at 200 mm centres are required.

(c) Support 3B/A (and 3C/D). Design shear force at distance d beyond the face of the support, V_D, is

$$V_D = V_{3B/A} - 0.7\omega = 279.5 - 0.7 \times 54.8 = 241.2 \text{ kN}$$

Shear stress, $v = \dfrac{V_D}{bd} = \dfrac{241.2 \times 10^3}{300 \times 485} = 1.66$ N/mm^2

Assume the tension steel is 4T25. Hence $A_s = 1960$ mm^2 and $\dfrac{100A_s}{bd} = \dfrac{100 \times 1960}{300 \times 485} = 1.35$

From *Table 3.8*, $v_c = \sqrt[3]{\dfrac{35}{25}} \times 0.69 = 0.77 \text{ N/mm}^2$

Since $v > (v_c + 0.4)$ provide design links according to

$\dfrac{A_{sv}}{s_v} \geqslant \dfrac{b(v - v_c)}{0.95 f_{yv}} = \dfrac{300(1.66 - 0.77)}{0.95 \times 250} = 1.124$. Select R10 at 125 centres ($A_{sv}/s_v = 1.256$).

To determine the number of links required assume that R10 links at 175 mm centres ($A_{sv}/s_v = 0.895$) are to be provided between the minimum links in span 3*A/B* (i.e. R10@300) and the design links at support 3*B* (R10@125). The shear resistance of R10 links at 175 centres and concrete, V_2, is

$$V_2 = \left(\left(\dfrac{A_{sv}}{s_v}\right)0.95 f_y + b v_c\right)d$$

$$= (0.895 \times 0.95 \times 250 + 300 \times 0.77)485 \times 10^{-3} = 215.1 \text{ kN}$$

This shear force occurs at $x_2 = (V_{3B/A} - V_2)/\omega = (279.5 - 215.1)/54.8 = 1.18$ m. Assuming that the first link is fixed 75 mm from the front face of column 3*B*, then nine R10 links at 125 mm centres are required. Note that since $x_2 < 2125 - d = 1640$ mm, the tension steel is 4T25 as assumed. Furthermore, the shear force at $x = 1.64$ m is

$$V_{x=1.64} = 279.5 - 1.64 \times 54.8 = 189.6 \text{ kN}$$

Here the minimum tension steel is 2T25 and from above $v_c = 0.6 \text{ N/mm}^2$. The shear resistance of R10 links at 175 centres and concrete at $x = 1.64$ m, V_R, is

$$V_R = \left(\left(\dfrac{A_{sv}}{s_v}\right)0.95 f_y + b v_c\right)d$$

$$= (0.895 \times 0.95 \times 250 + 300 \times 0.6)498 \times 10^{-3} = 195.5 \text{ kN} > 189.6 \text{ kN}$$

Therefore R10 links at 175 mm centres are suitable as assumed. The actual number of links required can be assessed using the procedure outlined above. Thus, shear resistance of the minimum links in span 3*A/B*, i.e. R10@300 ($A_{sv}/s_v = 0.523$), plus concrete, V_{21}, is

$$V_{21} = \left(\left(\dfrac{A_{sv}}{s_v}\right)0.95 f_y + b v_c\right)d$$

$$= (0.523 \times 0.95 \times 250 + 300 \times 0.6)498 \times 10^{-3} = 151.5 \text{ kN}$$

This shear force occurs at $x_{21} = (279.5 - 151.5)/54.8 = 2.34$ m (*Fig. 3.41*). Therefore provide seven R10 links at 175 mm centres and fifteen R10 links at 300 centres arranged as shown in *Fig. 3.42*.

(d) **Support 3*B/C* (and 3*C/B*).** Shear force at distance d from support 3*B/C*, V_D, is

$$V_D = V_{3B/A} - 0.7\omega = 256.2 - 0.7 \times 54.8 = 217.9 \text{ kN}$$

Shear stress, $v = \dfrac{V_D}{bd} = \dfrac{217.9 \times 10^3}{300 \times 485} = 1.50 \text{ N/mm}^2$

Again, assume the tension steel is 4T25. Hence $A_s = 1960 \text{ mm}^2$, $\dfrac{100 A_s}{bd} = 1.35$ and $v_c = 0.77 \text{ N/mm}^2$

Since $v > (v_c + 0.4)$ provide design links according to

$\dfrac{A_{sv}}{s_v} \geqslant \dfrac{b(v - v_c)}{0.95 f_{yv}} = \dfrac{300(1.50 - 0.77)}{0.95 \times 250} = 0.922$. Select R10 at 150 centres ($A_{sv}/s_v = 1.047$).

To determine the number of links required assume that R10 links at 225 mm centres ($A_{sv}/s_v = 0.698$) are to be provided between the minimum links in span 3*B/C* and the design links at support 3*B*. Shear resistance of R10 links at 225 and concrete, V_3, is

$$V_3 = \left(\left(\frac{A_{sv}}{s_v}\right)0.95f_y + bv_c\right)d$$

$$= (0.698 \times 0.95 \times 250 + 300 \times 0.77)485 \times 10^{-3} = 192.4 \text{ kN}$$

This shear force occurs at $x_3 = (V_{3B/A} - V_3)/\omega = (256.2 - 192.4)/54.8 = 1.17$ m. Assuming that the first link is fixed 75 mm from the front face of column 3B, then seven R10 links at 150 mm centres are required. Note that since $x_3 < 2125 - d = 1640$ mm, the tension steel is 4T25 as assumed. Furthermore, the shear force at $x = 1.64$ m is

$$V_{x=1.64} = 256.3 - 1.64 \times 54.8 = 166.4 \text{ kN}$$

Again $v_c = 0.6 \text{ N/mm}^2$ since $A_s = 980 \text{ mm}^2$ and the shear resistance of R10 links at 225 centres plus concrete, V_R, is

$$V_R = \left(\left(\frac{A_{sv}}{s_v}\right)0.95f_y + bv_c\right)d$$

$$= (0.698 \times 0.95 \times 250 + 300 \times 0.6)498 \times 10^{-3} = 172.2 \text{ kN} > 166.4 \text{ kN}$$

Therefore R10 links at 225 centres are suitable as assumed. The number of links can be assessed as noted above. The shear resistance of the minimum links in span 3B/C, i.e. R10@300 ($A_{sv}/s_v = 0.523$), plus concrete, V_{31}, is

$$V_{31} = \left(\left(\frac{A_{sv}}{s_v}\right)0.95f_y + bv_c\right)d$$

$$= (0.523 \times 0.95 \times 250 + 300 \times 0.6)498 \times 10^{-3} = 151.5 \text{ kN}$$

This shear force occurs at $x_{31} = (V_{3B/A} - V_{31})/\omega = (256.3 - 151.5)/54.8 = 1.91$ m. Therefore provide four R10 links at 225 mm centres and fourteen R10 at 300 centres arranged as shown in *Fig. 3.42*.

Fig. 3.42 shows the main reinforcement requirements for spans *3A/B* and *3B/C*. Note that in the above calculations the serviceability limit state of cracking has not been considered. For this reason as well as reasons of buildability, the actual reinforcement details may well be slightly different to those indicated in the figure.

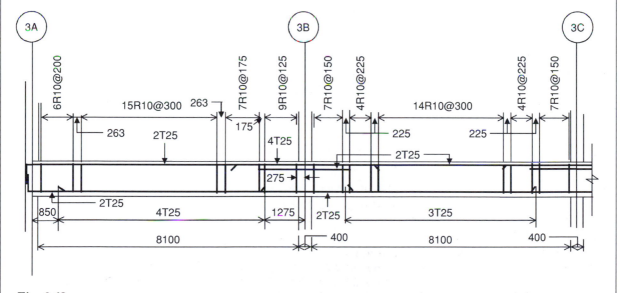

Fig. 3.42

Deflection

$$\text{Actual } \frac{\text{span}}{\text{effective depth}} = \frac{8500}{498} = 17$$

By inspection, exterior span is critical.

$$\frac{b_w}{b} = \frac{300}{1490} = 0.2 < 0.3 \Rightarrow \text{basic span/effective depth ratio of beam} = 20.8 \quad (\textit{Table 3.9})$$

Service stress, f_s, is

$$f_s = \frac{2}{3} f_y \frac{A_{s,req}}{A_{s,prov}} = \frac{2}{3} \times 460 \times \frac{1723}{1960} = 270 \text{ N/mm}^2$$

$$\text{modification factor} = 0.55 + \frac{447 - f_s}{120\left(0.9 + \dfrac{M}{bd^2}\right)} = 0.55 + \frac{477 - 270}{120\left(0.9 + \dfrac{356.3 \times 10^6}{1490 \times 498^2}\right)} = 1.47$$

Therefore,

$$\text{permissible } \frac{\text{span}}{\text{effective depth}} = \text{basic ratio} \times \text{mod.factor} = 20.8 \times 1.47 = 30.6 > \text{actual} \quad \text{OK}$$

PRIMARY BEAM ON GRID LINE B

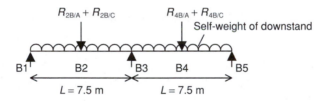

Fig. 3.43

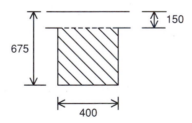

Loading

Design load on beam = uniformly distributed load from self weight of downstand
+ reactions at B2 and B4 from beams 2A/B, 2B/C, 4A/B and 4B/C,
i.e. $R_{2B/A}$, $R_{2B/C}$, $R_{4B/A}$ and $R_{4B/C}$.

Uniform loads

Dead load from self weight of downstand,

$$g_k = 0.4 \times (0.675 - 0.15) \times 24 = 5.04 \text{ kN/m}$$

$$\equiv 5.04 \times 7.5 = 37.8 \text{ kN on each span.}$$

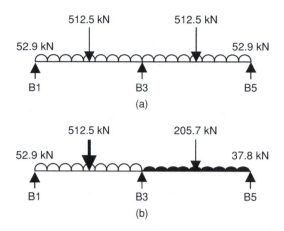

Fig. 3.44 *Load cases: (a) load case 1 (b) load case 2.*

Imposed load, $q_k = 0$

Point loads. Dead load from reactions $R_{2B/A}$ and $R_{2B/C}$ (and $R_{4B/A}$ and $R_{4B/C}$), G_k, is

$$G_k = 22 \times (0.6 \times 8.5) + 22 \times 4.25 = 205.7 \text{ kN}$$

Imposed load from reactions $R_{2B/A}$ and $R_{2B/C}$ (and $R_{4B/A}$ and $R_{4B/C}$), Q_k, is

$$Q_k = 15 \times (0.6 \times 8.5) + 15 \times 4.25 = 140.3 \text{ kN}$$

Load cases

Since the beam does not satisfy the conditions in clause 3.4.3, the coefficients in *Table 3.16* cannot be used to estimate the design moments and shear forces. They can be obtained using techniques such as moment distribution, as discussed in *section 3.9.3.1*. As noted in *section 3.6.2* for continuous beams, two load cases must be considered: (1) maximum design load on all spans (*Fig. 3.44(a)*) and (2) maximum and minimum design loads on alternate spans (*Fig. 3.44(b)*). Assume for the sake of simplicity that the ends of beam B1/5 are simple supports.

Maximum design load = uniform load($W' = 1.4g_k + 1.6q_k = 1.4 \times 37.8 + 0 = 52.9$ kN)
+ point load($W'' = 1.4G_k + 1.6Q_k = 1.4 \times 205.7 + 1.6 \times 140.3 = 512.5$ kN)

Minimum design load = uniform load($W''' = 1.0g_k = 1.0 \times 37.8 = 37.8$ kN)
+ point load($W'''' = 1.0G_k = 1.0 \times 205.7 = 205.7$ kN)

Load case 1

Fixed end moments From *Table 3.17*

$$M_{B1/3} = M_{B3/5} = -\frac{W'L}{12} - \frac{W''L}{8} = -\frac{52.9 \times 7.5}{12} - \frac{512.5 \times 7.5}{8} = -513.5 \text{ kN m}$$

$$M_{B3/1} = M_{B5/3} = \frac{W'L}{12} + \frac{W''L}{8} = 513.5 \text{ kN m}$$

Stiffness and distribution factors. Referring to Example 3.7 it can be seen that the stiffness factors for members B3/1 and B3/5 are both $(3/4)4EI/7.5$. Therefore the distribution factor at end B3/1 and end B3/5 are both 0.5.

End moments

Joint	B1	B3		B5
End	B1/3	B3/1	B3/5	B5/3
Distribution factors		0.5	0.5	
FEM (kN m)	−513.5	513.5	−513.5	513.5
Release B1 & B5	+513.5			−513.5
Carry over		256.8	−256.8	
Sum (kN m)	0	770.3	−770.3	0

Span moments and reactions The support reactions and mid-span moments are obtained using statics.

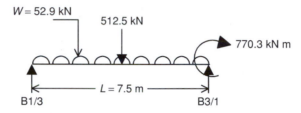

Taking moments about end B3/1 obtains the reaction at end B1/3, $R_{B1/3}$, as follows

$$7.5R_{B1/3} = 512.5 \times (7.5/2) + 52.9 \times (7.5/2) - 770.3$$

$$= 1350 \text{ kN m} \Rightarrow R_{B1/3} = 180 \text{ kN}$$

Reaction at end B3/1, $R_{B3/1} = 512.5 + 52.9 - 180 = 385.4$ kN. Since the beam is symmetrically loaded, the reaction at end B3/5 is 385.4 kN and end B5/3 is 180 kN.

By inspection, the maximum sagging moment occurs at the point load, i.e. $x = 3.75$ m, and is given by

$$M_{x=3.75m} = 180x - (52.9/7.5)x^2/2 = 625.4 \text{ kN m}$$

Load case 2

Fixed end moments

$$M_{B1/3} = -M_{B3/1} = -\frac{W'L}{12} - \frac{W''L}{8} = -\frac{52.9 \times 7.5}{12} - \frac{512.5 \times 7.5}{8} = -513.5 \text{ kN m}$$

$$M_{B3/5} = -M_{B5/3} = -\frac{W'''L}{12} - \frac{W''''L}{8} = -\frac{37.8 \times 7.5}{12} - \frac{205.7 \times 7.5}{8} = -216.5 \text{ kN m}$$

Stiffness and distribution factors. The stiffness and distribution factors are unchanged from the values calculated above.

End moments

Joint	B1	B3		B5
End	B1/3	B3/1	B3/5	B5/3
Distribution factors		0.5	0.5	
FEM (kN m)	−513.5	513.5	−216.5	216.5
Release B1 & B5	+513.5			−216.5
Carry over		256.8	−108.3	
Release B3		−222.8	−222.8	
Sum (kN m)	0	547.5	−547.6	0

Span moments and reactions. The support reactions and mid-span moments are obtained using statics.

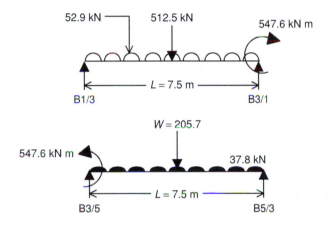

Taking moments about end B3/1 of beam B1/3, the reaction at end B1/3, $R_{B1/3}$, is

$$7.5R_{B1/3} = 512.5 \times (7.5/2) + 52.9 \times (7.5/2) − 547.6 = 1572.7 \text{ kN m}$$

$$R_{B1/3} = 209.7 \text{ kN}$$

Hence, reaction at end B3/1, $R_{B3/1} = 512.5 + 52.9 − 209.7 = 355.7$ kN.
Similarly for beam B3/5, the reaction at end B5/3, $R_{B5/3}$, is

$$7.5R_{B5/3} = 205.7 \times (7.5/2) + 37.8 \times (7.5/2) − 547.6 = 365.5 \text{ kN m}$$

$$R_{B5/3} = 48.7 \text{ kN}$$

and

$$R_{B3/5} = 205.7 + 37.8 − 48.7 = 194.8 \text{ kN}$$

By inspection, the maximum sagging moment in span B1/3 occurs at the point load and is given by

$$M_{x=3.75} = 209.7x − (52.9/7.5)x^2/2 = 736.8 \text{ kN m}$$

Similarly, the maximum sagging moment in span B3/5 is

$$M_{x=3.75} = 194.8x − (37.8/7.5)x^2/2 − 547.6 = 147.5 \text{ kN m}$$

Design moments and shear forces

The bending moment and shear force distributions along the beam are shown below. It can be seen that the design sagging moment is 736.8 kN m and the design hogging moment is 770.3 kN m. The design shear force at supports B1 and B5 is 209.7 kN and at support B3 is 385.4 kN.

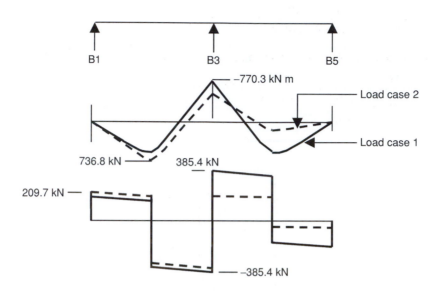

Steel reinforcement

Middle of span B1/3 (and B3/5). Assume the main steel consists of two layers of 25 mm diameter bars, diameter of links, $\phi' = 12$ mm and nominal cover, $c = 30$ mm. Hence effective depth, d, is

$$d = h - (\phi + \phi' + c) = 675 - (25 + 12 + 30) = 608 \text{ mm}$$

From clause 3.4.1.5 of BS 8110 the effective width of beam, b, is

$$b = b_w + 0.7\ell/5 = 400 + 0.7 \times 7500/5 = 1450 \text{ mm}$$

$$K = \frac{M}{f_{cu}bd^2} = \frac{736.8 \times 10^6}{35 \times 1450 \times 608^2} = 0.0393$$

$$z = d\left(0.5 + \sqrt{(0.25 - K/0.9)}\right) \leq 0.95d = 0.95 \times 608 = 578 \text{ mm} \quad \text{(critical)}$$

$$= 608\left(0.5 + \sqrt{(0.25 - 0.0393/0.9)}\right) = 608 \times 0.954 = 580 \text{ mm}$$

$x = (d - z)/0.45 = (608 - 578)/0.45 = 67$ mm $<$ flange thickness (= 150 mm). Hence

Area of tension steel, $A_s = \dfrac{M}{0.95f_y z} = \dfrac{736.8 \times 10^6}{0.95 \times 460(0.95 \times 608)} = 2919 \text{ mm}^2$

Provide 6T25 ($A_s = 2950$ mm²).

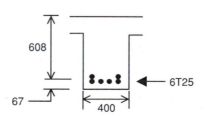

At support B3. Again, assuming that the main steel consists of two layers of 25 mm diameter bars, diameter of links, $\phi' = 12$ mm and nominal cover, $c = 30$ mm, implies that the effective depth, $d = 608$ mm

Since the beam is in hogging, effective width of beam, $b = b_w = 400$ mm

$$K = \frac{M}{f_{cu}bd^2} = \frac{770.3 \times 10^6}{35 \times 400 \times 608^2} = 0.1488$$

$$z = d\left(0.5 + \sqrt{(0.25 - K/0.9)}\right) = 608\left(0.5 + \sqrt{(0.25 - 0.1488/0.9)}\right) = 481 \text{ mm}$$

$$A_s = \frac{M}{0.95 f_y z} = \frac{770.3 \times 10^6}{0.95 \times 460 \times 481} = 3665 \text{ mm}^2$$

Provide 8T25 ($A_s = 3930$ mm^2).

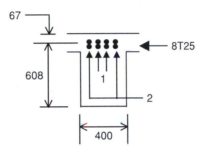

Note that, in practice, it would be difficult to hold bars 1 in place and a spacer bar would be needed between the two layers of reinforcement. This would reduce the value of the effective depth, but this aspect has been ignored in the calculations.

The simplified rules for curtailment of bars in continuous beams illustrated in *Fig. 3.26* do not apply here since the loading on the beam is not predominantly uniformly distributed. The general rule given in clause 3.12.9.1 can be used, however, to establish the curtailment length of bars as discussed below with reference to bar marks 1 and 2.

The theoretical position along the beam where mark 1 bars (i.e. 2T25 from the inner layer of reinforcement) can be stopped off is where the moment of resistance of the section considering only the continuing bars (see sketch), M_r, is equal to the design moment, M.

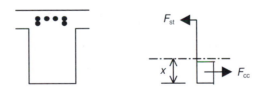

For equilibrium

$$F_{cc} = F_{st}$$

$$\frac{0.67 f_{cu}}{\gamma_{mc}} 0.9xb = 0.95 f_y A_s$$

$$\frac{0.67 \times 35}{1.5} 0.9 \times x \times 400 = 0.95 \times 460 \times 2950$$

Hence $x = 229$ mm

Also, $z = d - 0.9x/2 = 608 - 0.9 \times 229/2 = 505$ mm

Moment of resistance of section, M_r, is

$$M_r = F_{cc}z = \left(\frac{0.67 f_{cu}}{\gamma_{mc}} 0.9xb\right)z = \left(\frac{0.67 \times 35}{1.5} 0.9 \times 229 \times 400\right)505 \times 10^{-6} = 650 \text{ kN m}$$

The design moment along the beam from end B3/1, M, is

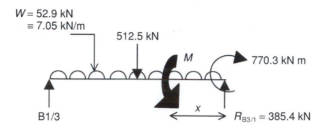

$M = 770.3 - 385.4x + \dfrac{7.05x^2}{2}$. Solving for $M = 650$ kN m implies that the theoretical cut-off point of mark 1 bars from the centre-line of support B3 is 0.31 m.

According to clause 3.12.9.1, the actual cut-off point of bars in the tension zone is obtained by extending the bars an anchorage length ($= 38\phi$ from *Table 3.27* of BS 8110 assuming the use of grade 35 concrete and grade 460, deformed type 2 bars) beyond the theoretical cut-off point (i.e. 310 mm $+ 38 \times 25 = 1260$ mm) or where other bars continuing past provide double the area required to resist the moment at the section i.e. where the design moment is $\frac{1}{2}M = 325$ kN m. The latter is obtained by solving the above expression for x assuming $M = 325$ kN m. This implies that the actual cut-off point of the bars is 1.16 m. Hence the 2T25 bars can be stopped off at, say, 1.3 m from support B3.

The cut-off point of mark 2 bars can be similarly evaluated. Here A_s is 1960 mm^2. Hence $x = 152$ mm, $z = 551.6$ mm assuming $d = 620$ mm and $M_r = 471.9$ kN m. The theoretical cut-off point of the bars is 0.78 m from the centre-line of support B3. The actual cut-off point is then either $780 + 38 \times 25 = 1730$ mm or where the design moment is 235.9 kN m i.e. 1.406 m. Thus it can be assumed the mark 2 bars can be stopped off at, say, 1.8 m from the centre-line of support B3.

Repeating the above procedure will obtain the cut-off points of the remaining sets of bars. Not all bars will need to be curtailed however. Some should continue through to supports as recommended in the simplified rules for curtailment of reinforcement for beams. Also the anchorage length of bars that continue to end supports or are stopped off in the compression zone will vary and the reader is referred to the provisions in clause 3.12.9.1 for further details. *Fig. 3.45* shows a sketch of the bending reinforcement for the beam.

Shear

Support B1 (and B5)

Shear stress, $v = \dfrac{V}{bd} = \dfrac{209.7 \times 10^3}{400 \times 620} = 0.85 \text{ N/mm}^2 < 0.8\sqrt{f_{cu}} = 0.8\sqrt{35} = 4.7 \text{ N/mm}^2$ OK

From *Fig. 3.45* it can be seen that the area of tension steel is 4T25. Hence $A_s = 1960$ mm^2 and

$$\frac{100A_s}{bd} = \frac{100 \times 1960}{400 \times 620} = 0.79$$

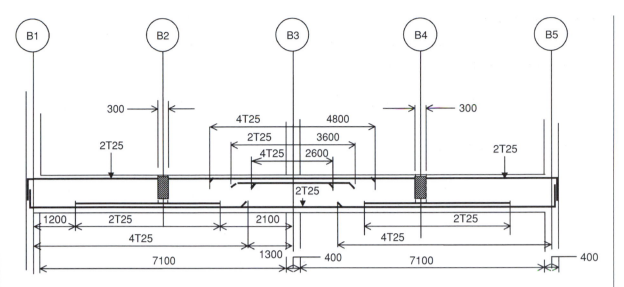

Fig. 3.45

From *Table 3.8*, $v_c = \sqrt[3]{\dfrac{35}{25}} \times 0.58 = 0.65 \text{ N/mm}^2$

Since $v < (v_c + 0.4)$ minimum links are required according to

$$\frac{A_{sv}}{s_v} = \frac{0.4b}{0.95f_{yv}} = \frac{0.4 \times 400}{0.95 \times 250} = 0.674$$

Provide R12–300 ($A_{sv}/s_v = 0.753$)

Support B3

Shear stress, $v = \dfrac{V}{bd} = \dfrac{385.4 \times 10^3}{400 \times 620} = 1.55 \text{ N/mm}^2 < 0.8\sqrt{f_{cu}} = 0.8\sqrt{35} = 4.7 \text{ N/mm}^2$ OK

From *Fig. 3.45* it can be seen that the minimum tension steel at any point between B3 and B2 is 2T25. Hence $A_s = 980 \text{ mm}^2$ and

$$\frac{100A_s}{bd} = \frac{100 \times 980}{400 \times 620} = 0.40$$

From *Table 3.8*, $v_c = \sqrt[3]{\dfrac{35}{25}} \times 0.46 = 0.52 \text{ N/mm}^2$

Since $v > (v_c + 0.4)$ provide design links according to

$$\frac{A_{sv}}{s_v} \geqslant \frac{b(v - v_c)}{0.95f_{yv}} = \frac{400(1.58 - 0.52)}{0.95 \times 250} = 1.785$$

Provide R12–125 ($A_{sv}/s_v = 1.808$)

Figure 3.46 shows the main reinforcement requirements for beam B1/5. Note that in the above calculations the serviceability limit state of cracking has not been considered. For this reason as well as reasons of buildability, the actual reinforcement details may well be slightly different to those shown.

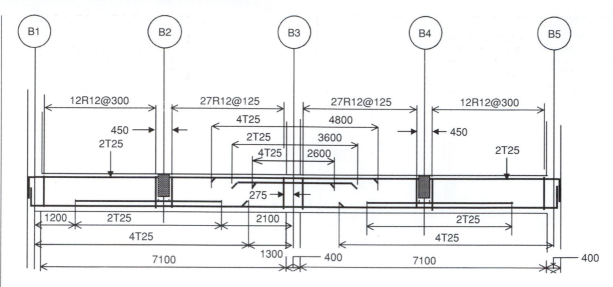

Fig. 3.46

Deflection

$$\text{Actual } \frac{\text{span}}{\text{effective depth}} = \frac{7500}{608} = 12.3$$

$\dfrac{b_w}{b} = \dfrac{400}{1450} = 0.276 < 0.3 \Rightarrow$ basic span/effective depth ratio of beam = 20.8   (*Table 3.9*)

Service stress, f_s, is

$$f_s = \frac{2}{3}f_y \frac{A_{s,req}}{A_{s,prov}} = \frac{2}{3} \times 460 \times \frac{2919}{2950} = 303 \text{ N/mm}^2$$

$$\text{Modification factor} = 0.55 + \frac{477 - f_s}{120\left(0.9 + \dfrac{M}{bd^2}\right)} = 0.55 + \frac{477 - 303}{120\left(0.9 + \dfrac{736.8 \times 10^6}{1450 \times 608^2}\right)} = 1.19$$

$$\text{Therefore, permissible } \frac{\text{span}}{\text{effective depth}} = \text{basic ratio} \times \text{modification factor}$$

$$= 20.8 \times 1.19 = 24.8 > \text{actual} \quad \text{OK}$$

Once the end moments have been determined, it is a simple matter to calculate the bending moments and shear forces in individual spans using statics as discussed in *Chapter 2*.

3.9.4 SUMMARY FOR BEAM DESIGN
Figure 3.47 shows the basic steps that should be followed in order to design reinforced concrete beams.

3.10 Slabs

If a series of very wide, shallow rectangular beams were placed side by side and connected transversely such that it was possible to share the load between adjacent beams, the combination of beams would act as a slab (*Fig. 3.48*).

Reinforced concrete slabs are used to form a variety of elements in building structures such as floors, roofs, staircases, foundations and some types

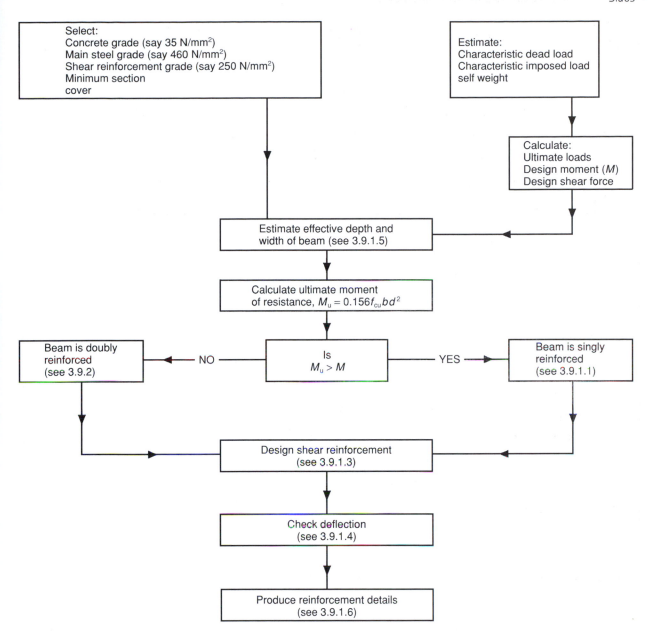

Fig. 3.47 *Beam design procedure.*

of walls (*Fig. 3.49*). Since these elements can be modelled as a set of transversely connected beams, it follows that the design of slabs is similar, in principle, to that for beams. The major difference is that in slab design the serviceability limit state of deflection is normally critical, rather than the ultimate limit states of bending and shear.

3.10.1 TYPES OF SLABS

Slabs may be solid, ribbed, precast or in situ and if in situ they may span two-ways. In practice, the choice of slab for a particular structure will largely depend upon economy, buildability, the loading conditions and the length of the span. Thus for short spans, generally less than 5 m, the most

Fig. 3.48 *Floor slab as a series of beams connected transversely.*

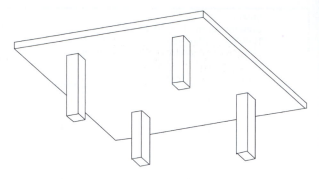

Fig. 3.51 *Flat slab.*

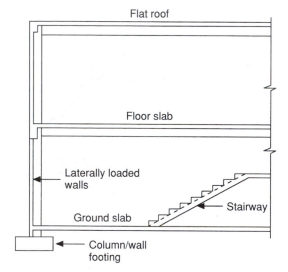

Fig. 3.49 *Various applications for slabs in reinforced concrete structures.*

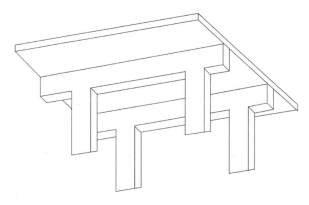

Fig. 3.50 *Solid slab.*

economical solution is to provide a solid slab of constant thickness supported on beams (*Fig. 3.50*).

With medium size spans from 5 to 9 m it is more economical to provide flat slabs since they are generally easier to construct (*Fig. 3.51*). The ease of construction chiefly arises from the fact that the floor has a flat soffit. This avoids having to erect complicated shuttering, thereby making possible speedier and cheaper construction. The use of flat slab construction offers a number of other advantages, absent from other flooring systems, including reduced storey heights, no restrictions on the positioning of partitions, windows can extend up to the underside of the slab and ease of installation of horizontal services. The main drawbacks with flat slabs are that they may deflect excessively and are vulnerable to punching failure. Excessive deflection of flat slabs can be avoided by using deep slabs or by thickening the slab near the columns, using drop panels.

Punching failure arises from the fact that high live loads results in high shear stresses at the supports which may allow the columns to punch through the slab unless appropriate steps are taken. Using deep slabs with large diameter columns, providing drop panels and/or by flaring column heads (*Fig. 3.52*), can avoid this problem. All these methods have drawbacks, however, and research effort has therefore been directed at finding alternative solutions. Use of shear hoops, ACI shear stirrups, shears ladders and stud rails (*Fig. 3.53*) are just a few of the solutions that have been proposed over recent years. All are designed to overcome the problem of fixing individual shear links, which is both labour intensive and a practical difficulty.

Shear hoops are prefabricated cages of shear reinforcement which are attached to the main steel. They are available in a range of diameters and are

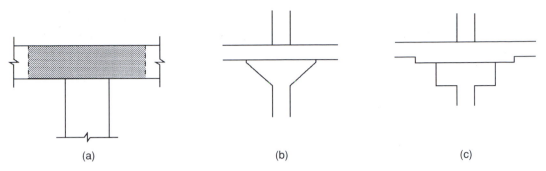

Fig. 3.52 *Methods of reducing shear stresses in flat slab construction: (a) deep slab and large column; (b) slab with flared column head; (c) slab with drop panel and column head.*

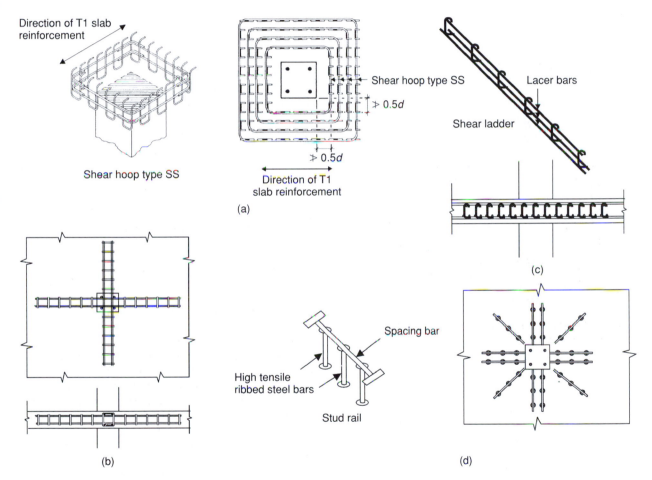

Fig. 3.53 *Prefabricated punching shear reinforcement for flat slabs: (a) shear hoops (b) ACI shear stirrups (c) shear ladders (d) stud rails. Typical arrangements for an internal column.*

suitable for use with internal and edge columns. Although superficially attractive, use of this system has declined significantly over recent years.

The use of ACI shear stirrups is potentially the simplest and cheapest method of preventing punch-ing shear in flat slabs. The shear stirrups are arrangements of conventional straight bars and links that form a +, **T** or **L** shape for an internal, edge or corner column respectively. The stirrups work in exactly the same way as conventional shear rein-

89

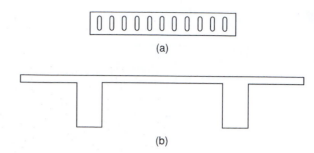

(a)

(b)

Fig. 3.54 Ribbed slab.

Fig. 3.55 Precast concrete floor units: (a) hollow core plank (b) double 'T' unit.

forcement but can simply be attached to the main steel via the straight bars.

Shear ladders are rows of traditional links that are welded to lacer bars. The links resist the shear stresses and the lacer bars anchor the links to the main steel. Whilst they are simple to design and use they can cause problems of congestion of reinforcement.

Stud rails are prefabricated high tensile ribbed headed studs, which are held at standard centres by a welded spacer bar. These rails are arranged in a radial pattern and held in position during the concrete pour by tying to either the top or bottom reinforcement. The studs work through direct mechanical anchorage provided by their heads. They are easy to install but quite expensive.

With medium to long spans and light to moderate live loads from 3 to 5 kN/m^2, it is more economical to provide ribbed slabs constructed using glass reinforced polyester, polypropylene or encapsulated expanded polystyrene moulds (*Fig. 3.54*). Such slabs have reduced self-weight compared to solid slabs since part of the concrete in the tension zone is omitted. However, ribbed slabs have higher form-work costs than the other slabs systems mentioned above and, generally, they are found to be economic in the range 8 to 12 m.

With the emphasis on speed of erection and economy of construction, the use of precast concrete floor slabs is now also popular with both clients and designers. *Figure 3.55* shows two types of precast concrete units that can be used to form floors. The hollow core planks are very common as they are economic over short, medium and long spans. If desired the soffit can be left exposed whereas the top is normally finished with a levelling screed

or appropriate flooring system. Cranage of large precast units, particularly in congested city centre developments, is the biggest obstacle to this type of floor construction.

The span ranges quoted above generally assume the slab is supported along two opposite edges i.e. it is one-way spanning (*Fig. 3.56*). Where longer in-situ concrete floor spans are required it is usually more economical to support the slab on all four sides. The cost of supporting beams or walls needs to be considered though. Such slabs are referred to as two-way spanning and are normally designed as two-dimensional plates provided the ratio of the length of the longer side to the length of the shorter side is equal to 2 or less (*Fig. 3.57*).

This book only discusses the design of one-way and two-way spanning solid slabs supporting uniformly distributed loads. The reader is referred to more specialised books on this subject for guidance on the design of the other slab types described above.

3.10.2 DESIGN OF ONE-WAY SPANNING SOLID SLAB

The general procedure to be adopted for slab design is as follows:

1. Determine a suitable depth of slab.
2. Calculate main and secondary reinforcement areas.
3. Check critical shear stresses.
4. Check detailing requirements.

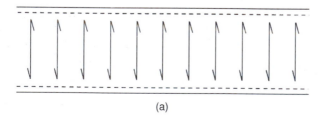

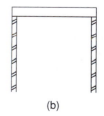

(a)

(b)

Fig. 3.56 One-way spanning solid slab: (a) plan; (b) elevation.

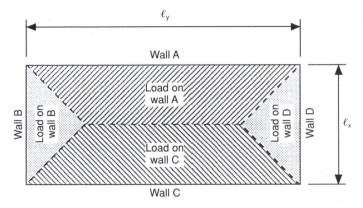

Fig. 3.57 *Plan of two-way spanning slab. l_x length of shorter side, l_y length of longer side. Provided $l_y/l_x \leqslant 2$ slab will span in two directions as indicated.*

3.10.2.1 Depth of slab (clause 3.5.7, BS 8110)

Solid slabs are designed as if they consist of a series of beams of 1 m width.

The effective span of the slab is taken as the smaller of

(a) the distance between centres of bearings, A, or
(b) the clear distance between supports, D, plus the effective depth, d, of the slab (*Fig. 3.58*).

The deflection requirements for slabs, which are the same as those for beams, will often control the depth of slab needed. The minimum effective depth of slab, d_{min}, can be calculated using

$$d_{min} = \frac{\text{span}}{\text{basic ratio} \times \text{modification factor}} \quad (3.20)$$

The basic (span/effective depth) ratios are given in *Table 3.11*. The modification factor is a function of the amount of reinforcement in the slab which is itself a function of the effective depth of the slab. Therefore, in order to make a first estimate of the effective depth, d_{min}, of the slab, a value of (say) 1.4 is assumed for the modification factor. The main steel areas can then be calculated (*section 3.10.2.2*), and used to determine the actual value of the modification factor. If the assumed value is slightly

greater than the actual value, the depth of the slab will satisfy the deflection requirements in BS 8110. Otherwise, the calculation must be repeated using a revised value of the modification factor.

3.10.2.2 Steel areas (clause 3.5.4, BS 8110)

The overall depth of slab, h, is determined by adding allowances for cover (*Table 3.5*) and half the (assumed) main steel bar diameter to the effective depth. The self weight of the slab together with the dead and live loads are used to calculate the design moment, M.

The ultimate moment of resistance of the slab, M_u, is calculated using equation 3.11, developed in *section 3.9.1.1*, namely

$$M_u = 0.156 f_{cu} b d^2$$

If $M_u \geqslant M$, which is the usual condition for slabs, compression reinforcement will not be required and the area of tensile reinforcement, A_s, is determined using equation 3.12 developed in *section 3.9.1.1*, namely

$$A_s = \frac{M}{0.95 f_y z}$$

where $z = d[0.5 + \sqrt{(0.25 - K/0.9)}]$ in which $K = M/f_{cu}bd^2$.

Secondary or distribution steel is required in the transverse direction and this is usually based on the minimum percentages of reinforcement ($A_{s\ min}$) given in *Table 3.25* of BS 8110:

$$A_{s\ min} = 0.24\% A_c \quad \text{when } f_y = 250 \text{ N/mm}^2$$

$$A_{s\ min} = 0.13\% A_c \quad \text{when } f_y = 460 \text{ N/mm}^2$$

where A_c is the total area of concrete.

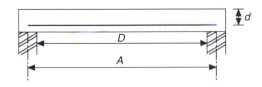

Fig. 3.58 *Effective span of simply supported slab.*

Table 3.18 Form and area of shear reinforcement in solid slabs (Table 3.16, BS 8110)

Values of v (N/mm^2)	Area of shear reinforcement to be provided
$v < v_c$	None required
$v_c < v < (v_c + 0.4)$	Minimum links in areas where $v > v_c$ $A_{sv} \geqslant 0.4 b s_v / 0.95 f_{yv}$
$(v_c + 0.4) < v < 0.8 \sqrt{f_{cu}}$ or 5 N/mm^2	Design links $A_{sv} \geqslant b s_v (v - v_c) / 0.95 f_{yv}$

3.10.2.3 Shear (clause 3.5.5 of BS 8110)

Shear resistance is generally not a problem in solid slabs subject to uniformly distributed loads and, in any case, shear reinforcement should not be provided in slabs less than 200 mm deep.

As discussed for beams in *section 3.9.1.3*, the design shear stress, v, is calculated from

$$v = \frac{V}{bd}$$

The ultimate shear resistance, v_c, is determined using *Table 3.8*. If $v < v_c$, no shear reinforcement is required. Where $v > v_c$, the form and area of shear reinforcement in solid slabs should be provided in accordance with the requirements contained in *Table 3.18*.

3.10.2.4 Reinforcement details (clause 3.12, BS 8110)

For reasons of durability the code specifies limits in respect of:

1. maximum and minimum reinforcement areas
2. spacing of reinforcement
3. maximum crack widths.

These are outlined below together with the simplified rules for curtailment of reinforcement.

1. Reinforcement areas (clause 3.12.5, BS 8110). The area of tension reinforcement, A_s, should exceed the following limits:

$$A_s \geqslant 0.24\% A_c \quad \text{when } f_y = 250 \text{ N/mm}^2$$
$$A_s \geqslant 0.13\% A_c \quad \text{when } f_y = 460 \text{ N/mm}^2$$

where A_c is the total area of concrete.

2. Spacing of reinforcement (clause 3.12.11.2.7, BS 8110). The clear distance between tension bars, s_b, should lie within the following limits: $h_{agg} + 5$ mm or bar diameter $\leqslant s_b \leqslant 3d$ or 750 mm whichever

is the lesser where h_{agg} is the maximum aggregate size. (See also below, section on crack widths.)

3. Crack width (clause 3.12.11.2.7, BS 8110). Unless the actual crack widths have been checked by direct calculation, the following rules will ensure that crack widths will not generally exceed 0.3 mm. This limiting crack width is based on considerations of appearance and durability.

(i) No further check is required on bar spacing if either:
   (a) $f_y = 250$ N/mm^2 and slab depth $\leqslant 250$ mm, or
   (b) $f_y = 460$ N/mm^2 and slab depth $\leqslant 200$ mm, or
   (c) the reinforcement percentage $(100 A_s / bd)$ $< 0.3\%$.

(ii) Where none of conditions (a), (b) or (c) apply and the percentage of reinforcement in the slab exceed 1 per cent, then the maximum clear distance between bars (s_{max}) given in *Table 3.28* of BS 8110 should be used, namely:

$$s_{max} \leqslant 300 \text{ mm} \quad \text{when } f_y = 250 \text{ N/mm}^2$$
$$s_{max} \leqslant 160 \text{ mm} \quad \text{when } f_y = 460 \text{ N/mm}^2$$

4. Curtailment of reinforcement (clause 3.12.10.3, BS 8110). Simplified rules for the curtailment of reinforcement are given in clause 3.12.10.3 of BS 8110. These are shown diagrammatically in *Fig. 3.59* for simply supported and continuous solid slabs.

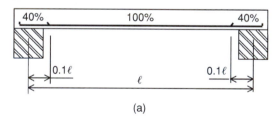

(a)

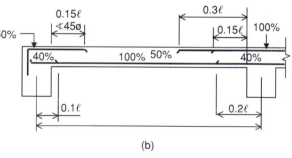

(b)

Fig. 3.59 *Simplified rules for curtailment of bars in slabs: (a) simply supported ends; (b) continuous slab (based on Fig. 3.25, BS 8110).*

Example 3.10 Design of a one–way spanning concrete floor (BS 8110)

A reinforced concrete floor subject to an imposed load of 3.5 kN/m^2 spans between brick walls as shown below. Design the floor for mild exposure conditions assuming the following material strengths:

$$f_{cu} = 35 \text{ N/mm}^2$$

$$f_y = 460 \text{ N/mm}^2$$

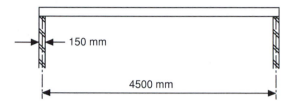

(I) DEPTH OF SLAB AND MAIN STEEL AREA

Overall depth of slab, h

$$\text{Minimum effective depth, } d_{min} = \frac{\text{span}}{\text{basic ratio} \times \text{modification factor}}$$

$$= \frac{4500}{20 \times (\text{say})1.4} = 161 \text{ mm}$$

Hence, assume effective depth of slab (d) = 165 mm. Assume diameter of main steel (Φ) = 10 mm. From *Table 3.5*, cover to all steel for mild conditions of exposure (c) = 20 mm.

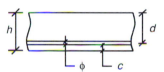

Overall depth of slab (h) = $d + \Phi/2 + c$

$$= 165 + 10/2 + 20 = 190 \text{ mm}$$

Loading

Dead. Self weight of slab (g_k) = 0.19 × 24 kN/m^3 = 4.56 kN/m^2

Imposed. Total imposed load (q_k) = 3.5 kN/m^2

Ultimate load. For 1 m width of slab total ultimate load is

$$= (1.4g_k + 1.6q_k)\text{width of slab} \times \text{span}$$

$$= (1.4 \times 4.56 + 1.6 \times 3.5)1 \times 4.5 = 53.93 \text{ kN}$$

Design moment

$$M = \frac{W\ell}{8} = \frac{53.93 \times 4.5}{8} = 30.34 \text{ kN m}$$

Ultimate moment

$$M_u = 0.156 f_{cu} b d^2$$
$$= 0.156 \times 35 \times 10^3 \times 165^2$$
$$= 148.6 \times 10^6 = 148.6 \text{ kN m}$$

Since $M < M_u$, no compression reinforcement is required.

Main steel

$$K = \frac{M}{f_{cu} b d^2} = \frac{30.34 \times 10^6}{35 \times 10^3 \times 165^2} = 0.0318$$

$$z = d[0.5 + \sqrt{(0.25 - K/0.9)}]$$

$$= 165[0.5 + \sqrt{(0.25 - 0.0318/0.9)}]$$

$$= 165 \times 0.963 \leqslant 0.95d \ (= 157 \text{ mm})$$

Hence $z = 157$ mm.

$$A_s = \frac{M}{0.95 f_y z} = \frac{30.34 \times 10^6}{0.95 \times 460 \times 157} = 443 \text{ mm}^2/\text{m width of slab}$$

For detailing purposes this area of steel has to be transposed into bars of a given diameter and spacing using steel area tables. Thus from *Table 3.19*, provide 10 mm diameter bars spaced at 150 mm, i.e. T10 at 150 centres ($A_s = 523$ mm^2/m).

Table 3.19 Cross-sectional area per metre width for various bar spacing (mm^2)

Bar size (mm)	Spacing of bars								
	50	75	100	125	150	175	200	250	300
6	566	377	283	226	189	162	142	113	94.3
8	1 010	671	503	402	335	287	252	201	168
10	1 570	1 050	785	628	523	449	393	314	262
12	2 260	1 510	1 130	905	754	646	566	452	377
16	4 020	2 680	2 010	1 610	1 340	1 150	1 010	804	670
20	6 280	4 190	3 140	2 510	2 090	1 800	1 570	1 260	1 050
25	9 820	6 550	4 910	3 930	3 270	2 810	2 450	1 960	1 640
32	16 100	10 700	8 040	6 430	5 360	4 600	4 020	3 220	2 680
40	25 100	16 800	12 600	10 100	8 380	7 180	6 280	5 030	4 190

Actual modification factor

The actual value of the modification can now be calculated using equations 7 and 8 given in *Table 3.13* (*section 3.9.1.4*).

$$\text{Design service stress}, f_s = \frac{2 f_y A_{s,req}}{3 A_{s,prov}} \quad (\text{equation 8, } Table\ 3.13)$$

$$= \frac{2 \times 460 \times 443}{3 \times 523} = 260 \text{ N/mm}^2$$

$$\text{Modification factor} = 0.55 + \frac{(477 - f_s)}{120\left(0.9 + \dfrac{M}{bd^2}\right)} \leq 2.0 \quad \text{(equation 7, Table 3.13)}$$

$$= 0.55 + \frac{(477 - 260)}{120\left(0.9 + \dfrac{30.34 \times 10^6}{10^3 \times 165^2}\right)} = 1.44$$

Hence,

$$\text{New } d_{\min} = \frac{4500}{20 \times 1.44} = 157 \text{ mm} < \text{assumed } d = 165 \text{ mm}$$

Minimum area of reinforcement, $A_{s\,\min}$, is equal to

$$A_{s\,\min} = 0.13\% bh = 0.13\% \times 10^3 \times 190 = 247 \text{ mm}^2/\text{m} < A_s$$

Therefore take $d = 165$ mm and provide T10 at 150 mm centres as main steel.

(II) SECONDARY STEEL
Based on minimum steel area = 247 mm²/m. Hence from *Table 3.19*, provide T8 at 200 mm centres ($A_s = 252$ mm²/m).

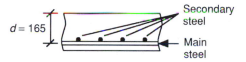

(III) SHEAR REINFORCEMENT

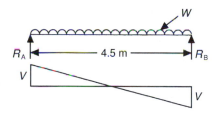

Design shear stress, v
Since slab is symmetrically loaded

$$R_A = R_B = W/2 = 27 \text{ kN}$$

Ultimate shear force (V) = 27 kN and design shear stress, v, is

$$v = \frac{V}{bd} = \frac{27 \times 10^3}{10^3 \times 165} = 0.16 \text{ N/mm}^2$$

Design concrete shear stress, v_c
Assuming that 50% of main steel is curtailed at the supports, $A_s = 523/2 = 262$ mm²/m

$$\frac{100 A_s}{bd} = \frac{100 \times 262}{10^3 \times 165} = 0.16$$

From *Table 3.8*, design concrete shear stress for grade 25 concrete is 0.42 N/mm². Hence

$$v_c = (35/25)^{1/3}0.42 = 0.47 \text{ N/mm}^2$$

From *Table 3.16*, since $v < v_c$, no shear reinforcement is required.

(IV) REINFORCEMENT DETAILS

The sketch below shows the main reinforcement requirements for the slab. For reasons of buildability the actual reinforcement details may well be slightly different.

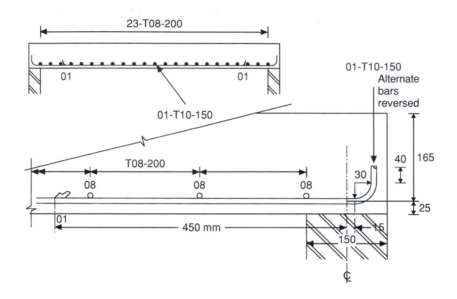

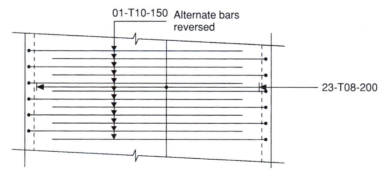

Check spacing between bars

Maximum spacing between bars should not exceed the lesser of $3d$ (= 495 mm) or 750 mm. Actual spacing = 150 mm main steel and 200 mm secondary steel.   OK

Maximum crack width

Since the slab depth does not exceed 200 mm, the above spacing between bars will automatically ensure that the maximum permissible crack width of 0.3 mm will not be exceeded.

Example 3.11 Analysis of a one–way spanning concrete floor (BS 8110)

A concrete floor reinforced with 10 mm diameter mild steel bars ($f_y = 250$ N/mm^2) at 125 mm centres spans between brick walls as shown in *Fig. 3.60*. Calculate the maximum uniformly distributed imposed load the floor can carry.

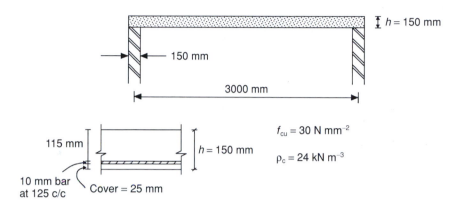

Fig. 3.60

(I) EFFECTIVE SPAN

Effective depth of slab, d, is

$$d = h - \text{cover} - \Phi/2$$

$$= 150 - 25 - 10/2 = 120 \text{ mm}$$

Effective span is the lesser of

(a) centre to centre distance between bearings = 3000 mm
(b) clear distance between supports plus effective depth = 2850 + 120 = 2970 mm

Hence effective span = 2970 mm

(II) MOMENT CAPACITY, *M*

Assume $z = 0.95d = 0.95 \times 120 = 114$ mm

$$A_s = \frac{M}{0.95 f_y z}$$

Hence

$$M = A_s \cdot 0.95 f_y z = 628 \times 0.95 \times 250 \times 114$$

$$= 17 \times 10^6 \text{ N mm} = 17 \text{ kN m/m width of slab}$$

(III) MAXIMUM UNIFORMLY DISTRIBUTED IMPOSED LOAD (q_k)

Loading

Dead load. Self weight of slab (g_k) = 0.15×24 kN/m^3 = 3.6 kN/m^2

Ultimate load
Total ultimate load (W) = $(1.4 g_k + 1.6 q_k)$span

$$= (1.4 \times 3.6 + 1.6 q_k)2.970$$

Imposed load

Design moment $(M) = \dfrac{W\ell}{8}$

From above, $M = 17$ kN m $= (5.04 + 1.6q_k)\dfrac{2.970^2}{8}$

Rearranging gives

$$q_k = \frac{17 \times 8/2.970^2 - 5.04}{1.6} = 6.4 \text{ kN/m}^2$$

Lever arm (z)

Check that assumed value of z is correct, i.e. $z = 0.95d$.

$$K = \frac{M}{f_{cu}bd^2} = \frac{17 \times 10^6}{30 \times 10^3 \times 120^2} = 0.0393$$

$$z = d[0.5 + \sqrt{(0.25 - K/0.9)}] \leqslant 0.95d$$

$$= d[0.5 + \sqrt{(0.25 - 0.0393/0.9)}] = 0.954d$$

Hence, assumed value of z is correct and the maximum uniformly distributed load that the floor can carry is 6.4 kN/m^2.

Table 3.20 Ultimate bending moments and shear forces in one-way spanning slabs with simple end supports (Table 3.12, BS 8110)

	End support	End span	Penultimate support	Interior span	Interior support
Moment	0	0.086Fℓ	−0.086Fℓ	0.063Fℓ	−0.063Fℓ
Shear	0.4F	−	0.6F	−	0.5F

$F = 1.4G_k + 1.6Q_k$; ℓ = effective span

3.10.3 DESIGN OF CONTINUOUS ONE-WAY SPANNING SOLID SLABS

The design of continuous one-way spanning slabs is similar to that outlined above for single-span slabs. The main differences are that (a) several loading arrangements may need to be considered and (b) such slabs are not statically determinate. Methods such as moment distribution can be used to determine the design moments and shear forces along the slab as discussed in *section 3.9.3.1*. Where the following conditions are met, however, the moments and shear forces can be calculated using the coefficients in *Table 3.12* of BS 8110. Values for slabs with simple end supports are reproduced here as *Table 3.20*.

1. There are three or more spans of approximately equal length.
2. The area of each bay exceeds 30 m^2 (*Fig. 3.61*).
3. The ratio of the characteristic imposed load to the characteristic dead load does not exceed 1.25.
4. The characteristic imposed load does not exceed 5 kN/m^2 excluding partitions.

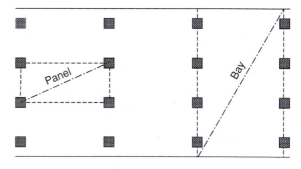

Fig. 3.61 *Definition of panels and bays (Fig. 3.7, BS 8110).*

Example 3.12 Continuous one-way spanning slab design (BS 8110)

Design the continuous one-way spanning slab in Example 3.9 assuming the cover to the reinforcement is 25 mm (*Fig. 3.62*).

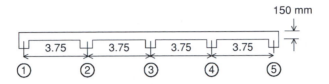

Fig 3.62

Loading

Dead load, g_k = self-weight of slab + finishes = $0.15 \times 24 + 1.5 = 5.1$ kN/m^2
Imposed load, $q_k = 4$ kN/m^2
  For a 1 m width of slab, total ultimate load, $F = (1.4g_k + 1.6q_k)$ width of slab $\times$ span = $(1.4 \times 5.1 + 1.6 \times 4)1 \times 3.75 = 50.8$ kN

Design moments and shear forces

Since area of each bay (= $8.5 \times 15 = 127.5$ m^2) > 30 m^2, q_k/g_k (= $4/5.1 = 0.78$) < 1.25 and q_k < 5 kN/m^2, the coefficients in Table 3.20 can be used to calculate the design moments and shear forces along the slab.

Position	Bending moments (kN m)	Shear forces (kN)
Supports 1 & 5	0	$0.4 \times 50.8 = 20.3$
Near middle of spans 1/2 & 4/5	$0.086 \times 50.8 \times 3.75 = 16.4$	
Supports 2 & 4	$-0.086 \times 50.8 \times 3.75 = -16.4$	$0.6 \times 50.8 = 30.5$
Middle of spans 2/3 & 3/4	$0.063 \times 50.8 \times 3.75 = 12$	
Support 3	$-0.063 \times 50.8 \times 3.75 = -12$	$0.5 \times 50.8 = 25.4$

Steel reinforcement

Middle of span 1/2 (and 4/5)

Assume diameter of main steel, $\phi = 10$ mm
Effective depth, $d = h_s - (\phi/2 + c) = 150 - (10/2 + 25) = 120$ mm

$$K = \frac{M}{f_{cu}bd^2} = \frac{16.4 \times 10^6}{35 \times 1000 \times 120^2} = 0.0325$$

$$z = d\left(0.5 + \sqrt{(0.25 - K/0.9)}\right) = 120\left(0.5 + \sqrt{(0.25 - 0.0325/0.9)}\right) = 115.5 \leqslant 0.95d = 114 \text{ mm}$$

$$A_s = \frac{M}{0.95 \times f_y \times z} = \frac{16.4 \times 10^6}{0.95 \times 460 \times 114} = 330 \text{ mm}^2 > A_{s,min} = 0.13\%bh = 195 \text{ mm}^2 \quad \text{OK}$$

From *Table 3.19*, provide T10@200 mm centres ($A_s = 393$ mm^2/m) in bottom face of slab.

Support 2 (and 4). Since the design moment is numerically the same as for the middle of span 1/2, provide T10@200 mm centres in top face of slab.

Middle of span 2/3 (and 3/4)
$M = 12$ kN m and $z = 0.95d = 114$ mm. Hence

$$A_s = \frac{M}{0.95 \times f_y \times z} = \frac{12 \times 10^6}{0.95 \times 460 \times 114} = 241 \text{ mm}^2 > A_{s,min} \quad \text{OK}$$

Provide T10@300 mm centres ($A_s = 262$ mm^2/m) in bottom face of slab.

Support 3. Since $M = -12$ kN m provide T10@300 mm centres in top face of slab.

Support 1 (and 5). According to clause 3.12.10.3.2 of BS 8110, although simple supports may have been assumed at end supports of continuous slabs for analysis, negative moments may arise which could lead to cracking. Therefore an amount of reinforcement equal to half the area of bottom steel at mid-span but not less than the minimum percentage of steel recommended in *Table 3.25* of BS 8110 should be provided in the top of the slab. Furthermore, this reinforcement should be anchored at the support and extend not less than 0.15ℓ or 45 times the bar size into the span.

From above, area of reinforcement at middle of span 1/2 is 330 mm^2/m. From *Table 3.25* of BS 8110, the minimum area of steel reinforcement is $0.13\%bh = 0.0013 \times 1000 \times 150 = 195$ mm^2/m. Hence provide T10 at 300 mm centres ($A_s = 262$ mm^2/m) in the top of the slab.

Distribution steel
Area of distribution steel in this case is equal to the minimum area of reinforcement for solid slabs i.e. 195 mm^2/m. Provide T10 at 350 centres ($A_s = 224$ mm^2/m).

Shear reinforcement

Support 2 (and 4)
Design shear force, $V = 30.5$ kN

$$v = \frac{V}{bd} = \frac{30.5 \times 10^3}{1000 \times 120} = 0.25 \text{ N/mm}^2$$

$$\frac{100A_s}{bd} = \frac{100 \times 393}{1000 \times 120} = 0.33$$

$$v_c = \sqrt[3]{\frac{35}{25}} \times 0.57 = 0.64 \text{ N/mm}^2 > v.$$

Hence, shear reinforcement is not required.

Support 3
Design shear force, $V = 25.4$ kN

$$v = \frac{V}{bd} = \frac{25.4 \times 10^3}{1000 \times 120} = 0.21 \text{ N/mm}^2$$

$$\frac{100A_s}{bd} = \frac{100 \times 262}{1000 \times 120} = 0.22$$

$$v_c = \sqrt[3]{\frac{35}{25}} \times 0.51 = 0.57 \text{ N/mm}^2 > v \quad \text{OK}$$

From *Table 3.18*, no shear reinforcement is necessary.

Deflection

$$\text{Actual } \frac{\text{span}}{\text{effective depth}} = \frac{3750}{120} = 31.25$$

Exterior spans. Steel service stress, f_s, is

$$f_s = \frac{2}{3}f_y\frac{A_{s,req}}{A_{s,prov}} = \frac{2}{3} \times 460 \times \frac{330}{393} = 257.5 \text{ N/mm}^2$$

$$\text{Modification factor} = 0.55 + \frac{477 - f_s}{120\left(0.9 + \dfrac{M}{bd^2}\right)} = 0.55 + \frac{477 - 257.5}{120\left(0.9 + \dfrac{16.4 \times 10^6}{10^3 \times 120^2}\right)} = 1.45$$

From *Table 3.11*, basic span to effective depth ratio is 26. Hence

$$\text{permissible} \ \frac{\text{span}}{\text{effective depth}} = \text{basic ratio} \times \text{mod.factor} = 26 \times 1.45 = 37.7 > 31.25 \quad \text{OK}$$

Interior spans. Steel service stress, f_s, is

$$f_s = \frac{2}{3}f_y\frac{A_{s,req}}{A_{s,prov}} = \frac{2}{3} \times 460 \times \frac{241}{262} = 282 \text{ N/mm}^2$$

$$\text{Modification factor} = 0.55 + \frac{477 - f_s}{120\left(0.9 + \dfrac{M}{bd^2}\right)} = 0.55 + \frac{477 - 282}{120\left(0.9 + \dfrac{12 \times 10^6}{10^3 \times 120^2}\right)} = 1.49$$

Hence permissible $\dfrac{\text{span}}{\text{effective depth}}$ = basic ratio × mod.factor = 26 × 1.49 = 38.7 > 31.25   OK

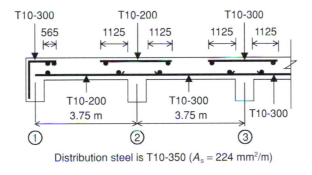

Distribution steel is T10-350 ($A_s = 224 \text{ mm}^2$/m)

3.10.4 DESIGN OF TWO-WAY SPANNING RESTRAINED SOLID SLABS

The design of two-way spanning restrained slabs (*Fig. 3.63*) supporting uniformly distributed loads is largely similar to that outlined above for one-way spanning slabs. The extra complication arises from the fact that it is rather difficult to determine the moments and shear forces in these plate-like elements. Fortunately BS 8110 contains tables of coefficients (β_{sx}, β_{sy}, β_{vx}, β_{vy}) that may assist in this task (*Tables 3.21* and *3.22*). Thus, the maximum design moments per unit width of rectangular slabs of shorter side ℓ_x and longer side ℓ_y are given by

$$m_{sx} = \beta_{sx}n\ell_x^2 \qquad (3.21)$$

$$m_{sy} = \beta_{sy}n\ell_x^2 \qquad (3.22)$$

where

m_{sx} maximum design ultimate moments either over supports or at mid-span on strips of unit width and span ℓ_x (*Fig. 3.64*)

m_{sy} maximum design ultimate moments either over supports or at mid-span on strips of unit width and span ℓ_y

n total design ultimate load per unit area $= 1.4g_k + 1.6q_k$

Similarly, the design shear forces at supports in the long span direction, v_{sy}, and short span direction, v_{sx}, may be obtained from the following expressions

$$v_{sy} = \beta_{vy}n\ell_x \qquad (3.23)$$

$$v_{sx} = \beta_{vx}n\ell_x \qquad (3.24)$$

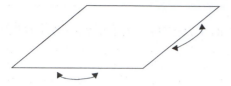

Fig. 3.63 Bending of two-way spanning slabs.

These moments and shears are considered to act over the middle three quarters of the panel width. The remaining edge strips, of width equal to one-eight of the panel width, may be provided with minimum tension reinforcement. In some cases, where there is a significant difference in the support moments calculated for adjacent panels, it may be necessary to modify the mid-span moments in accordance with the procedure given in BS 8110.

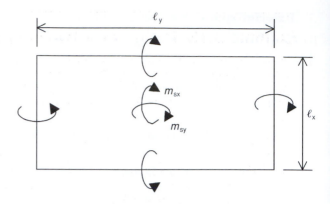

Fig. 3.64 Location of moments.

Table 3.21 Bending moment coefficients, β_{sx} and β_{sy}, for restrained slabs (based on Table 3.14, BS 8110)

Type of panel and moments considered	Short span coefficients, β_{sx} Values of ℓ_y/ℓ_x								Long span coefficients, β_{sy}, for all values of ℓ_y/ℓ_x
	1.0	1.1	1.2	1.3	1.4	1.5	1.75	2.0	
Interior panels									
Negative moment at continuous edge	0.031	0.037	0.042	0.046	0.050	0.053	0.059	0.063	0.032
Positive moment at mid-span	0.024	0.028	0.032	0.035	0.037	0.040	0.044	0.048	0.024
One long edge discontinuous									
Negative moment at continuous edge	0.039	0.049	0.056	0.062	0.068	0.073	0.082	0.089	0.037
Positive moment at mid-span	0.030	0.036	0.042	0.047	0.051	0.055	0.062	0.067	0.028
Two adjacent edges discontinuous									
Negative moment at continuous edge	0.047	0.056	0.063	0.069	0.074	0.078	0.087	0.093	0.045
Positive moment at mid-span	0.036	0.042	0.047	0.051	0.055	0.059	0.065	0.070	0.034

Table 3.22 Shear force coefficients, β_{vx} and β_{vy}, for restrained slabs (based on Table 3.15, BS 8110)

Type of panel and location	β_{vx} for values of ℓ_y/ℓ_x								β_{vy}
	1.0	1.1	1.2	1.3	1.4	1.5	1.75	2.0	
Four edges continuous									
Continuous edge	0.33	0.36	0.39	0.41	0.43	0.45	0.48	0.50	0.33
One long edge discontinuous									
Continuous edge	0.36	0.40	0.44	0.47	0.49	0.51	0.55	0.59	0.36
Discontinuous edge	0.24	0.27	0.29	0.31	0.32	0.34	0.36	0.38	–
Two adjacent edges discontinuous									
Continuous edge	0.40	0.44	0.47	0.50	0.52	0.54	0.57	0.60	0.40
Discontinuous edge	0.26	0.29	0.31	0.33	0.34	0.35	0.38	0.40	0.26

Example 3.13 Design of a two-way spanning restrained slab (BS 8110)

Fig. 3.65 shows a part plan of an office floor supported by monolithic concrete beams (not detailed), with individual slab panels continuous over two or more supports. The floor is to be designed to support an imposed load of 4 kN/m² and finishes plus ceiling loads of 1.25 kN/m². The characteristic strength of the concrete is 30 N/mm² and the steel reinforcement is 460 N/mm². The cover to steel reinforcement is 25 mm.

(a) Calculate the mid-span moments for panels AB2/3 and BC1/2 assuming the thickness of the floor is 180 mm.
(b) Design the steel reinforcement for panel BC2/3 (shown hatched) and check the adequacy of the slab in terms of shear resistance and deflection. Illustrate the reinforcement details on plan and elevation views of the panel.

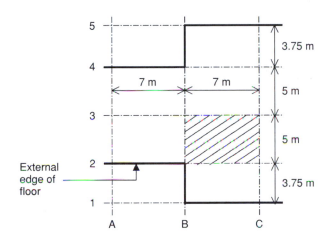

Fig. 3.65

MID-SPAN MOMENTS

Loading
Total dead load, g_k = finishes etc. + self weight of slab = 1.25 + 0.180 × 24 = 5.57 kN/m²
Imposed load, q_k = 4 kN/m²
Design load, $n = 1.4g_k + 1.6q_k = 1.4 \times 5.57 + 1.6 \times 4 = 14.2$ kN/m²

Panel AB2/3
By inspection, panel AB2/3 has a discontinuous long edge. Also $\ell_y/\ell_x = 7/5 = 1.4$
From *Table 3.21*,

short span coefficient for mid-span moment, $\beta_{sx} = 0.051$
long span coefficient for mid-span moment, $\beta_{sy} = 0.028$

Hence mid-span moment in the short span, $m_{sx} = \beta_{sx}n\ell_x^2 = 0.051 \times 14.2 \times 5^2 = 18.1$ kN m and mid-span moment in the long span, $m_{sy} = \beta_{sy}n\ell_x^2 = 0.028 \times 14.2 \times 5^2 = 9.9$ kN m

Panel BC1/2
By inspection, panel BC1/2 has two adjacent discontinuous edges and $\ell_y/\ell_x = 7/3.75 = 1.87$. From *Table 3.21*,

short span coefficient for mid-span moment, $\beta_{sx} = 0.0675$
long span coefficient for mid-span moment, $\beta_{sy} = 0.034$

Hence mid-span moment in the short span, $m_{sx} = \beta_{sx}n\ell_x^2 = 0.0675 \times 14.2 \times 3.75^2 = 13.5$ kN m and mid-span moment in the long span, $m_{sy} = \beta_{sy}n\ell_x^2 = 0.034 \times 14.2 \times 3.75^2 = 6.8$ kN m

PANEL BC2/3

Design moment
By inspection, panel BC2/3 is an interior panel. $\ell_y/\ell_x = 7/5 = 1.4$
From *Table 3.21*,

short span coefficient for negative (i.e. hogging) moment at continuous edge, $\beta_{sx,n} = 0.05$
short span coefficient for positive (i.e. sagging) moment at mid-span, $\beta_{sx,p} = 0.037$

long span coefficient for negative moment at continuous edge, $\beta_{sy,n} = 0.032$ and
long span coefficient for positive moment at mid-span, $\beta_{sy,p} = 0.024$

Hence negative moment at continuous edge in the short span,
$$m_{sx,n} = \beta_{sx,n}n\ell_x^2 = 0.05 \times 14.2 \times 5^2 = 17.8 \text{ kN m};$$
positive moment at mid-span in the short span,
$$m_{sx,p} = \beta_{sx,p}n\ell_x^2 = 0.037 \times 14.2 \times 5^2 = 13.1 \text{ kN m};$$
negative moment at continuous edge in the long span,
$$m_{sy,n} = \beta_{sy,n}n\ell_x^2 = 0.032 \times 14.2 \times 5^2 = 11.4 \text{ kN m};$$
and positive moment at mid-span in the long span,
$$m_{sy,p} = \beta_{sy,p}n\ell_x^2 = 0.024 \times 14.2 \times 5^2 = 8.5 \text{ kN m}.$$

Steel reinforcement

Continuous supports. At continuous supports the slab resists hogging moments in both the short-span and long-span directions. Therefore two layers of reinforcement will be needed in the top face of the slab. Comparison of design moments shows that the moment in the short span (17.8 kN m) is greater than the moment in the long span (11.4 kN m) and it is appropriate therefore that the steel in the short span direction (i.e. main steel) be placed at a greater effective depth than the steel in the long-span direction (i.e. secondary steel) as shown.

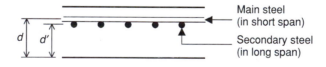

Assume diameter of main steel, $\phi = 10$ mm and nominal cover, $c = 25$ mm. Hence,

$$\text{Effective depth of main steel, } d = h - \frac{\phi}{2} - c = 180 - \frac{10}{2} - 25 = 150 \text{ mm}$$

Assume diameter of secondary steel, $\phi' = 10$ mm. Hence,

$$\text{Effective depth of secondary steel, } d' = h - \phi - \frac{\phi'}{2} - c = 180 - 10 - \frac{10}{2} - 25 = 140 \text{ mm}$$

(I) MAIN STEEL

$$K = \frac{m_{sx,n}}{f_{cu}bd^2} = \frac{17.8 \times 10^6}{30 \times 10^3 \times 150^2} = 0.0264$$

$$z = d\left(0.5 + \sqrt{(0.25 - K/0.9)}\right) \leqslant 0.95d = 0.95 \times 150 = 142.5 \text{ mm}$$

$$= 150\left(0.5 + \sqrt{(0.25 - 0.0264/0.9)}\right) = 150 \times 0.97 = 146 \text{ mm}$$

$$A_s = \frac{M}{0.95f_y z} = \frac{17.8 \times 10^6}{0.95 \times 460(0.95 \times 150)} = 286 \text{ mm}^2/\text{m} > 0.13\%bh = 234 \text{ mm}^2/\text{m} \quad \text{OK}$$

Provide T10@250 centres ($A_s = 314$ mm^2/m) in short span direction.

(II) SECONDARY STEEL

$$K = \frac{m_{sx,n}}{f_{cu}bd^2} = \frac{11.4 \times 10^6}{30 \times 10^3 \times 140^2} = 0.0194$$

$$z = d\left(0.5 + \sqrt{(0.25 - K/0.9)}\right) \leqslant 0.95d = 0.95 \times 140 = 133 \text{ mm}$$

$$= 140\left(0.5 + \sqrt{(0.25 - 0.0194/0.9)}\right) = 140 \times 0.98 = 137 \text{ mm}$$

(Note that for slabs generally, $z = 0.95d$)

$$A_s = \frac{m_{sy,n}}{0.95 \times f_y \times z} = \frac{11.4 \times 10^6}{0.95 \times 460 \times (0.95 \times 140)} = 196 \text{ mm}^2/\text{m}$$

$$\geqslant 0.13\%bh = 234 \text{ mm}^2/\text{m}$$

$$\therefore \text{ provide T10@300 centres } (A_s = 262 \text{ mm}^2/\text{m}) \text{ in long span direction.}$$

Mid-span. At mid-span the slab resists sagging moments in both the short-span and long-span directions, necessitating two layers of reinforcement in the bottom face of the slab too. Comparison of mid-span moments shows that the moment in the short span (14.4 kN m) is greater than the moment in the long span (9.4 kN m) and it is again appropriate therefore that the steel in the short span direction (main steel) be placed at a greater effective depth than the steel in the long span direction (secondary steel) as shown.

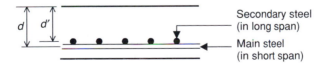

Assume diameter of main steel, $\phi = 10$ mm and nominal cover, $c = 25$ mm. Hence

$$\text{Effective depth of main steel, } d = h - \frac{\phi}{2} - c = 180 - \frac{10}{2} - 25 = 150 \text{ mm}$$

Assuming diameter of secondary steel, $\phi' = 10$ mm. Hence

$$\text{Effective depth of secondary steel, } d' = h - \phi - \frac{\phi'}{2} - c = 180 - 10 - \frac{10}{2} - 25 = 140 \text{ mm}$$

(I) MAIN STEEL

$$A_s = \frac{m_{sx,p}}{0.95 f_y z} = \frac{13.1 \times 10^6}{0.95 \times 460 \,(0.95 \times 150)} = 210 \text{ mm}^2/\text{m}$$

$$\geqslant A_{s,min} = 0.13\% bh = 234 \text{ mm}^2/\text{m}$$

Provide T10@300 centres ($A_s = 262$ mm^2/m) in short span direction.

(II) SECONDARY STEEL

$$A_s = \frac{m_{sy,p}}{0.95 \times f_y \times z} = \frac{8.5 \times 10^6}{0.95 \times 460 \times (0.95 \times 140)} = 146 \text{ mm}^2/\text{m}$$

$$\geqslant A_{s,min} = 0.13\% bh = 234 \text{ mm}^2/\text{m}$$

Provide T10@300 centres ($A_s = 262$ mm^2/m) in long span direction.

SHEAR
From *Table 3.22*,

long span coefficient, $\beta_{vy} = 0.33$ and
short span shear coefficient, $\beta_{vx} = 0.43$

Design load on beams B2/3 and C2/3, $v_{sy} = \beta_{vy} n \ell_x = 0.33 \times 14.2 \times 5 = 23.4$ kN/m
Design load on beams 2B/C and 3B/C, $v_{sx} = \beta_{vx} n \ell_x = 0.43 \times 14.2 \times 5 = 30.5$ kN/m   (critical)

Shear stress, $v = \dfrac{v_{sx}}{bd} = \dfrac{30.5 \times 10^3}{10^3 \times 150} = 0.20$ N/mm^2

$$\frac{100 A_s}{bd} = \frac{100 \times 314}{10^3 \times 150} = 0.21$$

From *Table 3.9*, $v_c = \sqrt[3]{\dfrac{30}{25}} \times 0.48 = 0.51$ N/mm$^2 > v$

Hence no shear reinforcement is required.

DEFLECTION
For two-way spanning slabs, the deflection check is satisfied provided the span/effective depth ratio in the shorter span does not exceed the appropriate value in *Table 3.11* multiplied by the modification factor obtained via equations 7 and 8 of *Table 3.13*

$$\text{Actual} \frac{\text{span}}{\text{effective depth}} = \frac{5000}{150} = 33.3$$

Service stress, f_s, is

$$f_s = \frac{2}{3} f_y \frac{A_{s,req}}{A_{s,prov}} = \frac{2}{3} \times 460 \times \frac{210}{262} = 246 \text{ N/mm}^2$$

$$\text{Modification factor} = 0.55 + \frac{477 - f_s}{120\left(0.9 + \dfrac{m_{sx,p}}{bd^2}\right)} = 0.55 + \frac{477 - 246}{120\left(0.9 + \dfrac{13.1 \times 10^6}{10^3 \times 150^2}\right)} = 1.85$$

$$\text{Permissible } \frac{\text{span}}{\text{effective depth}} = 26 \times 1.85 = 48 > \text{actual}\quad \text{OK}$$

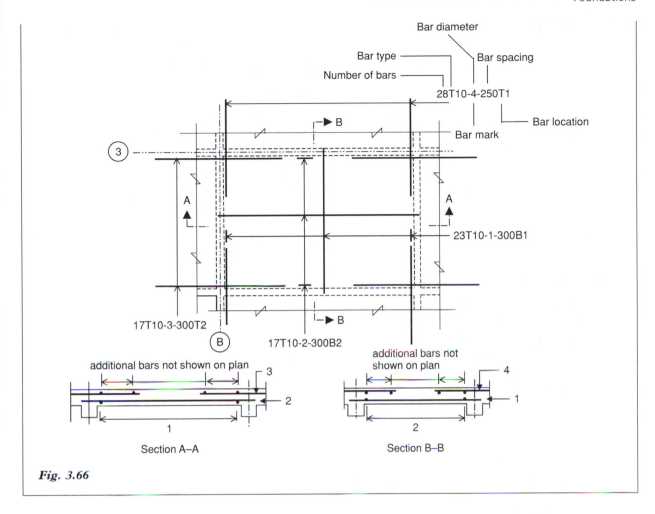

Section A–A   Section B–B

Fig. 3.66

3.11 Foundations

Foundations are required primarily to carry the dead and imposed loads due to the structure's floors, beams, walls, columns etc. and transmit and distribute the loads safely to the ground (*Fig. 3.67*). The purpose of distributing the load is to avoid the safe bearing capacity of the soil being exceeded otherwise excessive settlement of the structure may occur.

Foundation failure can produce catastrophic effects on the overall stability of a structure so that it may slide or even overturn (*Fig. 3.68*). Such failures are likely to have tremendous financial and safety implications. It is essential, therefore, that much attention is paid to the design of this element of a structure.

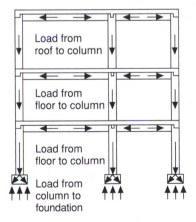

Foundation loads resisted by ground

Fig. 3.67 *Loading on foundations.*

107

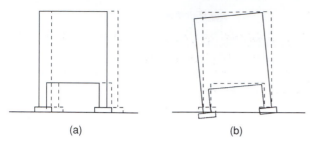

Fig. 3.68 *Foundation failures: (a) sliding failure; (b) overturning failure.*

3.11.1 FOUNDATION TYPES

There are many types of foundations which are commonly used, namely strip, pad and raft. The foundations may bear directly on the ground or be supported on piles. The choice of foundation type will largely depend upon (1) ground conditions (i.e. strength and type of soil) and (2) type of structure (i.e. layout and level of loading).

Pad footings are usually square or rectangular slabs and used to support a single column (*Fig. 3.69*). The pad may be constructed using mass concrete or reinforced concrete depending on the relative size of the loading. Detailed design of pad footings is discussed in *section 3.11.2.1*.

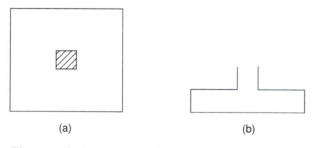

Fig. 3.69 *Pad footing: (a) plan; (b) elevation.*

Continuous strip footings are used to support loadbearing walls or under a line of closely spaced columns (*Fig. 3.70*). Strip footings are designed as pad footings in the transverse direction and in the longitudinal direction as an inverted continuous beam subject to the ground bearing pressure.

Where the ground conditions are relatively poor, a raft foundation may be necessary in order to distribute the loads from the walls and columns over a large area. In its simplest form this may consist of a flat slab, possibly strengthened by upstand or downstand beams for the more heavily loaded structures (*Fig. 3.71*).

Where the ground conditions are so poor that it is not practical to use strip or pad footings but better quality soil is present at lower depths, the use of pile foundations should be considered (*Fig. 3.72*).

The piles may be made of precast reinforced concrete, prestressed concrete or in-situ reinforced concrete. Loads are transmitted from the piles to the surrounding strata by end bearing and/or friction. End bearing piles derive most of their carrying capacity from the penetration resistance of the soil at the toe of the pile, while friction piles rely on the adhesion or friction between the sides of the pile and the soil.

3.11.2 FOUNDATION DESIGN

Foundation failure may arise as a result of (a) allowable bearing capacity of the soil being exceeded, or (b) bending and/or shear failure of the base. The first condition allows the plan-area of the base to be calculated, being equal to the design load divided by the bearing capacity of the soil, i.e.

$$\frac{\text{Ground}}{\text{pressure}} = \frac{\text{design load}}{\text{plan area}} > \frac{\text{bearing}}{\text{capacity of soil}} \quad (3.25)$$

Since the settlement of the structure occurs during its working life, the design loadings to be considered when calculating the size of the base

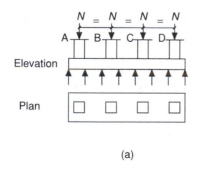

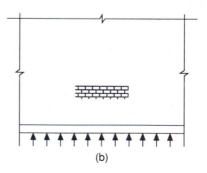

Fig. 3.70 *Strip footings: (a) footing supporting columns; (b) footing supporting wall.*

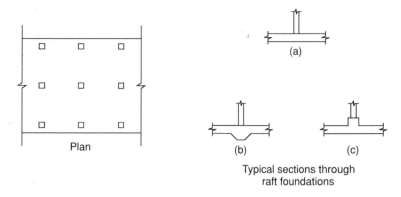

Fig. 3.71 *Raft foundations. Typical sections through raft foundations: (a) flat slab; (b) flat slab and downstand; (c) flat slab and upstand.*

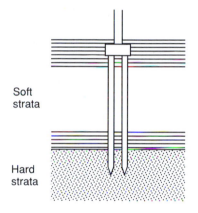

Fig. 3.72 *Piled foundations.*

foundation types discussed above. However, it should be born in mind that in most cases the design process would be similar to that for beams and slabs.

3.11.2.1 Pad footing

The general procedure to be adopted for the design of pad footings is as follows:

1. Calculate the plan area of the footing using serviceability loads.
2. Determine the reinforcement areas required for bending using ultimate loads (*Fig. 3.73*).
3. Check for punching and transverse shear failures (*Fig. 3.74*).

should be taken as those for the **serviceability limit state** (i.e. $1.0G_k + 1.0Q_k$). The calculations to determine the thickness of the base and the bending and shear reinforcement are, however, based on **ultimate loads** (i.e. $1.4G_k + 1.6Q_k$). The design of a pad footing only will be considered here. The reader is referred to other more specialised books on this subject for the design of the other

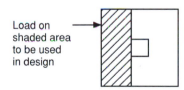

Fig. 3.73 *Critical section for bending.*

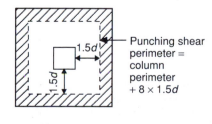

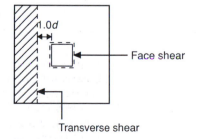

Fig. 3.74 *Critical sections for shear. (Load on shaded areas to be used in design.)*

Example 3.14 Design of a pad footing (BS 8110)

A 400 mm square column carries a dead load (G_k) of 1050 kN and imposed load (Q_k) of 300 kN. The safe bearing capacity of the soil is 170 kN/m^2. Design a square pad footing to resist the loads assuming the following material strengths:

$$f_{cu} = 35 \text{ N/mm}^2$$

$$f_y = 460 \text{ N/mm}^2$$

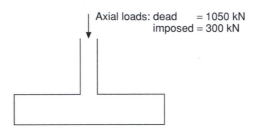

Axial loads: dead = 1050 kN
imposed = 300 kN

(I) PLAN AREA OF BASE

Loading

Dead load. Assume a footing weight of 130 kN

$$\text{Total dead load } (G_k) = 1050 + 130 = 1180 \text{ kN}$$

Serviceability load

Design axial load $(N) = 1.0G_k + 1.0Q_k$

$$= 1.0 \times 1180 + 1.0 \times 300 = 1480 \text{ kN}$$

Plan area

$$\text{Plan area of base} = \frac{N}{\text{bearing capacity of soil}} = \frac{1480}{170} = 8.70 \text{ m}^2$$

Hence provide a 3 m square base (plan area = 9 m^2)

(II) BENDING REINFORCEMENT

Self weight of footing

Assume the overall depth of footing (h) = 600 mm

Self weight of footing = area × h × density of concrete

$$= 9 \times 0.6 \times 24 = 129.6 \text{ kN} < \text{assumed } (130 \text{ kN})$$

Design moment, M

Total ultimate load $(W) = 1.4G_k + 1.6Q_k$

$$= 1.4 \times 1050 + 1.6 \times 300 = 1950 \text{ kN}$$

$$\text{Earth pressure } (p_s) = \frac{W}{\text{plan area of base}} = \frac{1950}{9} = 217 \text{ kN/m}^2$$

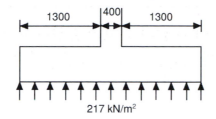

Maximum design moment occurs at face of column $(M) = \dfrac{p_s \ell^2}{2} = \dfrac{217 \times 1.300^2}{2}$

$$= 183 \text{ kN m/m width of slab}$$

Ultimate moment

Effective depth. Base to be cast against blinding, hence cover (c) to reinforcement = 40 mm (see clause 3.3.1.4, BS 8110). Assume 20 mm diameter (Φ) bars will be needed as bending reinforcement in both directions.

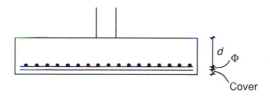

Hence, average effective depth of reinforcement, d, is

$$d = h - c - \Phi = 600 - 40 - 20 = 540 \text{ mm}$$

Ultimate moment

$$M_u = 0.156 f_{cu} b d^2 = 0.156 \times 35 \times 10^3 \times 540^2$$

$$= 1592 \times 10^6 \text{ N mm} = 1592 \text{ kN m}$$

Since $M_u > M$ no compression reinforcement is required.

Main steel

$$K = \frac{M}{f_{cu} b d^2} = \frac{183 \times 10^6}{35 \times 1000 \times 540^2} = 0.018$$

$$z = d[0.5 + \sqrt{(0.25 - K/0.9)}]$$

$$= d[0.5 + \sqrt{(0.25 - 0.018/0.9)}]$$

$$= 0.981d \leqslant 0.95d = 0.95 \times 540 = 513 \text{ mm}$$

$$A_s = \frac{M}{0.95 f_y z} = \frac{183 \times 10^6}{0.95 \times 460 \times 513} = 816 \text{ mm}^2/\text{m}$$

Minimum steel area is

$$0.13\%bh = 780 \text{ mm}^2/\text{m} < A_\text{s} \quad \text{OK}$$

Hence from *Table 3.19*, provide T20 at 300 mm centres ($A_\text{s} = 1050$ mm^2/m) distributed uniformly across the full width of the footing parallel to the x–x and y–y axis (see clause 3.11.3.2, BS 8110).

(III) CRITICAL SHEAR STRESSES

Punching shear

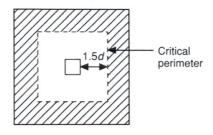

Critical perimeter, p_crit, is

$$= \text{column perimeter} + 8 \times 1.5d$$

$$= 4 \times 400 + 8 \times 1.5 \times 540 = 8080 \text{ mm}$$

Area within perimeter is

$$(400 + 3d)^2 = (400 + 3 \times 540)^2 = 4.08 \times 10^6 \text{ mm}^2$$

Ultimate punching force, V, is

$$V = \text{load on shaded area} = 217 \times (9 - 4.08) = 1068 \text{ kN}$$

Design punching shear stress, v, is

$$v = \frac{V}{p_\text{crit}d} = \frac{1068 \times 10^3}{8080 \times 540} = 0.24 \text{ N/mm}^2$$

$$\frac{100A_\text{s}}{bd} = \frac{100 \times 1050}{10^3 \times 540} = 0.19$$

Hence from *Table 3.8*, design concrete shear stress, v_c, is

$$v_\text{c} = (35/25)^{1/3} \times 0.39 = 0.40 \text{ N/mm}^2$$

Since $v_\text{c} > v$, punching failure is unlikely and a 600 mm depth of slab is acceptable.

Face shear
Maximum shear stress (v_max) occurs at face of column. Hence

$$v_\text{max} = \frac{W}{\text{column perimeter} \times d} = \frac{1950 \times 10^3}{(4 \times 400) \times 540}$$

$$= 2.3 \text{ N/mm}^2 < \text{permissible} (= 0.8\sqrt{35} = 4.73 \text{ N/mm}^2)$$

Transverse shear

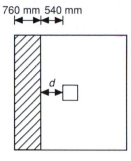

760 mm   540 mm

Ultimate shear force (V) = load on shaded area

$$= p_s \times \text{area} = 217(3 \times 0.760) = 495 \text{ kN}$$

Design shear stress, v, is

$$v = \frac{V}{bd} = \frac{495 \times 10^3}{3 \times 10^3 \times 540} = 0.31 \text{ N/mm}^2 < v_c$$

Hence no shear reinforcement is required.

(IV) REINFORCEMENT DETAILS

The sketch below shows the main reinforcement requirements for the pad footing.

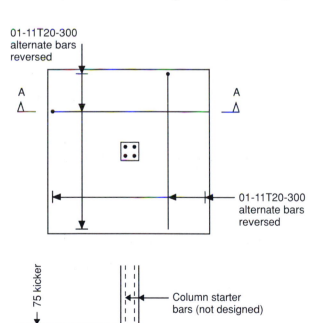

01-11T20-300
alternate bars
reversed

A

01-11T20-300
alternate bars
reversed

75 kicker

Column starter
bars (not designed)

01   01

01   01   01   01

Section A–A

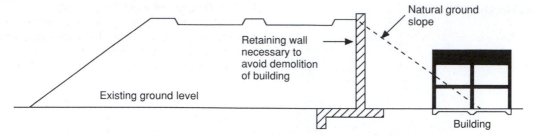

Fig. 3.75 Section through road embankment incorporating a retaining wall.

3.12 Retaining walls

Sometimes it is necessary to maintain a difference in ground levels between adjacent areas of land. Typical examples of this include road and railway embankments, reservoirs and ramps. A common solution to this problem is to build a natural slope between the two levels. However, this is not always possible because slopes are very demanding of space. An alternative solution which allows an immediate change in ground levels to be effected is to build a vertical wall which is capable of resisting the pressure of the retained material. These structures are commonly referred to as retaining walls (*Fig. 3.75*). Retaining walls are important elements in many building and civil engineering projects and the purpose of the following sections is to briefly describe the various types of retaining walls available and outline the design procedure associated with one common type, namely cantilever retaining walls.

3.12.1 TYPES OF RETAINING WALLS
Retaining walls are designed on the basis that they are capable of withstanding all horizontal pressures and forces without undue movement arising from deflection, sliding or overturning. There are two main categories of concrete retaining walls (a) gravity walls and (b) flexible walls.

3.12.1.1 Gravity walls
Where walls up to 2 m in height are required, it is generally economical to choose a gravity retaining wall. Such walls are usually constructed of mass concrete with mesh reinforcement in the faces to reduce thermal and shrinkage cracking. Other construction materials for gravity walls include masonry and stone (*Fig. 3.76*).

Gravity walls are designed so that the resultant force on the wall due to the dead weight and the earth pressures is kept within the middle third of the base. A rough guide is that the width of the base should be about a third of the height of the retained material. It is usual to include a granular layer behind the wall and weep holes near the base to minimise hydrostatic pressure behind the wall. Gravity walls rely on their dead weight for strength and stability. The main advantages with this type of wall are simplicity of construction and ease of maintenance.

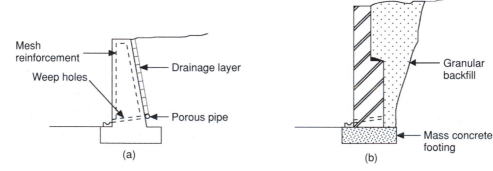

Fig. 3.76 Gravity retaining walls: (a) mass concrete wall; (b) masonry wall.

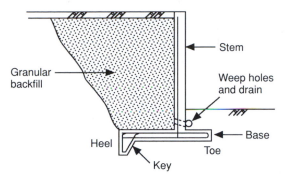

Fig. 3.77 *Cantilever wall.*

3.12.1.2 Flexible walls
These retaining walls may be of two basic types, namely (i) cantilever and (ii) counterfort.

(i) Cantilever walls. Cantilevered reinforced concrete retaining walls are suitable for heights up to about 7 m. They generally consist of a uniform vertical stem monolithic with a base slab (*Fig. 3.77*). A key is sometimes incorporated at the base of the wall in order to prevent sliding failure of the wall. The stability of these structures often relies on the weight of the structure and on the weight of backfill on the base. This is perhaps the most common type of wall and, therefore, the design of such walls is considered in detail in *section 3.12.2.*

(ii) Counterfort walls. In cases where a higher stem is needed, it may be necessary to design the wall as a counterfort (*Fig. 3.78*). Counterfort walls can be designed as continuous slabs spanning horizontally between vertical supports known as counterforts. The counterforts are designed as cantilevers and will normally have a triangular or trapezoidal shape. As with cantilever walls, stability is provided by the weight of the structure and earth on the base.

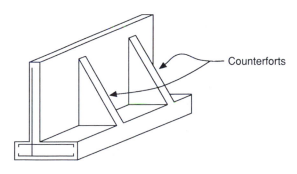

Fig. 3.78 *Counterfort retaining wall.*

3.12.2 Design of cantilever walls
Generally, the design process involves ensuring that the wall will not fail either due to foundation failure or structural failure of the stem or base. Specifically, the design procedure involves the following steps:

1. Calculate the soil pressures on the wall.
2. Check the stability of the wall.
3. Design the bending reinforcement.

As in the case of slabs, the design of retaining walls is usually based on a 1 m width of section.

3.12.2.1 Soil pressures
The method most commonly used for determining the soil pressures is based on Rankin's formula, which may be considered to be conservative but is straightforward to apply. The pressure on the wall resulting from the retained fill has a destabilising effect on the wall and is normally termed active pressure (*Fig. 3.79*). The earth in front of the wall resists the destabilising forces and is termed passive pressure.

The active pressure (p_a) is given by

$$p_a = \rho k_a z \tag{3.26}$$

where
ρ = unit weight of soil (kN/m³)
k_a = coefficient of active pressure
z = height of retained fill.

Here k_a is calculated using

$$k_a = \frac{1 - \sin \phi}{1 + \sin \phi} \tag{3.27}$$

where ϕ is the internal angle of friction of retained soil. Typical values of ρ and ϕ for various soil types are shown in *Table 3.23*.

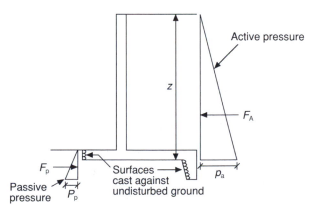

Fig. 3.79 *Active and passive pressure acting on a wall.*

Table 3.23 Values of ρ and ϕ

Material	ρ (kN/m^3)	ϕ
Sandy gravel	17–22	35°–40°
Loose sand	15–16	30°–35°
Crushed rock	12–22	35°–40°
Ashes	9–10	35°–40°
Broken brick	15–16	35°–40°

The passive pressure (p_p) is given by

$$p_p = \rho k_p z \tag{3.28}$$

where
ρ = unit weight of soil
z = height of retained fill
k_p = coefficient of passive pressure and is calculated using

$$k_p = \frac{1 + \sin \phi}{1 - \sin \phi} = \frac{1}{k_a} \tag{3.29}$$

3.12.2.2 Stability

Failure of the wall may arise due to (a) sliding or (b) rotation. Sliding failure will occur if the active pressure force (F_A) exceeds the passive pressure force (F_P) plus the friction force (F_F) arising at the base/ground interface (*Fig. 3.80(a)*) where

$$F_A = 0.5 p_a h_1 \tag{3.30}$$

$$F_P = 0.5 p_p h_2 \tag{3.31}$$

$$F_F = \mu W_t \tag{3.32}$$

The factor of safety against this type of failure occurring is normally taken to be at least 1.5:

$$\frac{F_F + F_P}{F_A} \geqslant 1.5 \tag{3.33}$$

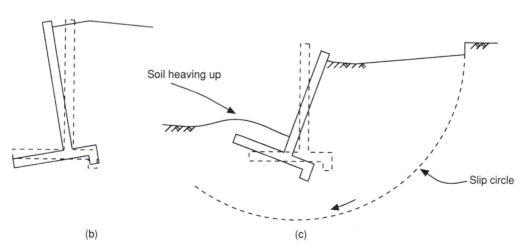

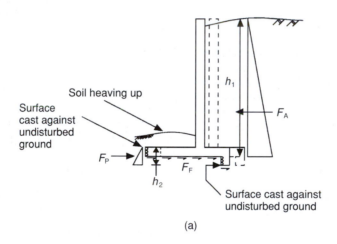

Fig. 3.80 *Modes of failure: (a) sliding; (b) overturning; (c) slip circle.*

Rotational failure of the wall may arise due to:

1. the overturning effect of the active pressure force (*Fig. 3.80(b)*);
2. bearing pressure of the soil being exceeded which will display similar characteristics to (1); or
3. failure of the soil mass surrounding the wall (*Fig. 3.80(c)*).

Failure of the soil mass (type (3)) will not be considered here but the reader is referred to any standard book on soil mechanics for an explanation of the procedure to be followed to avoid such failures.

Failure type (1) can be checked by taking moments about the toe of the foundation (A) as shown in *Fig. 3.81* and ensuring that the ratio of sum of restoring moments (ΣM_{res}) and sum of overturning moments (ΣM_{over}) exceeds 2.0, i.e.

$$\frac{\Sigma M_{res}}{\Sigma M_{over}} \geq 2.0 \qquad (3.34)$$

Failure type (2) can be avoided by ensuring that the ground pressure does not exceed the allowable bearing pressure for the soil. The ground pressure under the toe (p_{toe}) and the heel (p_{heel}) of the base can be calculated using

$$p_{toe} = \frac{N}{D} + \frac{6M}{D^2} \qquad (3.35)$$

$$p_{heel} = \frac{N}{D} - \frac{6M}{D^2} \qquad (3.36)$$

provided that the load eccentricity lies within the middle third of the base, that is

$$M/N \leq D/6 \qquad (3.37)$$

where
M = moment about centre line of base
N = total vertical load (W_t)
D = width of base

3.12.2.3 Reinforcement areas
Structural failure of the wall may arise if the base and stem are unable to resist the vertical and horizontal forces due to the retained soil. The area of steel reinforcement needed in the wall can be calculated by considering the ultimate limit states of bending and shear. As was pointed out at the beginning of this chapter, cantilever retaining walls can be regarded for design purposes as three cantilever beams (*Fig. 3.2*) and thus the equations developed in *section 3.9* can be used here.

The areas of main reinforcement (A_s) can be calculated using

$$A_s = \frac{M}{0.95 f_y z}$$

where
M = design moment
f_y = reinforcement grade
$z = d[0.5 + \sqrt{(0.25 - K/0.9)}]$
$K = M/f_{cu}bd^2$

The area of distribution steel is based on the minimum steel area (A_s) given in *Table 3.25* of BS 8110, i.e.

$$A_s = 0.24\%A_c \quad \text{when } f_y = 250 \text{ N/mm}^2$$
$$A_s = 0.13\%A_c \quad \text{when } f_y = 460 \text{ N/mm}^2$$

where A_c is the total cross-sectional area of concrete.

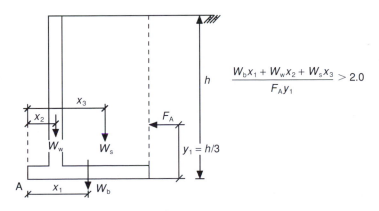

$$\frac{W_b x_1 + W_w x_2 + W_s x_3}{F_A y_1} > 2.0$$

Fig. 3.81

Example 3.15 Design of a cantilever retaining wall (BS 8110)

The cantilever retaining wall shown below is backfilled with granular material having a unit weight, ρ, of 19 kN/m³ and an internal angle of friction, ϕ, of 30°. Assuming that the allowable bearing pressure of the soil is 120 kN/m², the coefficient of friction is 0.4 and the unit weight of reinforced concrete is 24 kN/m³

1. Determine the factors of safety against sliding and overturning.
2. Calculate ground bearing pressures.
3. Design the wall and base reinforcement assuming $f_{cu} = 35$ N/mm², $f_y = 460$ N/mm² and the cover is 35 mm.

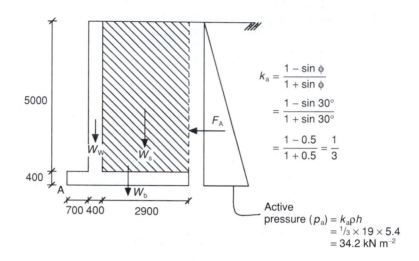

$$k_a = \frac{1 - \sin \phi}{1 + \sin \phi}$$
$$= \frac{1 - \sin 30°}{1 + \sin 30°}$$
$$= \frac{1 - 0.5}{1 + 0.5} = \frac{1}{3}$$

Active pressure $(p_a) = k_a \rho h$
$= \frac{1}{3} \times 19 \times 5.4$
$= 34.2$ kN m⁻²

(I) SLIDING

Consider the forces acting on a 1 m length of wall. Horizontal force on wall due to backfill, F_A, is

$$F_A = 0.5 p_a h = 0.5 \times 34.2 \times 5.4 = 92.34 \text{ kN}$$

and

Weight of wall (W_w) = 0.4 × 5 × 24 = 48.0 kN

Weight of base (W_b) = 0.4 × 4 × 24 = 38.4 kN

Weight of soil (W_s) = 2.9 × 5 × 19 = 275.5 kN

Total vertical force (W_t) = 361.9 kN

Friction force, F_F, is

$$F_F = \mu W_t = 0.4 \times 361.9 = 144.76 \text{ kN}$$

Assume passive pressure force $(F_P) = 0$. Hence factor of safety against sliding is

$$\frac{144.76}{92.34} = 1.56 > 1.5 \quad \text{OK}$$

(II) OVERTURNING

Taking moments about point A (see above), sum of overturning moments (M_{over}) is

$$\frac{F_A \times 5.4}{3} = \frac{92.34 \times 5.4}{3} = 166.2 \text{ kN m}$$

Sum of restoring moments (M_{res}) is

$$M_{res} = W_w \times 0.9 + W_b \times 2 + W_s \times 2.55$$
$$= 48 \times 0.9 + 38.4 \times 2 + 275.5 \times 2.55 = 822.5 \text{ kN m}$$

Factor of safety against overturning is

$$\frac{822.5}{166.2} = 4.9 > 2.0 \quad \text{OK}$$

(III) GROUND BEARING PRESSURE

Moment about centre line of base (M) is

$$M = \frac{F_A \times 5.4}{3} + W_w \times 1.1 - W_s \times 0.55$$

$$= \frac{92.34 \times 5.4}{3} + 48 \times 1.1 - 275.5 \times 0.55 = 67.5 \text{ kN m}$$

$$N = 361.9 \text{ kN}$$

$$\frac{M}{N} = \frac{67.5}{361.9} = 0.187 \text{ m} < \frac{D}{6} = \frac{4}{6} = 0.666 \text{ m}$$

Therefore, the maximum ground pressure occurs at the toe, p_{toe}, which is given by

$$p_{toe} = \frac{361.9}{4} + \frac{6 \times 67.5}{4^2} = 116 \text{ kN/m}^2 < \text{allowable } (120 \text{ kN/m}^2)$$

Ground bearing pressure at the heel, p_{heel}, is

$$p_{heel} = \frac{361.9}{4} - \frac{6 \times 67.5}{4^2} = 65 \text{ kN/m}^2$$

(IV) BENDING REINFORCEMENT

Wall

Height of stem of wall, $h_s = 5$ m. Horizontal force on stem due to backfill, F_s, is

$$F_s = 0.5 k_a \rho h_s^2$$
$$= 0.5 \times \frac{1}{3} \times 19 \times 5^2$$
$$= 79.17 \text{ kN/m width}$$

Design moment at base of wall, M, is

$$M = \frac{\gamma_f F_s h_s}{3} = \frac{1.4 \times 79.17 \times 5}{3} = 184.7 \text{ kN m}$$

Effective depth

Assume diameter of main steel (Φ) = 20 mm.
Hence effective depth, d, is

$$d = 400 - \text{cover} - \Phi/2$$
$$= 400 - 35 - 20/2 = 355 \text{ mm}$$

Ultimate moment of resistance

$$M_u = 0.156 f_{cu} b d^2 = 0.156 \times 35 \times 10^3 \times 355^2 \times 10^{-6} = 688 \text{ kN m}$$

Since $M_u > M$, no compression reinforcement is required.

Steel area

$$K = \frac{M}{f_{cu}bd^2} = \frac{184.7 \times 10^6}{35 \times 10^3 \times 355^2} = 0.0419$$

$$z = d[0.5 + \sqrt{(0.25 - K/0.9)}]$$

$$= 355[0.5 + \sqrt{(0.25 - 0.0419/0.9)}] = 337 \text{ mm}$$

$$A_s = \frac{M}{0.95f_yz} = \frac{184.7 \times 10^6}{0.95 \times 460 \times 337} = 1254 \text{ mm}^2/\text{m}$$

Hence from *Table 3.19*, provide T20 at 200 mm centres ($A_s = 1570$ mm^2/m) in near face (NF) of wall. Steel is also required in the front face (FF) of wall in order to prevent excessive cracking. This is based on the minimum steel area, i.e.

$$= 0.13\%bh = 0.13\% \times 10^3 \times 400 = 520 \text{ mm}^2/\text{m}$$

Hence, provide T12 at 200 centres ($A_s = 566$ mm^2)

Base

(a) Heel

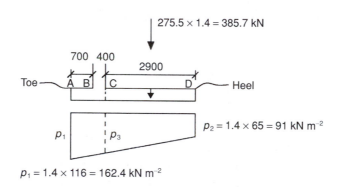

$$p_3 = 91 + \frac{2.9(162.4 - 91)}{4} = 142.8 \text{ kN/m}^2$$

Design moment at point C, M_c, is

$$\frac{385.7 \times 2.9}{2} + \frac{2.9 \times 38.4 \times 1.4 \times 1.45}{4} - \frac{91 \times 2.9^2}{2} - \frac{51.8 \times 2.9 \times 2.9}{2 \times 3} = 160.5 \text{ kN m}$$

$$K = \frac{160.5 \times 10^6}{35 \times 10^3 \times 355^2} = 0.036$$

$$z = 355[0.5 + \sqrt{(0.25 - 0.036/0.9)}] \leqslant 0.95d = 337 \text{ mm}$$

$$A_s = \frac{M}{0.95f_yz} = \frac{160.5 \times 10^6}{0.95 \times 460 \times 337} = 1091 \text{ mm}^2/\text{m}$$

Hence from *Table 3.19*, provide T20 at 200 mm centres ($A_s = 1570$ mm^2/m) in top face (T) of base.

(b) Toe. Design moment at point B, M_B, is given by

$$M_B \approx \frac{162.4 \times 0.7^2}{2} - \frac{0.7 \times 38.4 \times 1.4 \times 0.7}{4 \times 2} = 36.5 \text{ kN m}$$

$$A_s = \frac{36.5 \times 1091}{160.5} = 248 \text{ mm}^2/\text{m} \not< \text{ minimum steel area} = 520 \text{ mm}^2/\text{m}$$

Hence provide T12 at 200 mm centres ($A_s = 566$ mm^2/m), in bottom face (B) of base and as distribution steel in base and stem of wall.

(V) REINFORCEMENT DETAILS

The sketch below shows the main reinforcement requirements for the retaining wall. For reasons of buildability the actual reinforcement details may well be slightly different.

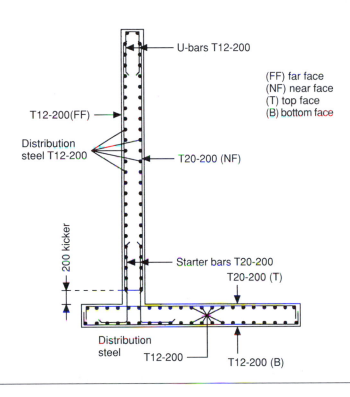

3.13 Design of short–braced columns

The function of columns in a structure is to act as vertical supports to suspended members such as beams and roofs and to transmit the loads from these members down to the foundations (*Fig. 3.82*). Columns are primarily compression members although they may also have to resist bending moments transmitted by beams.

Columns may be classified as short or slender, braced or unbraced, depending on various dimensional and structural factors which will be discussed

below. However, due to limitations of space, the study will be restricted to the design of the most common type of column found in building structures, namely short-braced columns.

3.13.1 COLUMN SECTIONS

Some common column cross-sections are shown in *Fig. 3.83*. Any section can be used, however, provided that the greatest overall cross-sectional dimension does not exceed four times its smaller dimension (i.e. $h \leqslant 4b$, *Fig. 3.83(c)*). With sections where $h > 4b$ the member should be regarded as a wall for design purposes (clause 1.2.4.1, BS 8110).

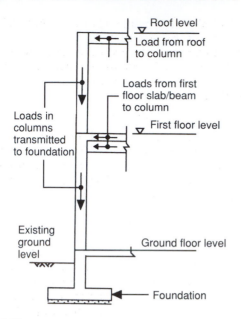

Roof level
Load from roof to column

Loads from first floor slab/beam to column

First floor level

Loads in columns transmitted to foundation

Existing ground level

Ground floor level

Foundation

Fig. 3.82

Fig. 3.84

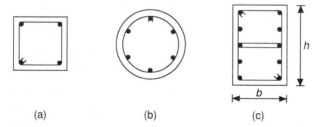

(a)   (b)   (c)

Fig. 3.83 *Column cross-sections.*

Fig. 3.85

3.13.2 SHORT AND SLENDER COLUMNS

Columns may fail due to one of three mechanisms:

1. compression failure of the concrete/steel reinforcement (*Fig. 3.84*);
2. buckling (*Fig. 3.85*);
3. combination of buckling and compression failure.

For any given cross-section, failure mode (1) is most likely to occur with columns which are short and stocky, while failure mode (2) is probable with columns which are long and slender. It is important, therefore, to be able to distinguish between columns which are short and those which are slender since the failure mode and hence the design procedures for the two column types are likely to be different.

Clause 3.8.1.3 of BS 8110 classifies a column as being short if

$$\frac{\ell_{ex}}{h} < 15 \quad \text{and} \quad \frac{\ell_{ey}}{b} < 15$$

where

ℓ_{ex} effective height of the column in respect of the major axis (i.e. x–x axis)
ℓ_{ey} effective height of the column in respect of the minor axis
b width of the column cross-section
h depth of the column cross-section.

It should be noted that the above definition only applies to columns which are braced, rather than unbraced. This distinction is discussed more fully in *section 3.13.3*. Effective heights of columns is covered in *section 3.13.4*.

3.13.3 BRACED AND UNBRACED COLUMNS (CLAUSE 3.8.1.5, BS 8110)

A column may be considered braced if the lateral loads, due to wind for example, are resisted by walls or some other form of bracing rather than by the column. For example, all the columns in the reinforced concrete frame shown in *Fig. 3.86* are braced in the y–y direction. A column may be considered to be unbraced if the lateral loads are resisted by the bending action of the column. For

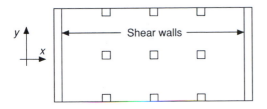

Fig. 3.86 *Columns braced in y direction and unbraced in the x direction.*

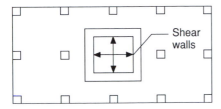

Fig. 3.87 *Columns braced in both directions.*

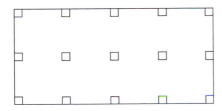

Fig. 3.88 *Columns unbraced in both directions.*

example, all the columns shown in *Fig. 3.86* are unbraced in the x–x direction.

Depending upon the layout of the structure, it is possible for the columns to be braced or unbraced in both directions as shown in *Fig. 3.87* and *3.88* respectively.

3.13.4 EFFECTIVE HEIGHT

The effective height (ℓ_e) of a column in a given plane is obtained by multiplying the clear height between lateral restraints (ℓ_o) by a coefficient (β) which is a function of the fixity at the column ends and is obtained from *Table 3.24*.

$$\ell_e = \beta\ell_o \qquad (3.38)$$

End condition 1 signifies that the column end is fully restrained. End condition 2 signifies that the column end is partially restrained and end condition 3 signifies that the column end is nominally restrained. In practice it is possible to infer the degree of fixity at the column ends simply by reference to the diagrams shown in *Fig. 3.89*.

Table 3.24 Values of β for braced columns (Table 3.19, BS 8110)

End condition of top	End condition at bottom		
	1	*2*	*3*
1	0.75	0.80	0.90
2	0.80	0.85	0.95
3	0.90	0.95	1.00

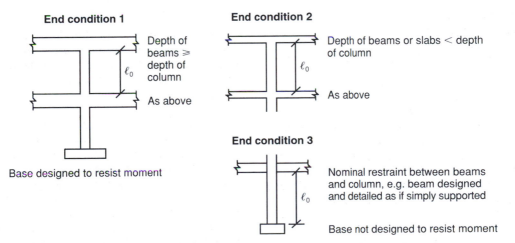

Fig. 3.89 *Column end restraint conditions.*

Example 3.16 Classification of a concrete column (BS 8110)

Determine if the column shown in *Fig. 3.90* is short.

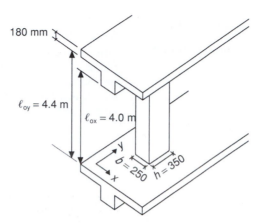

Fig. 3.90

For bending in the y direction: end condition at top of column = 1, end condition at bottom of column = 1. Hence from *Table 3.24*, $\beta_x = 0.75$.

$$\frac{\ell_{ex}}{h} = \frac{\beta_x \ell_{ox}}{h} = \frac{0.75 \times 4000}{350} = 8.57$$

For bending in the x direction: end condition at top of column = 2, end condition at bottom of column = 2. Hence from *Table 3.24*, $\beta_y = 0.85$

$$\frac{\ell_{ey}}{b} = \frac{\beta_y \ell_{oy}}{b} = \frac{0.85 \times 4400}{250} = 14.96$$

Since both ℓ_{ex}/h and ℓ_{ey}/b are both less than 15, the column is short.

3.13.5 SHORT BRACED COLUMN DESIGN

For design purposes, BS 8110 divides short-braced columns into three categories. These are:

1. columns resisting axial loads only;
2. columns supporting an approximately symmetrical arrangement of beams;
3. columns resisting axial loads and uniaxial or biaxial bending.

Referring to the floor plan shown in *Fig. 3.91*, it can be seen that column B2 supports beams which are equal in length and symmetrically arranged. Provided the floor is uniformly loaded, column B2 will resist an axial load and is an example of category 1.

Column C2 supports a symmetrical arrangement of beams but which are unequal in length.

Column C2 will, therefore, resist an axial load and moment. However, provided that (a) the loadings on the beams are uniformly distributed and (b) the beam spans do not differ by more than 15% of the longer, the moment will be small. As such, column C2 belongs to category 2 and it can safely be designed by considering the axial load only but using slightly reduced values of the design stresses in the concrete and steel reinforcement (*section 3.13.5.2*).

Columns belong to category 3 if conditions (a) and (b) are not satisfied. The moment here becomes significant and the column may be required to resist an axial load and uni-axial bending e.g. columns A2, B1, B3, C1, C3 and D2, or an axial load and biaxial bending, e.g. A1, A3, D1 and D3.

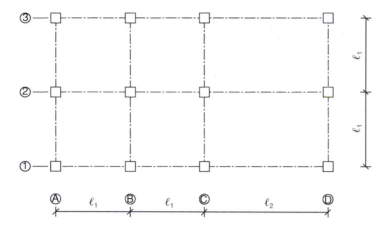

Fig. 3.91 *Floor plan.*

The design procedures associated with each of these categories are discussed in the subsection below.

3.13.5.1 Axially loaded columns
(clause 3.8.4.3, BS 8110)

Consider a column having a net cross-sectional area of concrete A_c and a total area of longitudinal reinforcement A_{sc} (*Fig. 3.92*).

As discussed in section 3.7, the design stresses for concrete and steel in compression are $0.67f_{cu}/1.5$ and $f_y/1.05$, respectively, i.e.

$$\text{Concrete design stress} = \frac{0.67f_{cu}}{1.5} = 0.45f_{cu}$$

$$\text{Reinforcement design stress} = \frac{f_y}{1.05} = 0.95f_y$$

Both the concrete and reinforcement assist in carrying the load. Thus, the ultimate load N which can be supported by the column is the sum of the loads carried by the concrete (F_c) and the reinforcement (F_s), i.e.

$$N = F_c + F_s$$

$$F_c = \text{stress} \times \text{area} = 0.45f_{cu}A_c$$

$$F_s = \text{stress} \times \text{area} = 0.95f_yA_{sc}$$

$$\text{Hence, } N = 0.45f_{cu}A_c + 0.95f_yA_{sc} \qquad (3.39)$$

Equation 3.39 assumes that the load is applied perfectly axially to the column. However, in practice, perfect conditions never exist. To allow for a small eccentricity, i.e. BS 8110 reduces the design stresses in equation 3.39 by about 10%, giving the following expression:

$$N = 0.4f_{cu}A_c + 0.8f_yA_{sc} \qquad (3.40)$$

This is equation 38 in BS 8110 which can be used to design short-braced axially loaded columns.

3.13.5.2 Columns supporting an approximately symmetrical arrangement of beams
(clause 3.8.4.4, BS 8110)

Where the column is subject to an axial load and 'small' moment (*section 3.13.5*), the latter is taken into account simply by decreasing the design stresses in equation 3.40 by around 10%, giving the following expression for the load carrying capacity of the column:

$$N = 0.35f_{cu}A_c + 0.7f_yA_{sc} \qquad (3.41)$$

This is equation 39 in BS 8110 and can be used to design columns supporting an approximately symmetrical arrangement of beams provided (a) the loadings on the beams are uniformly distributed and (b) the beam spans do not differ by more than 15% of the longer.

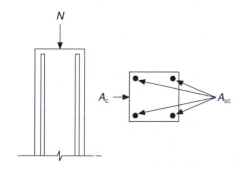

Fig. 3.92

Example 3.17 Sizing a concrete column (BS 8110)

A short-braced column in which $f_{cu} = 30$ N/mm^2 and $f_y = 460$ N/mm^2 is required to support an ultimate axial load of 2000 kN. Determine a suitable section for the column assuming that the area of longitudinal steel, A_{sc}, is of the order of 3% of the gross cross-sectional area of column, A_{col}.

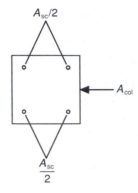

Since the column is axially loaded use equation 3.40

$$N = 0.4f_{cu}A_c + 0.8f_yA_{sc}$$

$$2000 \times 10^3 = 0.4 \times 30\left(A_{col} - \frac{3A_{col}}{100}\right) + 0.8 \times 460 \times \frac{3A_{col}}{100}$$

$$A_{col} = 88\,184 \text{ mm}^2$$

Assuming that the column is square,

$$b = h \approx \sqrt{88\,184} = 297 \text{ mm}.$$

Hence a 300 mm square column constructed of grade 30 concrete would be suitable.

Equations 3.40 and 3.41 are not only used to determine the load carrying capacities of short-braced columns predominantly supporting axial loads but can also be used for initial sizing of these elements as illustrated in Example 3.17.

3.13.5.3 Columns resisting axial forces and moments

The area of longitudinal steel for columns resisting axial loads and uniaxial or biaxial bending, A_{sc}, is normally determined using the design charts in Part 3 of BS 8110. These charts are available for columns having a rectangular cross-section and a symmetrical arrangement of reinforcement. BSI issued these charts when the partial factor of safety for steel reinforcement was 1.15 and not 1.05 and, therefore, use of these charts will lead to conservative estimates of A_{sc}. *Figure 3.93* presents

a modified version of chart 27 which incorporates the new partial safety factor for reinforcement.

It should be noted that each chart is particular for a selected

1. characteristic strength of concrete, f_{cu};
2. characteristic strength of reinforcement, f_y;
3. d/h ratio.

Design charts are available for concrete grades 25, 30, 35, 40, 45 and 50 and reinforcement grade 460. For a specified concrete and steel strength there is a series of charts for different d/h ratios in the range 0.75 to 0.95 in 0.05 increments.

The construction of these charts can best be illustrated by considering how the axial load and moment capacity of an existing column section is assessed. The solution to this problem is somewhat simpler than normal column design as many

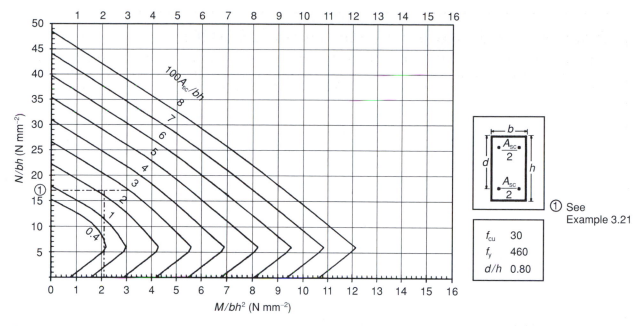

Fig. 3.93 *Column design chart (based on chart 27, BS 8110: Part 3).*

of the design parameters, e.g. grades of materials and area and location of the reinforcement, are predefined. Nonetheless, both rely on an iterative method for solution. Determining the load capacity

of an existing section involves investigating the relationship between the depth of neutral axis of the section and its axial load and co-existent moment capacity. For a range of neutral axis depths

Example 3.18 Analysis of a column section (BS 8110)

Determine whether the column section shown in *Fig. 3.94* of grade 35 concrete and grade 460 steel can support an axial load of 200 kN and a moment about the x–x axis of 200 kN m by calculating the load and moment capacity of the section when the depth of neutral axis of the section, $x = \infty$, 200 and 350 mm.

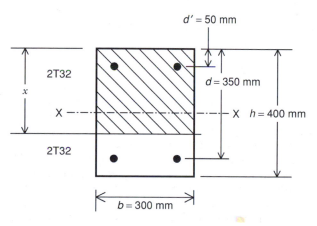

Fig. 3.94

127

(I) LOAD AND MOMENT CAPACITY OF COLUMN WHEN $X = \infty$

Assuming the simplified stress block for concrete, the stress and strain distributions in the section will be as shown in *Fig. 3.95*.

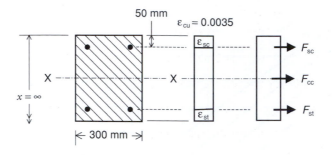

Fig. 3.95

The compressive force in the concrete, F_{cc}, neglecting the area displaced by the reinforcement is

$$F_{cc} = \left(\frac{0.67 f_{cu}}{\gamma_{mc}}\right) bh$$

$$= \left(\frac{0.67 \times 35}{1.5}\right) \times 300 \times 400$$

$$= 1876 \times 10^3 \text{ N}$$

By inspection, $\varepsilon_{sc} = \varepsilon_{st} = \varepsilon_{cu} = 0.0035 > \varepsilon_y \ (= 0.0022)$. Hence

$$F_{sc} = F_{st} = 0.95 f_y (A_{sc}/2)$$

$$= 0.95 \times 460 \times (3216/2) = 702\ 696 \text{ N}$$

Axial load capacity of column, N, is

$$N = F_{cc} + F_{sc} + F_{st}$$

$$= 1876 \times 10^3 + 2 \times 702\ 696$$

$$= 3\ 281\ 392 \text{ N}$$

The moment capacity of the section, M, is obtained by taking moments about the centreline of the section. By inspection, it can be seen that $M = 0$.

(II) LOAD AND MOMENT CAPACITY OF COLUMN WHEN $X = 200$ MM

Figure 3.96 shows the stress and strain distributions when $x = 200$ mm.

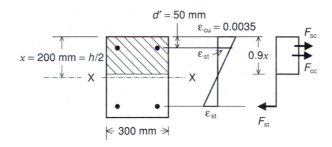

Fig. 3.96

From similar triangles

$$\frac{\varepsilon_{cu}}{x} = \frac{\varepsilon_{st}}{d-x}$$

$$\frac{0.0035}{200} = \frac{\varepsilon_{st}}{350-200}$$

$$\Rightarrow \varepsilon_{st} = 2.625 \times 10^{-3} > \varepsilon_y \ (= 0.0022)$$

By inspection $\varepsilon_{sc} = \varepsilon_{st}$. Hence the tensile force, F_{st}, and compressive force in the steel, F_{sc}, is

$$F_{sc} = F_{st} = 0.95f_y(A_{sc}/2)$$

$$= 0.95 \times 460 \times (3216/2)$$

$$= 702\ 696\ \text{N}$$

The compressive force in the concrete, F_{cc}, is

$$F_{cc} = \left(\frac{0.67f_{cu}}{\gamma_{mc}}\right)0.9xb$$

$$= \left(\frac{0.67 \times 35}{1.5}\right) \times 0.9 \times 200 \times 300$$

$$= 844\ 200\ \text{N}$$

Axial load capacity of column, N, is

$$N = F_{cc} + F_{sc} - F_{st} = 844\ 200\ \text{N}$$

The moment capacity of the section, M, is again obtained by taking moments about the centreline of the section:

$$M = F_{cc}(h/2 - 0.9x/2) + F_{st}(d - h/2) + F_{sc}(h/2 - d')$$

$$= 844\ 200(400/2 - 0.9 \times 200/2) + 702\ 696(350 - 400/2) + 702\ 696(400/2 - 50)$$

$$= 303.7 \times 10^6\ \text{N mm}$$

(III) LOAD AND MOMENT CAPACITY OF COLUMN WHEN $X = 350$ MM
Figure 3.97 shows the stress and strain distributions when $x = 350$ mm.

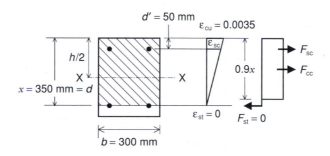

Fig. 3.97

As before

$$\frac{\varepsilon_{cu}}{x} = \frac{\varepsilon_{st}}{d-x}$$

$$\frac{0.0035}{350} = \frac{\varepsilon_{st}}{350-350} \Rightarrow \varepsilon_{st} = 0 \text{ and } F_{st} = 0$$

Similarly, $\varepsilon_{sc} = \varepsilon_{cu}(x-d')/x = 0.0035(350-50)/350 = 0.003 > \varepsilon_y$. Hence, the compressive force in the steel, F_{sc}, is

$$F_{sc} = 0.95f_y(A_{sc}/2)$$

$$= 0.95 \times 460 \times (3216/2)$$

$$= 702\ 696 \text{ kN}$$

$$F_{cc} = \left(\frac{0.67f_{cu}}{\gamma_{mc}}\right)0.9xb$$

$$= \left(\frac{0.67 \times 35}{1.5}\right) \times 0.9 \times 350 \times 300$$

$$= 1\ 477\ 350 \text{ kN}$$

Axial load capacity of column, N, is

$$N = F_{cc} + F_{sc} - F_{st}$$

$$= 1\ 477\ 350 + 702\ 696 - 0$$

$$= 2\ 180\ 046 \text{ N}$$

The moment capacity of the section, M, is

$$M = F_{cc}(h/2 - 0.9x/2) + F_{st}(d - h/2) + F_{sc}(h/2 - d')$$

$$= 1\ 477\ 350(400/2 - 0.9 \times 350/2) + 0 + 702\ 696(400/2 - 50)$$

$$= 168.2 \times 10^6 \text{ N mm}$$

(IV) CHECK SUITABILITY OF PROPOSED SECTION

By dividing the axial loads and moments calculated in (i)–(iii) by bh and bh^2 respectively, the following values obtain:

x (mm)	∞	200	350
N/bh (N/mm^2)	27.3	7.0	18.2
M/bh^2 (N/mm^2)	0	6.3	3.5

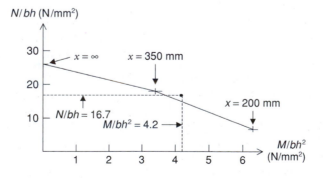

Fig. 3.98

Figure 3.98 shows a plot of the results. By calculating N/bh and M/bh^2 ratios for the design axial load and moment (respectively 16.7 and 4.2) and plotting on *Fig. 3.98* the suitability of the section can be determined.

The results show that the column section is incapable of supporting the design loads. Readers may like to confirm that if two 20 mm diameter bars (i.e. one in each face) were added to the section, the column would then have sufficient capacity.

Comparison of *Figs 3.93* and *3.98* shows that the curves in both cases are similar and, indeed, if the area of longitudinal steel in the section analysed in Example 3.18 was varied between 0.4% and 8%, a chart of similar construction to that shown in *Fig. 3.93* would result. The slight differences in the two charts arise from the fact that the d/h ratio and f_{cu} are not the same.

---

the tensile and compressive forces acting on the section are calculated. The size of these forces can be evaluated using the assumptions previously outlined in connection with the analysis of beam sections *(3.9.1.1)*, namely:

1. Sections that are plane before loading remain plane after loading.
2. The tensile and compressive stresses in the steel reinforcement are derived from *Fig. 3.9*.
3. The compressive stresses in concrete are based on either the rectangular-parabolic stress block for concrete *(Fig. 3.7)* or the equivalent rectangular stress block *(Fig. 3.15(e))*.
4. The tensile strength of concrete is zero.

Once the magnitude and position of the tensile and compressive forces have been determined, the axial load and moment capacity of the section can be evaluated. Example 3.18 illustrates how the results can be used to assess the suitability of the section to resist a particular axial load and moment.

(i) Uniaxial bending. Returning to the original problem of design, the area of longitudinal steel in a column resisting an axial load, N, and uni-axial moment, M, is normally determined using design charts. The procedure simply involves plotting the N/bh and M/bh^2 ratios on the appropriate chart and reading off the corresponding area of reinforcement as a percentage of the gross-sectional area of concrete ($100A_{sc}/bh$) (Example 3.21). Where the actual d/h ratio for the section being designed lies between two charts, both charts may be read and the longitudinal steel area found by linear interpolation.

(ii) Biaxial bending (clause 3.8.4.5, BS 8110). Where the column is subject to biaxial bending, the problem is reduced to one of uniaxial bending simply by increasing the moment about one of the axes using the procedure outlined below. Referring

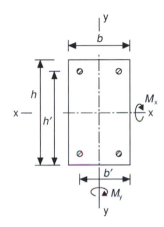

Fig. 3.99

to *Fig. 3.99*, if $M_x/M_y \geqslant h'/b'$ the enhanced design moment, about the x–x axis, M'_x, is

$$M'_x = M_x + \frac{\beta h'}{b'}M_y \qquad (3.42)$$

If $M_x/M_y < h'/b'$, the enhanced design moment about the y–y axis, M'_y, is

$$M'_y = M_y + \frac{\beta b'}{h'}M_x \qquad (3.43)$$

where
b' and h' are the effective depths (*Fig. 3.99*)
β is the enhancement coefficient for biaxial bending obtained from *Table 3.25*.

Table 3.25 Values of the coefficient β (Table 3.22, BS 8110)

$\dfrac{N}{bhf_{cu}}$	0	0.1	0.2	0.3	0.4	0.5	$\geqslant 0.06$
β	1.00	0.88	0.77	0.65	0.53	0.42	0.30

131

The area of longitudinal steel can then be determined using the ultimate axial load (N) and enhanced moment (M'_x or M'_y) in the same way as that described for uniaxial bending.

3.13.6 REINFORCEMENT DETAILS

In order to ensure structural stability, durability and practicability of construction BS 8110 lays down various rules governing the minimum size, amount and spacing of (i) longitudinal reinforcement and (ii) links. These are discussed in the subsections below.

3.13.6.1 Longitudinal reinforcement

(a) Size and minimum number of bars (clause 3.12.5.4, BS 8110). Columns with rectangular cross-sections should be reinforced with a minimum of four longitudinal bars; columns with circular cross-sections should be reinforced with a minimum of six longitudinal bars. Each of the bars should not be less than 12 mm in diameter.

(b) Reinforcement areas (clause 3.12.5, BS 8110). The code recommends that for columns with a gross cross-sectional area A_{col}, the area of longitudinal reinforcement (A_{sc}) should lie within the following limits:

$$0.4\% A_{col} \leq A_{sc} \leq 6\% A_{col} \text{ in a vertically cast}$$
$$\text{column and}$$

$$0.4\% A_{col} \leq A_{sc} \leq 8\% A_{col} \text{ in a horizontally}$$
$$\text{cast column.}$$

At laps the maximum area of longitudinal reinforcement may be increased to 10% of the gross cross-sectional area of the column for both types of columns.

(b) Spacing of reinforcement. The minimum distance between adjacent bars should not be less than the diameter of the bars or $h_{agg} + 5$ mm, where h_{agg} is the maximum size of the coarse aggregate. The code does not specify any limitations with regards to the maximum spacing of bars, but for practical reasons it should not normally exceed 250 mm.

3.13.6.2 Links (clause 3.12.7, BS 8110)

The axial loading on the column may cause buckling of the longitudinal reinforcement and subsequent cracking and spalling of the adjacent concrete cover (*Fig. 3.100*). In order to prevent such a situation from occurring, the longitudinal steel is normally

Axial load may cause buckling of longitudinal reinforcement

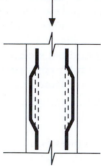

Fig. 3.100

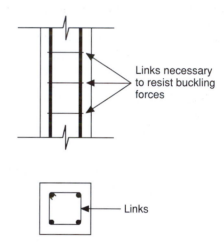

Links necessary to resist buckling forces

Links

Fig. 3.101

laterally restrained at regular intervals by links passing round the bars (*Fig. 3.101*).

(a) Size and spacing of links. Links should be at least one quarter of the size of the largest longitudinal bar or 6 mm, whichever is the greater. However, in practice 6 mm bars may not be freely available and a minimum size of 8 mm is preferable.

Links should be provided at a maximum spacing of 12 times the size of the smallest longitudinal bar or the smallest cross-sectional dimension of the column. The latter condition is not mentioned in BS 8110 but was referred to in CP114 and is still widely observed in order to reduce the risk of diagonal shear failure of columns.

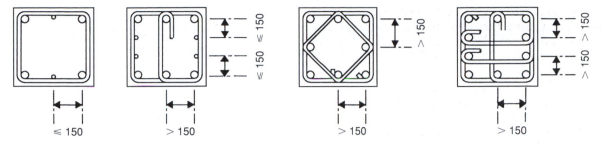

Fig. 3.102 *Arrangement of links in columns.*

(b) Arrangement of links. The code further requires that links should be so arranged that every corner and alternate bar in an outer layer of reinforcement is supported by a link passing around the bar and having an included angle of not more than 135°. All other bars should be within 150 mm of a restrained bar (*Fig. 3.102*).

Example 3.19 Design of an axially loaded column (BS 8110)

Design the longitudinal steel and links for a 350 mm square, short-braced column which supports the following axial loads:

$$G_k = 1000 \text{ kN} \quad Q_k = 1000 \text{ kN}$$

Assume $f_{cu} = 40 \text{ N/mm}^2$, $f_{yv} = 250 \text{ N/mm}^2$ and $f_y = 460 \text{ N/mm}^2$.

(I) LONGITUDINAL STEEL
Since column is axially loaded, use equation 3.40, i.e.

$$N = 0.4f_{cu}A_c + 0.8f_yA_{sc}$$

Total ultimate load $(N) = 1.4G_k + 1.6Q_k = 1.4 \times 1\,000 + 1.6 \times 1\,000 = 3\,000 \text{ kN}$
Substituting this into the above equation for N gives

$$3\,000 \times 10^3 = 0.4 \times 40 \times (350^2 - A_{sc}) + 0.8 \times 460A_{sc}$$

$$A_{sc} = 2\,955 \text{ mm}^2$$

Hence from *Table 3.7*, provide 4T32 ($A_{sc} = 3\,220 \text{ mm}^2$)

(II) LINKS
Diameter of links is one-quarter times the diameter of the largest longitudinal bar, that is, $^1/_4 \times 32$ = 8 mm, but not less than 8 mm diameter. The spacing of links is the lesser of (a) 12 times the diameter of the smallest longitudinal bar, that is, $12 \times 32 = 384$ mm, or (b) the smallest cross-sectional dimension of the column (= 350 mm).

Hence, provide R8 links at 350 mm centres.

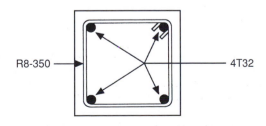

R8-350   4T32

Example 3.20 Column supporting an approximately symmetrical arrangement of beams (BS 8110)

An internal column in a braced two-storey building supporting an approximately symmetrical arrangement of beams (350 mm wide × 600 mm deep) results in characteristic dead and imposed loads each of 1000 kN being applied to the column. The column is 350 mm square and has a clear height of 4.5 m as shown in *Fig. 3.103*. Design the longitudinal reinforcement and links assuming

$$f_{cu} = 40 \text{ N/mm}^2, \quad f_y = 460 \text{ N/mm}^2 \quad \text{and} \quad f_{yv} = 250 \text{ N/mm}^2$$

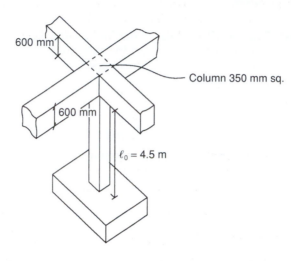

Fig. 3.103

(I) CHECK IF COLUMN IS SHORT

Effective height

Depth of beams (600 mm) > depth of column (350 mm), therefore end condition at top of column = 1. Assuming that the pad footing is not designed to resist any moment, end condition at bottom of column = 3. Therefore, from *Table 3.24*, β = 0.9.

$$\ell_{ex} = \ell_{ey} = \beta\ell_o = 0.9 \times 4\,500 = 4\,050 \text{ mm}$$

Short or slender

$$\frac{\ell_{ex}}{h} = \frac{\ell_{ey}}{b} = \frac{4\,050}{350} = 11.6$$

Since both ratios are less than 15, the column is short.

(II) LONGITUDINAL STEEL

Since column supports an approximately symmetrical arrangement of beams equation 3.41, i.e.

$$N = 0.35f_{cu}A_c + 0.7f_yA_{sc}$$

Total axial load, N, is

$$N = 1.4G_k + 1.6Q_k$$
$$= 1.4 \times 1\,000 + 1.6 \times 1\,000 = 3\,000 \text{ kN}$$

Substituting this into the above equation for N

$$3\,000 \times 10^3 = 0.35 \times 40(350^2 - A_{sc}) + 0.7 \times 460A_{sc}$$

$$\Rightarrow A_{sc} = 4172 \text{ mm}^2$$

Hence from *Table 3.7*, provide 4T32 and 4T25

$$(A_{sc} = 3\,220 + 1\,960 = 5\,180 \text{ mm}^2)$$

(III) LINKS

The diameter of links is one-quarter times the diameter of the largest longitudinal bar, that is $^1/_4 \times 32 = 8$ mm, but not less than 8 mm diameter. The spacing of the links is the lesser of (a) 12 times the diameter of the smallest longitudinal bar, that is, $12 \times 25 = 300$ mm, or (b) the smallest cross-sectional dimension of the column (= 350 mm). Provide R8 links at 300 mm centres.

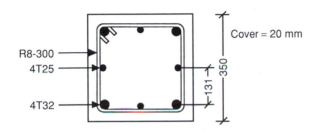

Example 3.21 Columns resisting an axial load and bending (BS 8110)

Design the longitudinal and shear reinforcement for a 275 mm square, short-braced column which supports either

(a) an ultimate axial load of 1280 kN and a moment of 62.5 kN m about the x–x axis or
(b) an ultimate axial load of 1280 kN and bending moments of 35 kN m about the x–x axis and 25 kN m about the y–y axis.

Assume $f_{cu} = 30$ N/mm^2, $f_y = 460$ N/mm^2 and cover to all reinforcement is 35 mm.

LOAD CASE (A)

(i) Longitudinal steel

$$\frac{M}{bh^2} = \frac{62.5 \times 10^6}{275 \times 275^2} = 3$$

$$\frac{N}{bh} = \frac{1280 \times 10^3}{275 \times 275} = 17$$

Assume diameter of longitudinal bars (Φ) = 20 mm, diameter of links (Φ') = 8 mm

$$d = h - \text{cover} - \Phi' - \Phi/2 = 275 - 35 - 8 - 20/2 = 222 \text{ mm}$$

$$d/h = 222/275 = 0.8$$

From *Fig. 3.93*, $100A_{sc}/bh = 3 \Rightarrow A_{sc} = 3 \times 275 \times 275/100 = 2269 \text{ mm}^2$
Provide 8T20 ($A_{sc} = 2510 \text{ mm}^2$, *Table 3.7*).

(ii) Links

The diameter of the links is one-quarter times the diameter of the largest longitudinal bar, that is, $^1/_4 \times 20 = 5$ mm, but not less than 8 mm diameter. The spacing of the links is the lesser of (a) 12 times the diameter of the smallest longitudinal bar, that is, $12 \times 20 = 240$ mm, or (b) the smallest cross-sectional dimension of the column (= 275 mm). Provide R8 links at 240 mm centres

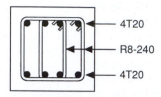

4T20

R8-240

4T20

LOAD CASE (B)

(i) Longitudinal steel

Assume diameter of longitudinal bars (Φ) = 25 mm, diameter of links (Φ') = 8 mm

$$b' = h' = h - \Phi/2 - \Phi' - \text{cover}$$

$$= 275 - 25/2 - 8 - 35 = 220 \text{ mm}$$

$$M_x/M_y = 35/25 = 1.4 > h'/b' = 1$$

$$\frac{N}{bhf_{cu}} = \frac{1280 \times 10^3}{275 \times 275 \times 30} = 0.56$$

Hence $\beta = 0.35$ (*Table 3.25*)

Enhanced design moment about x–x axis, M'_x, is

$$M'_x = M_x + \frac{\beta h'}{b'} M_y$$

$$= 35 + \frac{0.35 \times 220}{220} \times 25 = 43.8 \text{ kN m}$$

$$\frac{M'_x}{bh^2} = \frac{43.8 \times 10^6}{275 \times 275^2} = 2.1$$

$$\frac{N}{bh} = \frac{1280 \times 10^3}{275 \times 275} = 17$$

$$d/h = 220/275 = 0.8$$

From *Fig. 3.93*, $100A_{sc}/bh = 2.2 \Rightarrow A_{sc} = 2.2 \times 275 \times 275/100 = 1664 \text{ mm}^2$
Provide 4T25 ($A_{sc} = 1960 \text{ mm}^2$, *Table 3.7*).

(ii) Links

The diameter of the links is one-quarter times the diameter of the largest longitudinal bar, that is, $^1/_4 \times 25 \approx 6$ mm, but not less than 8 mm diameter. The spacing of the links is the lesser of (a) 12 times the diameter of the smallest longitudinal bar, that is, $12 \times 25 = 300$ mm, or (b) the smallest cross-sectional dimension of the column (= 275 mm). Provide R8 links at 275 mm centres.

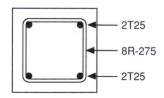

2T25

8R-275

2T25

3.14 Summary

This chapter has considered the design of a number of reinforced concrete elements to BS 8110: *Structural Use of Concrete*. The elements considered primarily either resist bending or support compressive loading. The latter category includes columns subject to axial loading and axial loading and small moments. Elements that fall into the first category include beams, slabs, retaining walls and foundations. They are generally designed for the ultimate limit states of bending and shear and checked for the serviceability limit states of deflection and cracking. The notable exception to this is slabs where the deflection requirements will usually be used to determine the depth of the member. Furthermore, where slabs resist point loads e.g. pad foundations or in flat slab construction, checks on punching shear will also be required.

Questions

1. (a) Derive from first principles the following equation for the ultimate moment of resistance (M_u) of a singly reinforced section assuming a rectangular stress-block distribution

$$M_u = 0.156 f_{cu} bd^2$$

   (b) Draw a fully annotated sketch of the assumed stress block used in the derivation and list any assumptions/limitations of the formula.

2. (a) Design the bending reinforcement for a rectangular concrete beam whose breadth and effective depth are 400 mm and 650 mm respectively to resist an ultimate bending moment of 700 kN m. The characteristic strengths of the concrete and steel reinforcement may be taken as 40 N/mm^2 and 460 N/mm^2.

   (b) Calculate the increase in moment of resistance of the beam in (a) assuming that two 25 mm diameter high yield bars are introduced into the compression face at an effective depth of 50 mm from the extreme compression face of the section.

3. (b) Explain the difference between M and M_u.

   (b) Design the bending and shear reinforcement for the beam shown below using the following information

   $f_{cu} = 30$ N/mm^2 $f_y = 460$ N/mm^2
   $f_{yv} = 250$ N/mm^2 $b = 300$ mm
   span/depth ratio = 12

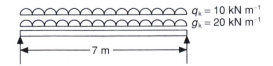

4. (a) Discuss how shear failure can arise in reinforced concrete members and how such failures can be avoided.

   (b) Describe the measures proposed in BS 8110 regarding design for durability of structures.

5. (a) A simply supported T-beam of 7 m clear span carries uniformly distributed dead (including self-weight) and imposed loads of 10 kN/m and 15 kN/m respectively. The beam is reinforced with two 25 mm diameter high yield steel bars as shown below.

   (i) Discuss the factors that influence the shear resistance of a concrete member without shear reinforcement. Assuming that the beam is made from grade 30 concrete, calculate the beam's shear resistance.

   (ii) Design the shear reinforcement for the beam. Assume $f_{yv} = 250$ N/mm^2.

   (b) Confirm whether the span/depth ratio is acceptable for deflection. (Assume $A_{s,req} = 935$ mm^2.)

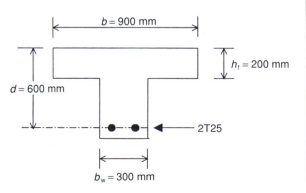

6. Redesign the slab in Example 3.7 assuming that the characteristic strength of the reinforcement is 250 N/mm^2. Comment on your results.

7. (a) Explain the difference between columns which are SHORT and SLENDER and those which are BRACED and UNBRACED.

   (b) Calculate the ultimate axial load capacity of a short-braced column supporting an approximately symmetrical arrangement of beams assuming that it is 500 mm square and is reinforced with 8T 20 mm diameter bars. Assume that $f_{cu} = 40$ N/mm^2, $f_y = 460$ N/mm^2 and the concrete cover is 20 mm. Design the shear reinforcement for the column.

8. (a) A braced column which is 300 mm square is restrained such that it has an effective height of 4.5 m. Classify the column as short or slender.

   (b) The column supports:
       (1) an ultimate axial load of 1300 kN and a bending moment of 200 kN m, or
       (2) an ultimate axial load of 1600 kN and bending moments of 75 kN m about the x-axis and 50 kN m about the y-axis.

   The characteristic strengths are $f_{cu} = 40$ N/mm^2 for the concrete and $f_y = 460$ N/mm^2 for the reinforcement. The cover is 35 mm. Determine the longitudinal steel required for both loading cases by constructing suitable design charts assuming a rectangular stress block distribution for concrete.

9. An internal column in a multi-storey building supporting an approximately symmetrical arrangement of beams carries an ultimate load of 2 000 kN. The storey height is 5.2 m and the effective height factor is 0.85, $f_{cu} = 30$ N/mm^2 and $f_y = 460$ N/mm^2.

   Assuming that the column is square, short and braced, calculate:
   1. a suitable cross-section for the column;
   2. the area of the longitudinal reinforcement;
   3. the size and spacing of the links.

   Sketch the reinforcement detail in cross-section.

Chapter 4

Design of structural steelwork elements to BS 5950

This chapter is concerned with the design of structural steelwork elements and composite construction to British Standard 5950. The chapter describes with the aid of a number of fully worked examples the design of the following elements: beams and joists, struts and columns, composite floors and beams, and bolted and welded connections. The section on beams and joists covers the design of members that are fully laterally restrained as well as members subject to lateral torsional buckling. The section on columns includes the design of cased columns and column base plates.

4.1 Introduction

Several codes of practice are currently in use in the UK for design in structural steelwork. For buildings the older, but still quite popular, permissible stress code BS 449 has now been largely superseded by BS 5950, which is a limit state code introduced in 1985. For steel bridges the limit state code BS 5400: Part 3 is used. Since the primary aim of this book is to give guidance on the design of structural elements, this is best illustrated by considering the contents of BS 5950.

BS 5950 is divided into the following nine parts:

Part 1: *Code of practice for design – Rolled and welded sections.*
Part 2: *Specification for materials, fabrication and erection – Rolled and welded sections.*
Part 3: *Design in composite construction – Section 3.1: Code of practice for design of simple and continuous composite beams.*
Part 4: *Code of practice for design of composite slabs with profiled steel sheeting.*
Part 5: *Code of practice for design of cold formed thin gauge sections.*
Part 6: *Code of practice for design of light gauge profiled steel sheeting.*
Part 7: *Specification for materials, fabrication and erection – Cold formed sections and sheeting.*

Part 8: *Code of practice for fire resistant design.*
Part 9: *Code of practice for stressed skin design.*

Part 1 covers most of the material required for everyday design. Since the majority of this chapter is concerned with the contents of Part 1, it should be assumed that all reference to BS 5950 refer to Part 1 exclusively. Part 3: Section 3.1 and Part 4 deal with, respectively, the design of composite beams and composite floors and will be discussed briefly in *section 4.10*. The remaining parts of the code generally either relate to specification or to specialist types of construction, which are not relevant to the present discussion and will not be mentioned further.

4.2 Iron and steel

4.2.1 MANUFACTURE

Iron has been produced for several thousands of years but it was not until the eighteenth century that it began to be used as a structural material. The first cast-iron bridge by Darby was built in 1779 at Coalbrookdale in Shropshire. Fifty years later wrought iron chains were used in Thomas Telford's Menai Straits suspension bridge. However, it was not until 1898 that the first steel-framed building was constructed. Nowadays most construction is carried out using steel which combines the best properties of cast and wrought iron.

Cast iron is basically remelted pig iron which has been cast into definite useful shapes. Charcoal was used for smelting iron but in 1740 Abraham Darby found a way of converting coal into coke which revolutionized the iron-making process. A later development to this process was the use of limestone to combine with the impurities in the ore and coke and form a slag which could be run off independently of the iron. Nevertheless, the pig iron contained many impurities which made the material brittle and weak in tension.

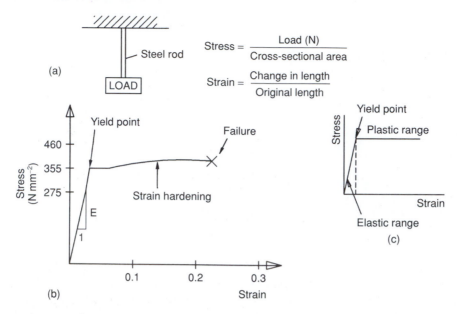

Fig. 4.1 *Stress–strain curves for structural steel: (a) schematic arrangements of test; (b) actual stress-strain curain curve from experiment; (c) idealized stress-strain relationship.*

In 1784, a method known as 'puddling' was developed which could be used to convert pig iron into a tough and ductile metal known as wrought iron. Essentially this process removed many of the impurities present in pig iron, e.g. carbon, manganese, silicon and phosphorus, by oxidation. However, it was difficult to remove the wrought iron from the furnace without contaminating it with slag. Thus although wrought iron was tough and ductile, it was rather soft and was eventually superseded by steel.

Bessemer discovered a way of making steel from iron. This involved filling a large vessel, lined with calcium silicate bricks, with molten pig iron which was then blown from the bottom to remove impurities. The vessel is termed a converter and the process is known as acid Bessemer. However, this converter was unable to remove the phosphorus from the molten iron which resulted in the metal being weak and brittle and completely unmalleable. Later in 1878 Gilchrist Thomas proposed an alternative lining for the converter in which dolomite was used instead of silica and which overcame the problems associated with the acid Bessemer. His process became known as basic Bessemer. The resulting metal, namely steel, was found to be superior to cast iron and wrought iron as it has the same high strength in tension and compression and was ductile.

4.2.2 PROPERTIES

If a rod of steel is subjected to a tensile test (*Fig. 4.1(a)*), and the stress in the rod (load/cross sectional area in N/mm^2) is plotted against the strain (change in length/original length), as the load is applied, a graph similar to that shown in *Fig. 4.1(b)* would be obtained. The stress-strain curve for all structural steels would be linear (**elastic**) up to a certain value, known as the yield point. After this point the steel yields without an increase in load (**plastic**), although there is significant 'strain hardening' as the bar continues to strain towards failure. In the elastic range the bar will return to it's original length if unloaded. However, once past the yield point, in the plastic range, the bar will be permanently strained after unloading. To simplify analysis of steel structures, the idealized stress-strain curve indicated in *Fig. 4.1(c)* is normally used. The slope of the stress-strain curve in the elastic region is referred to as the modulus of elasticity or Young's modulus and is denoted by the letter E. It indicates the stiffness of the material and is used to calculate deflections under load. All structural steels have a modulus of elasticity of 205 kN/mm^2.

4.3 Structural steel and steel sections

Structural steel is manufactured in three basic grades: S275, S355 and S460. Grade S460 is the strongest, but the lower strength grade S275 is the most commonly used in structural applications, for reasons that will later become apparent. In this

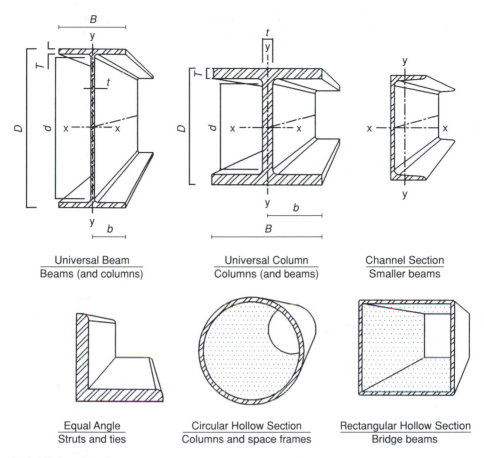

Fig. 4.2 *Standard rolled steel sections.*

classification system 'S' stands for structural and the number indicates the yield strength of the material in N/mm^2.

Figure 4.2 shows in end view the shapes of some commonly used steel elements of structure and their most usual usages. Depending on the size and the demand for a particular shape, some sections may be rolled into shape directly at a steel rolling mill, while others may be fabricated in a welding shop or on site using (usually) electric arc welding.

In each case these sections are designed to achieve economy of material while maximising strength, particularly bending strength. The material tends to be concentrated at the extremities of the section, where it can sustain the tensile and compressive stresses associated with bending. The most commonly used sections are still Universal Beams (UB's) and Universal Columns (UC's). While boxes and tubes have some popularity in specialist applications, they are more expensive to make and more difficult to maintain, particularly in small sizes.

The geometric properties of these steel sections, including the principal dimensions, area, second moment of area, radius of gyration and elastic and plastic section moduli have been tabulated in a booklet entitled *Structural Sections to BS4: Part 1: 1993 and BS EN10056: 1999* which is published by Corus Construction and Industrial. *Appendix B* contains extracts from these tables for UB's and UC's, and will be frequently referred to. The axes x–x and y–y shown in *Fig. 4.2* and referred to in the steel tables in *Appendix B* denote the strong and weak bending axes respectively. Note that the symbols used to identify particular dimensions of universal sections in the booklet are not consistent with the notation used in BS 5950 and have therefore been changed in *Appendix B* to conform with BS 5950.

In this chapter we will concentrate on the design of UB's and UC's and their connections to suit various applications. Irrespective of the element being designed, the designer will need a basic

141

understanding of the following aspects which are discussed next.

(a) symbols
(b) general principles and design methods
(c) loadings
(d) design strengths.

4.4 Symbols

For the purpose of this chapter, the following symbols have been used. These have largely been taken from BS 5950.

GEOMETRIC PROPERTIES

A	area of section
A_g	gross sectional area of steel section
B	breadth of section
b	outstand of flange
D	depth of section
d	depth of web
I_x, I_y	second moment of area about the major and minor axes
$\mathcal{J}$	torsion constant of section
L	length of section
r	root radius
r_x, r_y	radius of gyration of a member about its major and minor axes
S_{eff}	effective plastic modulus
S_x, S_y	plastic modulus about the major and minor axes
T	thickness of flange
t	thickness of web
u	buckling parameter of the section
x	torsional index of section
Z_{eff}	effective elastic modulus
Z_x, Z_y	elastic modulus about major and minor axes

BENDING

A_v	shear area
b_1	stiff bearing length
E	modulus of elasticity
F_t	tensile force
F_v	shear force
L	actual length
L_E	effective length
M	design moment *or* large end moment
M_c	moment capacity
M_b	buckling resistance moment
M_x	maximum major axis moment
P_{bw}	bearing resistance of an unstiffened web
P_s	bearing resistance of stiffener

P_x	buckling resistance of an unstiffened web
P_{xs}	buckling resistance of stiffener
k	$= T + r$
m_{LT}	equivalent uniform moment factor for lateral torsional buckling
P_{cs}	contact stress
P_v	shear capacity of a section
p_c	compressive strength of steel
p_b	bending strength of steel
p_y	design strength of steel
v	slenderness factor for a beam
β	ratio of smaller to larger end moment
β_w	a ratio for lateral torsional buckling
γ_f	overall load factor
γ_m	material strength factor
δ	deflection
ε	constant $= (275/p_y)^{1/2}$
λ	slenderness ratio
λ_{LT}	equivalent slenderness

COMPRESSION

A_{be}	effective area of baseplate
A_g	gross sectional area of steel section
c	largest perpendicular distance from the edge of the effective portion of the baseplate to the face of the column cross-section
F_c	ultimate applied axial load
L	actual length
L_E	effective length
M_b	buckling resistance moment
M_{cx}, M_{cy}	moment capacity of section about the major and minor axes in the absence of axial load
M_{ex}, M_{ey}	eccentric moment about the major and minor axes
M_x, M_y	applied moment about the major and minor axes
M_{LT}	maximum major axis moment in the segment between restraints against lateral torsional buckling
m	equivalent uniform moment factor
P_c	compression resistance of column
P_s	squash load of column
P_{cs}	compression resistance of short strut
P_E	Euler load
p_c	compressive strength
p_{yp}	design strength of the baseplate
t_p	thickness of baseplate
ω	pressure under the baseplate
λ	slenderness ratio
λ_{LT}	equivalent slenderness

CONNECTIONS

a	effective throat size of weld
α_e	effective net area
α_g	gross area
α_n	net area of plate
A_s	effective area of bolt
A_t	tensile stress area of a bolt
d_b	diameter of bolt
D_h	diameter of bolt hole
e_1	edge distance
e_2	end distance
p	pitch
t	thickness of part
F_s	applied shear force
F_t	applied tension force
P_{bb}	bearing capacity of a bolt
P_{bg}	bearing capacity of the parts connected by friction grip fasteners
P_{bs}	bearing capacity of connected parts
P_s	shear capacity of a bolt
P_{sL}	slip resistance provided by a preloaded bolt
P_t	tension capacity of a member or bolt
P_o	minimum shank tension
p_{bb}	bearing strength of a bolt
p_{bs}	bearing strength of connected parts
p_s	shear strength of a bolt
p_t	tension strength of a bolt
p_w	design strength of a fillet-weld
s	leg length of a fillet-weld
K_e	coefficient = 1.2 for S275 steel
K_s	coefficient = 1.0 for clearance holes
μ	slip factor

COMPOSITES

A_{cv}	mean cross-sectional area of concrete
A_{sv}	cross-sectional area of steel reinforcement
α_e	modular ratio
B_e	effective breadth of concrete flange
D_p	depth of profiled metal decking
D_s	depth of concrete flange
δ	deflection
F_v	shear force
Q	strength of shear studs
R_c	compression resistance of concrete flange
R_f	tensile resistance of steel flange
R_s	tensile resistance of steel beam
s	longitudinal spacing of studs
v	longitudinal shear stress per unit length
y_p	depth of neutral axis

4.5 General principles and design methods

As stated at the outset, BS 5950 is based on limit state philosophy. *Table 1* of BS 5950, reproduced as *Table 4.1*, outlines typical limit states appropriate to steel structures.

As this book is principally about the design of structural elements we will be concentrating on the ultimate limit state of strength (1), and the serviceability limit state of deflection (5). Stability (2) is an aspect of complete structures or sub-structures that will not be examined at this point, except to say that structures must be robust enough not to overturn or sway excessively under wind or other sideways loading. Fatigue (3) is generally taken account of by the provision of adequate safety factors to prevent occurrence of the high stresses associated with fatigue. Brittle fracture (4) can be avoided by selecting the correct grade of steel for the expected ambient conditions. Avoidance of excessive vibration (6) and oscillations (7), is an aspect of structural dynamics and is beyond the scope of this book. Corrosion can be a serious problem for exposed steelwork, but correct preparation and painting of the steel will ensure maximum durability (8) and minimum maintenance during the life of the structure. Alternatively, the use of weather resistant steels should be considered.

Table 4.1 Limit states (Table 1, BS 5950)

Ultimate	Serviceability
1 Strength (including general yielding, rupture, buckling and forming a mechanism)	5 Deflection
2 Stability against overturning and sway stability	6 Vibration
3 Fracture due to fatigue	7 Wind induced oscillation
4 Brittle fracture	8 Durability

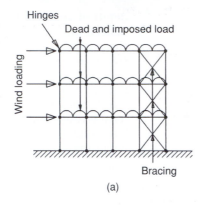

Fig. 4.3 *(a) Simple and (b) continuous design.*

Although BS 5950 does not specifically mention fire resistance, this is an important aspect that fundamentally affects steel's economic viability compared to it's chief rival, concrete. Exposed structural steelwork does not perform well in a fire. The high conductivity of steel together with the thin sections used causes high temperatures to be quickly reached in steel members, resulting in premature failure due to softening at around 600°C. Structural steelwork has to be insulated to provide adequate fire resistance in multi-storey structures. Insulation can be a sprayed treatment, intumescent coatings, concrete surround or boxing with plasterboard. Although it is expensive to provide suitable fire resistance, under certain conditions structural advantage can be taken of the concrete used for fire protection as discussed in *section 4.9.4*. Guidance on the design of fire protection for members in steel framed buildings can be found in Part 8 of BS 5950.

For steel structures three principal methods of design are identified in clause 2.1.2 of BS 5950:

1. *Simple design.* The structure is regarded as having pinned joints, and significant moments are not developed at connections (*Fig. 4.3(a)*). The structure is prevented from becoming a mechanism by appropriate bracing using shear walls for instance. This apparently rather unrealistic but conservative assumption is a very popular method of design, even in this era of powerful design computing.
2. *Continuous design.* The joints in the structure are assumed to be able to fully transfer the forces and moments in the members which they attach (*Fig. 4.3(b)*). Analysis of the structure may be by elastic or plastic methods, and will be more complex than simple design. However the increasing use of micro-computers has made this

method more viable. In theory a more economic design can be achieved by this method, but unless the joints are truly rigid the analysis will give an upper bound (unsafe) solution.
3. *Semi-continuous design.* The joints in the structure are assumed to have some degree of strength and stiffness but not provide complete restraint as in the case of continuous design. The actual strength and stiffness of the joints should be determined experimentally. Guidance on the design of semi-continuous frames can be found in the following Steel Construction Institute publications:
   (i) *Wind-moment Design of Unbraced Composite Frames*, SCI-P264, 2000.
   (ii) *Design of Semi-continuous Braced Frames*, SCI-P183, 1997.

4.6 Loading

As for structural design in other media, the designer needs to estimate the loading to which the structure may be subject during its design life.

The characteristic dead and imposed loads can be obtained from BS 6399: Parts 1 and 3. Wind loads should be determined from BS 6399: Part 2 or CP3: Chapter V: Part 2. In general, a characteristic load is expected to be exceeded in only 5% of instances, or for 5% of the time, but in the case of wind loads it represents a gust expected only once every 50 years.

To obtain design loading at **ultimate limit state** for strength and stability calculations the characteristic load is multiplied by a load factor obtained from *Table 2* of BS 5950, part of which is reproduced as *Table 4.2*. Generally, dead load is multiplied by 1.4 and imposed vertical (or live) load by 1.6, except when the load case considers wind load

Table 4.2 Partial factors for loads (Table 2, BS 5950)

Type of load and load combinations	Factor, γ_f
Dead load	1.4
Dead load acting together with wind and imposed load	1.2
Dead load whenever it counters the effects of other loads	1.0
Dead load when restraining sliding, overturning or uplift	1.0
Imposed load	1.6
Imposed load acting together with wind load	1.2
Wind load	1.4
Storage tanks, including contents	1.4
Storage tanks, empty, when restraining sliding, overturning or uplift	1.0

also, in which case, dead, imposed and wind loads are all multiplied by 1.2.

Several loading cases may be specified to give a 'worst case' envelope of forces and moments around the structure. In the design of buildings without cranes, the following load combinations should normally be considered (clause 2.4.1.2, BS 5950):

1. dead plus imposed
2. dead plus wind
3. dead, imposed plus wind.

Table 4.3 Design strengths p_y (Table 9, BS 5950)

Steel grade	Thickness, less than or equal to (mm)	Design strength, p_y (N/mm²)
S275	16	275
	40	265
	63	255
	80	245
	100	235
	150	225
S355	16	355
	40	345
	63	335
	80	325
	100	315
	150	295
S460	16	460
	40	440
	63	430
	80	410
	100	400

To obtain design loading at **serviceability limit state** for calculation of deflections the most adverse realistic combination of **unfactored** characteristic **imposed** loads is usually used. In the case of wind loads acting together with imposed loads, only 80% of the full specified values need to be considered.

4.7 Design strengths

In BS 5950 no distinction is made between characteristic and design strength. In effect the material safety factor $\gamma_m = 1.0$. Structural steel used in the UK is specified by BS 5950: Part 2, and strengths of the more commonly used steels are given in *Table 9* of BS 5950, reproduced here as *Table 4.3*. As a result of the residual stresses locked into the metal during the rolling process, the thicker the material, the lower the design strength.

Having discussed these more general aspects relating to structural steelwork design, the following sections will consider the detailed design of beams and joists (*section 4.8*), struts and columns (*section 4.9*), composite floors and beams (*section 4.10*) and connections (*section 4.11*).

4.8 Design of steel beams and joists

The design of beams and joists primarily involves predicting the strength of the member. This requires the designer to imagine all ways in which the member may fail during its design life. It would be useful at this point, therefore, to discuss some of the more important modes of failure associated with beams and joists.

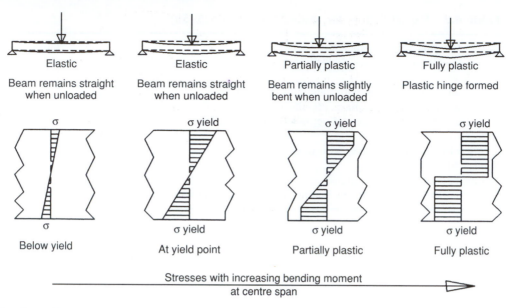

Fig. 4.4 *Bending failure of a beam.*

4.8.1 MODES OF FAILURE

4.8.1.1 Bending

As a result of bending, longitudinal stresses are set up in the beam. These stresses are tensile in one half of the beam and compressive in the other. As the bending moment increases yield is eventually reached. Failure takes place when steel yields in tension and/or compression across the entire cross section of the beam. When all of the beam cross-section has become plastic in this way the beam fails by formation of a plastic hinge at the point of maximum imposed moment. *Figure 4.4* reviews this process. Chapter 2 summarises how classical beam theory is derived from these considerations.

4.8.1.2 Local buckling

During the bending process outlined above, if the compression flange, or the part of the web subject to compression, is too thin the plate may actually fail by buckling, or rippling, as shown in *Fig. 4.5*, before the full plastic moment is reached.

4.8.1.3 Shear

Due to excessive shear forces, usually adjacent to supports, the beam may fail in shear. The beam web, which resists shear forces, may fail as shown in *Fig. 4.6(a)*, as steel yields in tension and compression in the shaded zones. The formation of plastic hinges in the flanges accompanies this process.

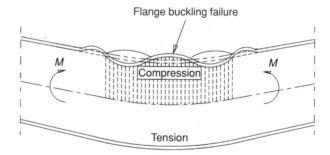

Fig. 4.5 *Local flange buckling failure.*

4.8.1.4 Shear buckling

During the shearing process described above, if the web is too thin it will fail by buckling or rippling in the shear zone, as shown in *Fig. 4.6(b)*.

4.8.1.5 Web bearing and buckling

Due to high vertical stresses directly over a support or under a concentrated load, the beam web may actually crush, or buckle as a result of these stresses, as illustrated in *Fig. 4.7*.

4.8.1.6 Lateral torsional buckling

When the beam has a higher bending stiffness in the vertical plane compared to the horizontal plane, the beam can twist sideways under the load. This is perhaps best visualised by loading a scale rule on

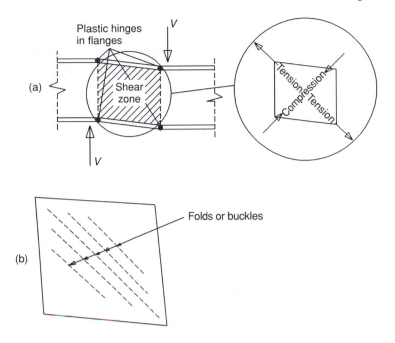

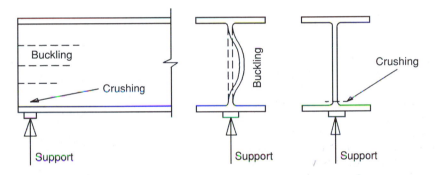

Fig. 4.6 *Shear and shear buckling failures: (a) shear failure; (b) shear buckling.*

Fig. 4.7 *Web buckling and web bearing failures.*

its edge, as it is held as a cantilever – it will tend to twist and deflect sideways. This is illustrated in *Fig. 4.8*. Where a beam is not prevented from moving sideways, by a floor, for instance, or the beam is not nominally torsionally restrained at supports, it is necessary to check that it is laterally stable under load. Nominal torsional restraint may be assumed to exist if web cleats, partial depth end plates or fin plates, for example, are present (*Fig. 4.9*).

4.8.1.7 Deflection
Although a beam cannot fail as a result of excessive deflection alone, it is necessary to ensure that deflections are not excessive under unfactored imposed loading. Excessive deflections are those

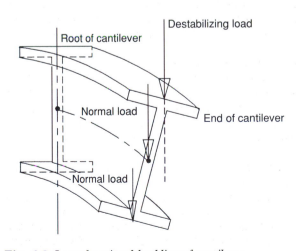

Fig. 4.8 *Lateral torsional buckling of cantilever.*

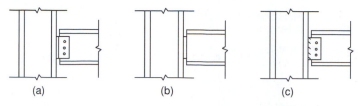

Fig. 4.9 *Nominal torsional restraint at beam support supplied by (a) web cleats (b) end plate (c) fin plate.*

resulting in severe cracking in finishes which would render the building unserviceable.

4.8.2 SUMMARY OF DESIGN PROCESS
The design process for a beam can be summarised as follows:

1. determination of design shear forces, F_v, and bending moments, M, at critical points on the element (see *Chapter 2*);
2. selection of UB or UC;
3. classification of section;
4. check shear strength; if unsatisfactory return to (2);
5. check bending capacity; if unsatisfactory return to (2);
6. check deflection; if unsatisfactory return to (2);
7. check web bearing and buckling at supports or concentrated load; if unsatisfactory provide web stiffener;
8. check lateral torsional buckling (*section 4.8.11*); if unsatisfactory return to (2) or provide lateral and torsional restraints;
9. summarise results.

4.8.3 INITIAL SECTION SELECTION
It is perhaps most often the case in the design of skeletal building structures, that **bending** is the critical mode of failure, and so beam bending theory can be used to make an initial selection of section. Readers should refer to Chapter 2 for more clarification on bending theory if necessary.

To avoid bending failure, it is necessary to ensure that the design moment, M, does not exceed the moment capacity of the section, M_c, i.e.

$$M < M_c \qquad (4.1)$$

Generally, the moment capacity for a steel section is given by

$$M_c = p_y S \qquad (4.2)$$

where
p_y is the assumed design strength of the steel
S is the plastic modulus of the section

Combining the above equations gives an expression for S:

$$S > M/p_y \qquad (4.3)$$

This can be used to select suitable universal beam sections from steel tables (*Appendix B*) with the plastic modulus of section S greater than the calculated value.

4.8.4 CLASSIFICATION OF SECTION
Having selected a suitable section from steel tables, or proposed a suitable section fabricated by welding, it must be classified.

4.8.4.1 Strength classification
In making the initial choice of section, a steel strength will have been assumed. If grade S275 steel is to be used, for example, it may have been assumed that the strength is 275 N/mm^2. Now by referring to the flange thickness T from the steel tables, the design strength can be obtained from *Table 9* of BS 5950, reproduced as *Table 4.3*.

If the section is fabricated from welded plate, the strength of the web and flange may be taken separately from *Table 9* of BS 5950 as that for the web thickness t and flange thickness T respectively.

4.8.4.2 Section classification
As previously noted, the bending strength of the section depends on how the section performs in bending. If the section is stocky, i.e. has thick flanges and web, for instance, it can sustain the formation of a plastic hinge. On the other hand, a slender section, i.e. with thin flanges and web, will fail by local buckling before the yield stress can be reached. Four classes of section are identified in clause 3.5.2 of BS 5950.

Class 1 Plastic cross sections are those in which a plastic hinge can be developed with significant rotation capacity (*Fig. 4.10*). If the plastic design method is used in the structural analysis, all members **must** be of this type.

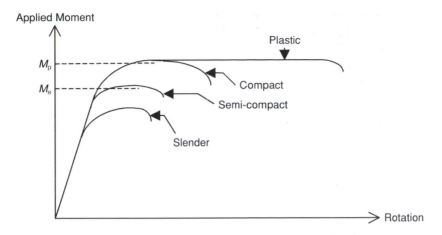

Fig. 4.10 *Typical moment/rotation characteristics of different classes of section.*

Class 2 Compact cross sections are those in which the full plastic moment capacity can be developed, but local buckling may prevent development of a plastic hinge with sufficient rotation capacity to permit plastic design.

Class 3 Semi-Compact cross sections can develop their elastic moment capacity, but local buckling may prevent the development of the full plastic moment.

Class 4 Slender cross sections contain slender elements subject to compression due to moment or axial load. Local buckling may prevent the full elastic moment capacity from being developed.

Limiting width to thickness ratios for elements for the above classes are given in *Table 11* of BS 5950, part of which is reproduced as *Table 4.4*. (Refer to *Fig. 4.2* for details of UB and UC dimensions.) Once the section has been classified, the various strength checks can be carried out to assess its suitability as discussed next.

4.8.5 SHEAR

According to clause 4.2.3 of BS 5950, the shear force, F_v, should not exceed the shear capacity, P_v, i.e.

$$F_v \leqslant P_v \tag{4.5}$$

where

$$P_v = 0.6 p_y A_v \tag{4.6}$$

in which A_v is the shear area ($= tD$ for rolled I-, H- and channel sections). Equation (4.6) assumes that the web carries the shear force alone.

Clause 4.2.3 also states that when the buckling ratio (d/t) of the web exceeds 70ε (see equation 4.4), then the web should be additionally checked for shear buckling. However, no British universal beam section, no matter what the grade, is affected.

Table 4.4 Limiting width to thickness ratios (elements which exceed these limits are to be taken as class 4, slender cross sections.) (based on Table 11, BS 5950)

Type of element (all rolled sections)	Class of section		
	(1) Plastic	(2) Compact	(3) Semi-comp
Outstand element of compression flange	$\dfrac{b}{T} \leqslant 9\varepsilon$	$\dfrac{b}{T} \leqslant 10\varepsilon$	$\dfrac{b}{T} \leqslant 15\varepsilon$
Web with neutral axis at mid-depth	$\dfrac{d}{t} \leqslant 80\varepsilon$	$\dfrac{d}{t} \leqslant 100\varepsilon$	$\dfrac{d}{t} \leqslant 120\varepsilon$
Web where the whole cross-section is subject to axial compression only	n/a	n/a	$\dfrac{d}{t} \leqslant 40\varepsilon$

Note. $\varepsilon = (275/p_y)^{1/2}$ (4.4)

4.8.6 LOW SHEAR AND MOMENT CAPACITY

As stated in clause 4.2.1.1 of BS 5950, at critical points the combination of (i) maximum moment and co-existent shear and (ii) maximum shear and co-existent moment should be checked.

If the shear force F_v does not exceed $0.6P_v$, then this is a **low shear load**. Otherwise, if $0.6P_v < F_v < P_v$, then it is a **high shear load**.

When the shear load is low, the moment capacity of the section is calculated according to clause 4.2.5.2 of BS 5950 as follows:

For **Class 1 Plastic** or **Class 2 Compact** sections, the moment capacity

$$M_c = p_y S \leqslant 1.2 p_y Z \qquad (4.7)$$

where

p_y design strength of the steel
S plastic modulus
Z section modulus

The additional check ($M_c \leqslant 1.2 p_y Z$) is to guard against plastic deformations under serviceability loads and is applicable to **simply supported and cantilever beams**. For other beam types this limit is $1.5 p_y Z$.

For **Class 3 Semi-compact** sections

$$M_c = p_y Z \qquad (4.8)$$

or alternatively $M_c = p_y S_{eff} \leqslant 1.2 p_y Z \qquad (4.9)$

where S_{eff} is the effective plastic modulus (clause 3.5.6 of BS 5950) and the other symbols are as defined for equation 4.7.

Note that whereas equation 4.8 provides a conservative estimate of the moment capacity of class 3 compact sections, use of equation 4.9 is more efficient but requires additional computational effort.

For **Class 4 Slender** sections

$$M_c = p_y Z_{eff} \qquad (4.10)$$

where Z_{eff} is the effective section modulus (clause 3.6.2 of BS 5950).

In practice the above considerations do not prove to be much of a problem. Nearly all sections in grade S275 steel are plastic, and only a few sections in higher strength steel are semi-compact. No British rolled universal beam sections in pure bending, no matter what the strength class, are slender or have plastic or compact flanges and semi-compact webs.

Example 4.1 Selection of a beam section in grade S275 steel (BS 5950)

A rolled universal beam section in grade S275 steel is to span 10 metres simply supported, and support 5 kN/m uniformly distributed dead load and 5 kN/m uniformly distributed imposed load, as well as a central imposed point load of 30 kN as shown in *Fig. 4.11*. Assuming the beam is fully laterally restrained and there is nominal torsional restraint at supports, select a suitable universal beam section to satisfy bending and shear considerations.

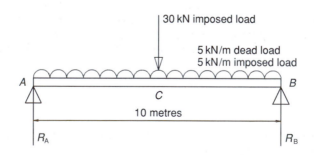

Fig. 4.11 *Loading for example 4.1.*

DESIGN BENDING MOMENT AND SHEAR FORCE

Total loading $= (30 \times 1.6) + (5 \times 1.4 + 5 \times 1.6)10$
$\qquad\qquad = 48 + 15 \times 10 = 198$ kN

Because the structure is symmetrical $R_A = R_B = 198/2 = 99$ kN. The central bending moment, M, is

$$M = \frac{W\ell}{4} + \frac{\omega\ell^2}{8}$$

$$= \frac{48 \times 10}{4} + \frac{15 \times 10^2}{8}$$

$$= 120 + 187.5 = 307.5 \text{ kN m}$$

Shear force and bending moment diagrams are shown in *Fig. 4.12*.

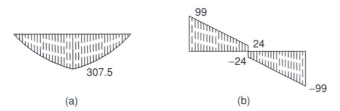

      (a)   (b)

Fig. 4.12 *B.M. and S.F. diagrams.*

INITIAL SECTION SELECTION
Assuming $p_y = 275$ N/mm^2

$$S_x > \frac{M}{p_y} = \frac{307.5 \times 10^6}{275} = 1.118 \times 10^6 \text{ mm}^3 = 1118 \text{ cm}^3$$

From steel tables (*Appendix B*), suitable sections are:

1. $356 \times 171 \times 67$ UB: $S = 1210$ cm^3;
2. $406 \times 178 \times 60$ UB: $S = 1190$ cm^3;
3. $457 \times 152 \times 60$ UB: $S = 1280$ cm^3.

The above illustrates how steel beam sections are specified. For section 1, for instance, **356 × 171** represents the serial size in the steel tables; **67** represents the mass per metre in kilograms; and UB stands for **universal beam**.

    All the above sections give a value of plastic modulus about axis x–x, S_x, just greater than that required. Whichever one is selected will depend on economic and engineering considerations. For instance, if lightness were the primary consideration, perhaps section 3 would be selected, which is also the strongest (largest S_x). However if minimising the depth of the member were the main consideration then section 1 would be chosen. Let us choose the compromise candidate, section 2.

CLASSIFICATION

Strength Classification
Because the flange thickness $T = 12.8$ mm (< 16 mm), then $p_y = 275$ N/mm^2 (as assumed) from *Table 4.3* and $\varepsilon = (275/p_y)^{1/2} = 1$ (*Table 4.4*).

Section classification
$b/T = 6.95$ which is less than $9\varepsilon = 9$. Hence from *Table 4.4*, flange is plastic. Also $d/t = 46.2$ which is less than $80\varepsilon = 80$. Hence from *Table 4.4*, web is plastic. Therefore $406 \times 178 \times 60$ UB section is class 1 plastic.

SHEAR STRENGTH

As $d/t = 46.2 < 70\varepsilon$, shear buckling need not be considered. Shear capacity of section, P_v, is

$$P_v = 0.6p_ytD = 0.6 \times 275 \times 7.8 \times 406.4$$

$$= 523 \times 10^3 \text{ N} = 523 \text{ kN}$$

Now, as $F_v(99 \text{ kN}) < 0.6P_v = 0.6 \times 523 = 314 \text{ kN}$ (low shear load).

BENDING MOMENT

Moment capacity of section, M_c, is

$$M_c = p_yS = 275 \times 1\,190 \times 10^3$$

$$= 327 \times 10^6 \text{ N mm} = 327 \text{ kN m}$$

$$\leqslant 1.2p_yZ = 1.2 \times 275 \times 1060 \times 10^3$$

$$= 349.8 \times 10^6 \text{ N mm} = 349.8 \text{ kN m} \quad \text{OK}$$

Moment M due to imposed loading $= 307.5$ kN m. Extra moment due to self weight, M_{sw}, is

$$M_{sw} = 1.4 \times (60 \times 9.81/10^3)\frac{10^2}{8} = 10.3 \text{ kN m}$$

Total imposed moment $M_t = 307.5 + 10.3 = 317.8$ kN m < 327 kN m. Hence proposed section is suitable.

Example 4.2 Selection of a beam section in grade S460 steel (BS 5950)

Repeat the above design in grade S460 steel.

INITIAL SECTION SELECTION

Since section is of grade S460 steel, assume $p_y = 460$ N/mm^2

$$S_x > \frac{M}{p_y} = \frac{307.5 \times 10^6}{460} = 668.5 \times 10^3 \text{ mm}^3 = 668.5 \text{ cm}^3$$

Suitable sections (with classifications) are:

1. $305 \times 127 \times 48$ UB: $S = 706$ cm^3, $p_y = 460$, plastic;
2. $305 \times 165 \times 46$ UB: $S = 723$ cm^3, $p_y = 460$, compact;
3. $356 \times 171 \times 45$ UB: $S = 774$ cm^3, $Z = 687$ cm^3, $p_y = 460$, semi-compact;
4. $406 \times 140 \times 39$ UB: $S = 721$ cm^3, $Z = 627$ cm^3, $p_y = 460$, semi-compact.

Section 4 must be discounted. As it is semi-compact and Z is much less than 669 cm^3 it will fail in bending. The same is probably true of section 3. Section 1 is plastic and S is sufficiently large to take care of it's own self weight. However, section 2 looks the better choice, being the lighter of 1 and 2, and with the greater strength.

SHEAR STRENGTH

As $d/t < 70\varepsilon$, no shear buckling check is required. Shear capacity of section, P_v, is

$$P_v = 0.6p_ytD = 0.6 \times 460 \times 6.7 \times 307.1$$

$$= 567.9 \times 10^3 \text{ N} = 567.9 \text{ kN}$$

Now, as $F_v(99 \text{ kN}) < 0.6P_v = 340.7$ kN (low shear load).

BENDING MOMENT
Moment capacity of section subject to low shear load, M_c, is

$$M_c = p_y S = 460 \times 723 \times 10^3$$

$$= 332.6 \times 10^6 \text{ N mm} = 332.6 \text{ kN m}$$

$$\leqslant 1.2 p_y Z = 1.2 \times 460 \times 648 \times 10^3$$

$$= 357.7 \times 10^6 \text{ N mm} = 357.7 \text{ kN m} \quad \text{OK}$$

Moment M due to imposed loading = 307.5 kN m. Extra moment due to self weight, M_{sw}, is

$$M_{sw} = 1.4 \times (46 \times 9.81/10^3) \frac{10^2}{8} = 7.9 \text{ kN m}$$

Total imposed moment $M_t = 307.5 + 7.9 = 315.4$ kN m, which is less than the section's moment of resistance $M_c = 332.6$ kN m. This section is satisfactory.

4.8.7 HIGH SHEAR AND MOMENT CAPACITY
When the shear load is high, i.e. $F_v > 0.6P_v$, the moment-carrying capacity of the section is reduced. This is because the web cannot take the full tensile or compressive stresses associated with the bending moment as well as a substantial shear stress due to the shear force. According to clause 4.2.5.3 of BS 5950, the moment capacity of UB and UC sections, M_c, should be calculated as follows:

For class 1 plastic and compact sections

$$M_c = p_y(S - \rho S_v) \tag{4.11}$$

For class 3 semi-compact sections

$$M_c = p_y(S - \rho S_v/1.5) \quad \text{or alternatively}$$

$$M_c = p_y(S_{eff} - \rho S_v) \tag{4.12}$$

For class 4 slender sections

$$M_c = p_y(Z_{eff} - \rho S_v/1.5) \tag{4.13}$$

where $\rho = [2(F_v/P_v) - 1]^2$ and S_v for sections with equal flanges, is the plastic modulus of the shear area of section equal to $tD^2/4$. The other symbols are as previously defined in *section 4.8.6*.

Note the effect of the ρ factor is to reduce the moment-carrying capacity of the web as the shear load rises from 50 to 100% of the web's shear capacity. However, the resulting reduction in moment capacity is negligible when $F_v < 0.6P_v$.

4.8.8 DEFLECTION
A check should be carried out on the maximum deflection of the beam due to the most adverse

Example 4.3 Selection of a cantilever beam section (BS 5950)

A proposed cantilever beam 1 metre long is to be built into a concrete wall as shown in *Fig. 4.13*. It supports characteristic dead and imposed loading of 450 kN/m and 270 kN/m respectively. Select a suitable universal beam section in grade S275 steel to satisfy bending and shear criteria only.

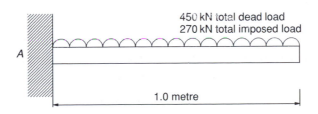

450 kN total dead load
270 kN total imposed load

A

1.0 metre

Fig. 4.13

DESIGN BENDING MOMENT AND SHEAR FORCE
Shear force F_v at A is

$$(450 \times 1.4) + (270 \times 1.6) = 1\,062 \text{ kN}$$

Bending Moment M at A is

$$\frac{W\ell}{2} = \frac{1\,062 \times 1}{2} = 531 \text{ kN m}$$

INITIAL SECTION SELECTION
Assuming $p_y = 275 \text{ N/mm}^2$

$$S_x > \frac{M}{p_y} = \frac{531 \times 10^6}{275} = 1931 \times 10^3 \text{ mm}^3 = 1931 \text{ cm}^3$$

Suitable sections (with classifications) are:

1. $457 \times 191 \times 89$ UB: $S = 2010 \text{ cm}^3$, $p_y = 275$, plastic;
2. $533 \times 210 \times 82$ UB: $S = 2060 \text{ cm}^3$, $p_y = 275$, plastic.

SHEAR STRENGTH
Shear capacity of $533 \times 210 \times 82$ UB section, P_v, is

$$P_v = 0.6 p_y t D = 0.6 \times 275 \times 9.6 \times 528.3$$

$$= 837 \times 10^3 \text{ N} = 837 \text{ kN} < 1062 \text{ kN} \quad \text{Not OK}$$

In this case, because the length of the cantilever is so short, the selection of section will be determined from the shear strength, which is more critical than bending. Try a new section:

$$610 \times 229 \times 113 \text{ UB}: S_x = 3290 \text{ cm}^3, \quad p_y = 265, \text{ plastic}$$

$$P_v = 0.6 \times 265 \times 607.3 \times 11.2 = 1081 \times 10^3 \text{ N}$$

$$= 1081 \text{ kN} \quad \text{OK but high shear load}$$

BENDING MOMENT
From above

$$\rho = [2(F_v/P_v) - 1]^2 = [2(1062/1081) - 1]^2 = 0.93$$

$$M_c = p_y(S_x - \rho S_v)$$

$$= 265(3290 \times 10^3 - 0.93[11.2 \times 607.3^2/4])$$

$$= 617 \times 10^6 = 617 \text{ kN m}$$

$$M_{sw} = 1.4 \times (113 \times 9.81/10^3)\frac{1^2}{2} = 0.8 \text{ kN m}$$

$$M_t = M + M_{sw} = 531 + 0.8 = 532 \text{ kN m} < M_c$$

This section is satisfactory.

realistic combination of **unfactored imposed serviceability** loading. In BS 5950 this is covered by clause 2.5.2 and *Table 8*, part of which is reproduced below as *Table 4.5*. *Table 8* outlines recommended limits to these deflections, compliance with which should avoid significant damage to the structure and finishes.

Calculation of deflections from first principles has to be done using the Area-Moment Method, Macaulay's Method, or some other similar ap-

Table 4.5 Suggested vertical deflection limits on beams due to imposed load (based on Table 8, BS 5950)

Cantilevers	Length/180
Beams carrying plaster or other brittle finish	Span/360
Other beams (except purlins and sheeting rails)	Span/200

proach, a subject which is beyond the scope of this book.

The reader is referred to a suitable structural analysis text for more detail on this subject. However, many calculations of deflection are carried out using formulae for standard cases, which can be combined as necessary to give the answer for more complicated situations. *Figure 4.14* summarises some of the more useful formulae.

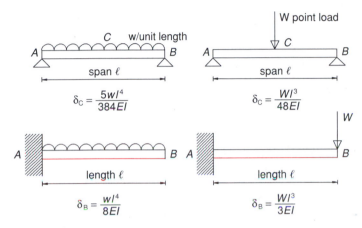

Fig. 4.14 *Deflections for standard cases. E = elastic modulus of steel (205 kN/mm²) and I = second moment of area (x–x) of section.*

Example 4.4 Deflection checks on steel beams (BS 5950)

Carry out a deflection check for Examples 4.1–4.3 above.

FOR EXAMPLE 4.1

$$\delta_C = \frac{5\omega\ell^4}{384EI} + \frac{W\ell^3}{48EI}$$

$$= \frac{5 \times 5 \times 10^4}{384 \times 205 \times 10^6 \times 21\,500 \times 10^{-8}} + \frac{30 \times 10^3}{48 \times 205 \times 10^6 \times 21\,500 \times 10^{-8}}$$

$$= 0.0148 + 0.0142 = 0.029 \text{ m} = 29 \text{ mm}$$

From *Table 4.5*, the recommended maximum deflection for beams carrying plaster is span/360 = 10 000/360 = 27.8 mm, and for other beams span/200 = 10 000/200 = 50 mm. Therefore if the beam was carrying a plaster finish one might consider choosing a larger section.

FOR EXAMPLE 4.2

$$\delta_C = \frac{5\omega\ell^4}{384EI} + \frac{W\ell^3}{48EI}$$

$$= \frac{5 \times 5 \times 10^4}{384 \times 205 \times 10^6 \times 9950 \times 10^{-8}} + \frac{30 \times 10^3}{48 \times 205 \times 10^6 \times 9950 \times 10^{-8}}$$

$$= 0.0319 + 0.0306 = 0.0625 \text{ m} = 62.5 \text{ mm}$$

The same recommended limits from *Table 4.5* apply as above. This grade S460 beam therefore fails the deflection test. This partly explains why the higher strength steel beams are not particularly popular.

FOR EXAMPLE 4.3

$$\delta_B = \frac{\omega \ell^4}{8EI} = \frac{270 \times 1^4}{8 \times 205 \times 10^6 \times 87\,400 \times 10^{-8}}$$

$$= 0.00019 \text{ m} = 0.19 \text{ mm}$$

The recommended limit from *Table 4.5* is

$$\text{Length}/180 = 5.6 \text{ mm}.$$

So deflection is no problem for this particular beam.

4.8.9 WEB BEARING AND WEB BUCKLING

Clause 4.5 of BS 5950 covers all aspects of web bearing, web buckling and stiffener design. Usually most critical at the position of a support or concentrated load is the problem of web buckling. The buckling resistance of a web is obtained via the web bearing capacity as discussed next.

4.8.9.1 Web bearing

The ways in which concentrated loads are transmitted through the flange/web connection in the span, and at supports when the distance to the end of the member from the end of the stiff bearing is zero is shown in *Fig. 4.15(a)*. Clause 4.5.2 of BS 5950 shows how the bearing resistance P_{bw} is calculated at the flange/web connection:

$$P_{bw} = (b_1 + nk)tp_{yw} \qquad (4.15)$$

where
b_1 is the stiff bearing length
n is as shown in Figure 4.15(a): $n = 5$ except at the end of a member and $n = 2 + 0.6b_e/k$ $\leqslant 5$ at the end of the member
b_e is the distance to the end of the member from the end of the stiff bearing (*Fig. 4.15(b)*)
k $= (T + r)$ for rolled I- or H-sections
T is the thickness of the flange
t is the web thickness
p_{yw} is the design strength of the web

This is essentially a simple check which ensures that the stress at the critical point on the flange/web connection does not exceed the strength of the steel.

It should also be checked that the contact stresses between load or support and flange do not exceed p_y.

4.8.9.2 Web buckling

According to clause 4.5.3.1 of BS 5950, provided the distance α_e from the concentrated load or reaction to the nearer end of the member is at least 0.7d, and if the flange through which the load or reaction is applied is effectively restrained against both

(a) rotation relative to the web
(b) lateral movement relative to the other flange (*Fig. 4.16*),

the buckling resistance of an unstiffened web is given by

$$P_x = \frac{25\varepsilon t}{\sqrt{(b_1 + nk)d}} P_{bw} \qquad (4.16)$$

The reader is referred to Appendix C for the background and derivation of this equation.

Alternatively, when $\alpha_e < 0.7d$, the buckling resistance of an unstiffened web is given by

$$P_x = \frac{\alpha_e + 0.7d}{1.4d} \frac{25\varepsilon t}{\sqrt{(b_1 + nk)d}} P_{bw} \qquad (4.17)$$

If the flange is not restrained against rotation and/or lateral movement the buckling resistance of the web is reduced to P_{xr}, given by

$$P_{xr} = \frac{0.7d}{L_E} P_x \qquad (4.18)$$

in which L_E is the effective length of the web determined in accordance with *Table 22* of BS 5950.

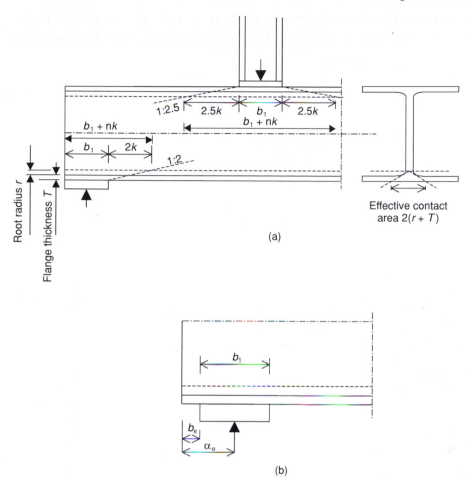

(a)

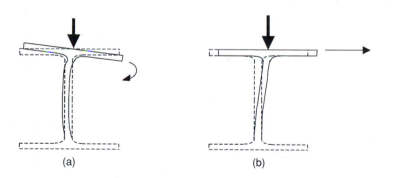

(b)

Fig. 4.15 *Web bearing.*

Fig. 4.16

Example 4.5 Checks on web buckling and bearing for steel beams (BS 5950)

Check web bearing and buckling for *Example 4.1*, assuming the beam sits on 100 mm bearings at each end.

WEB BEARING AT SUPPORTS

$$P_{bw} = (b_1 + nk)tp_{yw}$$
$$= (100 + 2 \times 23)7.8 \times 275$$
$$= 313 \times 10^3 \text{ N} = 313 \text{ kN} > 99 \text{ kN} \quad \text{OK}$$

where
$k = T + r = 12.8 + 10.2 = 23$ mm
$n = 2 + 0.6b_e/k = 2$ (since $b_e = 0$)

CONTACT STRESS AT SUPPORTS

$$P_{cs} = (b_1 \times 2(r + T))p_y = (100 \times 46) \times 275$$
$$= 1\,265 \times 10^3 \text{ N} = 1\,265 \text{ kN} > 99 \text{ kN} \quad \text{OK}$$

WEB BUCKLING AT SUPPORTS
Since α_e (= 50 mm) $< 0.7d = 0.7 \times 360.5 = 252$ mm, buckling resistance of the web is

$$P_x = \frac{\alpha_e + 0.7d}{1.4d} \frac{25\varepsilon t}{\sqrt{(b_1 + nk)d}} P_{bw}$$

$$= \frac{50 + 252}{1.4 \times 360.5} \times \frac{25 \times 1 \times 7.8}{\sqrt{(100 + 2 \times 23)360.5}} \, 313 = 159 \text{ kN} > 99 \text{ kN} \quad \text{OK}$$

So no web stiffeners are required at supports.
    A web check at the concentrated load should also be carried out, but readers can confirm that this aspect is not critical in this case.

4.8.10 STIFFENER DESIGN
If it is found that the web fails in buckling or bearing, it is not necessary to select another section; larger supports can be designed, or load-carrying stiffeners can be locally welded between the flanges and the web. Clause 4.5 of BS 5950 gives guidance on stiffener designer and the following example illustrates this.

Example 4.6 Design of a steel beam with web stiffeners (BS 5950)

A simply supported beam is to span 5 metres and support uniformly distributed characteristic loads of 200 kN/m (dead) and 100 kN/m (imposed). The beam sits on 150 mm long bearings at supports, and both flanges are laterally and torsionally restrained (*Fig. 4.17*). Select a suitable universal beam section to satisfy bending, shear and deflection criteria. Also check web bearing and buckling at supports, and design stiffeners if they are required.

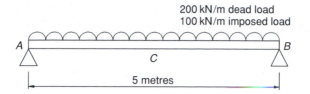

Fig. 4.17

DESIGN SHEAR FORCE AND BENDING MOMENT

Factored loading $= (200 \times 1.4) + (100 \times 1.6) = 440$ kN/m

Reactions $R_A = R_B = 440 \times 5 \times 0.5 \qquad = 1100$ kN

Bending moment, $M = \dfrac{\omega \ell^2}{8} = \dfrac{440 \times 5^2}{8} = 1375$ kN m

INITIAL SECTION SELECTION

Assuming $p_y = 275$ N/mm^2

$$S_x > \frac{M}{p_y} = \frac{1375 \times 10^6}{275} = 5000 \times 10^3 \text{ mm}^3 = 5000 \text{ cm}^3$$

From *Appendix B*, suitable sections (with classifications) are:

1. $610 \times 305 \times 179$ UB: $S = 5520$ cm^3, $p_y = 265$ plastic;
2. $686 \times 254 \times 170$ UB: $S = 5620$ cm^3, $p_y = 265$ plastic;
3. $762 \times 267 \times 173$ UB: $S = 6200$ cm^3, $p_y = 265$ plastic;
4. $838 \times 292 \times 176$ UB: $S = 6810$ cm^3, $p_y = 265$ plastic.

Section 2 looks like a good compromise; it is the lightest section, and it's depth is not excessive.

SHEAR STRENGTH

$$P_v = 0.6 p_y t D = 0.6 \times 265 \times 14.5 \times 692.9$$

$$= 1597 \times 10^3 \text{ N} = 1\,597 \text{ kN} > 1\,100 \text{ kN} \quad \text{OK}$$

However as F_v (= 1100 kN) $> 0.6 P_v$ (= 958 kN), high shear load.

BENDING MOMENT

As the shear force is zero at the centre, the point of maximum bending moment, M_c, is obtained from

$$M_c = p_y S = 265 \times 5620 \times 10^3$$

$$= 1489.3 \times 10^6 \text{ N mm} = 1489.3 \text{ kN m}$$

$$\leqslant 1.2 p_y Z = 1.2 \times 265 \times 4910 \times 10^3$$

$$= 1561.3 \times 10^6 \text{ N mm} = 1561.3 \text{ kN m} \quad \text{OK}$$

$$M_{sw} = 1.4 \times \left(170 \times 9.81/10^3\right)\frac{5^2}{8} = 7.3 \text{ kN m}$$

Total imposed moment $M_t = 1375 + 7.3 = 1382.3$ kN m < 1489.3 kN m. OK

DEFLECTION

Calculated deflection, δ_C, is

$$\delta_C = \frac{5\omega \ell^4}{384 EI} = \frac{5 \times 100 \times 5^4}{384 \times 205 \times 10^6 \times 170\,000 \times 10^{-8}}$$

$$= 0.0023 \text{ m} = 2.3 \text{ mm}$$

Maximum recommended deflection limit for a beam carrying plaster from *Table 4.5* = span/360 = 13.8 mm OK

WEB BEARING AT SUPPORTS

$$k = T + r = 23.7 + 15.2 = 38.9 \text{ mm}$$

$$n = 2 + 0.6b_e/k = 2 \text{ (since } b_e = 0)$$

$$P_{bw} = (b_1 + nk)t \cdot p_{yw}$$

$$= (150 + 2 \times 38.9)14.5 \times 275$$

$$= 908 \times 10^3 \text{ N} = 908 \text{ kN}$$

The web's bearing resistance P_{bw} (= 908 kN) < R_A (= 1100 kN), and so load carrying stiffeners are required. BS 5950 stipulates that load-carrying stiffeners should be checked for both bearing and buckling.

STIFFENER BEARING CHECK

Let us propose 12 mm thick stiffeners each side of the web and welded continuously to it. Since the width of section B = 255.8 mm, the stiffener outstand is effectively limited to 120 mm.

The actual area of stiffener in contact with the flange, if a 15 mm fillet is cut out for the root radius, $A_{s,net}$ = (120 − 15)2 × 12 = 2520 mm^2

The bearing capacity of the stiffener, P_s, is given by (clause 4.5.2.2)

$$P_s = A_{s,net}p_y = 2520 \times 275$$

$$= 693 \times 10^3 \text{ N} > \text{external reaction} = (R_A - P_{bw} = 1100 - 908 =) 192 \text{ kN}$$

WEB BUCKLING AT SUPPORTS

Since α_e (= 75 mm) < 0.7d = 0.7 × 692.9 = 485 mm, buckling resistance of the web is

$$P_x = \frac{\alpha_e + 0.7d}{1.4d} \frac{25\varepsilon t}{\sqrt{(b_1 + nk)d}} P_{bw}$$

$$= \frac{75 + 485}{1.4 \times 692.9} \times \frac{25 \times 1 \times 14.5}{\sqrt{(150 + 2 \times 38.9)692.9}} 908 = 478.3 \text{ kN} < 1100 \text{ kN}$$

Therefore, web stiffeners capable of resisting an external (buckling) load, F_x, of $F_x = (F_v - P_x) =$ 1100 − 478.3 = 621.7 kN are required.

STIFFENER BUCKLING CHECK

The buckling resistance of a stiffener, P_{xs}, is given by

$$P_{xs} = A_s p_c$$

The plan of the web and stiffener at the support position is shown in *Fig. 4.18*. Note that a length of web on each side of the centre line of the stiffener not exceeding 15 times the web thickness should be included in calculating the buckling resistance (clause 4.5.3.3). Hence, second moment of area (I_s) of the effective section (shown cross-hatched in *Fig. 4.18*) (based on $bd^3/12$) for stiffener buckling about the z–z axis is

$$I_s = \frac{12 \times (120 + 120 + 14.5)^3}{12} + \frac{(217.5 + 69) \times 14.5^3}{12} = 16.56 \times 10^6 \text{ mm}^4$$

Effective area of buckling section, A_s, is

$$A_s = 12 \times (120 + 120 + 14.5) + (217.5 + 69) \times 14.5$$

$$= 7208 \text{ mm}^2$$

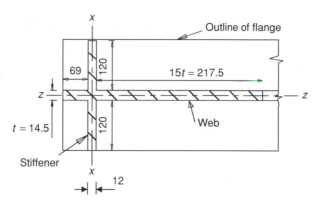

Fig. 4.18 *Plan of web and stiffener.*

Radius of gyration $r = (I_s/A_s)^{0.5}$

$$= (16.56 \times 10^6/7208)^{0.5}$$

$$= 47.9 \text{ mm}$$

According to Clause 4.5.3.3 of BS 5950, the effective length (L_E) of load carrying stiffeners when the compression flange is laterally restrained = 0.7L, where L = length of stiffener (= d)

$$= 0.7 \times 615.1$$

$$= 430.6 \text{ mm}$$

$$\lambda = L_E/r = 430.6/47.9 = 9$$

Then from *Table 24(c)* (*Table 4.14*) $p_c = p_y = 275 \text{ N/mm}^2$

Then $P_{xs} = A_s p_c = 7208 \times 275$

$$= 1982 \times 10^3 \text{ N}$$

$$= 1982 \text{ kN} > F_x = 621.7 \text{ kN}$$

Hence the stiffener is also adequate in buckling.

4.8.11 LATERAL TORSIONAL BUCKLING

If the loaded flange of a beam is not effectively restrained against lateral movement relative to the other flange, by a concrete floor fixed to the beam, for instance, and against rotation relative to the web, by web cleats or fin plates, for example, it is possible for the beam to twist sideways under a load less than that which would cause the beam to fail in bending, shear or deflection. This is called **lateral torsional buckling** which is covered by Clause 4.3 of BS 5950. It is illustrated by *Fig. 4.19* for a cantilever, but readers can experiment with this phenomenon very easily using a scale rule or ruler loaded on it's edge. They will see that although the problem occurs more readily with a cantilever, it also applies to beams supported at each end. Keen experimenters will also find that the more restraint they provide at the beam supports, the higher the load required to make the beam twist sideways. This effect is taken into account by using the concept of 'effective length', as discussed below.

4.8.11.1 Effective length

The concept of **effective length** is introduced in clause 4.3.5 of BS 5950. For a beam supported at its ends only, with no intermediate lateral restraint, and standard restraint conditions at supports, the effective length is equal to the actual length between supports. When a greater degree of lateral and torsional restraint is provided at supports, the effective length is less than the actual length, and vice-versa. The effective length appropriate to different end restraint conditions is specified in Table 13 of BS 5950, reproduced below as *Table 4.6*,

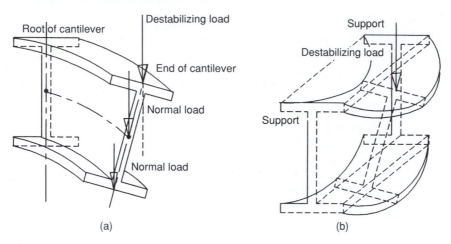

Fig. 4.19 *Lateral torsional buckling: (a) cantilever beam; (b) simply supported beam.*

Table 4.6   Effective length, L_E, for beams (Table 13, BS 5950)

Conditions of restraint at supports	Loading conditions	
	Normal	*Destab.*
Comp. flange laterally restrained		
Nominal torsional restraint against rotation about longitudinal axis		
Both flanges fully restrained against rotation on plan	$0.7L_{LT}$	$0.85L_{LT}$
Compression flange fully restrained against rotation on plan	$0.75L_{LT}$	$0.9L_{LT}$
Both flanges partially restrained against rotation on plan	$0.8L_{LT}$	$0.95L_{LT}$
Compression flange partially restrained against rotation on plan	$0.85L_{LT}$	$1.0L$
Both flanges free to rotate on plan	$1.0L_{LT}$	$1.2L_{LT}$
Comp. flange laterally unrestrained		
Both flanges free to rotate on plan		
Partial torsional restraint against rotation about longitudinal axis provided by connection of bottom flange to supports	$1.0L_{LT} + 2D$	$1.2L_{LT} + 2D$
Partial torsional restraint against rotation about longitudinal axis provided only by pressure of bottom flange onto supports	$1.2L_{LT} + 2D$	$1.4L_{LT} + 2D$

and illustrated in *Fig. 4.20*. The table gives different values of effective length depending on whether the load is **normal** or **destabilising**. This is perhaps best explained using *Fig. 4.19*, in which it can readily be seen that a load applied to the top of the beam will cause it to twist further, thus worsening the situation. If the load is applied below the centroid of the section however, it has a slightly restorative effect, and is conservatively assumed to be normal.

Determining the effective length for real beams, when it is difficult to define the actual conditions of restraint, and whether the load is normal or destabilising, is perhaps one of the greatest problems in the design of steelwork. It is an aspect that

was amended in the 1990 edition of BS 5950 (and once again in the 2000 edition) in order to relate more to practical circumstances.

4.8.11.2 Lateral torsional buckling resistance
Checking of lateral torsional buckling for rolled universal beam sections can be carried out in two different ways. Firstly there is a slightly conservative, but quite simple check in clause 4.3.7 of BS 5950, which only applies to equal flanged rolled sections. The approach is similar to that for struts (*section 4.9.1*), and involves determining the bending strength for the section, p_b, from Table 20 of BS 5950, reproduced as *Table 4.7*, via the slenderness value $(\beta_w)^{0.5}L_E/r_y$ and D/T.

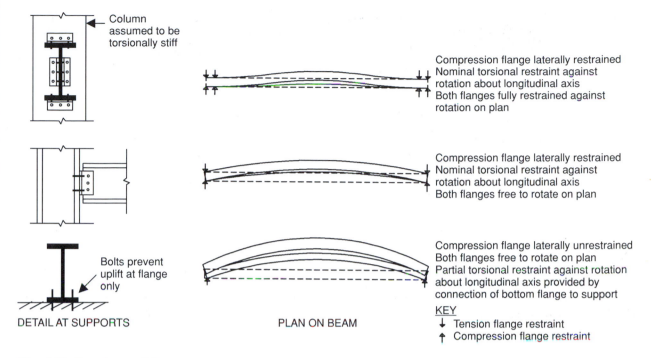

Fig. 4.20 *Restraint conditions.*

Table 4.7 Bending strength p_b (N/mm^2) for rolled sections with equal flanges (Table 20, BS 5950)

	1) Grade S275 steel $\leq$ 16 mm (p_y = 275 N/mm^2)									
$(\beta_w)^{0.5}L_E/r_y$					D/T					
	5	10	15	20	25	30	35	40	45	50
30	275	275	275	275	275	275	275	275	275	275
35	275	275	275	275	275	275	275	275	275	275
40	275	275	275	275	274	273	272	272	272	272
45	275	275	269	266	264	263	263	262	262	262
50	275	269	261	257	255	253	253	252	252	251
55	275	263	254	248	246	244	243	242	241	241
60	275	258	246	240	236	234	233	232	231	230
65	275	252	239	232	227	224	223	221	221	220
70	274	247	232	223	218	215	213	211	210	209
75	271	242	225	215	209	206	203	201	200	199
80	268	237	219	208	201	196	193	191	190	189
85	265	233	213	200	193	188	184	182	180	179
90	262	228	207	193	185	179	175	173	171	169
95	260	224	201	186	177	171	167	164	162	160
100	257	219	195	180	170	164	159	156	153	152
105	254	215	190	174	163	156	151	148	146	144
110	252	211	185	168	157	150	144	141	138	136
115	250	207	180	162	151	143	138	134	131	129
120	247	204	175	157	145	137	132	128	125	123
125	245	200	171	152	140	132	126	122	119	116
130	242	196	167	147	135	126	120	116	113	111
135	240	193	162	143	130	121	115	111	108	106

Table 4.7 (*cont'd*)

$(\beta_w)^{0.5}L_E/r_y$	D/T									
	5	10	15	20	25	30	35	40	45	50
140	238	190	159	139	126	117	111	106	103	101
145	236	186	155	135	122	113	106	102	99	96
150	233	183	151	131	118	109	102	98	95	92
155	231	180	148	127	114	105	99	94	91	88
160	229	177	144	124	111	101	95	90	87	84
165	227	174	141	121	107	98	92	87	84	81
170	225	171	138	118	104	95	89	84	81	78
175	223	169	135	115	101	92	86	81	78	75
180	221	166	133	112	99	89	83	78	75	72
185	219	163	130	109	96	87	80	76	72	70
190	217	161	127	107	93	84	78	73	70	67
195	215	158	125	104	91	82	76	71	68	65
200	213	156	122	102	89	80	74	69	65	63
210	209	151	118	98	85	76	70	65	62	59
220	206	147	114	94	81	72	66	62	58	55
230	202	143	110	90	78	69	63	58	55	52
240	199	139	106	87	74	66	60	56	52	50
250	195	135	103	84	72	63	57	53	50	47

2) Grade S275 steel > 16 mm ≤ 40 mm ($p_y = 265$ N/mm^2)

$(\beta_w)^{0.5}L_E/r_y$	D/T									
	5	10	15	20	25	30	35	40	45	50
30	265	265	265	265	265	265	265	265	265	265
35	265	265	265	265	265	265	265	265	265	265
40	265	265	265	265	265	264	264	264	263	263
45	265	265	261	258	256	255	254	254	254	254
50	265	261	253	249	247	246	245	244	244	244
55	265	255	246	241	238	236	235	235	234	234
60	265	250	239	233	229	227	226	225	224	224
65	265	245	232	225	221	218	216	215	214	214
70	265	240	225	217	212	209	207	205	204	204
75	263	235	219	210	204	200	198	196	195	194
80	260	230	213	202	196	191	189	187	185	184
85	257	226	207	195	188	183	180	178	176	175
90	254	222	201	188	180	175	171	169	167	166
95	252	217	196	182	173	167	163	160	158	157
100	249	213	190	176	166	160	156	153	150	149
105	247	209	185	170	160	153	148	145	143	141
110	244	206	180	164	154	147	142	138	136	134
115	242	202	176	159	148	140	135	132	129	127
120	240	198	171	154	142	135	129	125	123	121
125	237	195	167	149	137	129	124	120	117	115
130	235	191	163	144	132	124	119	114	111	109
135	233	188	159	140	128	119	114	109	106	104
140	231	185	155	136	124	115	109	105	102	99
145	229	182	152	132	120	111	105	101	97	95
150	227	179	148	129	116	107	101	97	93	91
155	225	176	145	125	112	103	97	93	89	87
160	223	173	142	122	109	100	94	89	86	83
165	221	170	139	119	106	97	91	86	83	80
170	219	167	136	116	103	94	88	83	80	77
175	217	165	133	113	100	91	85	80	77	74
180	215	162	130	110	97	88	82	77	74	71
185	213	160	128	108	95	86	79	75	71	69

Table 4.7 (*cont'd*)

$(\beta_w)^{0.5}L_E/r_y$	D/T									
	5	10	15	20	25	30	35	40	45	50
190	211	157	125	105	92	83	77	73	69	66
195	209	155	123	103	90	81	75	70	67	64
200	207	153	120	101	88	79	73	68	65	62
210	204	148	116	96	84	75	69	64	61	58
220	200	144	112	93	80	71	65	61	58	55
230	197	140	108	89	77	68	62	58	54	52
240	194	136	104	86	74	65	59	55	52	49
250	190	132	101	83	71	63	57	52	49	47

For **Class 1 Plastic** or **Class 2 Compact** sections, the buckling resistance moment, M_b, is obtained from

$$M_b = p_b S_x \qquad (4.19)$$

For **Class 3 Semi-compact** sections, M_b, is given by

$$M_b = p_b Z_x \qquad (4.20)$$

where
S_x plastic section modulus about the major axis
Z_x section modulus about the major axis
β_w a ratio equal to 1 for class 1 plastic or class 2 compact sections and Z_x/S_x for class 3 semi-compact sections
D depth of the section
r_y radius of gyration about the y–y axis
T flange thickness

Example 4.7 Design of a laterally unrestrained steel beam (simple method) (BS 5950)

Assuming the beam in Example 4.1 is not laterally restrained, determine whether the selected section is stable, and if not, select one which is. Assume the compression flange is laterally restrained and the beam fully restrained against torsion at supports, but both flanges are free to rotate on plan. The loading is normal.

EFFECTIVE LENGTH
Since beam is pinned at both ends, from *Table 4.6*, $L_E = 1.0L_{LT} = 10$ m

BUCKLING RESISTANCE
Using the conservative approach of clause 4.3.7

$$\left(\beta_w\right)^{0.5}\frac{L_E}{r_y} = \left(1.0\right)^{0.5}\frac{10\,000}{39.7} = 252$$

$$\frac{D}{T} = \frac{406.4}{12.8} = 31.75$$

From *Table 4.7*, $p_b = 60$ N/mm^2 (approx).

$$M_b = p_b S_x = 60 \times 1190 \times 10^3$$

$$= 71.4 \times 10^6 \text{ N mm} = 71.4 \text{ kN m} \ll 317.8 \text{ kN m}$$

As the buckling resistance moment is much less than the actual imposed moment, this beam would fail by lateral torsional buckling. Using trial and error, a more suitable section can be found

Try $305 \times 305 \times 137$ UC; $p_y = 265$, plastic.

$$\left(\beta_w\right)^{0.5} \frac{L_E}{r_y} = \left(1.0\right)^{0.5} \frac{10\,000}{78.2} = 128$$

$$\frac{D}{T} = \frac{320.5}{21.7} = 14.8$$

Then from *Table 4.7*, $p_b = 165.7$ N/mm^2

$$M_b = p_b S_x = 165.7 \times 2300 \times 10^3$$
$$= 381 \times 10^6 \text{ N mm} = 381 \text{ kN m}$$

$$M_{sw} = 1.4 \left(137 \times 9.81/10^3\right) \frac{10^2}{8} = 23.5 \text{ kN m}$$

Total imposed moment $M_t = 307.5 + 23.5 = 331$ kN m < 381 kN m   OK

So this beam, actually a column section, is suitable. Readers may like to check that there are not any lighter UB sections. Because the column has a greater r_y, it is laterally stiffer than a UB section of the same weight and is more suitable than a beam section in this particular situation.

4.8.11.3 Equivalent slenderness and uniform moment factors

The more rigorous approach for calculating values of M_b is covered by clauses 4.3.6.2 to 4.3.6.9 of BS 5950. It involves calculating an equivalent slenderness ratio, λ_{LT}, given by

$$\lambda_{LT} = uv\lambda \sqrt{\beta_w} \qquad (4.21)$$

in which

$$\lambda = \frac{L_E}{r_y}$$

where

L_E effective length for lateral torsional buckling
r_y radius of gyration about the minor axis
u buckling parameter = 0.9 for rolled *I*- and *H*-sections
v slenderness factor from Table 19 of BS 5950, part of which is reproduced as *Table 4.8*, given in terms of λ/x in which x torsional index = D/T where D is the depth of the section and T is the flange thickness
β_w = 1.0 for class 1 plastic or class 2 compact sections;
= Z_x/S_x for class 3 semi-compact sections if $M_b = p_b Z_x$;
= $S_{x,eff}/S_x$ for class 3 semi-compact sections if $M_b = p_b S_{x,eff}$;
= $Z_{x,eff}/S_x$ for class 4 slender cross-sections.

Knowing λ_{LT} and the design strength, p_y, the bending strength, p_b, is then read from *Table 16* for rolled sections, reproduced as *Table 4.9*. The buck-

Table 4.8 Slenderness factor v for beams with equal flanges (based on Table 19, BS 5950)

$\dfrac{\lambda}{x}$	$N = 0.5$
0.5	1.00
1.0	0.99
1.5	0.97
2.0	0.96
2.5	0.93
3.0	0.91
3.5	0.89
4.0	0.86
4.5	0.84
5.0	0.82
5.5	0.79
6.0	0.77
6.5	0.75
7.0	0.73
7.5	0.72
8.0	0.70
8.5	0.68
9.0	0.67
9.5	0.65
10.0	0.64
11.0	0.61
12.0	0.59
13.0	0.57
14.0	0.55
15.0	0.53
16.0	0.52
17.0	0.50
18.0	0.49
19.0	0.48
20.0	0.47

Table 4.9 Bending strength p_b (N/mm^2) for rolled sections (Table 16, BS 5950)

λ_{LT}	Steel grade and design strength, p_y (N/mm^2)														
	S275					S355					S460				
	235	245	255	265	275	315	325	335	345	355	400	410	430	440	460
25	235	245	255	265	275	315	325	335	345	355	400	410	430	440	460
30	235	245	255	265	275	315	325	335	345	355	395	403	421	429	446
35	235	245	255	265	273	307	316	324	332	341	378	386	402	410	426
40	229	238	246	254	262	294	302	309	317	325	359	367	382	389	404
45	219	227	235	242	250	280	287	294	302	309	340	347	361	367	381
50	210	217	224	231	238	265	272	279	285	292	320	326	338	344	356
55	199	206	213	219	226	251	257	263	268	274	299	305	315	320	330
60	189	195	201	207	213	236	241	246	251	257	278	283	292	296	304
65	179	185	190	196	201	221	225	230	234	239	257	261	269	272	279
70	169	174	179	184	188	206	210	214	218	222	237	241	247	250	256
75	159	164	168	172	176	192	195	199	202	205	219	221	226	229	234
80	150	154	158	161	165	178	181	184	187	190	201	203	208	210	214
85	140	144	147	151	154	165	168	170	173	175	185	187	190	192	195
90	132	135	138	141	144	153	156	158	160	162	170	172	175	176	179
95	124	126	129	131	134	143	144	146	148	150	157	158	161	162	164
100	116	118	121	123	125	132	134	136	137	139	145	146	148	149	151
105	109	111	113	115	117	123	125	126	128	129	134	135	137	138	140
110	102	104	106	107	109	115	116	117	119	120	124	125	127	128	129
115	96	97	99	101	102	107	108	109	110	111	115	116	118	118	120
120	90	91	93	94	96	100	101	102	103	104	107	108	109	110	111
125	85	86	87	89	90	94	95	96	96	97	100	101	102	103	104
130	80	81	82	83	84	88	89	90	90	91	94	94	95	96	97
135	75	76	77	78	79	83	83	84	85	85	88	88	89	90	90
140	71	72	73	74	75	78	78	79	80	80	82	83	84	84	85
145	67	68	69	70	71	73	74	74	75	75	77	78	79	79	80
150	64	64	65	66	67	69	70	70	71	71	73	73	74	74	75
155	60	61	62	62	63	65	66	66	67	67	69	69	70	70	71
160	57	58	59	59	60	62	62	63	63	63	65	65	66	66	67
165	54	55	56	56	57	59	59	59	60	60	61	62	62	62	63
170	52	52	53	53	54	56	56	56	57	57	58	58	59	59	60
175	49	50	50	51	51	53	53	53	54	54	55	55	56	56	56
180	47	47	48	48	49	50	51	51	51	51	52	53	53	53	54
185	45	45	46	46	46	48	48	48	49	49	50	50	50	51	51
190	43	43	44	44	44	46	46	46	46	47	48	48	48	48	48
195	41	41	42	42	42	43	44	44	44	44	45	45	46	46	46
200	39	39	40	40	40	42	42	42	42	42	43	43	44	44	44
210	36	36	37	37	37	38	38	38	39	39	39	40	40	40	40
220	33	33	34	34	34	35	35	35	35	36	36	36	37	37	37
230	31	31	31	31	31	32	32	33	33	33	33	33	34	34	34
240	28	29	29	29	29	30	30	30	30	30	31	31	31	31	31
250	26	27	27	27	27	28	28	28	28	28	29	29	29	29	29
λ_{L0}	37.1	36.3	35.6	35	34.3	32.1	31.6	31.1	30.6	30.2	28.4	28.1	27.4	27.1	26.5

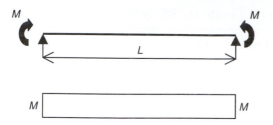

Fig. 4.21 *Standard case for buckling.*

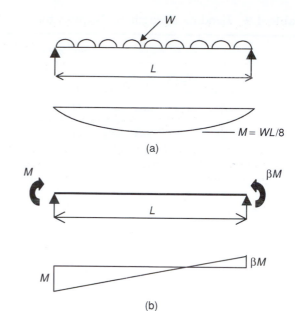

Fig. 4.22 *Load cases less susceptible to buckling.*

ling resistance moment, M_b, is obtained from the following:

For **Class 1 Plastic** or **Class 2 Compact** sections

$$M_b = p_b S_x \qquad (4.22)$$

For **Class 3 Semi-compact** sections, M_b, is given by

$$M_b = p_b Z_x \qquad (4.23)$$

or alternatively $\qquad M_b = p_b S_{x,eff} \qquad (4.24)$

For **Class 4 slender** sections

$$M_b = p_b Z_{x,eff} \qquad (4.25)$$

where
S_x plastic section modulus about the major axis
Z_x section modulus about the major axis
$S_{x,eff}$ effective plastic modulus about the major axis (see clause 3.5.6)
$Z_{x,eff}$ effective section modulus about the major axis (see clause 3.6.2)

This value of the buckling moment assumes the beam is acted on by a uniform, single curvature moment (*Fig. 4.21*), which is the most severe arrangement in terms of lateral stability. However, where the beam is acted on by variable moments (*Fig. 4.22(a)*) or unequal end moments (*Fig. 4.22(b)*), the maximum moment about the major axis, M_x, must satisfy the following conditions:

$$M_x \leqslant M_b/m_{LT} \text{ and } M_x \leqslant M_{cx} \qquad (4.26)$$

where
m_{LT} equivalent uniform moment factor for lateral torsional buckling read from *Table 18* of BS 5950, reproduced as *Table 4.10*. For the destabilizing loading condition $m_{LT} = 1$
M_{cx} moment capacity of the section about the major axis

Example 4.8 Design of a laterally unrestrained beam – rigorous method (BS 5950)

Repeat Example 4.7 using the more rigorous method.

As previously noted the 305 × 305 × 137 UC section is Class 1 plastic has a design strength, $p_y = 265$ N/mm². From steel tables (*Appendix B*), $r_y = 78.2$ mm, $D = 320.5$ mm and $T = 21.7$ mm. The slenderness ratio, λ, is

$$\lambda = \frac{L_E}{r_y} = \frac{10\,000}{78.2} = 127.9$$

$$\frac{\lambda}{x} = \frac{127.9}{320.5/21.7} = 8.7$$

From *Table 4.8*, $v = 0.678$. Since the section is Class 1 plastic, $\beta_w = 1.0$ and the equivalent slenderness ratio, λ_{LT}, is

$$\lambda_{LT} = uv\lambda \sqrt{\beta_w} = 0.9 \times 0.678 \times 127.9 \times \sqrt{1.0} = 78$$

From *Table 4.9*, $p_b = 165.4$ N/mm^2, hence

$$M_b = p_b S_x = 165.4 \times 2300 \times 10^3$$
$$= 380.4 \times 10^6 \text{ N mm} = 380.4 \text{ kN m}$$

m_{LT} from *Table 4.10* = 0.89 (approx.), by interpolation between 0.85 and 0.925

$$\frac{M_b}{m_{LT}} = \frac{380.4}{0.89} = 427.4 \text{ kN m} < M_{cx} \ (= p_y S_x = 265 \times 2300 \times 10^{-3}) = 609.5 \text{ kN m}$$

This gives, in this case, a buckling moment approximately 10% greater than in *Example 4.7*. This may enable a lighter member to be selected, but for rolled sections it may not be really worth the additional effort.

Table 4.10 Equivalent uniform moment factor m_{LT} for lateral torsional buckling (based on Table 18, BS 5950)

Segments with end moments only	β	m_{LT}
β positive	1.0	1.00
	0.9	0.96
	0.8	0.92
	0.7	0.88
	0.6	0.84
	0.5	0.80
	0.4	0.76
	0.3	0.72
	0.2	0.68
x Lateral restraint	0.1	0.64
	0.0	0.60
β negative	−0.1	0.56
	−0.2	0.52
	−0.3	0.48
	−0.4	0.46
	−0.5	0.44
	−0.6	0.44
	−0.7	0.44
	−0.8	0.44
	−0.9	0.44
	−1.0	0.44

Specific cases (no intermediate lateral restraints)

$m_{LT} = 0.850$ $m_{LT} = 0.925$

$m_{LT} = 0.925$ $m_{LT} = 0.744$

Table 4.11 Effective length, L_E, for cantilevers without intermediate constraint (based on Table 14, BS 5950)

Restraint conditions at supports	Restraint conditons at tip	Loading conditions	
		Normal	Destabilising
Restrained laterally, torsionally and against rotation on plan	1) Free	0.8L	1.4L
	2) Lateral restraint to top flange	0.7L	1.4L
	3) Torsional restraint	0.6L	0.6L
	4) Lateral and torsional restraint	0.5L	0.5L

Example 4.9 Checking for lateral instability in a cantilever steel beam (BS 5950)

Continue *Example 4.3* to determine whether the cantilever is laterally stable. Assume the load is destabilising.

From *Table 4.11*, $L_E = 1.4L = 1.4$ metres

For $610 \times 229 \times 113$ UB

$$\lambda = \frac{L_E}{r_y} = \frac{1400}{48.8} = 28.7$$

$$\frac{\lambda}{x} = \frac{28.7}{607.3/17.3} = 0.82 \Rightarrow v = 0.99 \quad (Table\ 4.8)$$

$$\lambda_{LT} = uv\lambda\sqrt{\beta_w} = 0.9 \times 0.99 \times 28.7 \times \sqrt{1.0} = 28.4$$

From *Table 4.9*, $p_b = 265$ N/mm$^2 = p_y$, and so lateral torsional buckling is not a problem with this cantilever.

4.8.11.4 Cantilever beams

For cantilevers, the effective length is given in clear diagrammatical form in Table 14 of BS 5950, part of which is reproduced as *Table 4.11*.

4.8.11.5 Summary of design procedures

The two alternative methods for checking lateral torsional buckling of beams can be summarized as follows.

Conservative method

1. Calculate the design shear force, F_v, and bending moment, M_x, at critical points along the beam.
2. Select and classify UB or UC section.
3. Check shear and bending capacity of section. If unsatisfactory return to (2).
4. Determine the effective length of the beam, L_E, using *Table 4.6*.

5. Determine S_x, r_y, D and T from steel tables (*Appendix B*).
6. Calculate the slenderness value, $\lambda = (\beta)^{0.5} L_E/r_y$.
7. Determine the bending strength, p_b, from *Table 4.7* using λ and D/T.
8. Calculate the buckling resistance moment, M_b, via equation 4.18 or 4.19.
9. Check $M_x \leq M_b$. If unsatisfactory return to (2).

Rigorous methods
1. Steps (1)–(4) as for conservative method.
2. Determine S_x, r_y, u and x from steel tables.
3. Calculate slenderness ratio, $\lambda = L_E/r_y$.
4. Determine v from *Table 4.8* using λ/x and $N = 0.5$.
5. Calculate equivalent slenderness ratio, $\lambda_{LT} = uv\lambda\sqrt{\beta_w}$.
6. Determine p_b from *Table 4.9* using λ_{LT}.
7. Calculate M_b via equations 4.22–4.25.
8. Obtain the equivalent uniform moment factor, m_{LT}, from *Table 4.10*.
9. Check $M_x \leq M_b/m_{LT}$. If unsatisfactory return to (2).

4.9 Design of compression members

4.9.1 STRUTS

Steel compression members, commonly referred to as stanchions, include struts and columns. A strut is a member subject to direct compression only. A column, on the other hand, refers to members subject to a combination of compressive load and bending. Although most columns in real structures resist compressive load and bending, the strut is a convenient starting point.

Struts (and columns) differ fundamentally in their behaviour under axial load depending on whether they are **slender** or **stocky**. Most real struts and columns can neither be regarded as slender nor stocky, but as something in between, but let us look at the behaviour of stocky and slender struts first.

Stocky struts will fail by crushing or squashing of the material. For stocky struts the 'squash load' is given by the simple formula

$$P_s = p_y A_g \qquad (4.27)$$

where
p_y design strength of steel
A_g gross cross sectional area of the section

Slender struts will fail by buckling. For elastic slender struts pinned at each end, the 'Euler

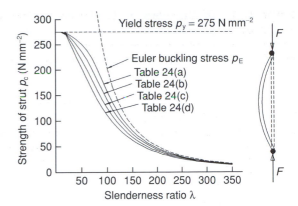

Fig. 4.23

buckling load', at which a perfect strut buckles elastically is given by

$$P_E = \frac{\pi^2 EI}{L^2} = \frac{\pi^2 EA_g r_y^2}{L^2} = \frac{\pi^2 EA_g}{\lambda^2} \qquad (4.28)$$

using $r = \sqrt{\dfrac{I}{A}}$ and $\lambda = \dfrac{L}{r}$.

If the compressive strength, p_c, which is given by

$$p_c = P_s/A_g \quad \text{(for stocky struts)} \qquad (4.29)$$

and

$$p_c = P_E/A_g \quad \text{(for slender struts)} \qquad (4.30)$$

are plotted against λ (*Fig. 4.23*), the area above the two dotted lines represents an impossible situation in respect of these struts. In this area, the strut has either buckled or squashed. Struts which fall below the dotted lines are theoretically able to withstand the applied load without either buckling or squashing. In reality, however, this tends not to be the case because of a combination of manufacturing and practical considerations. For example, struts are never completely straight, nor are subject to exactly concentric loading. During manufacture, stresses are locked into steel members which effectively reduce their load-carrying capacities. As a result of these factors, failure of a strut will not be completely due to buckling or squashing, but perhaps a combination, with partially plastic stresses appearing across the member section. These non-ideal factors or imperfections are found in practice, and laboratory tests have confirmed that in fact the failure line for real struts lies along a series of lines such as a,b,c and d in *Fig. 4.23*. These are derived from the Perry-Robertson equation that includes allowances for the various imperfections.

Whichever of the lines a–d is used depends on the shape of section and the axis of buckling. Table 23 of BS 5950, part of which is reproduced as *Table 4.12*, specifies which of the lines is appropriate for the shape of section, and Tables 24(a), (b), (c) and (d) enable values of p_c to be read off appropriate to the section used. (Tables 24(b) and (c) of BS 5950 have been reproduced as *Tables 4.13* and *4.14* respectively.) Alternatively, Appendix C of BS 5950 gives the actual Perry-Robertson equations which may be used in place of the tables if considered necessary.

Table 4.12 Strut table selection (based on Table 23, BS 5950)

Type of section	Thickness[a]	Axis of buckling	
		x–x	y–y
Hot-rolled structural hollow section		24(a)	24(a)
Rolled *I*-section		24(a)	24(b)
Rolled *H*-section	Up to 40 mm	24(b)[b]	24(c)[c]
	Over 40 mm	24(c)	24(d)

Notes. [a] For thicknesses between 40 and 50 mm the value of p_c may be taken as the average of the values for thicknesses up to 40 mm and over 40 mm.
[b] Reproduced as Table 4.13.
[c] Reproduced as Table 4.14.

Table 4.13 Compressive strength, p_c (N/mm^2) with $\lambda < 110$ for strut curve b (Table 24(b), BS 5950)

λ	Steel grade and design strength p_y (N/mm^2)														
	S275					S355					S460				
	235	245	255	265	275	315	325	335	345	355	400	410	430	440	460
15	235	245	255	265	275	315	325	335	345	355	399	409	428	438	457
20	234	243	253	263	272	310	320	330	339	349	391	401	420	429	448
25	229	239	248	258	267	304	314	323	332	342	384	393	411	421	439
30	225	234	243	253	262	298	307	316	325	335	375	384	402	411	429
35	220	229	238	247	256	291	300	309	318	327	366	374	392	400	417
40	216	224	233	241	250	284	293	301	310	318	355	364	380	388	404
42	213	222	231	239	248	281	289	298	306	314	351	359	375	383	399
44	211	220	228	237	245	278	286	294	302	310	346	354	369	377	392
46	209	218	226	234	242	275	283	291	298	306	341	349	364	371	386
48	207	215	223	231	239	271	279	287	294	302	336	343	358	365	379
50	205	213	221	229	237	267	275	283	290	298	330	337	351	358	372
52	203	210	218	226	234	264	271	278	286	293	324	331	344	351	364
54	200	208	215	223	230	260	267	274	281	288	318	325	337	344	356
56	198	205	213	220	227	256	263	269	276	283	312	318	330	336	347
58	195	202	210	217	224	252	258	265	271	278	305	311	322	328	339
60	193	200	207	214	221	247	254	260	266	272	298	304	314	320	330
62	190	197	204	210	217	243	249	255	261	266	291	296	306	311	320
64	187	194	200	207	213	238	244	249	255	261	284	289	298	302	311
66	184	191	197	203	210	233	239	244	249	255	276	281	289	294	301
68	181	188	194	200	206	228	233	239	244	249	269	273	281	285	292
70	178	185	190	196	202	223	228	233	238	242	261	265	272	276	282
72	175	181	187	193	198	218	223	227	232	236	254	257	264	267	273
74	172	178	183	189	194	213	217	222	226	230	246	249	255	258	264
76	169	175	180	185	190	208	212	216	220	223	238	241	247	250	255
78	166	171	176	181	186	203	206	210	214	217	231	234	239	241	246

Table 4.13 (cont'd)

λ	Steel grade and design strength p_y (N/mm²)														
	S275					S355					S460				
	235	245	255	265	275	315	325	335	345	355	400	410	430	440	460
80	163	168	172	177	181	197	201	204	208	211	224	226	231	233	237
82	160	164	169	173	177	192	196	199	202	205	217	219	223	225	229
84	156	161	165	169	173	187	190	193	196	199	210	212	216	218	221
86	153	157	161	165	169	182	185	188	190	193	203	205	208	210	213
88	150	154	158	161	165	177	180	182	185	187	196	198	201	203	206
90	146	150	154	157	161	172	175	177	179	181	190	192	195	196	199
92	143	147	150	153	156	167	170	172	174	176	184	185	188	189	192
94	140	143	147	150	152	162	165	167	169	171	178	179	182	183	185
96	137	140	143	146	148	158	160	162	164	165	172	173	176	177	179
98	134	137	139	142	145	153	155	157	159	160	167	168	170	171	173
100	130	133	136	138	141	149	151	152	154	155	161	162	164	165	167
102	127	130	132	135	137	145	146	148	149	151	156	157	159	160	162
104	124	127	129	131	133	141	142	144	145	146	151	152	154	155	156
106	121	124	126	128	130	137	138	139	141	142	147	148	149	150	151
108	118	121	123	125	126	133	134	135	137	138	142	143	144	145	147
110	115	118	120	121	123	129	130	131	133	134	138	139	140	141	142
112	113	115	117	118	120	125	127	128	129	130	134	134	136	136	138
114	110	112	114	115	117	122	123	124	125	126	130	130	132	132	133
116	107	109	111	112	114	119	120	121	122	122	126	126	128	128	129
118	105	106	108	109	111	115	116	117	118	119	122	123	124	124	125
120	102	104	105	107	108	112	113	114	115	116	119	119	120	121	122
122	100	101	103	104	105	109	110	111	112	112	115	116	117	117	118
124	97	99	100	101	102	106	107	108	109	109	112	112	113	114	115
126	95	96	98	99	100	103	104	105	106	106	109	109	110	111	111
128	93	94	95	96	97	101	101	102	103	103	106	106	107	107	108
130	90	92	93	94	95	98	99	99	100	101	103	103	104	105	105
135	85	86	87	88	89	92	93	93	94	94	96	97	97	98	98
140	80	81	82	83	84	86	87	87	88	88	90	90	91	91	92
145	76	77	78	78	79	81	82	82	83	83	84	85	85	86	86
150	72	72	73	74	74	76	77	77	78	78	79	80	80	80	81
155	68	69	69	70	70	72	72	73	73	73	75	75	75	76	76
160	64	65	65	66	66	68	68	69	69	69	70	71	71	71	72
165	61	62	62	62	63	64	65	65	65	65	66	67	67	67	68
170	58	58	59	59	60	61	61	61	62	62	63	63	63	64	64
175	55	55	56	56	57	58	58	58	59	59	60	60	60	60	60
180	52	53	53	53	54	55	55	55	56	56	56	57	57	57	57
185	50	50	51	51	51	52	52	53	53	53	54	54	54	54	54
190	48	48	48	48	49	50	50	50	50	50	51	51	51	51	52
195	45	46	46	46	46	47	47	48	48	48	49	49	49	49	49
200	43	44	44	44	44	45	45	45	46	46	46	46	47	47	47
210	40	40	40	40	41	41	41	41	42	42	42	42	42	43	43
220	36	37	37	37	37	38	38	38	38	38	39	39	39	39	39
230	34	34	34	34	34	35	35	35	35	35	35	36	36	36	36
240	31	31	31	31	32	32	32	32	32	32	33	33	33	33	33
250	29	29	29	29	29	30	30	30	30	30	30	30	30	30	30
260	27	27	27	27	27	27	28	28	28	28	28	28	28	28	28
270	25	25	25	25	25	26	26	26	26	26	26	26	26	26	26
280	23	23	23	23	24	24	24	24	24	24	24	24	24	24	24
290	22	22	22	22	22	22	22	22	22	22	23	23	23	23	23
300	20	20	21	21	21	21	21	21	21	21	21	21	21	21	21
310	19	19	19	19	19	20	20	20	20	20	20	20	20	20	–
320	18	18	18	18	18	18	18	19	19	19	19	19	19	19	19
330	17	17	17	17	17	17	17	17	17	18	18	18	18	18	18
340	16	16	16	16	16	16	16	16	17	17	17	17	17	17	17
350	15	15	15	15	15	16	16	16	16	16	16	16	16	16	16

Table 4.14 Compressive strength, p_c (N/mm^2) for strut curve c (Table 24(c), BS 5950)

λ	Steel grade and design strength p_y (N/mm^2)														
	S275				S355					S460					
	235	*245*	*255*	*265*	*275*	*315*	*325*	*335*	*345*	*355*	*400*	*410*	*430*	*440*	*460*
15	235	245	255	265	275	315	325	335	345	355	398	408	427	436	455
20	233	242	252	261	271	308	317	326	336	345	387	396	414	424	442
25	226	235	245	254	263	299	308	317	326	335	375	384	402	410	428
30	220	228	237	246	255	289	298	307	315	324	363	371	388	396	413
35	213	221	230	238	247	280	288	296	305	313	349	357	374	382	397
40	206	214	222	230	238	270	278	285	293	301	335	343	358	365	380
42	203	211	219	227	235	266	273	281	288	296	329	337	351	358	373
44	200	208	216	224	231	261	269	276	284	291	323	330	344	351	365
46	197	205	213	220	228	257	264	271	279	286	317	324	337	344	357
48	195	202	209	217	224	253	260	267	274	280	311	317	330	337	349
50	192	199	206	213	220	248	255	262	268	275	304	310	323	329	341
52	189	196	203	210	217	244	250	257	263	270	297	303	315	321	333
54	186	193	199	206	213	239	245	252	258	264	291	296	308	313	324
56	183	189	196	202	209	234	240	246	252	258	284	289	300	305	315
58	179	186	192	199	205	229	235	241	247	252	277	282	292	297	306
60	176	183	189	195	201	225	230	236	241	247	270	274	284	289	298
62	173	179	185	191	197	220	225	230	236	241	262	267	276	280	289
64	170	176	182	188	193	215	220	225	230	235	255	260	268	272	280
66	167	173	178	184	189	210	215	220	224	229	248	252	260	264	271
68	164	169	175	180	185	205	210	214	219	223	241	245	252	256	262
70	161	166	171	176	181	200	204	209	213	217	234	238	244	248	254
72	157	163	168	172	177	195	199	203	207	211	227	231	237	240	246
74	154	159	164	169	173	190	194	198	202	205	220	223	229	232	238
76	151	156	160	165	169	185	189	193	196	200	214	217	222	225	230
78	148	152	157	161	165	180	184	187	191	194	207	210	215	217	222
80	145	149	153	157	161	176	179	182	185	188	201	203	208	210	215
82	142	146	150	154	157	171	174	177	180	183	195	197	201	203	207
84	139	142	146	150	154	167	169	172	175	178	189	191	195	197	201
86	135	139	143	146	150	162	165	168	170	173	183	185	189	190	194
88	132	136	139	143	146	158	160	163	165	168	177	179	183	184	187
90	129	133	136	139	142	153	156	158	161	163	172	173	177	178	181
92	126	130	133	136	139	149	152	154	156	158	166	168	171	173	175
94	124	127	130	133	135	145	147	149	151	153	161	163	166	167	170
96	121	124	127	129	132	141	143	145	147	149	156	158	160	162	164
98	118	121	123	126	129	137	139	141	143	145	151	153	155	157	159
100	115	118	120	123	125	134	135	137	139	140	147	148	151	152	154
102	113	115	118	120	122	130	132	133	135	136	143	144	146	147	149
104	110	112	115	117	119	126	128	130	131	133	138	139	142	142	144
106	107	110	112	114	116	123	125	126	127	129	134	135	137	138	140
108	105	107	109	111	113	120	121	123	124	125	130	131	133	134	136
110	102	104	106	108	110	116	118	119	120	122	126	127	129	130	132
112	100	102	104	106	107	113	115	116	117	118	123	124	125	126	128
114	98	100	101	103	105	110	112	113	114	115	119	120	122	123	124
116	95	97	99	101	102	108	109	110	111	112	116	117	118	119	120
118	93	95	97	98	100	105	106	107	108	109	113	114	115	116	117
120	91	93	94	96	97	102	103	104	105	106	110	110	112	112	113

Table 4.14 (*cont'd*)

λ	*Steel grade and design strength* p_y *(N/mm²)*														
	S275					S355					S460				
	235	245	255	265	275	315	325	335	345	355	400	410	430	440	460
122	89	90	92	93	95	99	100	101	102	103	107	107	109	109	110
124	87	88	90	91	92	97	98	99	100	100	104	104	106	106	107
126	85	86	88	89	90	94	95	96	97	98	101	102	103	103	104
128	83	84	86	87	88	92	93	94	95	95	98	99	100	100	101
130	81	82	84	85	86	90	91	91	92	93	96	96	97	98	99
135	77	78	79	80	81	84	85	86	87	87	90	90	91	92	92
140	72	74	75	76	76	79	80	81	81	82	84	85	85	86	87
145	69	70	71	71	72	75	76	76	77	77	79	80	80	81	81
150	65	66	67	68	68	71	71	72	72	73	75	75	76	76	76
155	62	63	63	64	65	67	67	68	68	69	70	71	71	72	72
160	59	59	60	61	61	63	64	64	65	65	66	67	67	67	68
165	56	56	57	58	58	60	60	61	61	61	63	63	64	64	64
170	53	54	54	55	55	57	57	58	58	58	60	60	60	60	61
175	51	51	52	52	53	54	54	55	55	55	56	57	57	57	58
180	48	49	49	50	50	51	52	52	52	53	54	54	54	54	55
185	46	46	47	47	48	49	49	50	50	50	51	51	52	52	52
190	44	44	45	45	45	47	47	47	47	48	49	49	49	49	49
195	42	42	43	43	43	45	45	45	45	45	46	46	47	47	47
200	40	41	41	41	42	43	43	43	43	43	44	44	45	45	45
210	37	37	38	38	38	39	39	39	40	40	40	40	41	41	41
220	34	34	35	35	35	36	36	36	36	36	37	37	37	37	38
230	31	32	32	32	32	33	33	33	33	34	34	34	34	34	35
240	29	29	30	30	30	30	31	31	31	31	31	31	32	32	32
250	27	27	27	28	28	28	28	28	29	29	29	29	29	29	29
260	25	25	26	26	26	26	26	26	27	27	27	27	27	27	27
270	23	24	24	24	24	24	25	25	25	25	25	25	25	25	25
280	22	22	22	22	22	23	23	23	23	23	23	24	24	24	24
290	21	21	21	21	21	21	21	22	22	22	22	22	22	22	22
300	19	19	20	20	20	20	20	20	20	20	21	21	21	21	21
310	18	18	18	19	19	19	19	19	19	19	19	19	19	19	20
320	17	17	17	17	18	18	18	18	18	18	18	18	18	18	18
330	16	16	16	16	17	17	17	17	17	17	17	17	17	17	17
340	15	15	15	16	16	16	16	16	16	16	16	16	16	16	16
350	15	15	15	15	15	15	15	15	15	15	15	15	15	15	15

4.9.2 EFFECTIVE LENGTH

As mentioned in *section 4.9.1*, the compressive strength of struts is primarily related to their slenderness ratio. The slenderness ratio, λ, is given by

$$\lambda = \frac{L_E}{r} \qquad (4.31)$$

where

L_E effective length of the member
r radius of gyration obtained from steel tables.

The concept of effective length was discussed in *section 4.8.11.1*, in the context of lateral torsional buckling, and is similarly applicable to the design of struts and columns. The effective length is simply a function of the actual length of the member and the restraint at the member ends.

The formulae in Appendix C of BS 5950 and the graph in *Fig. 4.23* relate to standard restraint conditions in which each end is pinned. In reality each end of the strut may be free, pinned, partially fixed,

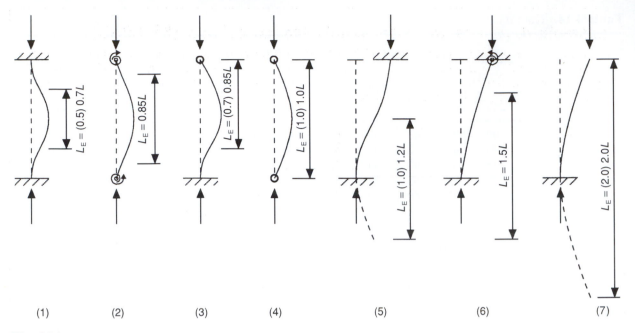

(1) (2) (3) (4) (5) (6) (7)

Fig. 4.24

Table 4.15   Nominal effective length, L_E, for a compression member (Table 22, BS 5950)

a) non-sway mode

Restraint (in the plane under consideration) by other parts of the structure		L_E
Effectively held in position at both ends	Effectively restrained in direction at both ends (1)	0.7L
	Partially restrained in direction at both ends (2)	0.85L
	Restrained in direction at one end (3)	0.85L
	Not restrained in direction at either end (4)	1.0L

b) sway mode

One end	Other end		L_E
Effectively held in position and restrained in direction	Not held in position	Effectively restrained in direction (5)	1.2L
		Partially restrained in direction (6)	1.5L
		Not restrained in direction (7)	2.0L

or fully fixed (rotationally). Also, whether or not the top of the strut is allowed to move laterally with respect to the bottom end is important. *Figure 4.24* summarises these restraints, and Table 22 of BS 5950, reproduced above as *Table 4.15* stipulates conservative assumptions of effective length L_E from which the slenderness λ can be calculated. Note that the design effective lengths are greater than the theoretical values where one or both ends of the member are partially or wholly restrained. This is because, in practice, it is difficult if not impossible to guarantee that some rotation of the member will not take place. Furthermore, the effective lengths are always less than the actual length of the compression member except when the structure is unbraced.

Example 4.10 Design of an axially loaded column (BS 5950)

A proposed 5 metre long internal column in a 'rigid' jointed steel structure is to be loaded concentrically with 1000 kN dead and 1000 kN imposed load (*Fig. 4.25*). Assuming that fixity at the top and bottom of the column gives effective rotational restraints, design column sections assuming the structure will be (a) braced and (b) unbraced.

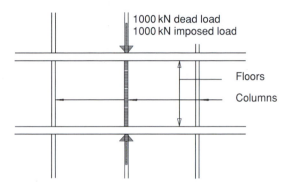

1000 kN dead load
1000 kN imposed load

Floors

Columns

Fig. 4.25

BRACED COLUMN

Design axial loading
Factored loading, $F_c = (1.4 \times 1000) + (1.6 \times 1000) = 3000$ kN

Effective length
For the braced case the column is assumed to be effectively held in position at both ends, and restrained in direction at both ends. It will buckle about the weak (y–y) axis. From *Table 4.15* therefore, the effective length, L_E, is

$$L_E = 0.7L = 0.7 \times 5 = 3.5 \text{ m}$$

Section selection
This aspect of column design can only really be done by trial and error.
Initial trial. Try 254 × 254 × 107 UC:

$$p_y = 265 \text{ N/mm}^2 \quad r_y = 65.7 \text{ mm} \quad A_g = 13\,700 \text{ mm}^2 \quad b/T = 6.3 \quad d/t = 15.4$$

$$\lambda = L_E/r_y = 3500/65.7 = 53$$

From *Table 4.12*, use Table 24(c) of BS 5950 (i.e. *Table 4.14*), from which $p_c = 208$ N/mm². Since UC section is not slender since $b/T < 15\varepsilon = 15 \times (275/265)^{0.5} = 15.28$ and $d/t < 40\varepsilon = 40.74$ (*Table 4.4*). From clause 4.7.4 of BS 5950, compression resistance of column, P_c, is

$$P_c = A_g p_c = 13\,700 \times 208/10^3 = 2850 \text{ kN} < 3000 \text{ kN} \quad \text{Not OK}$$

Second trial. Try 305 × 305 × 118 UC:

$$p_y = 265 \text{ N/mm}^2 \quad r_y = 77.5 \text{ mm} \quad A_g = 15\,000 \text{ mm}^2 \quad b/T = 8.20 \quad d/t = 20.7$$

$$\lambda = L_E/r_y = 3500/77.5 = 45$$

Then from Table 24(c) of BS 5950 (i.e. *Table 4.14*), $p_c = 222$ N/mm². Since UC section is not slender

$$P_c = A_g p_c = 15\,000 \times 222/10^3 = 3330 \text{ kN} > 3000 \text{ kN} \quad \text{OK}$$

UNBRACED COLUMN

For the unbraced case, $L_E = 1.2L = 6.0$ metres from *Table 4.15*, and the most economic member would appear to be $305 \times 305 \times 158$ UC:

$$p_y = 265 \text{ N/mm}^2 \quad r_y = 78.9 \text{ mm} \quad A_g = 20\,100 \text{ mm}^2 \quad b/T = 6.21 \quad d/t = 15.7$$

$$\lambda = L_E/r_y = 6000/78.9 = 76$$

Then from *Table 4.14*, $p_c = 165$ N/mm^2. Since section is not slender

$$P_c = A_g p_c = 20\,100 \times 165/10^3 = 3317 \text{ kN} > 3000 \text{ kN} \quad \text{OK}$$

Hence, it can immediately be seen that for a given axial load, a bigger steel section will be required if the column is unbraced.

4.9.3 COLUMNS WITH BENDING MOMENTS

As noted earlier, most columns in steel structures are subject to both axial load and bending. According to clause 4.8.3.1 of BS 5950, the cross-sectional capacity of such members should be checked (for local yielding or buckling) at the points of greatest bending moment and axial load, which usually occurs at the member ends. In addition, the buckling resistance of the member as a whole should be checked.

4.9.3.1 Cross-section capacity check

The purpose of this check is to ensure that nowhere across the section does the steel stress exceed yield. Generally, except for class 4 slender cross-sections, clause 4.8.3.2 states that the following relationship should be satisfied:

$$\frac{F_c}{A_g p_y} + \frac{M_x}{M_{cx}} + \frac{M_y}{M_{cy}} \leq 1 \quad (4.32)$$

where
F_c axial compression load
A_g gross cross-sectional area of section
p_y design strength of steel
M_x applied major axis moment
M_{cx} moment capacity about the major axis in the absence of axial load
M_y applied minor axis moment
M_{cy} moment capacity about the minor axis in the absence of axial load

When the section is slender the expression in clause 4.8.3.2 (c) should be used. Note that paragraph (b) of this clause gives an alternative expression for calculating the (local) capacity of compression members of class 1 plastic or class 2 compact cross-section, which yields a more exact estimate of member strength.

4.9.3.2 Buckling resistance check

Buckling due to imposed axial load, lateral torsional buckling due to imposed moment, or a combination of buckling and lateral torsional buckling are additional possible modes of failure in most practical columns in steel structures.

Clause 4.8.3.3.1 of BS 5950 gives a simplified approach for calculating the buckling resistance of columns which involves checking that the following relationships are both satisfied:

$$\frac{F_c}{P_c} + \frac{m_x M_x}{p_y Z_x} + \frac{m_y M_y}{p_y Z_y} \leq 1 \quad (4.33)$$

$$\frac{F_c}{P_{cy}} + \frac{m_{LT} M_{LT}}{M_b} + \frac{m_y M_y}{p_y Z_y} \leq 1 \quad (4.34)$$

where
P_c smaller of P_{cx} and P_{cy}
P_{cx} compression resistance of member considering buckling about the major axis
P_{cy} compression resistance of member considering buckling about the minor axis
m_x equivalent uniform moment factor for major axis flexural buckling obtained from Table 26 of BS 5950, reproduced as *Table 4.16*
m_y equivalent uniform moment factor for minor axis flexural buckling obtained from Table 26 of BS 5950, reproduced as *Table 4.16*
m_{LT} equivalent uniform moment factor for lateral torsional buckling obtained from Table 18 of BS 5950, reproduced as *Table 4.10*
M_{LT} maximum major axis moment in the segment length L_x governing P_{cx}
M_b buckling resistance moment
Z_x section modulus about the major axis
Z_y section modulus about the minor axis and the other symbols are as above.

Table 4.16 Equivalent uniform moment factor m for flexural buckling (Table 26, BS 5950)

Segments with end moments only		β	m
β positive $\qquad$ β negative		1.0	1.00
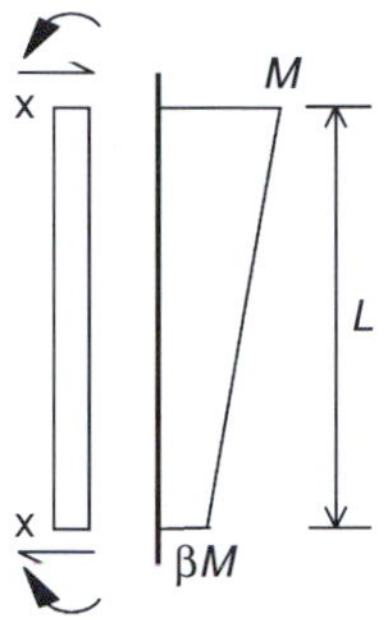	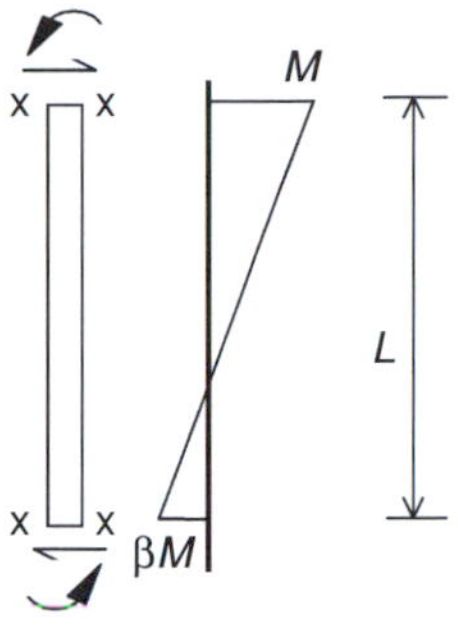	0.9	0.96
		0.8	0.92
		0.7	0.88
		0.6	0.84
		0.5	0.80
		0.4	0.76
		0.3	0.72
		0.2	0.68
		0.1	0.64
		0.0	0.60
		−0.1	0.58
		−0.2	0.56
		−0.3	0.54
		−0.4	0.52
		−0.5	0.50
		−0.6	0.48
		−0.7	0.46
		−0.8	0.44
		−0.9	0.42
		−1.0	0.40

Segments between intermediate lateral restraints

Specific cases

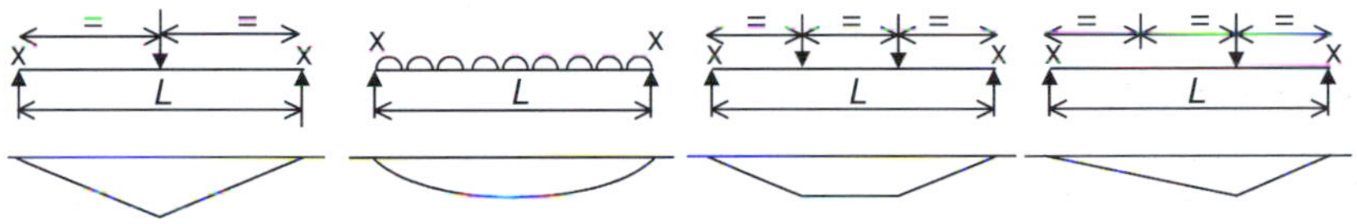

$m = 0.90$	$m = 0.95$	$m = 0.95$	$m = 0.80$

Example 4.11 Column resisting an axial load and bending (BS 5950)

Select a suitable column section in grade S275 steel to support a factored axial concentric load of 2000 kN and factored bending moments of 100 kN m about the major axis, and 20 kN m about the minor axis (*Fig. 4.26*), both applied at each end. The column is 10 metres long and is fully fixed against rotation at top and bottom, and the floors it supports are braced against sway.

INITIAL SECTION SELECTION
$305 \times 305 \times 118$ UC:

$p_y = 265$ N/mm^2 plastic $\qquad$ $S_x = 1950$ cm^3 $\qquad$ $Z_x = 1760$ cm^3
$A_g = 150$ cm^2 $\qquad\qquad\quad$ $S_y = 892$ cm^3 $\qquad$ $Z_y = 587$ cm^3
$r_y = 7.75$ cm $\qquad\qquad\quad$ $x = D/T = 314.5/18.7$ $\qquad$ $u = 0.9$
$\qquad\qquad\qquad\qquad\qquad\qquad = 16.8$

Note: In this case the classification procedure is slightly different in respect of web classification. Initially, it is conservatively assumed that the web is wholly in compression, the neutral axis lying in the flange. From Table 11 of BS 5950 (*Table 4.4*), for the member to be plastic, compact or semi-compact $d/t \leqslant 40\varepsilon$.

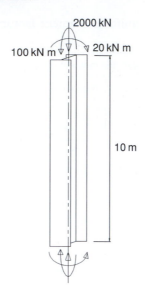

Fig. 4.26

$$M_{cx} = p_y S_x = 265 \times 1\,950 \times 10^{-3} = 516.75 \text{ kN m}$$

$$M_{cy} = p_y S_y = 265 \times 892 \times 10^{-3} = 236.38 \text{ kN m}$$

CROSS-SECTION CAPACITY CHECK

Substituting into $\dfrac{F_c}{A_g p_y} + \dfrac{M_x}{M_{cx}} + \dfrac{M_y}{M_{cy}}$ gives

$$\frac{2000 \times 10^3}{150 \times 10^2 \times 265} + \frac{100}{516.75} + \frac{20}{236.38} = 0.503 + 0.193 + 0.085 = 0.781 < 1 \quad \text{OK}$$

BUCKLING RESISTANCE CHECK

Bi–axial bending

From *Table 4.15*, effective length $L_E = 0.7L = 7$ m

$$\lambda = L_E/r_y = 7000/77.5 = 90$$

From *Table 4.12*, for buckling about the x–x axis and y–y axis use, respectively, Table 24(b) of BS 5950 (*Table 4.13*) from which $p_{cx} = 157$ N/mm^2, and Table 24(c) of BS 5950 (*Table 4.14*) from which $p_{cy} = 139$ N/mm^2. Then

$$P_{cx} = A_g p_{cx} = 150 \times 10^2 \times 157 \times 10^{-3} = 2355 \text{ kN}$$

$$P_{cy} = A_g p_{cy} = 150 \times 10^2 \times 139 \times 10^{-3} = 2085 \text{ kN} = P_c$$

Ratio of end moments about both x–x and y–y axes, $\beta = 1$. Hence from *Table 4.16*, $m_x = m_y = 1$

Substituting into $\dfrac{F_c}{P_c} + \dfrac{m_x M_x}{p_y Z_x} + \dfrac{m_y M_y}{p_y Z_y}$ gives

$$\frac{2000}{2085} + \frac{1 \times 100}{265 \times 1760 \times 10^{-3}} + \frac{1 \times 20}{265 \times 587 \times 10^{-3}} = 0.96 + 0.21 + 0.13 = 1.30 > 1 \quad \text{Not OK}$$

Buckling

$$\left(\beta_w\right)^{0.5} \frac{L_E}{r_y} = \left(1.0\right)^{0.5} \frac{7000}{77.5} = 90 \quad \text{and} \quad D/T = 16.8$$

From *Table 4.7*, $p_b = 196$ N/mm^2

$$M_b = p_b S_x = 196 \times 1950 \times 10^{-3} = 382.2 \text{ kN m}$$

Ratio of end moments about both major axis, $\beta = 1$. Hence from *Table 4.10*, $m_{LT} = 1$

Substituting into $\dfrac{F_c}{P_{cy}} + \dfrac{m_{LT}M_{LT}}{M_b} + \dfrac{m_y M_y}{p_y Z_y}$ gives

$$\frac{2000}{2085} + \frac{1 \times 100}{382.2} + \frac{1 \times 20}{265 \times 587 \times 10^{-3}} = 0.96 + 0.26 + 0.13 = 1.35 > 1 \quad \text{Not OK}$$

Hence, a bigger section should be selected.

SECOND SECTION SELECTION
Try 356 × 368 × 177 UC:

$$p_y = 265 \text{ N/mm}^2, \text{ plastic} \qquad S_x = 3460 \text{ cm}^3 \qquad Z_x = 3100 \text{ cm}^3$$
$$A_g = 226 \text{ cm}^2 \qquad\qquad S_y = 1670 \text{ cm}^3 \qquad Z_y = 1100 \text{ cm}^3$$
$$r_y = 9.52 \text{ cm}^3 \qquad\qquad x = D/T = 368.3/23.8 \qquad u = 0.9$$
$$= 15.5$$

$$M_{cx} = p_y S_x = 265 \times 3460 \times 10^{-3} = 916.9 \text{ kN m}$$

$$M_{cy} = p_y S_y = 265 \times 1670 \times 10^{-3} = 442.55 \text{ kN m}$$

CROSS-SECTION CAPACITY CHECK
Again using $\dfrac{F_c}{A_g p_y} + \dfrac{M_x}{M_{cx}} + \dfrac{M_y}{M_{cy}}$ gives

$$\frac{2000 \times 10^3}{226 \times 10^2 \times 265} + \frac{100}{916.9} + \frac{20}{442.55} = 0.33 + 0.11 + 0.05 = 0.49 < 1 \quad \text{OK}$$

BUCKLING RESISTANCE CHECK
Bi-axial bending
From *Table 4.15*, effective length $L_E = 0.7L = 7$ m

$$\lambda = L_E/r_y = 7000/95.2 = 73.5$$

From *Table 4.12*, for buckling about the x–x axis and y–y axis use, respectively, Table 24(b) of BS 5950 (*Table 4.13*) from which $p_{cx} = 190$ N/mm^2, and Table 24(c) of BS 5950 (*Table 4.14*) from which $p_{cy} = 169$ N/mm^2. Then

$$P_{cx} = A_g p_{cx} = 226 \times 10^2 \times 190 \times 10^{-3} = 4294 \text{ kN}$$

$$P_{cy} = A_g p_{cy} = 226 \times 10^2 \times 169 \times 10^{-3} = 3819.4 \text{ kN} = P_c$$

Ratio of end moments about both x–x and y–y axes, $\beta = 1$. Hence from *Table 4.16*, $m_x = m_y = 1$

Substituting into $\dfrac{F_c}{P_c} + \dfrac{m_x M_x}{p_y Z_x} + \dfrac{m_y M_y}{p_y Z_y}$ gives

$$\frac{2000}{3819.4} + \frac{1 \times 100}{265 \times 3100 \times 10^{-3}} + \frac{1 \times 20}{265 \times 1100 \times 10^{-3}} = 0.52 + 0.12 + 0.07 = 0.71 < 1 \quad \text{OK}$$

Buckling

$$(\beta_w)^{0.5}\frac{L_E}{r_y} = (1.0)^{0.5}\frac{7000}{95.2} = 73.5 \quad \text{and} \quad D/T = 15.5$$

From *Table 4.7*, $p_b = 220$ N/mm^2

$$M_b = p_b S_x = 220 \times 3460 \times 10^{-3} = 761.2 \text{ kN m}$$

Ratio of end moments about both major axis, $\beta = 1$. Hence from *Table 4.10*, $m_{LT} = 1$

Substituting into $\dfrac{F_c}{P_{cy}} + \dfrac{m_{LT}M_{LT}}{M_b} + \dfrac{m_y M_y}{p_y Z_y} \leqslant 1$ gives

$$\frac{2000}{3819.4} + \frac{1 \times 100}{761.2} + \frac{1 \times 20}{265 \times 1100 \times 10^{-3}} = 0.52 + 0.13 + 0.07 = 0.72 < 1 \quad \text{OK}$$

Hence a $356 \times 368 \times 177$ UC section is satisfactory.

Clause 4.8.3.3.2 of BS 5950 gives a more exact approach, but as in practice most designers tend to use the simplified approach, the more exact method is not discussed here.

4.9.4 COLUMN DESIGN IN 'SIMPLE' CONSTRUCTION

At first sight it would appear that columns in so-called 'simple construction' are not subject to moments, as the beams are all joined at connections which allow no moment to develop. In fact, in most cases there is a bending moment due to the eccentricity of the shear load from the beam. This is summarised in Clause 4.7.7 of BS 5950 and illustrated in *Fig. 4.27*. Note that where a beam sits on a column cap plate, for example at A (*Fig. 4.27(c)*), it can be assumed that the reaction from the beam acts at the face of the column. However, where the beam is connected to a column by means of a 'simple' connection, e.g. using web cleats, the reaction from the beam can be assumed to act 100 mm from the column (web or flange) face as illustrated in *Fig. 4.27(b)*.

When a roof truss is supported on a column cap plate (*Fig. 4.28*), and the connection is unable to develop significant moments, it can be assumed that the load from the truss is transmitted concentrically to the column.

In simple structures it is not necessary to check the cross-sectional capacity of columns as this

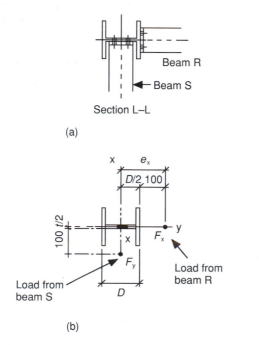

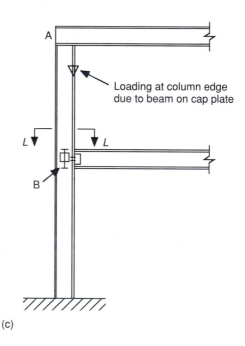

Fig. 4.27 *Load eccentricity for columns in simple construction.*

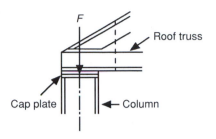

Fig. 4.28 *Column supporting a roof truss.*

cross-sectional area of the section and p_c is the compressive strength, see *section 4.9.1*

M_x nominal major axis moment
M_y nominal minor axis moment
M_{bs} buckling resistance moment for 'simple' columns
p_y design strength of steel
Z_y section modulus about the minor axis.

should not be critical. However, it will still be necessary to check for buckling, which involves satisfying the following relationship:

$$\frac{F_c}{P_c} + \frac{M_x}{M_{bs}} + \frac{M_y}{p_y Z_y} \leq 1 \qquad (4.35)$$

where
F_c axial compressive load
P_c $= A_g p_c$ – for all classes except class 4 (clause 4.7.4 of BS 5950) – in which A_g is the gross

Note that this expression is similar to that used for checking the buckling resistance of columns in continuous structures (equation 4.34), but with all equivalent moment factors m taken as 1.0. For *I*- and *H*-sections $M_{bs} = M_b$ determined as discussed in *section 4.8.11.3* but using the equivalent slenderness of the column, λ_{LT}, given by

$$\lambda_{LT} = 0.5L/r_y \qquad (4.36)$$

where
L is the distance between levels at which the column is laterally restrained in both directions
r_y is the radius of gyration about the minor axis.

Example 4.12 Design of a steel column in 'simple' construction (BS 5950)

Select a suitable short column section in grade S275 steel to support the ultimate loads from beams A and B shown in *Fig. 4.29*. Assume the column is 7 m long and is effectively held in position at both ends but only restrained in direction at the bottom.

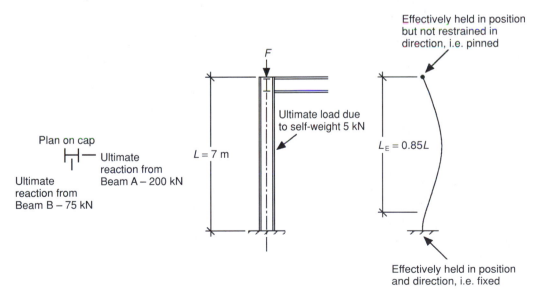

Fig. 4.29

SECTION SELECTION
This can only be done by trial and error. Therefore, try a 203 × 203 × 52 UC section.

DESIGN LOADING AND MOMENTS
Ultimate reaction from beam A, $R_A = 200$ kN; ultimate reaction from beam B, $R_B = 75$ kN; assume self-weight of column = 5 kN. Ultimate axial load, F, is

$$F = R_A + R_B + \text{self-weight of column}$$
$$= 200 + 75 + 5 = 280 \text{ kN}$$

Load eccentricity for beam A,

$$e_x = D/2 + 100 = 206.2/2 + 100 = 203.1 \text{ mm}$$

Load eccentricity for beam B,

$$e_y = t/2 + 100 = 8/2 + 100 = 104 \text{ mm}$$

Moment due to beam A,

$$M_x = R_A e_x = 200 \times 10^3 \times 203.1 = 40.62 \times 10^6 \text{ N mm}$$

Moment due to beam B,

$$M_y = R_B e_y = 75 \times 10^3 \times 104 = 7.8 \times 10^6 \text{ N mm}$$

EFFECTIVE LENGTH
From *Table 4.15*, effective length coefficient = 0.85. Hence, effective length is

$$L_E = 0.85L = 0.85 \times 7000 = 5950 \text{ mm}$$

BENDING STRENGTH
From *Table 4.12*, relevant compressive strength values for buckling about the x–x axis are obtained from Table 24(b) (*Table 4.13*) and from Table 24(c) (*Table 4.14*) for bending about the y–y axis.

$$\lambda_x = L_E/r_x = 5950/89 = 66.8$$

From *Table 4.13*, $p_c = 208$ N/mm².

$$\lambda_y = L_E/r_y = 5950/51.6 = 115.3$$

From *Table 4.14*, $p_c = 103$ N/mm². Hence critical compressive strength of column is 103 N/mm².

BUCKLING RESISTANCE

$$\lambda_{LT} = 0.5L/r_y = 0.5 \times 7000/51.6 = 67.8$$

From *Table 4.9*, $p_b = 193$ N/mm². Buckling resistance moment capacity of column, M_{bs}, is given by

$$M_{bs} = M_b = p_b S_x = 193 \times 568 \times 10^3 = 109.6 \times 10^6 \text{ N mm}$$

Hence for stability,

$$\frac{F_c}{P_c} + \frac{M_x}{M_{bs}} + \frac{M_y}{p_y Z_y} \leq 1$$

$$\frac{280 \times 10^3}{66.4 \times 10^2 \times 103} + \frac{40.6 \times 10^6}{109.6 \times 10^6} + \frac{7.8 \times 10^6}{275 \times 174 \times 10^3} = 0.41 + 0.37 + 0.16 = 0.94 < 1$$

Therefore, the 203 × 203 × 52 UC section is suitable.

4.9.5 SUMMARY OF DESIGN PROCEDURES FOR COMPRESSION MEMBERS

Axially loaded
1. Calculate the design axial force, F_c.
2. Select universal section and check that it is non-slender.
3. Determine r_x, r_y and A_g from steel tables.
4. Determine the major and minor axes effective lengths, L_{EX} and L_{EY}, using *Table 4.15*.
5. Calculate the major and minor axes slenderness ratios, λ_{EX} $(= L_{EX}/r_x)$ and λ_{EY} $(= L_{EY}/r_y)$.
6. Select appropriate strut tables from *Table 4.12*.
7. Determine the critical compressive strength of the section, p_c, using *Table 4.13, 4.14* or similar.
8. Calculate the compression resistance of the section, $P_c = A_g p_c$.
9. Check $F_c \leqslant P_c$. If unsatisfactory return to 2.

Axial load and bending
1. Calculate the design axial force, F_c, and major and minor axes bending moments, M_x and M_y.
2. Select universal section and check that it is non-slender.
3. Calculate the major and minor axes moment capacities of the section, M_{cx} and M_{cy}. If either $M_x > M_{cx}$ or $M_y > M_{cy}$ return to 2.
4. Obtain A_g from steel tables and check the cross-section capacity of the section via equation 4.32. If unsatisfactory return to 2.
5. Determine the major and minor axes effective lengths, L_{EX} and L_{EY}, using *Table 4.15*.
6. Calculate the major and minor axes slenderness ratios, λ_{EX} $(= L_{EX}/r_x)$ and λ_{EY} $(= L_{EY}/r_y)$.
7. Select appropriate strut tables from *Table 4.12*.
8. Determine the major and minor axes compressive strengths, p_{cx} and p_{cy}, using *Table 4.13, 4.14* or similar.
9. Calculate the major and minor axes compressive resistances, P_{cx} $(= p_{cx}A_g)$ and P_{cy} $(= p_{cy}A_g)$.
10. Calculate the buckling resistance of the section, M_b.
11. Determine the equivalent uniform moment factors for major and minor axes flexural buckling, m_x and m_y, using *Table 4.16*.
12. Check the buckling resistance of the section using equations (4.33) and (4.34). If unsatisfactory return to 2.

Compression members in simple construction
1. Calculate the design axial force, F_c, and major and minor axes bending moments, M_x and M_y.
2. Select universal section and check that it is non-slender.

3. Determine the major and minor axes effective lengths, L_{EX} and L_{EY}, using *Table 4.15*.
4. Calculate the major and minor axes slenderness ratios, λ_{EX} $(= L_{EX}/r_x)$ and λ_{EY} $(= L_{EY}/r_y)$.
6. Select appropriate strut tables from *Table 4.12*.
7. Determine the critical compressive strength of the section, p_c, using *Table 4.13, 4.14* or similar.
8. Calculate the compression resistance of the section, $P_c = A_g p_c$. If $P_c < F_c$ return to 2.
9. Calculate the effective slenderness ratio, $\lambda_{LT} = 0.5L/r_y$.
10. Calculate the buckling moment resistance of the section, $M_{bs} = M_b = p_b S_x$.
11. Check the buckling resistance of the section using equation (4.35). If unsatisfactory return to (2).

4.9.6 DESIGN OF CASED COLUMNS

As discussed in *section 4.5*, steel columns are sometimes cased in concrete for fire protection. However, the concrete also increases the strength of the section, a fact which can be used to advantage in design provided that the conditions stated in Clause 4.14.1 of BS 5950 are met. Some of these conditions are illustrated in *Fig. 4.30*.

BS 5950 gives guidance on the design of UC sections encased in concrete for the following loading conditions which are discussed below.

(i) axially loaded columns
(ii) columns subject to axial load and bending.

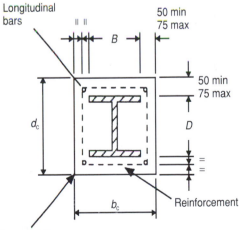

Characteristic strength of concrete $\geqslant 20$ N/mm²

Reinforcement: steel fabric type D98 (BS 4483) or $\geqslant 5$ mm diameter longitudinal bars and links at a maximum spacing of 200 mm

Fig. 4.30 *Cased UC section.*

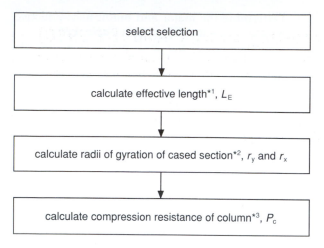

Fig. 4.31 *Design procedure for axially loaded cased columns.*

4.9.6.1 Axially loaded columns
The design procedure for this case is shown in *Fig. 4.31*.

Notes to Fig. 4.31
1 The effective length, L_E, is taken as the lesser of:
   (i) $40b_c$
   (ii) $\dfrac{100b_c^2}{d_c}$
   (iii) $250r$

   where
   b_c is the minimum width of solid casing within the depth of the steel section (*Fig. 4.30*)
   d_c is the minimum depth of solid casing within the width of the steel section
   r is the minimum radius of gyration of the uncased section i.e. r_y.
2 The radius of gyration of the cased section about the y–y axis, r_y, is taken as $0.2b_c$ but not more than $0.2(B + 150)$ mm and not less than that of the steel section alone.

   The radius of gyration of the cased section about the x–x axis, r_x, taken as the radius of gyration of the uncased steel section.
3 The compression resistance of the cased section, P_c, is given by

$$P_c = \left(A_g + \frac{0.45f_{cu}A_c}{p_y} \right) p_c \qquad (4.37)$$

However, this should not be greater than the short strut capacity of the section, P_{cs}, which is given by:

$$P_{cs} = \left(A_g + \frac{0.25f_{cu}A_c}{p_y} \right) p_y \qquad (4.38)$$

where
A_c is the gross sectional area of the concrete but neglecting any casing in excess of 75 mm from the overall dimensions of the UC section or any applied finish
A_g is the gross sectional area of the UC section
f_{cu} is the characteristic strength of the concrete which should not be greater than 40 N/mm²
p_c is the compressive strength of the UC section determined as discussed for uncased columns (*section 4.9.1*), using r_x and r_y for the cased section (see note 2) and taking $p_y \leqslant 355$ N/mm²
p_y design strength of the UC section which should not exceed 355 N/mm².

4.9.6.2 Cased columns subject to axial load and moment
The design procedure here is similar to that when the column is axially loaded but also involves checking the members cross-section capacity and buckling resistance using the following relationships:

1. *Cross-section capacity check*

$$\frac{F_c}{P_{cs}} + \frac{M_x}{M_{cx}} + \frac{M_y}{M_{cy}} \leqslant 1 \qquad (4.39)$$

where
F_c axial compression load
P_{cs} short strut capacity (equation 4.38)
M_x applied moment about major axis
M_{cx} major axis moment capacity of steel section
M_y applied moment about minor axis
M_{cy} minor axis moment capacity of steel section

2. *Buckling resistance check*

Major axis $\dfrac{F_c}{P_c} + \dfrac{m_x M_x}{p_y Z_x} + \dfrac{m_y M_y}{p_y Z_y} \leqslant 1 \qquad (4.40)$

Minor axis $\dfrac{F_c}{P_{cy}} + \dfrac{m_{LT} M_{LT}}{M_b} + \dfrac{m_y M_y}{p_y Z_y} \leqslant 1 \ (4.41)$

where
F_c maximum compressive axial force
P_c smaller of P_{cx} and P_{cy} (equation 4.37)
P_{cx} compression resistance of member considering buckling about the major axis

Example 4.13 Cased steel column resisting an axial load (BS 5950)

Calculate the compression resistance of a $305 \times 305 \times 118$ kg/m UC column if it is encased in concrete of compressive strength 20 N/mm² in the manner shown below. Assume that the effective length of the column about both axes is 3.5 m.

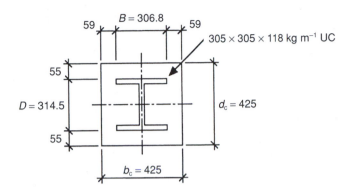

PROPERTIES OF UC SECTION

Area of UC section (A_g) = 15 000 mm²   (*Appendix B*)
Radius of gyration (r_x) = 136 mm
Radius of gyration (r_y) = 77.5 mm
Design strength (p_y) = 265 N/mm⁻² (since $T = 18.7$ mm)
Effective length (L_E) = 3.5 m

EFFECTIVE LENGTH

Check that the effective length of column (= 3500 mm) does not exceed the least of:

(i) $40b_c = 40 \times 425 = 17\,000$ mm

(ii) $\dfrac{100b_c^2}{d_c} = \dfrac{100 \times 425^2}{425} = 42\,500$ mm

(iii) $250r_y = 250 \times 77.5 = 19\,375$ mm   OK

RADII OF GYRATION FOR THE CASED SECTION

For the cased section r_x is the same as for UC section = 136 mm
For the cased section $r_y = 0.2b_c = 0.2 \times 425 = 85$ mm $\not> 0.2(B + 150) = 0.2(306.8 + 150) = 91.36$ mm
but not less than that for the uncased section (= 77.5 mm)
Hence $r_y = 85$ mm and $r_x = 136$ mm

COMPRESSION RESISTANCE

Slenderness ratio

$$\lambda_x = \frac{L_E}{r_x} = \frac{3500}{136} = 25.7$$

$$\lambda_y = \frac{L_E}{r_y} = \frac{3500}{85} = 41.2$$

Compressive strength

From *Table 4.12*, relevant compressive strength values for buckling about the x–x axis are obtained from Table 24(b) of BS 5950 (*Table 4.13*) and from Table 24(c) of BS 5950 (*Table 4.14*) for bending about the y–y axis.

For $\lambda_x = 25.7$ and $p_y = 265$ N/mm^2 compressive strength, $p_c = 257$ N/mm^2 (*Table 4.13*). For $\lambda_y = 41.2$ and $p_y = 265$ N/mm^2 compressive strength, $p_c = 228$ N/mm^2 (*Table 4.14*). Hence, p_c is equal to 228 N/mm^2.

Compression resistance

$$A_g = 15\,000 \text{ mm}^2$$

$$A_c = d_c b_c = 425 \times 425 = 180\,625 \text{ mm}^2$$

$$p_y = 265 \text{ N/mm}^2 \quad (\text{since } T = 18.7 \text{ mm})$$

$$p_c = 228 \text{ N/mm}^2$$

$$f_{cu} = 20 \text{ N/mm}^2$$

Compression resistance of encased column, P_c, is given by

$$P_c = \left(A_g + \frac{0.45 f_{cu} A_c}{p_y} \right) p_c$$

$$P_c = \left(15\,000 + \frac{0.45 \times 20 \times 180\,625}{265} \right) 228 = 4.81 \times 10^6 \text{ N} = 4810 \text{ kN}$$

which should not be greater than the short strut capacity, P_{cs}, given by

$$P_{cs} = \left(A_g + \frac{0.25 f_{cu} A_c}{p_y} \right) p_y$$

$$P_{cs} = \left(15\,000 + \frac{0.25 \times 20 \times 180\,625}{265} \right) 265 = 4.878 \times 10^6 \text{ N} = 4878 \text{ kN} \quad \text{OK}$$

Hence the compression resistance of the encased column is 4.81×10^6 N. Comparing this with the compression resistance of the uncased column (*Example 4.10*) shows that the load capacity of the column has been increased from 3357 kN to 4810 kN, which represents an increase of approximately 45%.

P_{cy} compression resistance of member considering buckling about the minor axis

m_x, m_y equivalent uniform moment factors for major axis and minor axis buckling respectively obtained from Table 26 of BS 5950, reproduced as *Table 4.16*

m_{LT} equivalent uniform moment factor for lateral torsional buckling obtained from Table 18 of BS 5950, reproduced as *Table 4.10*

M_b buckling resistance moment of the cased column $= S_x p_b \leqslant 1.5 M_b$ for the uncased section. To determine p_b, r_y should be taken as the greater of r_y of the uncased section or $0.2(B + 100)$ mm (*Fig. 4.30*).

M_x, M_y maximum moment about the major and minor axes respectively

M_{LT} maximum major axis moment in the segment L governing M_b

Z_x, Z_y section modulus about the major and minor axes respectively.

Example 4.14 Cased steel column resisting an axial load and bending (BS 5950)

In Example 4.11 it was found that a 305 × 305 × 118 kg/m UC column was incapable of resisting the design load and moments below:

Design axial load = 2000 kN
Design moment about x–x axis = 100 kN m
Design moment about y–y axis = 20 kN m

Assuming that the same column is now encased in concrete as show below, determine its suitability. The effective length of the column about both axes is 7 m.

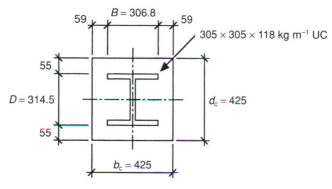

PROPERTIES OF UC SECTION

Area of UC section, A_g = 15 000 mm^2
Radius of gyration about x–x axis, r_x = 136 mm
Radius of gyration about y–y axis, r_y = 77.5 mm
Elastic modulus about x–x axis, Z_x = 1760 × 10^3 mm^3
Elastic modulus about y–y axis, Z_y = 587 × 10^3 mm^3
Plastic modulus about x–x axis, S_x = 1950 × 10^3 mm^3
Design strength, p_y = 265 N/mm^2 (since T = 18.7 mm)
Effective length, L_E = 7 m

LOCAL CAPACITY

Axial load, F_c = 2000 kN
Applied moment about x–x axis, M_x = 100 kN m
Applied moment about y–y axis, M_y = 20 kN m

Short strut capacity, P_{cs}, is given by

$$P_{cs} = \left(A_g + \frac{0.25 f_{cu} A_c}{p_y} \right) p_y$$

$$P_{cs} = \left(15\ 000 + \frac{0.25 \times 20 \times 425^2}{265} \right) 265 = 4.878 \times 10^6 \text{ N} = 4878 \text{ kN}$$

Moment capacity of column about the x–x axis, M_{cx}, is given by

$$M_{cx} = p_y Z_x = 265 \times 1760 \times 10^3 = 466.4 \times 10^6 \text{ N mm} = 466.4 \text{ kN m}$$

Moment capacity of column about the y–y axis, M_{cy}, is given by

$$M_{cy} = p_y Z_y = 265 \times 587 \times 10^3 = 155.6 \times 10^6 \text{ N mm} = 155.6 \text{ kN m}$$

$$\frac{F_c}{P_{cs}} + \frac{M_x}{M_{cx}} + \frac{M_y}{M_{cy}} = \frac{2000}{4878} + \frac{100}{466.4} + \frac{20}{155.6} = 0.41 + 0.21 + 0.13 = 0.75 < 1$$

Hence, local capacity of the section is satisfactory.

BUCKLING RESISTANCE

Radii of gyration for cased section
For the cased section, r_x, is the same as for the UC section = 136 mm
For the cased section $r_y = 0.2b_c = 0.2 \times 425 = 85$ mm $\not> 0.2(B + 150) = 0.2(306.8 + 150) = 91.36$ mm
but not less than that for the uncased section (= 77.5 mm)
Hence $r_y = 85$ mm and $r_x = 136$ mm.

Slenderness ratio

$$\lambda_x = \frac{L_E}{r_x} = \frac{7000}{136} = 51.5 \quad \lambda_y = \frac{L_E}{r_y} = \frac{7000}{85} = 82.4$$

Compressive strength
For $\lambda_x = 51.5$ and $p_y = 265$ N/mm^2 compressive strength, $p_c = 226$ N/mm^2 (*Table 4.13*). For $\lambda_y = 82.4$ and $p_y = 265$ N/mm^2 compressive strength, $p_c = 153$ N/mm^2 (*Table 4.14*).
Hence, p_c is equal to 153 N/mm^2.

Compression resistance

$$A_g = 15\,000 \text{ mm}^2$$
$$A_c = d_c b_c = 425 \times 425 = 180\,625 \text{ mm}^2$$
$$p_y = 265 \text{ N/mm}^2 \text{ (since } T = 18.7 \text{ mm)}$$
$$p_c = 153 \text{ N/mm}^2$$
$$f_{cu} = 20 \text{ N/mm}^2$$

Compression resistance of encased column, P_c, is given by

$$P_c = \left(A_g + \frac{0.45 f_{cu} A_c}{p_y} \right) p_c$$

$$P_c = \left(15\,000 + \frac{0.45 \times 20 \times 180\,625}{265} \right) 153 = 3.233 \times 10^6 \text{ N} = 3233 \text{ kN}$$

which should not be greater than the short strut capacity, $P_{cs} = 4878$ kN (see above)   OK

Buckling resistance
For the **uncased** section,

$$\lambda_y = \frac{L_E}{r_y} = \frac{7000}{77.5} = 90$$

$$\frac{\lambda_y}{x} = \frac{90}{314.5/18.7} = 5.4 \Rightarrow v = 0.79 \quad (\textit{Table 4.8})$$

$$\lambda_{LT} = uv\lambda_y \sqrt{\beta_w} = 0.851 \times 0.79 \times 90 = 61$$

From *Table 4.9*, $p_b = 205$ N/mm^2

$$M_b = S_x p_b = 1950 \times 10^3 \times 205 \times 10^{-6} = 400 \text{ kN m}$$

For the **cased** section,

$$\lambda_y = \frac{L_E}{r_y} = \frac{7000}{85} = 82.4$$

$$\frac{\lambda_y}{x} = \frac{82.4}{314.5/18.7} = 4.9 \Rightarrow v = 0.82 \quad (\textit{Table 4.8})$$

$$\lambda_{LT} = uv\lambda_y \sqrt{\beta_w} = 0.851 \times 0.82 \times 82.4 = 57.4$$

From *Table 4.9*, $p_b = 213 \text{ N/mm}^2$

$$M_b = S_x p_b = 1950 \times 10^3 \times 213 \times 10^{-6} = 415 \text{ kN m}$$

Hence, M_b (for cased UC section = 415 kN m) $< 1.5 M_b$ (for uncased UC section = $1.5 \times 400 = 600$ kN m).

Checking buckling resistance

$$\frac{F_c}{P_c} + \frac{m_x M_x}{p_y Z_x} + \frac{m_y M_y}{p_y Z_y}$$

$$\frac{2000}{3233} + \frac{1 \times 100}{265 \times 1760 \times 10^{-3}} + \frac{1 \times 20}{265 \times 587 \times 10^{-3}} = 0.62 + 0.21 + 0.13 = 0.96 < 1 \quad \text{OK}$$

$$\frac{F_c}{P_{cy}} + \frac{m_{LT} M_{LT}}{M_b} + \frac{m_y M_y}{p_y Z_y}$$

$$\frac{2000}{3233} + \frac{1 \times 100}{415} + \frac{1 \times 20}{265 \times 587 \times 10^{-3}} = 0.62 + 0.24 + 0.13 = 0.99 < 1 \quad \text{OK}$$

Hence, the section is now just adequate to resist the design axial load of 2000 kN and design moments about the x–x and y–y axes of 100 kN m and 20 kN m respectively.

4.9.7 DESIGN OF COLUMN BASEPLATES

Clause 4.13 gives guidance on the design of concentrically loaded column slab baseplates, which covers most practical design situations. The plan size of the baseplate is calculated by assuming

a) the nominal bearing pressure between the baseplate and support is uniform and
b) the applied load acts over a portion of the baseplate known as the effective area, the extent of which for UB and UCs is as indicated on *Fig. 4.32*.

For concrete foundations the bearing strength may be taken as 0.6 times the characteristic cube strength of the concrete base or the bedding material (i.e. $0.6 f_{cu}$), whichever is the lesser. The effective area of the baseplate, A_{be}, is then obtained from

$$A_{be} = \frac{\text{axial load}}{\text{bearing strength}} \quad (4.42)$$

In determining the overall plan size of the plate allowance should be made for the presence of holding bolts.

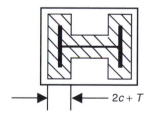
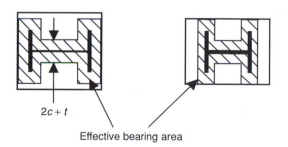

Effective bearing area

Fig. 4.32

The required minimum baseplate thickness, t_p, is given by

$$t_p = c[3\omega/p_{yp}]^{0.5} \qquad (4.43)$$

where

c is the largest perpendicular distance from the edge of the effective portion of the baseplate to the face of the column cross-section (*Fig. 4.32*)

ω pressure on the underside of the plate assuming a uniform distribution throughout the effective portion, but $\leq 0.6f_{cu}$

p_{yp} design strength of the baseplate which may be taken from *Table 4.3*

Example 4.15 Design of a column baseplate (BS 5950)

Design a suitable baseplate for the axially loaded column shown below assuming it is supported on concrete foundations and the characteristic strength of the bedding grout is 30 N/mm^2.

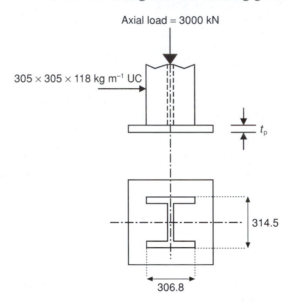

AREA OF BASEPLATE

Effective area

$$A_{be} \geqslant \frac{\text{axial load}}{\text{bearing strength}} = \frac{3000 \times 10^3}{0.6 \times 30} = 1.666 \times 10^5 \text{ mm}^2$$

Actual area

$$A_{be} = (B + 2c)(D + 2c) - 2\{(D - 2[T + c])([B + 2c] - [t + 2c])\}$$

$$1.666 \times 10^5 = (306.8 + 2c)(314.5 + 2c) - 2\{(314.5 - 2[18.7 + c])(306.8 - 11.9)\} \Rightarrow c = 84.6 \text{ mm}$$

Minimum length of baseplate = $D + 2c = 314.5 + 2 \times 84.6 = 483.7$ mm
Minimum width of baseplate = $B + 2c = 306.4 + 2 \times 84.6 = 476$ mm
Provide 500 × 500 mm baseplate in grade S275 steel.

BASEPLATE THICKNESS

Assuming a baseplate thickness of less than 40 mm the design strength $p_{yp} = 265$ N/mm^2. The actual baseplate thickness, t_p, is

$$t_p = c[3\omega/p_{yp}]^{0.5} = 84.6[3 \times (0.6 \times 30)/265]^{0.5} = 34.9 \text{ mm}$$

Hence, a 500 mm × 500 mm × 35 mm thick baseplate in grade S275 steel should be suitable.

4.10 Floor systems for steel framed structures

In temporary steel framed structures such as car parks and Bailey bridges the floor deck can be formed from steel plates. In more permanent steel framed structures the floors generally comprise:

- precast, prestressed concrete slabs
- in-situ reinforced concrete slabs
- composite metal deck floors.

Precast floors are normally manufactured using prestressed hollow core planks, which can easily span up to 6–8 m (*Fig. 4.33(a)*). The top surface can be finished with a levelling screed or, if a composite floor is required, with an in-situ concrete structural topping. This type of floor slab offers a number of advantages over other flooring systems including:

- elimination of shuttering and propping
- reduced floor depth by supporting the precast units on shelf angles *Fig. 4.33(b)*
- rapid construction since curing or strength development of the concrete is unnecessary.

Precast floors are generally designed to act non-compositely with the supporting steel floor beams. Nevertheless, over recent years, hollow core planks that act compositely with the floor beams have been developed and are slowly beginning to be specified. A major drawback of precast concrete slabs, which restricts their use in many congested city centre developments, is that the precast units are heavy and cranage may prove difficult. In such locations, in-situ concrete slabs have invariably been found to be more practical.

In-situ reinforced concrete floor slabs can be formed using conventional removable shuttering and are normally designed to act compositely with the steel floor beams. Composite action is achieved by welding steel studs to the top flange of the steel beams and embedding the studs in the concrete when cast (*Fig. 4.34*). The studs prevent slippage and also enable shear stresses to be transferred between the slab and supporting beams. This increases both the strength and stiffnesses of the beams, thereby allowing significant reductions in construction depth and weight of steel beams to be achieved (see *Example 4.16*). Composite construction not only reduces frame loadings but also results in

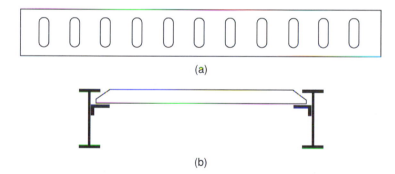

(a)

(b)

Fig. 4.33 *Precast concrete floor: (a) hollow core plank-section (b) precast concrete plank supported on shelf angles.*

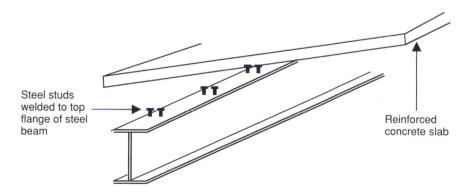

Steel studs welded to top flange of steel beam

Reinforced concrete slab

Fig. 4.34 *In-situ reinforced concrete slab.*

Example 4.16 Advantages of composite construction (BS5950)

Two simply supported, solid steel beams 250 mm wide and 600 mm deep are required to span 8 m. Both beams are manufactured using two smaller beams, each 250 mm wide and 300 mm deep, positioned one above the other. In beam A the two smaller beams are not connected but act independently whereas in beam B they are fully joined and act together as a combined section.

(a) Assuming the permissible strength of steel is 165 N/mm^2, determine the maximum uniformly distributed load that Beam A and Beam B can support.

(b) Calculate the mid-span deflections of both beams assuming that they are subjected to a uniformly distributed load of 140 kN/m.

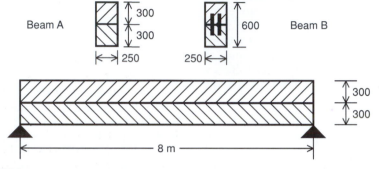

(A) LOAD CAPACITY

(i) Beam A

Elastic modulus of single 250×300 mm beam, Z_s, is

$$Z_s = \frac{I_s}{y} = \frac{bd^3/12}{d/2} = \frac{250 \times 300^2}{6} = 3.75 \times 10^6 \text{ mm}^3$$

Combined elastic modulus of two 250×300 mm beams acting separately, $Z_c = 2Z_s = 7.5 \times 10^6$ mm^3
Moment capacity of combined section, M, is

$$M = \sigma Z_c = 165 \times 7.5 \times 10^6 = 1.2375 \times 10^9 \text{ N mm}$$

Hence, load carrying capacity of Beam A, ω_A, is

$$M = \frac{\omega \ell^2}{8} = \frac{\omega_A(8 \times 10^3)^2}{8} = 1.2375 \times 10^9 \text{ N mm}$$

$$\Rightarrow \omega_A = 154.7 \text{ N/mm} = 154.7 \text{ kN/m}$$

(ii) Beam B

In beam B the two smaller sections act together and behave like a beam 250 mm wide and 600 mm deep. The elastic modulus of the combined section, Z, is

$$Z = \frac{I}{y} = \frac{bd^3/12}{d/2} = \frac{250 \times 600^2}{6} = 15 \times 10^6 \text{ mm}^3$$

Bending strength of beam, $M = \sigma Z = 165 \times 15 \times 10^6 = 2.475 \times 10^6$ N mm
Hence, load carrying capacity of Beam B, ω_B, is

$$M = \frac{\omega \ell^2}{8} = \frac{\omega_B(8 \times 10^3)^2}{8} = 2.475 \times 10^9 \text{ N mm}$$

$$\Rightarrow \omega_B = 309.4 \text{ N/mm} = 309.4 \text{ kN m}$$

(B) DEFLECTION

Mid-span deflection of beam A (for a notional load of $\omega = 140$ kN/m), δ_A, is

$$\delta_A = \frac{5\omega\ell^4}{384EI} = \frac{5 \times 140 \times (8 \times 10^3)^4}{384 \times 205 \times 10^3 \times 2(5.625 \times 10^8)} = 32.3 \text{ mm}$$

Mid-span deflection of beam B, δ_B, is

$$\delta_B = \frac{5\omega\ell^4}{384EI} = \frac{5 \times 140 \times (8 \times 10^3)^4}{384 \times 205 \times 10^3 \times 4.5 \times 10^9} = 8 \text{ mm}$$

Hence, it can be seen that the load capacity has been doubled and the stiffness quadrupled by connecting the two beams, a fact which will generally be found to hold for other composite sections.

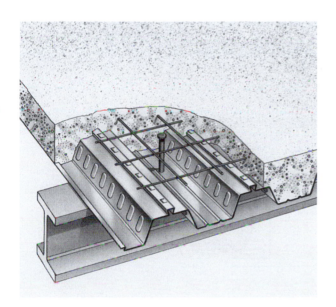

Fig. 4.35 *Composite metal deck floor.*

smaller and hence cheaper foundations. One drawback of this form of construction is that shuttering is needed and the slab propped until the concrete develops adequate strength.

A development of this approach, which can eliminate the need for tensile steel reinforcement and propping of the slab during construction, is to use profiled metal decking as permanent shuttering (*Fig. 4.35*). Three common types of metal decking used in composite slab construction are shown in *Fig. 4.36*. The metal decking is light and easy to work, which makes for simple and rapid construction. This system is most efficient for slab spans of between 3–4 m, and beam spans of up to around 12 m. Where longer spans and/or higher

loads are to be supported, the steel beams may be substituted with cellular beams or stub-girders (*Fig. 4.37*). In structures where there is a need to reduce the depth of floor construction, for example tall buildings, the Slimflor system developed by British Steel can be used. Floor spans are limited to 7.5 m using this system, which utilises stiff steel beams, fabricated from universal column sections welded to a steel plate, and deep metal decking that rests on the bottom flange of the beam (*Fig. 4.38*).

Nowadays, almost all composite floors in steel framed buildings are formed using profiled metal decking and the purpose of this section is to discuss the design of (a) composite slabs and (b) composite beams.

4.10.1 COMPOSITE SLABS

Composite slabs (i.e. metal decking plus concrete) are normally designed to BS 5950: Part 4. Although explicit procedures are given in the standard, these tend to be overly conservative when compared with the results of full-scale tests. Therefore, designers mostly rely on load/span tables produced by metal deck manufacturers in order to determine the thickness of slab and mesh reinforcement required for a given floor arrangement, fire rating, method of construction, etc. *Table 4.17* shows an example of a typical load/span table available from one supplier of metal decking. Concrete grades in the range 30–40 N/mm^2 are common. Slab depths may vary between 100 and 200 mm. *Example 4.20* illustrates the use of this table.

4.10.2 COMPOSITE BEAMS

Once the composite slab has been designed, design of the primary and secondary composite beams (i.e. steel beams plus slab) can begin. This is normally

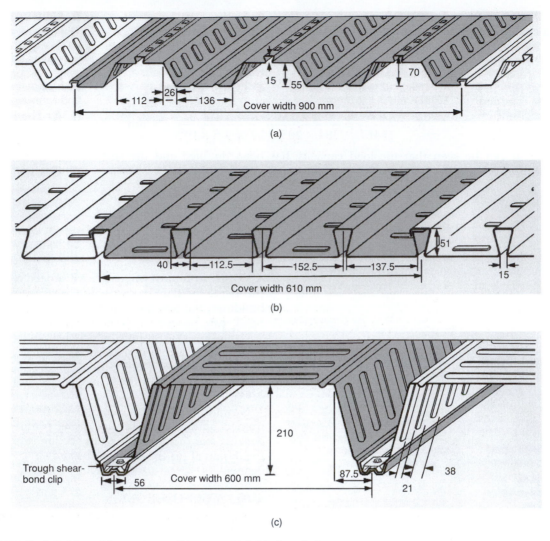

Fig. 4.36 *Steel decking: (a) re-entrant, (b) trapezoidal (c) deep deck.*

carried out in accordance with the recommendations in Part 3: Section 3.1 of BS 5950, hereafter referred to as BS 5950–3.1, and involves the following steps:

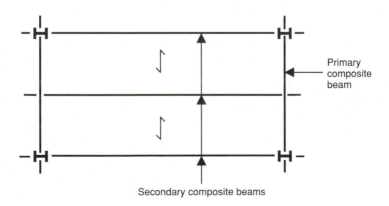

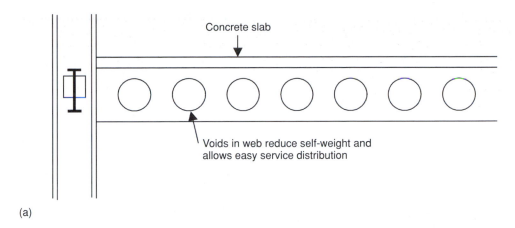

(a)

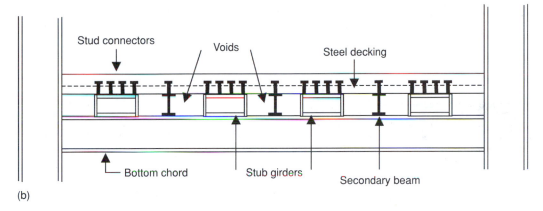

(b)

Fig. 4.37 *Long span structures: (a) cellular beams (b) stub-girder.*

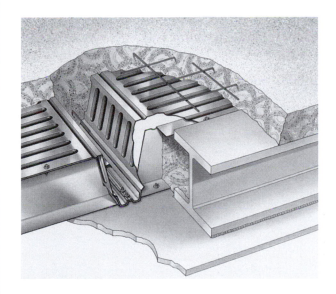

Fig. 4.38 *Slimflor system.*

1. Determine the effective breadth of the concrete slab.
2. Calculate the moment capacity of the section.
3. Evaluate the shear capacity of the section.
4. Design the shear connectors.
5. Assess the longitudinal shear capacity of the section.
6. Check deflection.

Each of these steps is explained below by reference to simply supported secondary beams of class 1 plastic UB section supporting a solid slab.

4.10.2.1 Effective breadth of concrete slab, B_e

According to BS 5950–3.1, the effective breadth of concrete slab, B_e, acting compositely with simply supported beams of length L should be taken as the lesser of $L/4$ and the sum of the effective breadths, b_e, of the portions of flange each side of the centreline of the steel beam (*Fig. 4.39*).

Table 4.17 Typical load/span table for design of unpropped double span slab and deck made of normal weight grade 35 concrete (PMF, CF70, Corus).

Fire Rating	Slab Depth (mm)	Mesh	Maximum span (m)					
			Deck thickness (mm)					
			0.9			1.2		
			Total applied load (kN/m²)					
			3.50	5.00	10.0	3.50	5.00	10.0
1 hr	125	A142	3.2	3.2	2.8	4.0	3.7	3.0
1½ hr	135	A193	3.1	3.1	2.7	3.9	3.5	2.8
2 hr	150	A193	2.9	2.9	2.5	3.5	3.2	2.6
	200	A393	2.5	2.5	2.5	3.5	3.5	3.5
	250	A393	2.1	2.1	2.1	3.2	3.2	3.2

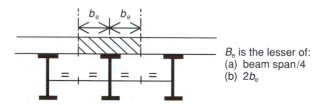

B_e is the lesser of:
(a) beam span/4
(b) $2b_e$

Fig. 4.39 *Effective breadth.*

4.10.2.2 Moment capacity
In the analysis of a composite section to determine its moment capacity the following assumptions can be made:

a) The stress block for concrete in compression at ultimate conditions is rectangular with a design stress of $0.45f_{cu}$
b) The stress block for steel in both tension and compression at ultimate conditions is rectangular with a design stress equal to p_y
c) The tensile strength of the concrete is zero

d) The ultimate moment capacity of the composite section is independent of the method of construction i.e. propped or unpropped.

The moment capacity of a composite section depends upon where the plastic neutral axis falls within the section. Three outcomes are possible, namely:

1. plastic neutral axis occurs within the concrete flange;
2. plastic neutral axis occurs within the steel flange;
3. plastic neutral axis occurs within the web (*Fig. 4.40*).

Only the first two cases will be discussed here.

(i) Case 1: $R_c > R_s$. *Figure 4.41* shows the stress distribution in a typical composite beam section when the plastic neutral axis lies within the concrete slab.

Since there is no resultant axial force on the section, the force in the concrete, R_c', must equal the force in the steel beam, R_s. Hence

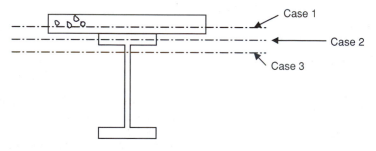

Case 1
Case 2
Case 3

Fig. 4.40 *Plastic neutral axis positions.*

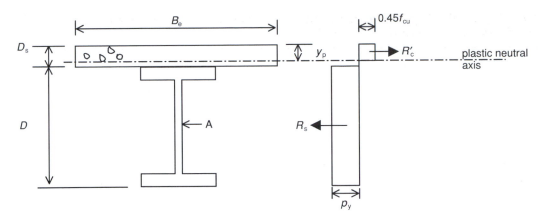

Fig. 4.41 *Stress distribution when plastic neutral axis lies within concrete flange.*

$$R'_c = R_s \qquad (4.44)$$

where
R'_c = design stress in concrete × area of concrete in compression
$$= (0.45f_{cu})(B_e y_p) \qquad (4.45)$$
R_s = steel design strength × area of steel section
$$= p_y A$$

The maximum allowable force in the concrete flange, R_c, is given by

$$R_c = \text{design stress in concrete} \times \text{area of concrete flange}$$
$$= (0.45f_{cu})(B_e D_s) \qquad (4.46)$$

Eliminating $0.45f_{cu}$ from equations 4.45 and 4.46 gives

$$R'_c = \frac{R_c y_p}{D_s} \qquad (4.47)$$

Combining equations 4.44 and 4.47 and rearranging obtains the following expression for depth of the plastic neutral axis, y_p

$$y_p = \frac{R_s}{R_c} D_s \leqslant D_s \qquad (4.48)$$

Taking moments about the top of the concrete flange and substituting for y_p, the moment capacity of the section, M_c, is given by

$$M_c = R_s\left(D_s + \frac{D}{2}\right) - R'_c\frac{y_p}{2}$$

$$= R_s\left(D_s + \frac{D}{2}\right) - R'_c\frac{R_s}{2R_c}D_s$$

$$= R_s\left(D_s + \frac{D}{2}\right) - \frac{R_s^2 D_s}{2R_c} \qquad (4.49)$$

where
D = depth of steel section
D_s = depth of concrete flange

Note that the equation for M_c does not involve y_p. Nonetheless it should be remembered that this equation might only be used to calculate M_c provided that $y_p < D_s$. This condition can be checked either via equation 4.48 or, since $R'_c = R_s$ (equation 4.44) and $R_c > R'_c$, by checking that $R_c > R_s$. Clearly if $R_c < R_s$, it follows that the plastic neutral axis will occur within the steel beam.

(ii) Case 2: $R_c < R_s$. *Figure 4.42a* shows the stress distribution in the section when the plastic neutral axis lies within the steel flange.

By equating horizontal forces, the depth of plastic neutral axis below the top of the steel flange, y, is obtained as follows

$$R_c + p_y(By) = R_s - p_y(By)$$

$$\Rightarrow y = \frac{R_s - R_c}{2Bp_y}$$

Resistance of the steel flange, $R_f = p_y(BT) \Rightarrow p_y$
$$= \frac{R_f}{BT}$$
Substituting into the above expression for y gives

$$y = \frac{R_s - R_c}{2R_f/T} \leqslant T \qquad (4.50)$$

The expression for moment capacity is derived using the equivalent stress distribution shown in *Fig. 4.42(b)*. Taking moments about the top of the steel flange, the moment capacity of the section is given by

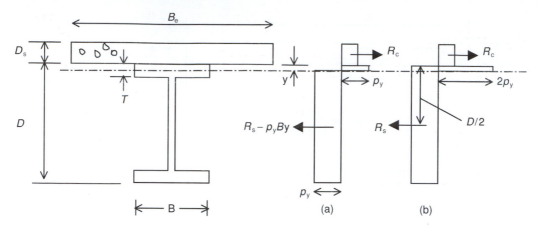

Fig. 4.42 Stress distribution when plastic neutral axis lies within steel flange.

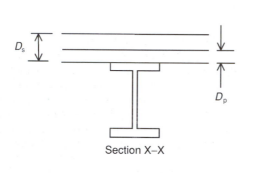

Fig. 4.43 Slab thickness and depth of metal decking.

$$M_c = R_s\frac{D}{2} + R_c\frac{D_s}{2} - (2p_yBy)\frac{y}{2}$$

$$= R_s\frac{D}{2} + R_c\frac{D_s}{2} - p_yBy^2$$

Substituting for p_y and y and simplifying gives

$$M_c = R_s\frac{D}{2} + R_c\frac{D_s}{2} - \frac{(R_s - R_c)^2}{4R_f}T \quad (4.51)$$

(iii) Moment capacity of composite beam incorporating metal decking. The moment capacity of composite beams incorporating profiled metal decking is given by the following:

Case 1: Plastic neutral axis is in the concrete flange

$$M_c = R_s\left[\frac{D}{2} + D_s - \frac{R_s}{R_c}\left(\frac{D_s - D_p}{2}\right)\right] \quad (4.52)$$

Case 2: Plastic neutral axis is in the steel flange

$$M_c = R_s\frac{D}{2} + R_c\left(\frac{D_s + D_p}{2}\right) - \frac{(R_s - R_c)^2}{R_f}\frac{T}{4} \quad (4.53)$$

These expressions are derived in the same way as for beams incorporating solid slabs but further assume that (a) the ribs of the metal decking run perpendicular to the beams and (b) the concrete within the depth of the ribs is ignored (*Fig. 4.43*). The stress distributions used to derive equations 4.52 and 4.53 are shown in *Fig. 4.44*. Note that the symbols in these equations are as previously defined except for R_c, which is given by

$$R_c = 0.45f_{cu}B_e(D_s - D_p) \quad (4.54)$$

where D_p is the overall depth of the profiled metal decking (*Fig. 4.43*).

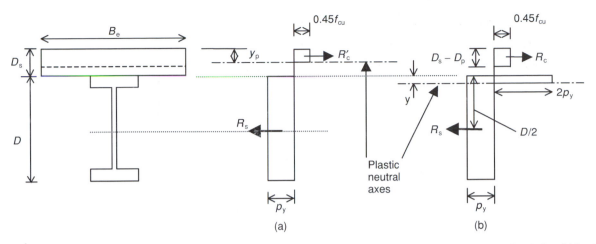

Fig. 4.44 *Stress distributions in composite beams incorporating profiled metal decking: (a) plastic neutral axis is within the concrete flange (b) plastic neutral axis is within the steel flange.*

4.10.2.3 Shear capacity

According to BS 5950–3.1, the steel beam should be capable of resisting the whole of the vertical shear force, F_v. As discussed in 4.8.5, the shear capacity, P_v, of a rolled I-section is given by

$$P_v = 0.6 p_y t D$$

Like BS 5950: Part 1, BS 5950–3.1 recommends that where the co-existent shear force exceeds $0.5 P_v$ the moment capacity of the section, M_{cv}, should be reduced in accordance with the following:

$$M_{cv} = M_c - (M_c - M_f)(2F_v/P_v - 1)^2 \qquad (4.55)$$

where

M_f is the plastic moment capacity of that part of the section remaining after deduction of the shear area A_v defined in Part 1 of BS 5950

P_v is the lesser of the shear capacity and the shear buckling resistance, both determined from Part 1 of BS 5950

4.10.2.4 Shear connectors

(i) Headed studs. For the steel beams and slab to act compositely and also to prevent separation of the two elements under load, they must be structurally tied. This is normally achieved by providing shear connectors in the form of headed studs as shown in *Fig. 4.45*. The shear studs are usually welded to the steel beams through the metal decking.

Shear studs are available in a range of diameters and lengths as indicated in *Table 4.18*. The 19 mm diameter by 100 mm high stud is by far the most common in buildings. In slabs comprising profiled metal decking and concrete, the heights of the studs should be at least 35 mm greater than the overall depth of the decking. Also, the centre-to-centre distance between studs along the beam should lie between 5ϕ and 600 mm or $4D_s$ if smaller, where ϕ is the shank diameter and D_s the depth of the concrete slab. Some of the other code recommendations governing the minimum size, transverse spacing and edge distances of shear connectors are shown in *Fig. 4.45*.

(ii) Design procedure. The shear strength of headed studs can be determined using standard push-out specimens consisting of a short section of beam and slab connected by two or four studs. *Table 4.18* gives the characteristic resistances, Q_k, of headed studs embedded in a solid slab of normal weight concrete. For positive (i.e. sagging) moments, the design strength, Q_p, should be taken as

$$Q_p = 0.8 Q_k \qquad (4.56)$$

A limit of 80% of the static capacity of the shear connector is deemed necessary for design to ensure that full composite action is achieved between the slab and the beam.

The capacity of headed studs in composite slabs with the ribs running perpendicular to the beam should be taken as their capacity in a solid slab

201

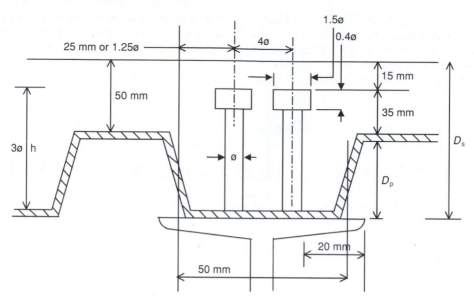

Fig. 4.45 *Geometrical requirements for placing of studs (CIRIA Report 99).*

Table 4.18   Characteristic resistance, Q_k, of headed studs in normal weight concrete (Table 5, BS 5950–3.1)

Shank diameter (mm)	Height (mm)	Characteristic strength (N/mm²)			
		25	30	35	40
25	100	146	154	161	168 kN
22	100	119	126	132	139 kN
19	100	95	100	104	109 kN
19	75	82	87	91	96 kN
16	75	70	74	78	82 kN
13	65	44	47	49	52 kN

multiplied by a reduction factor, k, given by the following expressions:

for one stud per rib
$$k = 0.85(b_r/D_p)\{(h/D_p) - 1\} \leqslant 1$$

for two studs per rib
$$k = 0.6(b_r/D_p)\{(h/D_p) - 1\} \leqslant 0.8$$

for three or more studs per rib
$$k = 0.5(b_r/D_p)\{(h/D_p) - 1\} \leqslant 0.6$$

where
b_r breadth of the concrete rib
D_p overall depth of the profiled steel sheet
h overall height of the stud but not more than $2D_p$ or $D_p + 75$ mm, although studs of greater height may be used

For full shear connection, the total number of studs, N_p, required over half the span of a simply supported beam in order to develop the positive moment capacity of the section can be determined using the following expression:

$$N_p = F_c/Q_p \qquad (4.57)$$

where
$F_c = Ap_y$ (if plastic neutral axis lies in the concrete flange)
$F_c = 0.45f_{cu}B_eD_s$ (if plastic neutral axis lies in the steel beam)
Q_p = design strength of shear studs = $0.8Q_k$ (in solid slab) and kQ_k (in a composite slab formed using ribbed profile sheeting aligned perpendicular to the beam)

In composite floors made with metal decking it may not always be possible to provide full shear connection between the beams and the slab because the top flange of the beam may be too narrow and/or the spacing of the ribs may be too great to accommodate the total number of studs required. In this case, the reader should refer to BS 5950–3.1: Appendix B, which gives alternative expressions for moment capacity of composite sections with partial shear connection.

Example 4.17 Moment capacity of a composite beam (BS 5950)

Determine the moment capacity of the section shown in *Fig. 4.46* assuming the universal beam is of grade S275 steel and the characteristic strength of the concrete is 35 N/mm^2.

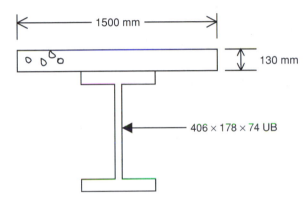

Fig. 4.46

(A) PLASTIC NEUTRAL AXIS OF COMPOSITE SECTION
Resistance of the concrete flange, R_c, is

$$R_c = (0.45f_{cu})B_eD_s = (0.45 \times 35)1500 \times 130 \times 10^{-3} = 3071.25 \text{ kN}$$

From *Table 4.3*, since $T(= 16 \text{ mm}) \leqslant 16$ mm and steel grade is S275, design strength of beam, $p_y = 275$ N/mm^2. Resistance of steel beam, R_s, is

$$R_s = Ap_y = 95 \times 10^2 \times 275 \times 10^{-3} = 2612.5 \text{ kN}$$

Since $R_c > R_s$, the plastic neutral axis falls within the concrete slab. Confirm this by calculating y_p:

$$y_p = \frac{Ap_y}{(0.45f_{cu})B_e} = \frac{2612.5 \times 10^3}{0.45 \times 35 \times 1500} = 110.6 \text{ mm} < D_s \quad \text{OK}$$

(B) MOMENT CAPACITY
Since $y_p < D_s$ use equation 4.49 to calculate the moment capacity of the section, M_c. Hence

$$M_c = Ap_y\left(D_s + \frac{D}{2} - \frac{R_s}{R_c}\frac{D_s}{2}\right) = 2612.5 \times 10^3\left(130 + \frac{412.8}{2} - \frac{2612.5}{3071.25} \times \frac{130}{2}\right)10^{-6} = 734.4 \text{ kN m}$$

Example 4.18 Moment capacity of a composite beam (BS 5950)

Repeat Example 4.17 assuming the beam is made of grade S355 steel. Also, design the shear connectors assuming the beam is 6 m long and that full composite action is to be provided.

(A) PLASTIC NEUTRAL AXIS OF COMPOSITE SECTION

As before, the resistance of the concrete flange, R_c, is

$$R_c = (0.45f_{cu})A_c = (0.45 \times 35)1500 \times 130 \times 10^{-3} = 3071.25 \text{ kN}$$

From *Table 4.3*, since T (= 16 mm) $\leqslant$ 16 mm and steel grade is S355, design strength of beam, $p_y = 355 \text{ N/mm}^2$. Resistance of steel beam, R_s, is

$$R_s = Ap_y = 95 \times 10^2 \times 355 \times 10^{-3} = 3372.5 \text{ kN}$$

Since $R_c < R_s$, the plastic neutral axis will lie within the steel beam. Confirm this by calculating y:

$$y = \frac{R_s - R_c}{2Bp_y} = \frac{3372.5 \times 10^3 - 3071.25 \times 10^3}{2 \times 179.7 \times 355} = 2.4 \text{ mm} \quad \text{OK}$$

(B) MOMENT CAPACITY

Since $y < T$ use equation 4.51 to calculate moment capacity of the section, M_c.

$$\text{Resistance of steel flange, } R_f = BTp_y = 179.7 \times 16 \times 355 = 1.02 \times 10^6 \text{ N}$$

Moment capacity of composite section, M_c, is

$$M_c = R_s\frac{D}{2} + R_c\frac{D_s}{2} - \frac{(R_s - R_c)^2}{R_f}\frac{T}{4}$$

$$= 3372.5 \times 10^3\frac{412.8}{2} + 3071.25 \times 10^3\frac{130}{2} - \frac{\left((3372.5 - 3071.25) \times 10^3\right)^2}{1.02 \times 10^6}\frac{16}{4}$$

$$= 895.4 \times 10^6 \text{ N mm} = 895.4 \text{ kN m}$$

(C) SHEAR STUDS

From *Table 4.18*, characteristic resistance, Q_k, of headed studs 19 mm diameter × 100 mm high embedded in grade 35 concrete is 104 kN.

Design strength of studs under positive moment, Q_p, is given by

$$Q_p = 0.8Q_k = 0.8 \times 104 = 83.2 \text{ kN}$$

Number of studs required $= \dfrac{R_c}{Q_p} = \dfrac{3071.25}{83.2} \geqslant 36.9$

Provide 38 studs, evenly arranged in pairs, in each half span of beam as shown.

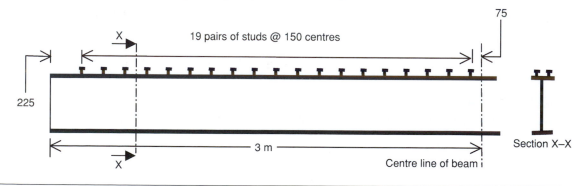

19 pairs of studs @ 150 centres

75

225

3 m

Centre line of beam

Section X–X

4.10.2.5 Longitudinal shear capacity

A solid concrete slab, which is continuous over supports, will need to be reinforced with top and bottom steel to resist the sagging and hogging moments due to the applied loading. Over beam supports, this steel will also be effective in transferring longitudinal forces from the shear connectors to the slab without splitting the concrete. Design involves checking that the applied longitudinal shear force per unit length, v, does not exceed the shear resistance of the concrete, v_r.

The total longitudinal shear force per unit length, v, is obtained using the following

$$v = NQ_p/s \qquad (4.58)$$

where

N number of shear connectors in a group
s longitudinal spacing centre-to-centre of groups of shear connectors
Q_p design strength of shear connectors

In a solid slab, the concrete shear resistance, v_r, is obtained using the following:

$$v_r = 0.7A_{sv}f_y + 0.03\eta A_{cv}f_{cu} \leqslant 0.8\eta A_{cv}\sqrt{f_{cu}} \qquad (4.59)$$

In slabs with profiled steel sheeting, v_r, is given by

$$v_r = 0.7A_{sv}f_y + 0.03\eta A_{cv}f_{cu} + v_p$$
$$\leqslant 0.8\eta A_{cv}\sqrt{f_{cu}} + v_p \qquad (4.60)$$

where

f_{cu} characteristic strength of the concrete $\leqslant 40$ N/mm^2
η 1.0 for normal weight concrete
A_{cv} mean cross-sectional area per unit length of the beam of the concrete surface under consideration

A_{sv} cross-sectional area, per unit length of the beam, of the combined top and bottom reinforcement crossing the shear surface (*Fig. 4.47*)
v_p contribution of the profiled steel decking. Assuming the decking is continuous across the top flange of the steel beam and that the ribs are perpendicular to the span of the beam, v_p is given by

$$v_p = t_p p_{yp} \qquad (4.61)$$

in which
t_p thickness of the steel decking
p_{yp} design strength of the steel decking obtained either from Part 4 of BS 5950 or manufacturer's literature

4.10.2.6 Deflection

The deflection experienced by composite beams will vary depending on the method of construction used. Thus where steel beams are unpropped during construction, the total deflection, δ_T, will be the sum of the dead load deflection, δ_D, due to the self weight of the slab and beam, based on the properties of the steel beam alone, plus the imposed load deflection, δ_I, based on the properties of the composite section:

$$\delta_T = \delta_D + \delta_I$$

For propped construction the total deflection is calculated assuming that the composite section supports both dead and imposed loads.

The mid-span deflection of a simply supported beam of length L subjected to a uniformly distributed load, ω, is given by

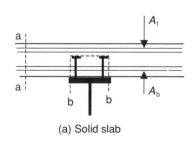

(a) Solid slab

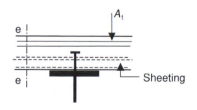

(b) Composite slab with the sheeting spanning perpendicular to the beam

Surface	A_{sv}
a–a	$A_b + A_t$
b–b	$2A_b$
e–e	A_t

Fig. 4.47

$$\delta = \frac{5\omega L^4}{384EI}$$

where
E elastic modulus = 205 kN/mm² for steel
I second moment of area of the section

Deflections of simply supported composite beams should be calculated using the gross value of the second moment of area of the uncracked section, I_g, determined using a modular ratio approach. The actual value of modular ratio, α_e, depends on the proportions of the loading which are considered to be long term and short term. Imposed loads on floors should be assumed to be 2/3 short term and 1/3 long term. On this basis, appropriate values of modular ratio for normal weight and lightweight concrete are 10 and 15, respectively. For composite beams with solid slabs I_g is given by

$$I_g = I_s + \frac{B_e D_s^3}{12\alpha_e} + \frac{AB_e D_s(D + D_s)^2}{4(A\alpha_e + B_e D_s)} \quad (4.62)$$

where I_s is the second moment of area of the steel section.

Equation 4.62 is derived assuming that the concrete flange is uncracked and unreinforced (see *Appendix D*). In slabs with profiled steel sheeting the concrete within the depth of the ribs may conservatively be omitted and I_g is then given by:

$$I_g = I_s + \frac{B_e(D_s - D_p)^3}{12\alpha_e}$$

$$+ \frac{AB_e(D_s - D_p)(D + D_s + D_p)^2}{4(A\alpha_e + B_e[D_s - D_p])} \quad (4.63)$$

The deflections under unfactored imposed loads should not exceed the limits recommended in BS 5590, as summarised in *Table 4.5*.

Example 4.19 Design of a composite floor (BS 5950)

Steel universal beams at 3.5 m centres with 9 m simple span are to support a 150 mm deep concrete slab of characteristic strength 30 N/mm² (*Fig. 4.48*). If the imposed load is 4 kN/m² and the weight of the partitions is 1 kN/m²

a) select a suitable universal beam section in grade S355 steel
b) check the shear capacity
c) determine the number and arrangement of 19 mm diameter × 100 mm long headed stud connectors required
d) assuming the slab is reinforced in both faces with T8@150 centres (A = 335 mm²/m), check the longitudinal shear capacity of the concrete
e) calculate the imposed load deflection of the beam.

Assume that weight of the finishes, and ceiling and service loads are 1.2 kN/m² and 1 kN/m² respectively. The density of normal weight reinforced concrete, ρ_c, can be taken as 24 kN/m³.

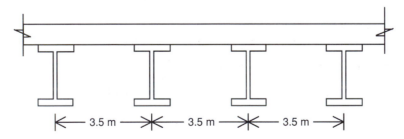

Fig. 4.48

(A) BEAM SELECTION

(i) Design moment
Beam span, L = 9 m
Slab span, ℓ = 3.5 m
Design load per beam, ω = 1.4($D_s\rho_c$ + finishes + ceiling/services)ℓ + 1.6(q_k + partition loading)ℓ
      = 1.4(0.15 × 24 + 1.2 + 1)3.5 + 1.6(4 + 1)3.5 = 56.4 kN/m

Design moment, $M = \dfrac{\omega L^2}{8} = \dfrac{56.4 \times 9^2}{8} = 571$ kN m

(ii) Effective width of concrete slab

B_e is the lesser of beam span/4 (= 9000/4 = 2250 mm) and beam spacing (= 3500 mm)
∴ B_e = 2250 mm

(iii) Moment capacity

Using trial and error, try $356 \times 171 \times 45$ UB in grade S355 steel
Resistance of the concrete flange, R_c, is

$$R_c = (0.45 f_{cu}) B_e D_s = (0.45 \times 30)\,2250 \times 150 \times 10^{-3} = 4556.3 \text{ kN}$$

From *Table 4.3*, since T (= 9.7 mm) < 16 mm and steel grade is S355, design strength of beam, $p_y = 355$ N/mm². Resistance of steel beam, R_s, is

$$R_s = A p_y = 57 \times 10^2 \times 355 \times 10^{-3} = 2023.5 \text{ kN}$$

Since $R_c > R_s$, the plastic neutral axis will lie in the concrete slab. Confirm this by substituting into the following expression for y_p

$$\Rightarrow y_p = \frac{A p_y}{(0.45 f_{cu}) B_e} = \frac{2023.5 \times 10^3}{0.45 \times 30 \times 2250} = 66.6 \text{ mm} < D_s \quad \text{OK}$$

Hence, moment capacity of composite section, M_c, is

$$M_c = A p_y \left(D_s + \frac{D}{2} - \frac{R_s}{R_c}\frac{D_s}{2} \right) = 2023.5 \times 10^3 \left(150 + \frac{352}{2} - \frac{2023.5}{4556.3} \times \frac{150}{2} \right) 10^{-6} = 592.3 \text{ kN m}$$

$$M_c > M + M_{sw} = 571 + (45 \times 9.8 \times 10^{-3})9^2/8 = 575.5 \text{ kN m} \quad \text{OK}$$

(B) SHEAR CAPACITY

Shear force, $F_v = \dfrac{1}{2}\omega L = \dfrac{1}{2} \times 56.4 \times 9 = 253.8$ kN

Shear resistance, $P_v = 0.6 p_y t D = 0.6 \times 355 \times 6.9 \times 352 \times 10^{-3} = 517$ kN $> F_v$ OK
At mid-span, F_v (= 0) < $0.5 P_v$ (= 258 kN) and therefore the moment capacity of the section calculated above is valid.

(C) SHEAR CONNECTORS

From *Table 4.18*, characteristic resistance, Q_k, of headed studs 19 mm diameter × 100 mm high is 100 kN.
Design strength of shear connectors, Q_p, is

$$Q_p = 0.8 Q_k = 0.8 \times 100 = 80 \text{ kN}$$

Longitudinal force that needs to be transferred, F_c, is 2023.5 kN

$$\text{Number of studs required} = \frac{R_c}{Q_p} = \frac{2023.5}{80} \geqslant 25.3$$

Provide 26 studs, evenly arranged in pairs, in each half span of beam.

Spacing $= \dfrac{4500}{12} = 375$ mm, say 350 mm centres

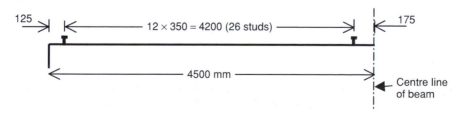

(D) LONGITUDINAL SHEAR
Longitudinal force, v, is

$$v = \frac{NQ_p}{s} = \frac{2 \times 80 \times 10^3}{350} = 457 \text{ N/mm}$$

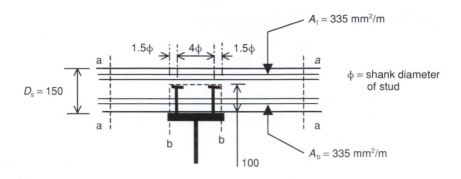

(i) Shear failure surface a–a
Length of failure surface = $2 \times 150 = 300$ mm
Cross-sectional area of failure surface per unit length of beam, $A_{cv} = 300 \times 10^3$ mm^2/m
Cross-sectional area of reinforcement crossing potential failure surface, A_{sv}, is

$$A_{sv} = A_t + A_b = 2 \times 335 = 670 \text{ mm}^2/\text{m}$$

Hence shear resistance of concrete, v_r, is

$$v_r = 0.7A_{sv}f_y + 0.03\eta A_{cv}f_{cu} \leqslant 0.8\eta A_{cv}\sqrt{f_{cu}} = \left(0.8 \times 1.0 \times (300 \times 10^3 \sqrt{30}\right)/10^3 = 1315 \text{ N/mm}$$
$$= \left(0.7 \times 670 \times 460 + 0.03 \times 1.0 \times 300 \times 10^3 \times 30\right)/10^3 = 485.7 \text{ N/mm} > v \quad \text{OK}$$

(ii) Shear failure surface b–b
Length of failure surface = 2 times stud height + 7ϕ = $2 \times 100 + 7 \times 19 = 333$ mm
Cross-sectional area of failure surface per unit length of beam, $A_{cv} = 333 \times 10^3$ mm^2/m
Cross-sectional area of reinforcement crossing potential failure surface, A_{sv}, is

$$A_{sv} = 2A_b = 2 \times 335 = 670 \text{ mm}^2/\text{mm}$$

Hence shear resistance of concrete, v_r, is

$$v_r = 0.7A_{sv}f_y + 0.03\eta A_{cv}f_{cu} \leqslant 0.8\eta A_{cv}\sqrt{f_{cu}} = \left(0.8 \times 1.0 \times (333 \times 10^3)\sqrt{30}\right)/10^3 \; 1459 \text{ N/m}$$
$$= \left(0.7 \times 670 \times 460 + 0.03 \times 1.0 \times 333 \times 10^3 \times 30\right)/10^3 = 515 \text{ N/mm} > v \quad \text{OK}$$

(E) DEFLECTION
Since beam is simply supported use the gross value of second moment of area, I_g, of the uncracked section to calculate deflections.

$$I_g = I_s + \frac{B_e D_s^3}{12\alpha_e} + \frac{AB_e D_s(D + D_s)^2}{4(A\alpha_e + B_e D_s)}$$

$$= 12\,100 \times 10^4 + \frac{2250 \times 150^3}{12 \times 10} + \frac{57 \times 10^2 \times 2250 \times 150(352 + 150)^2}{4(57 \times 10^2 \times 10 + 2250 \times 150)} = 49\,150 \times 10^4 \text{ mm}^4$$

Mid-span deflection of beam, δ, is

$$\delta = \frac{5\omega L^4}{384EI} = \frac{5 \times (5 \times 3.5 \times 9)9^3 \times 10^{12}}{384 \times 205 \times 10^3 \times 49\,150 \times 10^4}$$

$$= 14.8 \text{ mm} < \frac{L}{360} = \frac{9000}{360} = 25 \text{ mm} \quad \text{OK}$$

Therefore adopt $356 \times 171 \times 45$ UB in grade S355 steel.

Example 4.20 Design of a composite floor incorporating profiled metal decking (BS 5950)

Figure 4.49 shows a part plan of a composite floor measuring 9 m × 6 m. The slab is to be constructed using profiled metal decking and normal weight, grade 35 concrete and is required to have a fire resistance of 60 mins. The longitudinal beams are of grade S355 steel with a span of 9 m and spaced 3 m apart. Design the composite slab and internal beam A2–B2 assuming the floor loading is as follows:

$$\begin{aligned}
\text{imposed load} &= 4 \text{ kN/m}^2 \\
\text{partition load} &= 1 \text{ kN/m}^2 \\
\text{weight of finishes} &= 1.2 \text{ kN/m}^2 \\
\text{weight of ceiling and services} &= 1 \text{ kN/m}^2
\end{aligned}$$

The density of normal weight reinforced concrete can be taken to be 24 kN/m^3.

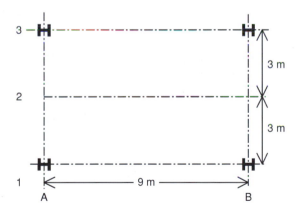

Fig. 4.49 *Part plan of composite floor.*

SLAB DESIGN

Assuming the slab is unpropped during construction, use *Table 4.17* to select suitable slab depth and deck gauge for required fire resistance of 1 hr. It can be seen that for a total imposed load of 7.2 kN/m^2 (i.e. occupancy, partition load, finishes, ceilings and services) and a slab span of 3 m, a 125 mm thick concrete slab reinforced with A142 mesh and formed on 1.2 mm gauge decking should be satisfactory.

A cross-section through the floor slab is shown below.

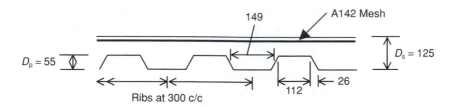

149   A142 Mesh

$D_p = 55$

$D_s = 125$

112   26

Ribs at 300 c/c

BEAM A2–B2

Beam selection

(i) Design moment

Beam span, L = 9 m
Beam spacing, $\ell = 3$ m

From manufacturers' literature effective slab thickness, $D_{ef} = D_s - 26 = 125 - 26 = 91$ mm

Design load, $\omega = 1.4(D_{ef}\rho_c + \text{finishes} + \text{ceiling/services})\ell + 1.6(q_k + \text{partition loading})\ell$

$= 1.4(0.091 \times 24 + 1.2 + 1)3 + 1.6(4 + 1)3 = 42.4$ kN/m

Design moment, $M = \dfrac{\omega L^2}{8} = \dfrac{42.4 \times 9^2}{8} = 429.3$ kN m

(ii) Effective width of concrete slab

B_e is the lesser of $L/4$ (= 9000/4 = 2250 mm) and beam spacing (= 3000 mm)
$\therefore B_e = 2250$ mm

(iii) Moment capacity

Using trial and error, try $305 \times 165 \times 40$ UB in grade S355 steel
Resistance of concrete flange, R_c, is

$R_c = \left(0.45 f_{cu}\right)B_e\left(D_s - D_p\right) = \left(0.45 \times 35\right) \times 2250 \times \left(125 \times 55\right) \times 10^{-3} = 2480.6$ kN

From *Table 4.3*, since T (= 10.2 mm) < 16 mm and steel grade is S355, design strength of beam, $p_y = 355$ N/mm². Resistance of steel beam, R_s, is

$R_s = Ap_y = 51.5 \times 10^2 \times 355 \times 10^{-3} = 1828.3$ kN

Since $R_c > R_s$, plastic neutral axis lies within the slab. Confirm this by substituting into the following expression for y_p

$y_p = \dfrac{Ap_y}{(0.45 f_{cu})B_e} = \dfrac{1828.3 \times 10^3}{0.45 \times 35 \times 2250} = 51.6$ mm $< D_s - D_p = 125 - 55 = 70$ mm   OK

Hence, moment capacity of composite section, M_c, is

$M_c = R_s\left[\dfrac{D}{2} + D_s - \dfrac{R_s}{R_c}\left(\dfrac{D_s - D_p}{2}\right)\right]$

$= 1828.3 \times 10^3\left[\dfrac{303.8}{2} + 125 - \dfrac{1828.3 \times 10^3}{2480.6 \times 10^3}\left(\dfrac{125 - 55}{2}\right)\right] \times 10^{-6} = 459$ kN m

$M_c > M + M_{sw} = 429.3 + 1.4(40 \times 9.8 \times 10^{-3})9^2/8 = 434.9$ kN m   OK

Shear capacity

Shear force, $F_v = \dfrac{1}{2}\omega L = \dfrac{1}{2} \times 42.4 \times 9 = 190.8$ kN

Shear resistance, $P_v = 0.6p_y tD = 0.6 \times 355 \times 6.1 \times 303.8 \times 10^{-3} = 394.7$ kN $> F_v$ OK

At mid-span $F_v = 0 < 0.5P_v$ (= 197.4 kN). Therefore moment capacity of section remains unchanged.

Shear connectors

Assume headed studs 19 mm diameter $\times$ 100 mm high are to be used as shear connectors. From *Table 4.18*, characteristic resistance of studs embedded in a solid slab, $Q_k = 104$ kN

Design strength of studs under positive (i.e. sagging) moment, Q_p, is

$$Q_p = 0.8Q_k = 0.8 \times 104 = 83.2 \text{ kN}$$

Design strength of studs embedded in a slab comprising profiled metal decking and concrete, Q'_p, is

$$Q'_p = kQ_p$$

Assuming two studs are to be provided per trough, the shear strength reduction factor, k, is given by

$$k = 0.6\left(b_r/D_p\right)\left\{\left(h/D_p\right) - 1\right\} \leqslant 0.8$$
$$= 0.6\left(149/55\right)\left\{\left(95/55\right) - 1\right\} = 1.2$$
$$\therefore k = 0.8$$
$$\Rightarrow Q'_p = 0.8 \times 83.2 = 66.6 \text{ kN}$$

Maximum longitudinal force in the concrete, $F_c = 1828.3$ kN

$$\text{Total number of studs required} = \frac{R_c}{Q_p} = \frac{1828.3}{66.6} = 27.5$$

Since the spacing of the deck troughs is 300 mm, fifteen (= 4500/300) trough positions are available for the positioning of the shear studs. Therefore, provide 30 studs in each half span of beam, i.e. two studs per trough.

Longitudinal shear

(i) Longitudinal shear stress
Longitudinal force, υ, is

$$\upsilon = \frac{NQ_p}{s} = \frac{2 \times 66.6 \times 10^3}{300} = 444 \text{ N/mm}$$

(ii) Shear failure surface e–e
Cross-sectional area of reinforcement crossing potential failure surface, A_{sv}, is

$$A_{sv} = A_t = 142 \text{ mm}^2/\text{m}$$

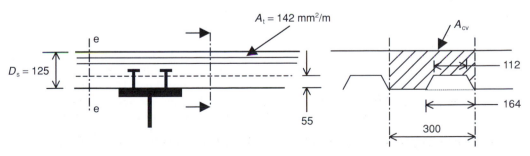

211

Mean cross-sectional area of shear surface, A_{cv}, is

$$[125 \times 300 - {}^1/_2(164 + 112)55]/0.3 = 99.7 \times 10^3 \text{ mm}^2/\text{m}$$

Hence shear resistance of concrete, v_r, is

$$v_r = 0.7A_{sv}f_y + 0.03\eta A_{cv}f_{cu} + v_p \leqslant 0.8\eta A_{cv}\sqrt{f_{cu}} + v_p$$

$$= \left(0.7 \times 142 \times 460 + 0.03 \times 1.0 \times 99.7 \times 10^3 \times 35 + 1.2 \times 10^3 \times 280\right)/10^3 = 486 \text{ N/mm}$$

$$\leqslant \left(0.8 \times 1.0 \times 99.7 \times 10^3 \sqrt{35} + 1.2 \times 10^3 \times 280\right)/10^3 = 808 \text{ N/mm} \quad \text{OK}$$

$$v_r = 486 \text{ N/mm} > v \quad \text{OK}$$

(Note $p_{yp} = 280 \text{ N/mm}^2$ from the steel decking manufacturer's literature.)

Deflection

Since beam is simply supported use the gross value of second moment of area, I_g, of the uncracked section to calculate deflection.

$$I_g = I_s + \frac{B_e\left(D_s - D_p\right)^3}{12\alpha_e} + \frac{AB_e\left(D_s - D_p\right)\left(D + D_s + D_p\right)^2}{4\left\{A\alpha_e + B_e\left(D_s - D_p\right)\right\}}$$

$$= 8520 \times 10^4 + \frac{2250(125 - 55)^3}{12 \times 10} + \frac{51.5 \times 10^2 \times 2250(125 - 55)(303.8 + 125 + 55)^2}{4\{51.5 \times 10^2 \times 10 + 2250(125 - 55)\}}$$

$$= 31\,873 \times 10^4 \text{ mm}^4$$

Mid-span deflection of beam, δ, is

$$\delta = \frac{5\omega L^4}{384EI} = \frac{5 \times (5 \times 3 \times 9)9^3 \times 10^{12}}{384 \times 205 \times 10^3 \times 31\,873 \times 10^4} = 19.6 \text{ mm} < \frac{L}{360} = \frac{9000}{360} = 25 \text{ mm} \quad \text{OK}$$

Therefore adopt $305 \times 165 \times 40$ UB in grade S355 steel.

4.11 Design of connections

There are two principal methods for connecting together steel elements of structure, and the various cleats, end plates etc., also required.

1. Bolting, using ordinary or high strength friction grip (HSFG) bolts, is the principal method of connecting together elements on site.
2. Welding, principally electric arc welding, is an alternative way of connecting elements on site, but most welding usually takes place in factory conditions. End plates and fixing cleats are welded to the elements in the fabrication yard. The elements are then delivered to site where they are bolted together in position.

Figure 4.50 shows some typical connections used in steel structures.

The aim of this section is to describe the design of some commonly used types of bolted and welded

connections in steel structures. However, at the outset, it is worthwhile reiterating some general points relating to connection design given in clause 6 of BS 5950.

The first couple of sentences are vitally important – 'Joints should be designed on the basis of realistic assumptions of the distribution of internal forces. These assumptions should correspond with direct load paths through the joint, taking account of the relative stiffnesses of the various components of the joint'. Before any detailed design is embarked upon therefore, a consideration of how forces will be transmitted through the joint is essential.

'The connections between members should be capable of withstanding the forces and moments to which they are subject . . . without invalidating the design assumptions'. If, for instance, the structure is designed in 'simple construction', the beam-column joint should be designed accordingly to accept rotations rather than moments. A rigid joint would be completely wrong in this situation, as it

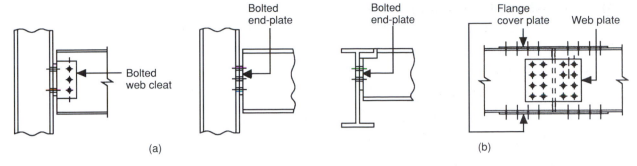

Fig. 4.50 Typical connections: (a) beam to column; (b) beam to beam.

would tend to generate a moment in the column for which it has not been designed.

'The ductility of steel assists in the distribution of forces generated within a joint.' This means that residual forces due to initial lack of fit, or due to bolt tightening do not normally have to be considered.

4.11.1 BOLTED CONNECTIONS

As mentioned above, two types of bolts commonly used in steel structures are ordinary (or black bolts) and HSFG bolts. Black bolts sustain a shear load by the shear strength of the bolt shank itself, whereas HSFG grip bolts rely on a high tensile strength to grip the joined parts together so tightly that they cannot slide.

There are three grades of ordinary bolts, namely 4.6, 8.8 and 10.9. HSFG bolts commonly used in structural connections conform to the general grade and may be parallel shank fasteners designed to be non-slip in service or waisted shank fasteners designed to be non-slip under factored loads. The preferred size of steel bolts are 12, 16, 20, 22, 24 and 30 mm in diameter. Generally, in structural connections, grade 8.8 bolts having a diameter not less than 12 mm are recommended. In any case, as far as possible, only one size and grade of bolt should be used on a project.

The nominal diameter of holes for ordinary bolts, D_h, is equal to the bolt diameter, d_b, plus 1 mm for 12 mm diameter bolts, 2 mm for bolts between 16 and 24 mm in diameter and 3 mm for bolts 27 mm or greater in diameter (*Table 33*: BS 5950):

$$D_h = d_b + 1 \text{ mm} \quad \text{for } d_b = 12 \text{ mm}$$

$$D_h = d_b + 2 \text{ mm} \quad \text{for } 16 \leqslant d_b \leqslant 24 \text{ mm}$$

$$D_h = d_b + 3 \text{ mm} \quad \text{for } d_b \geqslant 27 \text{ mm}$$

4.11.2 FASTENER SPACING AND EDGE/END DISTANCES

Clause 6.2 of BS 5950 contains various recommendations regarding the distance between fasteners and edge/end distances to fasteners, some of which are illustrated in *Fig. 4.51* and summarised below:

1. Spacing between centres of bolts, i.e. pitch (p), in the direction of stress and not exposed to corrosive influences should lie within the following limits:

$$2.5d_b \leqslant p \leqslant 14t$$

where d_b is the diameter of bolts and t the thickness of the thinner ply.

2. Minimum edge distance, e_1, and end distance, e_2, to fasteners should conform with the following limits:

Rolled, machine flame cut, sawn or planed edge/end $\geqslant 1.25D_h$

Sheared or hand flame cut edge/end $\geqslant 1.40D_h$

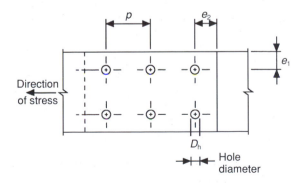

Fig. 4.51 Rules for fastner spacing and edge/end distances to fastners.

where D_h is the diameter of the bolt hole. Note that the edge distance, e_1, is the distance from the centre line of the hole to the outside edge of the plate at right angles to the direction of the stress, whereas the end distance, e_2, is the distance from the centre line to the edge of the plate in the direction of stress.

3. Maximum edge distance, e_1, should not exceed the following:

$$e_1 \leqslant 11t\varepsilon$$

where t is the thickness of the thinner part and $\varepsilon = (275/p_y)^{1/2}$.

4.11.3 STRENGTH CHECKS

Bolted connections may fail due to various mechanisms including shear, bearing, tension and combined shear and tension. The following sections describe these failure modes and outlines the associated design procedures for connections involving (a) ordinary bolts and (b) HSFG bolts.

4.11.3.1 Ordinary bolts

Shear and bearing. Referring to the connection detail shown in *Fig. 4.52*, it can be seen that the loading on bolt A between the web cleat and the column will be in shear, and that there are three principal ways in which the joint may fail. Firstly, the bolts can fail in shear, for example along surface x_1–y_1 (*Fig. 4.52(a)*). Secondly the bolts can fail in bearing as the web cleat cuts into the bolts (*Fig. 4.52(b)*). This can only happen when the bolts are softer than the metal being joined. Thirdly, the metal being joined, i.e. the cleat, can fail in bearing as the bolts cut into it (*Fig. 4.52(c)*). This is the converse of the above situation and can only happen when the bolts are harder than the metal being joined.

It follows, therefore, that the design shear strength of the connection should be taken as the least of:

1. Shear capacity of the bolt,

$$P_s = p_s A_s \qquad (4.64)$$

2. Bearing capacity of bolt,

$$P_{bb} = d_b t_p p_{bb} \qquad (4.65)$$

3. Bearing capacity of connected part,

$$P_{bs} = k_{bs} d_b t_p p_{bs} \leqslant 0.5 k_{bs} e t_p p_{bs} \qquad (4.66)$$

where
p_s shear strength of the bolts (*Table 4.19*)
p_{bb} bearing strength of the bolts (*Table 4.20*)

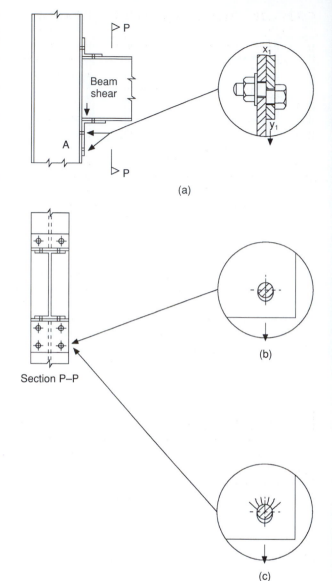

Section P–P

Fig. 4.52 *Failure modes of a beam-to-column connection: (a) single shear failure of bolt; (b) bearing failure of bolt; (c) bearing failure of cleat.*

p_{bs} bearing strength of the connected part (*Table 4.21*)
e end distance e_2
A_s effective area of bolts in shear, normally taken as the tensile stress area, A_t (*Table 4.22*)
t_p thickness of connected part
k_{bs} = 1.0 for bolts in standard clearance holes

Double shear. If a column supports two beams in the manner indicated in *Fig. 4.53*, the failure modes

Table 4.19 Shear strength of bolts (Table 30, BS 5950)

Bolt grade	Shear strength p_s (N/mm^2)
4.6	160
8.8	375
10.9	400
General grade HSFG $\leqslant$ M24	400
to BS 4395–1 $\geqslant$ M27	350
Higher grade HSFG to BS 4395–2	400
Other grades ($U_b \leqslant 1000$ N/mm^2)	$0.4U_b$

Note. U_b is the specified minimum tensile strength of the bolt.

Table 4.20 Bearing strength of bolts (Table 33, BS 5950)

Bolt grade	Bearing strength p_{bb} (N/mm^2)
4.6	460
8.8	1000
10.9	1300
General grade HSFG $\leqslant$ M24	1000
to BS 4395–1 $\geqslant$ M27	900
Higher grade HSFG to BS 4395–2	1300
Other grades ($U_b \leqslant 1000$ N/mm^2)	$0.7(U_b + Y_b)$

Note. U_b is the specified minimum tensile strength of the bolt and Y_b is the specified minimum yield strength of the bolt.

Table 4.21 Bearing strength of connected parts (Table 32)

Steel grade	S275	S355	S460	Other grades
Bearing strength p_{bs} (N/mm^2)	460	550	670	$0.67(U_b + Y_b)$

Note. U_b is the specified minimum tensile strength of the bolt and Y_b is the specified minimum yield strength of the bolt.

Table 4.22 Tensile stress area, A_t

Nominal size and thread diameter (mm)	Tensile stress area, A_t (mm^2)
12	84.3
16	157
20	245
22	303
24	353
27	459
30	561

essentially remain the same as for the previous case, except that the bolts (B) will be in 'double shear'. This means that failure of the bolts will only occur once surfaces x_2–y_2 and x_3–y_3 exceed the shear strength of the bolt (*Fig. 4.53(b)*).

The shear capacity of bolts in double shear, P_{sd}, is given by

$$P_{sd} = 2P_s \qquad (4.67)$$

Thus, double shear effectively doubles the shear strength of the bolt.

Tension. Tension failure may arise in simple connections as a result of excessive tension in the bolts (*Fig. 4.54(a)*) or cover plates (*Fig. 4.54(b)*). The tension capacity of ordinary bolts may be calculated using a simple or more exact approach. Only the simple method is discussed here as it is both easy to use and conservative. The reader is referred to clause 6.3.4.3 of BS 5950 for guidance on the more exact method.

According to the simple method, the nominal tension capacity of the bolt, P_{nom}, is given by

$$P_{nom} = 0.8p_t A_t \qquad (4.68)$$

where
p_t tension strength of the bolt (*Table 4.23*)
A_t tensile stress area of bolt (*Table 4.22*)

The tensile capacity of a flat plate is given by

$$P_t = \alpha_e p_y \qquad (4.69)$$

where effective net area, α_e, is

$$\alpha_e = K_e \alpha_n < \alpha_g \qquad (4.70)$$

in which
K_e = 1.2 for grade S275 steel plates
α_g gross area of plate = bt (*Fig. 4.54*)
α_n net area of plate = α_g – allowance for bolt holes (= $D_h t$, *Fig. 4.54(b)*).

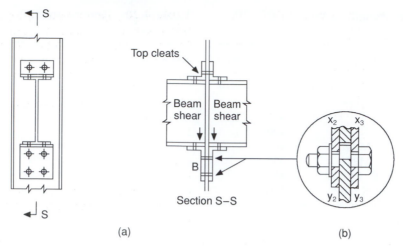

Fig. 4.53 *Double sheer failure.*

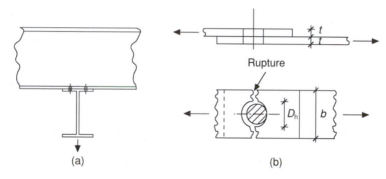

Fig. 4.54 *Typical tension: (a) bolts in tension; (b) cover plate in tension.*

Table 4.23 Tensile strength of bolts (Table 34, BS 5950)

Bolt grade		Tension strength p_t (N/mm^2)
4.6		240
8.8		560
10.9		700
General grade HSFG	$\leqslant$ M24	590
to BS 4395–1	$\geqslant$ M27	515
Higher grade HSFG to BS 4395–2		700
Other grades ($U_b \leqslant 1000$ N/mm^2)		$0.7U_b$ but $\leqslant Y_b$

Note. U_b is the specified minimum tensile strength of the bolt and Y_b is the specified minimum yield strength of the bolt.

Combined shear and tension. Where ordinary bolts are subject to combined shear and tension (*Fig. 4.55*), in addition to checking their shear and tension capacities separately, the following relationship should also be satisfied:

$$\frac{F_s}{P_s} + \frac{F_t}{P_{nom}} \leqslant 1.4 \qquad (4.71)$$

where

F_s applied shear
F_t applied tension
P_s shear capacity (equation 4.64)
P_{nom} tension capacity (equation 4.68).

Note that this expression should only be used when the bolt tensile capacity has been calculated using the simple method.

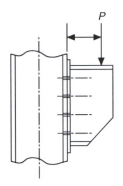

Fig. 4.55 *Bracket bolted to column.*

4.11.3.2 HSFG bolts

If parallel-shank or waisted-shank HSFG bolts, rather than ordinary bolts, were used in the connection detail shown in *Fig. 4.50*, failure of the connection would principally arise as a result of slip between the connected parts. All connections utilizing friction grip fasteners should be checked for slip resistance. Connections using parallel shank HSFG bolts designed to be non-slip in service, should additionally be checked for bearing capacity of the connected parts and shear capacity of the bolts after slip.

Slip resistance. According to clause 6.4.2, the slip resistance of HSFG bolts designed to be non-slip in service, P_{sL}, is given by

$$P_{sL} = 1.1 K_s \mu P_o \qquad (4.72)$$

and for HSFG bolts designed to be non-slip under factored loads by

$$P_{sL} = 0.9 K_s \mu P_o \qquad (4.73)$$

where
P_o minimum shank tension (proof load) (*Table 4.24*)
K_s = 1.0 for bolts in standard clearance holes
μ slip factor $\leqslant 0.5$

Table 4.24 Proof load of HSFG bolts, P_o

Nominal size and thread diameter (mm)	Minimum shank tension or proof load (kN)
12	49.4
16	92.1
20	144
22	177
24	207
27	234
30	286
36	418

The slip factor depends on the condition of the surfaces being joined. According to Table 35 of BS 5950, shot or grit blasted surfaces have a slip factor of 0.5 whereas wire brushed and untreated surfaces have slip factors of 0.3 and 0.2 respectively.

Bearing. The bearing capacity of connected parts after slip, P_{bg}, is given by

$$P_{bg} = 1.5 d_b t_p p_{bs} \leqslant 0.5 e t_p p_{bs} \qquad (4.74)$$

where
d_b bolt diameter
t_p thickness of connected part
p_{bs} bearing strength of connected parts (*Table 4.21*)
e end distance

Shear. As in the case of black bolts, the shear capacity of HSFG bolts, P_s, is given by

$$P_s = p_s A_s \qquad (4.75)$$

where
p_s shear strength of the bolts (*Table 4.19*)
A_s effective area of bolts in shear, normally taken as the tensile stress area, A_t (*Table 4.22*)

Combined shear and tension. When parallel shank friction grip bolts designed to be non-slip in service, are subject to combined shear and tension, then the following additional check should be carried out:

$$\frac{F_s}{P_{sL}} + \frac{F_{tot}}{1.1 P_o} \leqslant 1 \quad \text{but} \quad F_{tot} \leqslant A_t p_t \quad (4.76)$$

where
F_s applied shear
F_{tot} total applied tension in the bolt, including the calculated prying forces $= p_t A_t$ in which p_t is obtained from *Table 4.23* and A_t is obtained from *Table 4.22*
P_{sL} slip resistance (equation 4.72)
P_o minimum shank tension (*Table 4.24*).

4.11.4 BLOCK SHEAR

Bolted beam-to-column connection may fail as a result of block shear. Failure occurs in shear at a row of bolt holes parallel to the applied force, accompanied by tensile rupture along a perpendicular face. This type of failure results in a block of material being torn out by the applied shear force as shown in *Fig. 4.56*.

Block shear failure can be avoided by ensuring that the applied shear force, F_r, does not exceed the block shear capacity, P_r, given by

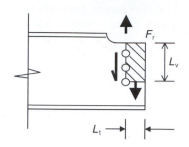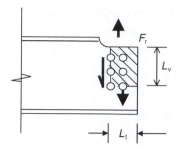

Fig. 4.56 *Block shear.*

$$P_r = 0.6p_y t[L_v + K_e(L_t - kD_t)] \quad (4.77)$$

where
D_t is the hole size for the tension face
t is the thickness

L_t and L_v are the dimensions shown in *Fig. 4.56*
K_e is the effective net area coefficient
k = 0.5 for a single line of bolts parallel to the applied shear
    = 2.5 for two lines of bolts parallel to the applied shear

---

Example 4.21 Analysis of a double angle web cleat beam-to-column connection (BS 5950)

Show that the double angle web cleat beam-to-column connection detail shown below is suitable to resist the design shear force, V, of 400 kN. Assume the steel is grade S275 and the bolts are M20, grade 8.8 in 2 mm clearance holes.

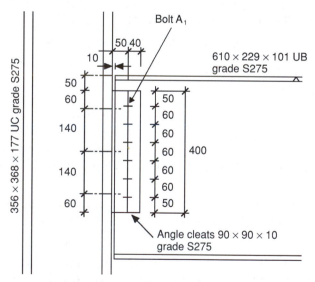

CHECK FASTENER SPACING AND EDGE/END DISTANCES

Diameter of bolt, d_b = 20 mm
Diameter of bolt hole, D_h = 22 mm
Pitch of bolt, p = 140 mm and 60 mm
Edge distance, e_1 = 40 mm
End distance, e_2 = 60 mm and 50 mm
Thickness of angle cleat, t_p = 10 mm

The following conditions need to be met:

$$\text{Pitch} \geq 2.5d_b = 2.5 \times 20 = 50 < 140 \text{ and } 60 \quad \text{OK}$$
$$\text{Pitch} \leq 14t_p = 14 \times 10 = 140 \leq 140 \text{ and } 60 \quad \text{OK}$$
$$\text{Edge distance } e_1 \geq 1.4D_h = 1.4 \times 22 = 30.8 < 40 \quad \text{OK}$$
$$\text{End distance } e_2 \geq 1.4D_h = 1.4 \times 22 = 30.8 < 60 \text{ and } 50 \quad \text{OK}$$
$$e_1 \text{ and } e_2 \leq 11t_p\varepsilon = 11 \times 10 \times 1 = 110 < 40, 50 \text{ and } 60 \quad \text{OK}$$

(For grade S275 steel with $t_p = 10$ mm, $p_y = 275$ N/mm^2, $\varepsilon = 1$.) Hence all fastener spacing and edge/end distances to fasteners are satisfactory.

CHECK STRENGTH OF BOLTS CONNECTING CLEATS TO SUPPORTING COLUMN

Shear
6 No., M20 grade 8.8 bolts. Hence $A_s = 245$ mm (*Table 4.22*) and $p_s = 375$ N/mm^2 (*Table 4.19*). Shear capacity of single bolt, P_s, is

$$P_s = p_sA_s = 375 \times 245 = 91.9 \times 10 = 91.9 \text{ kN}$$

Shear capacity of bolt group is

$$6P_s = 6 \times 91.9 = 551.4 \text{ kN} > V = 400 \text{ kN}$$

Hence bolts are adequate in shear.

Bearing
Bearing capacity of bolt, P_{bb}, is given by

$$P_{bb} = d_btp_{bb} = 20 \times 10 \times 1000 = 200 \times 10^3 = 200 \text{ kN}$$

Since thickness of angle cleat (= 10 mm) < thickness of column flange (= 23.8 mm), bearing capacity of cleat is critical. Bearing capacity of cleat, P_{bs}, is given by

$$P_{bs} = k_{bs}d_btp_{bs} = 1 \times 20 \times 10 \times 460 = 92 \times 10^3 \text{ N} = 92 \text{ kN}$$
$$\leq 0.5k_{bs}etp_{bs} = 0.5 \times 1 \times 60 \times 10 \times 460 = 138 \times 10^3 \text{ N} = 138 \text{ kN}$$

Bearing capacity of connection is

$$6 \times 92 = 552 \text{ kN} > V = 400 \text{ kN}$$

Therefore bolts are adequate in bearing.

CHECK STRENGTH OF BOLT GROUP CONNECTING CLEATS TO WEB OF SUPPORTED BEAM

Shear
6 No., *M*20 grade 8.8 bolts; from above, $A_s = 245$ mm^2 and $p_s = 375$ N/mm^2. Since bolts are in double shear, shear capacity of each bolt is $2P_s = 2 \times 91.9 = 183.8$ kN

Loads applied to the bolt group are vertical shear, $V = 400$ kN and moment, $M = 400 \times 50 \times 10^{-3} = 20$ kN m.

Outermost bolt (A_1) subject to greatest shear force which is equal to the resultant of the load due to the moment, $M = 20$ kN m and vertical shear force, $V = 400$ kN. Load on the outermost bolt due to moment, F_{mb}, is given by

$$F_{mb} = \frac{M}{Z}A = \frac{20 \times 10^3}{420A}A = 47.6 \text{ kN}$$

where A is the area of bolt and Z the modulus of the bolt group given by

$$\frac{I}{y} = \frac{63\,000}{150}A = 420A \text{ mm}^3$$

in which I is the inertia of the bolt group equal to

$$2A(30^2 + 90^2 + 150^2) = 63\,000A \text{ mm}^4$$

Load on outermost bolt due to shear, F_{vb}, is given by

$$F_{vb} = V/\text{No. of bolts} = 400/6 = 66.7 \text{ kN}$$

Resultant shear force of bolt, F_s, is

$$F_s = (F_{vb}^2 + F_{mb}^2)^{1/2} = (66.7^2 + 47.6^2)^{1/2} = 82 \text{ kN}$$

Since F_s (= 82 kN) $< 2P_s$ (= 183.8 kN) the bolts are adequate in shear.

Bearing
Bearing capacity of bolt, P_{bb}, is

$$P_{bb} = 200 \text{ kN (from above)} > F_s \quad \text{OK}$$

Bearing capacity of each cleat, P_{bs}, is

$$P_{bs} = 92 \text{ kN (from above)}$$

Bearing capacity of both cleats is

$$2 \times 92 = 184 \text{ kN} > F_s \quad \text{OK}$$

Bearing capacity of the web, P_{bs}, is

$$P_{bs} = k_{bs}d_bt_wp_{bs} = 1 \times 20 \times 10.6 \times 460 \times 10^{-3} = 97.52 \text{ kN} > F_s \quad \text{OK}$$

Hence bolts, cleats and beam web are adequate in bearing.

SHEAR STRENGTH OF CLEATS
Shear capacity of a single angle cleat, P_v, is

$$P_v = 0.6p_yA_v = 0.6 \times 275 \times 3600 \times 10^{-3} = 594 \text{ kN}$$

where

$$A_v = 0.9A_o \text{ (clause 4.2.3 of BS 5950)} = 0.9 \times \text{thickness of cleat } (t_p) \times \text{length of cleat } (\ell_p)$$

$$= 0.9 \times 10 \times 400 = 3600 \text{ mm}^2.$$

Since shear force $V/2$ (= 200 kN) $< P_v$ (= 594 kN) the angle is adequate in shear.

BENDING STRENGTH OF CLEATS

$$M = \frac{V}{2} \times 50 \times 10^{-3} = \frac{400}{2} \times 50 \times 10^{-3} = 10 \text{ kN m}$$

Assume moment capacity of one angle of cleat, M_c, is

$$M_c = p_yZ = 275 \times 266.7 \times 10^3 = 73.3 \times 10^6 \text{ N mm} = 73.3 \text{ kN m} > M$$

where $Z = \dfrac{t_p\ell_p^2}{6} = \dfrac{10 \times 400^2}{6} = 266\,667 \text{ mm}^3$. Angle cleat is adequate in bending.

LOCAL SHEAR STRENGTH OF THE BEAM
Shear capacity of the supported beam, P_v, is

$$P_v = 0.6p_yA_v = 0.6 \times 275 \times 6383.3 = 1053.2 \times 10^3 \text{ N} = 1053.2 \text{ kN} > V \text{ (= 400 kN)}$$

where $A_v = t_wD = 10.6 \times 602.2 = 6383.3 \text{ mm}^2$. Hence supported beam at the end is adequate in shear.

Example 4.22 Analysis of a bracket–to–column connection (BS 5950)

Show that the bolts in the bracket-to-column connection below are suitable to resist the design shear force of 200 kN. Assume the bolts are all $M16$, grade 8.8.

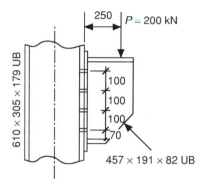

Since the bolts are subject to combined shear and tension, the bolts should be checked for shear, tension and combined shear and tension separately.

SHEAR

Design shear force, $P = 200$ kN
Number of bolts, N $= 8$
Shear force/bolt, F_s $= P/N = 200/8 = 25$ kN

Shear capacity of bolt, P_s, is

$$P_s = p_s A_s = 375 \times 157 = 58.9 \times 10^3 = 58.9 \text{ kN} > F_s \quad \text{OK}$$

TENSILE CAPACITY

Maximum bolt tension, F_t, is

$$F_t = Pey_1/2\Sigma y^2$$

$$= 200 \times 250 \times 370/2(70^2 + 170^2 + 270^2 + 370^2)$$

$$= 38 \text{ kN}$$

Tension capacity, P_{nom}, is

$$P_{nom} = 0.8 p_t A_t = 0.8 \times 560 \times 157 = 70.3 \times 10^3 \text{ N}$$

$$= 70.3 \text{ kN} > F_t \quad \text{OK}$$

COMBINED SHEAR AND TENSION

Combined check:

$$\frac{F_s}{P_s} + \frac{F_t}{P_{nom}} \leqslant 1.4$$

$$\frac{25}{58.9} + \frac{38}{70.3} = 0.96 \leqslant 1.4 \quad \text{OK}$$

Hence the $M16$, grade 8.8 bolts are satisfactory.

Example 4.23 Analysis of a beam splice connection (BS 5950)

Show that the splice connection shown below is suitable to resist a design bending moment, M, and shear force, F, of 270 kN m and 300 kN respectively. Assume the steel grade is S275 and the bolts are general grade, $M22$, parallel-shank HSFG bolts. The slip factor, μ, can be taken as 0.5.

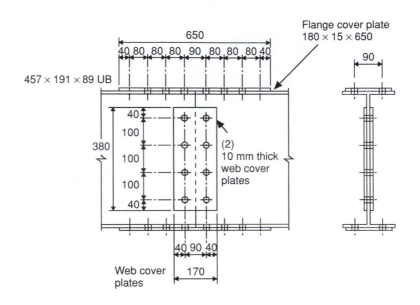

Assume that (1) flange cover plates resist the design bending moment, M and (2) web cover plates resist the design shear force, F, and the torsional moment (= Fe), where e is half the distance between the centroids of the bolt groups either side of the joint.

FLANGE SPLICE

Check bolts in flange cover plate
Single cover plate on each flange. Hence the bolts in the flange are subject to single shear and must resist a shear force equal to

$$\frac{\text{Applied moment}}{\text{Overall depth of beam}} = \frac{M}{D} = \frac{270 \times 10^3}{463.6} = 582.4 \text{ kN}$$

Slip resistance of single bolt in single shear, P_{sL}, is

$$P_{sL} = 1.1 K_s \mu P_o = 1.1 \times 1.0 \times 0.5 \times 177 = 97.3 \text{ kN} \quad (P_o = 177 \text{ kN, } Table\ 4.24)$$

Since thickness of beam flange (= 17.7 mm) > thickness of cover plate (= 15 mm), bearing in cover plate is critical. Bearing capacity, P_{bg}, of cover plate is

$$1.5 d_b t p_{bs} \leqslant 0.5 e t p_{bs}$$

$$1.5 d_b t p_{bs} = 1.5 \times 22 \times 15 \times 460 = 227.7 \times 10^3 \text{ N } (p_{bs} = 460 \text{ N/mm}^2, Table\ 4.21)$$

$$0.5 e t p_{bs} = 0.5 \times 40 \times 15 \times 460 = 138 \times 10^3 \text{ N}$$

Shear capacity of the bolts after slip, P_s, is

$$P_s = p_s A_s = 400 \times 303 \times 10^{-3} = 121.2 \text{ kN per bolt}$$

where $A_s = A_t = 303 \text{ mm}^2$ (*Table 4.22*)

Hence shear strength based on slip resistance of bolts. Slip resistance of 8 No., $M22$ bolts $= 8P_{sL}$ $= 8 \times 97.3 = 778.4$ kN $>$ shear force in bolts $= 582.4$ kN. Therefore the 8 No., $M22$ HSFG bolts provided are adequate in shear.

Check tension capacity of flange cover plate

Gross area of plate, α_g, is

$$\alpha_g = 180 \times 15 = 2700 \text{ mm}^2$$

Net area of plate, α_n, is

$$\alpha_n = \alpha_g - \text{area of bolt holes}$$
$$= 2700 - 2(15 \times 24) = 1980 \text{ mm}^2$$

Force in cover plate, F_t, is

$$F_t = \frac{M}{D + T_{fp}} = \frac{270 \times 10^3}{463.6 + 15} = 564.1 \text{ kN}$$

where T_{fp} is the thickness of flange cover plate $= 15$ mm.
Tension capacity of cover plate, P_t, is

$$P_t = \alpha_e p_y = 2376 \times 275 = 653.4 \times 10^3 \text{ N} = 653.4 \text{ kN}$$

where $\alpha_e = K_e \alpha_n = 1.2 \times 1980 = 2376 \text{ mm}^2$.
Since $P_t > F_t$, cover plate is OK in tension.

Check compressive capacity of flange cover plate

Top cover plate in compression will also be adequate provided the compression flange has sufficient lateral restraint.

WEB SPLICE

Check bolts in web splice

Vertical shear $F_v = 300$ kN
  Torsional moment $= F_v e_o$ (where e_o is half the distance between the centroids of the bolt groups either side of the joint in accordance with assumption (2) above)

$$= 300 \times 45 = 13\,500 \text{ kN mm}$$

Maximum resultant force F_R occurs on outermost bolts, e.g. bolt A, and is given by the following expression:

$$F_R = \sqrt{F_{vs}^2 + F_{tm}^2}$$

where F_{vs} is the force due to vertical shear and F_{tm} the force due to torsional moment.

$$F_{vs} = \frac{F_v}{N} = \frac{300}{4} = 75 \text{ kN}$$

$$F_{tm} = \frac{\text{Torsional moment} \times A_b}{Z_b}$$

where
A_b area of single bolt
Z_b modulus of the bolt group which is given by

$$\frac{\text{Inertia of bolt group}}{\text{Distance of furthest bolt}} = \frac{2A_b(50^2 + 150^2)}{150} = 333.33A_b$$

Hence

$$F_{tm} = \frac{13\,500A_b}{333.33A_b} = 40.5 \text{ kN}$$

and $$F_R = \sqrt{40.5^2 + 75^2} = 85.2 \text{ kN}$$

From above, slip resistance of $M22$ HSFG bolt, $P_{sL} = 97.3$ kN. Since two cover plates are present, slip resistance per bolt is

$$2P_{sL} = 2 \times 97.3 = 194.6 \text{ kN} > F_R$$

Therefore, slip resistance of bolts in web splice is adequate.
Similarly, shear capacity of bolt in double shear, $P_s = 2 \times 121.2 = 242.4$ kN $> F_R$
Therefore, shear capacity of bolts in web splice after slip is also adequate.

Check web of beam
The forces on the edge of the holes in the web may give rise to bearing failure and must therefore be checked. Here e = edge distance for web and

$$\tan\theta = \frac{40.5}{75} \Rightarrow \theta = 28.37°$$

$$e = \frac{45}{\sin\theta} = \frac{45}{0.475} = 94.7$$

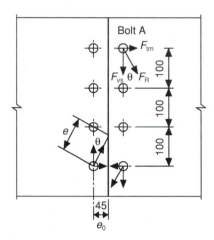

Bearing capacity of web, P_{bg}, is

$$= 1.5 d_b t p_{bs} \leqslant 0.5 e t p_{bs} = 0.5 \times 94.7 \times 10.6 \times 460 = 230.9 \times 10^3 \text{ N} = 203.9 \text{ kN}$$

$$= 1.5 \times 22 \times 10.6 \times 460 = 160.9 \times 10^3 \text{ N} = 160.9 \text{ kN} > F_R$$

Hence web of beam is adequate in bearing.

Check web cover plates
The force on the edge of the holes in the web cover plates may give rise to bearing failure and must therefore be checked. Here e = edge distance for web cover plates and $\theta = 28.37°$. Thus

$$e = \frac{40}{\cos\theta} = \frac{40}{0.88} = 45.5$$

Bearing capacity of one plate is

$$= 1.5 d_b t p_{bs} \leqslant 0.5 e t p_{bs} = 0.5 \times 45.5 \times 10 \times 460 = 104.7 \times 10^3 \text{ N} = 104.7 \text{ kN}$$

$$= 1.5 \times 22 \times 10 \times 460 = 151.8 \times 10^3 \text{ N} = 151.8 \text{ kN}$$

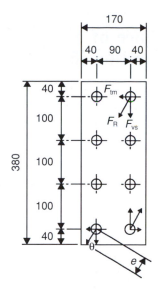

Thus bearing capacity of single web plate = 104.7 kN and web capacity of two web plates = 2×104.7 = 209 kN > F_R. Hence web cover plates are adequate in bearing.

Check web cover plates for shear and bending

Check shear capacity. Shear force applied to one plate is

$$F_v/2 = 300/2 = 150 \text{ kN}$$

Torsional moment applied to one plate is

$$150 \times 45 = 6750 \text{ kN mm} = 6.75 \text{ kN m}$$

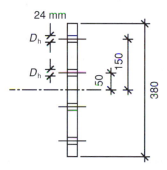

Shear capacity of one plate, P_v, is

$$P_v = 0.6p_y A_v = 0.6 \times 275 \times 0.9 \times 10(380 - 4 \times 24) = 421.7 \text{ kN} > 150 \text{ kN}$$

where $A_v = 0.9A_n$. Hence cover plate is adequate in shear.

Check moment capacity

Since applied shear force (= 150 kN) $< 0.6P_v$ = 253 kN, moment capacity, $M_c = p_y Z$ where $Z = I/y$.

Referring to the above diagram, I of plate = $10 \times 380^3/12 - 2(10 \times 24)50^2 - 2(10 \times 24)150^2$ = $337 \times 10^5 \text{ mm}^2$

$$Z = I/y = 337 \times 10^5/190 = 177 \times 10^3 \text{ mm}^2$$

$$M_c = p_y Z = 275 \times 177 \times 10^3 = 48.7 \times 10^6 \text{ N mm} = 48.7 \text{ kN m} > 6.75 \text{ kN m}$$

Hence web cover plates are also adequate for bending.

Example 4.24 Shear resistance of a welded end plate beam-to-column connection (BS 5950)

Calculate the design shear resistance of the connection shown below, assuming that the steel is grade S275 and the bolts are $M20$, grade 8.8 in 2 mm clearance holes.

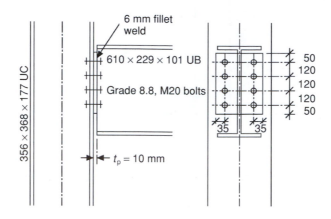

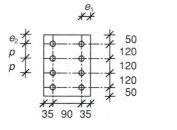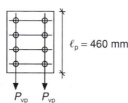

CHECK FASTENER SPACING AND EDGE/END DISTANCES

Diameter of bolt, d_b = 20 mm
Diameter of bolt hole, D_h = 22 mm
Pitch of bolt, p = 120 mm
Edge distance, e_1 = 35 mm
End distance, e_2 = 50 mm
Thickness of end plate, t_p = 10 mm

The following conditions need to be met:

Pitch $\geqslant 2.5d_b = 2.5 \times 20 = 50 < 120$ OK
Pitch $\leqslant 14t_p = 14 \times 10 = 140 > 120$ OK
Edge distance $e_1 \geqslant 1.4D_h = 1.4 \times 22 = 30.8 < 35$ OK
End distance $e_2 \geqslant 1.4D_h = 1.4 \times 22 = 30.8 < 50$ OK
e_1 and e_2 $\leqslant 11t_p\varepsilon = 11 \times 10 \times 1 = 110 > 35, 50$ OK

For grade S275 steel with $t_p = 10$ mm, $p_y = 275$ N/mm^2 (*Table 4.3*), $\varepsilon = 1$ (equation 4.4). Hence all fastener spacing and edge/end distances to fasteners are satisfactory.

BOLT GROUP STRENGTH

Shear

8 No., $M20$ grade 8.8 bolts; $A_s = 245$ mm (*Table 4.22*) and $p_s = 375$ N/mm (*Table 4.19*)
Shear capacity of single bolt, P_s, is

$$P_s = p_s A_s = 375 \times 245 = 91.9 \times 10^3 = 91.9 \text{ kN}$$

Shear capacity of bolt group is

$$8P_s = 8 \times 91.9 = 735 \text{ kN}$$

Bearing

Bearing capacity of bolt, P_{bb}, is given by

$$P_{bb} = d_b t p_{bb} = 20 \times 10 \times 1000 = 200 \times 10^3 = 200 \text{ kN}$$

End plate is thinner than column flange and will therefore be critical. Bearing capacity of end plate, P_{bs}, is given by

$$P_{bs} = k_{bs} d_b t_p p_{bs} = 1 \times 20 \times 10 \times 460 = 92 \times 10^3 \text{ N} = 92 \text{ kN}$$

$$\leqslant 0.5 k_{bs} e_2 t_p p_{bs} = 0.5 \times 1 \times 50 \times 10 \times 460 = 115 \times 10^3 \text{ N} = 115 \text{ kN}$$

Hence bearing capacity of connection = $8 \times 92 = 736$ kN.

END PLATE SHEAR STRENGTH

$$A_v = 0.9 A_n = 0.9 t_p (\ell_p - 4 D_h) = 0.9 \times 10 \times (460 - 4 \times 22) = 3348 \text{ mm}^2$$

Shear capacity assuming single plane of failure, P_{vp}, is

$$P_{vp} = 0.6 p_y A_v = 0.6 \times 275 \times 3348 = 552.4 \times 10^3 = 552.4 \text{ kN}$$

Shear capacity assuming two failure planes is

$$2P_{vp} = 2 \times 552.4 = 1104.8 \text{ kN}$$

WELD STRENGTH

(Readers should refer to *section 4.11.5* before performing this check.) Fillet weld of 6 mm provided. Hence

$$\text{Leg length, } s = 6 \text{ mm}$$

Design strength, $p_w = 220$ N/mm^2 (assuming electrode strength = 42 N/mm^2, *Table 4.25*)
Throat size, $a = 0.7s = 0.7 \times 6 = 4.2$ mm
Effective length of weld, ℓ_w, is

$$\ell_w = 2(\ell_p - 2s) = 2(460 - 2 \times 6) = 896 \text{ mm}$$

Hence design shear strength of weld, V_w, is

$$V_w = p_w a \ell_w = 220 \times 4.2 \times 896 = 828 \times 10^3 \text{ N} = 828 \text{ kN}$$

LOCAL SHEAR STRENGTH OF BEAM WEB AT THE END PLATE

$$P_{vb} = 0.6 p_y A_v = 0.6 \times 275 (0.9 \times 10.6 \times 460) \times 10^{-3} = 724 \text{ kN}$$

Hence, strength of connection is controlled by shear strength of beam web and is equal to 724 kN.

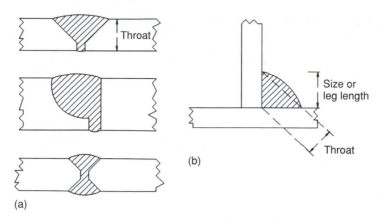

Fig. 4.57 *Types of weld; (a) butt weld; (b) fillet weld.*

4.11.5 WELDED CONNECTIONS

The two main types of welded joints are **fillet welds** and **butt welds**. Varieties of each type are shown in *Fig. 4.57*. Essentially the process of welding consists of heating and melting steel in and/or around the gap between the pieces of steel that are being welded together. Welding rods consist of a steel rod surrounded by a flux which helps the metal to melt and flow into the joint. Welding can be accomplished using oxy-acetylene equipment, but the easiest method uses electric arc welding.

4.11.5.1 Strength of welds

If welding is expertly carried out using the correct grade of welding rod, the resulting weld should be considerably stronger than the pieces held together. However, to allow for some variation in the quality of welds, it is assumed that the weld strength for fillet welds is as given in Table 37 of BS 5950, reproduced below as *Table 4.25*.

Table 4.25 Design strength of fillet welds, p_w (Table 37, BS 5950)

Steel grade	Electrode classification		
	35	*42*	*50*
	N/mm²	N/mm²	N/mm²
S275	220	220[a]	220[a]
S355	220[b]	250	250[a]
S460	220[b]	250[b]	280

Notes. [a] Over matching electrodes.
[b] Under matching electrodes. Not to be used for partial penetration butt welds.

Nevertheless, the design strength of the weld can be taken as the same as that for the parent metal if the joint is a butt weld, or alternatively a fillet weld satisfying the following conditions:

1. The weld is symmetrical as shown in *Fig. 4.58*.
2. It is made with suitable electrodes which will produce specimens at least as strong as the parent metal.
3. The sum of throat sizes (*Fig. 4.58*) is not less than the connected plate thickness.
4. The weld is principally subject to direct tension or compression (*Fig. 4.58*).

4.11.5.2 Design details

Figure 4.57 also indicates what is meant by the weld **leg length**, s, and the **effective throat size**, a, which should not be taken as greater than $0.7s$.

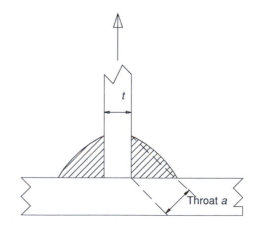

Fig. 4.58 *Special fillet weld.*

The **effective length** of a run of weld should be taken as the actual length, less one leg length for each end of the weld. Where the weld ends at a corner of the metal, it should be continued around the corner for a distance greater than 2*s*. In a lap joint, the minimum lap length should be not less than 4*t*, where *t* is the thinner of the pieces to be joined.

For fillet welds, the 'vector sum of the design stresses due to all forces and moments transmitted by the weld should not exceed the design strength, p_w'.

Example 4.25 Analysis of a welded beam-to-column connection (BS 5950)

A grade S275 steel 610 × 229 × 101 UB is to be connected, via a welded end plate onto a 356 × 368 × 177 UC. The connection is to be designed to transmit a bending moment of 500 kN m and a shear force of 300 kN. Show that the proposed welding scheme for this connection is adequate.

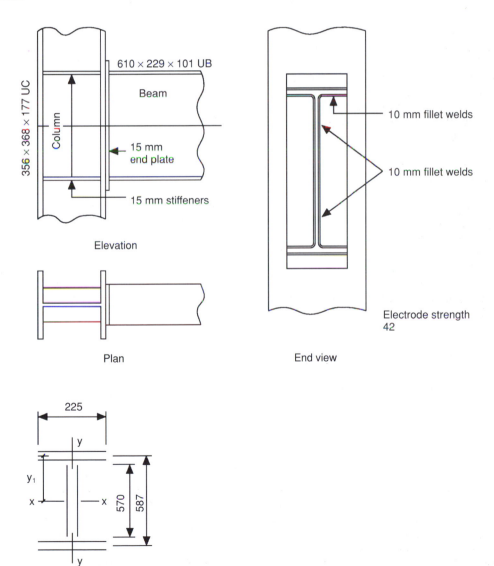

356 × 368 × 177 UC

Column

610 × 229 × 101 UB

Beam

15 mm end plate

15 mm stiffeners

Elevation

Plan

10 mm fillet welds

10 mm fillet welds

End view

Electrode strength 42

225

y

y_1

x — x

570

587

y

Leg length of weld, $s = 10$ mm. Effective length of weld is

$$4(\ell_w - 2s) + 2(h_w - 2s) = 4(225 - 2 \times 10) + 2(570 - 2 \times 10) = 1920 \text{ mm}$$

$$\text{Weld force} = \frac{\text{shear force}}{\text{effective length of weld}} = \frac{300}{1920} = 0.16 \text{ kN/mm}$$

Weld second moment of area, I_{xx}, is

$$I_{xx} = 4\left[1 \times \left(\ell_w - 2s\right)\right]y_1^2 + 2\left[1 \times \left(h_w - 2s\right)^3 / 12\right]$$

$$= 4\left[\left(205\right)\right]293.5^2 + 2\left[1 \times \left(550\right)^3 / 12\right]$$

$$= 98\ 365\ 812 \text{ mm}^4$$

Weld shear (moment) is

$$\frac{My}{I_{xx}} = \frac{500 \times 10^3 \times 293.5}{98\ 365\ 812} = 1.49 \text{ kN/mm}$$

where $y = D/2 = 587/2 = 293.5$ mm

Vectored force $= \sqrt{1.49^2 + 0.16^2} = 1.50$ kN/mm

Since the steel grade is S275 and the electrode strength is 42, the design strength of the weld is 220 N/mm^2 (*Table 4.25*).

Weld capacity of 10 mm fillet weld, p_{wc}, is

$$p_{wc} = ap_w = 0.7 \times 10 \times 220 \times 10^{-3} = 1.54 \text{ kN/mm} > 1.50 \text{ kN/mm} \quad \text{OK}$$

where throat thickness of weld, $a = 0.7s = 0.7 \times 10$ mm

Hence proposed welding scheme is adequate.

4.11 Summary

This chapter has considered the design of a number of structural steelwork elements, including beams, composite floors, columns and connections, to BS 5950: *Structural use of steel work in buildings*. The ultimate limit state of strength and the serviceability limit state of deflection principally influence the design of steel elements. Many steel structures are still analaysed by assuming that individual elements are simply supported at their ends. Steel sections are classified as plastic, compact, semi-compact or slender depending on how they perform in bending. The design of flexural members generally involves considering the limit states of bending, shear, web bearing/buckling, deflection and if the compression flanges are not fully restrained, lateral torsional buckling. Columns subject to axial load and bending are checked for cross-section capacity and overall buckling.

The two principal methods of connecting steel elements of a structure are bolting and welding. It is vitally important that the joints are designed to act in accordance with the assumptions made in the design. Design of bolted connections, using ordinary (or black) bolts, usually involves checking that neither the bolt nor the elements being joined exceed their shear, bearing or tension capacities. Where HSFG bolts are used, the slip resistance must also be determined. Welded connections are most often used to weld end plates or cleats to members, a task which is normally performed in the fabrication yard rather than on site.

Questions

1. A simply-supported beam spanning 8 m has central point dead and imposed loads of 200 kN and 100 kN respectively. Assuming the beam is fully laterally restrained select and check suitable universal beam sections in (a) grade S275 and (b) grade S460 steel.

2. A simply-supported beam spanning 8 m has uniformly distributed dead and imposed loads of 20 kN/m and 10 kN/m respectively. Assuming the beam is fully laterally restrained select and check suitable universal beam sections in (a) grade S275 and (b) grade S460 steel.

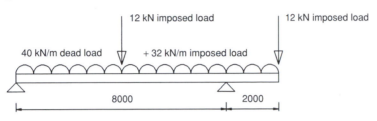

Fig. 4.59

3. For the fully laterally restrained beam shown in *Fig. 4.59* select and check a suitable universal beam section in grade S275 steel to satisfy shear, bending and deflection criteria.

4. If the above beam is laterally unrestrained, select and check a suitable section in grade S275 steel to additionally satisfy lateral torsional buckling, web bearing and buckling criteria. Assume that each support is 50 mm long and lateral restraint conditions at supports are as follows: –

   Compression flange laterally restrained.
   Beam fully restrained against torsion.
   Both flanges free to rotate on plan.
   Destabilizing load conditions.

5. If two discrete lateral restraints, one at mid-span, and one at the cantilever tip, are used to stabilise the above beam, select and check a suitable section in grade S275 steel to satisfy all the criteria in question 4.

6. Carry out designs for the beams shown in *Fig. 4.60*.

7. (a) List and discuss the merits of floor systems used in steel framed structures.
   (b) A floor consists of a series of beams 8.0 m span and 4.0 m apart, supporting a reinforced concrete slab 120 mm thick. The superimposed load is 4 kN/m² and the weight of finishes is 1.2 kN/m².

   (i) Assuming the slab and support beams are not connected, select and check a suitable UB section in grade S355 steel.
   (ii) Repeat the above design assuming the floor is of composite construction. Comment on your results. Assume the unit weight of reinforced concrete is 24 kN/m³.

8. (a) List and discuss the factors that influence the load carrying capacity of steel columns.
   (b) A 254 × 254 × 132 universal column in grade S275 steel is required to support an ultimate axial compressive load of 1400 kN and a major axis moment of 160 kN m applied at the top of the element. Assuming the column is 5 m long and effectively pin ended about both axes, check the suitability of the section.

9. A 305 × 305 × 137 universal column section extends through a height of 3.5 m.
   (a) Check that the section is suitable for plastic design using grade S275 and grade S355 steel.
   (b) Calculate the squash load for grade S275 steel.
   (c) Calculate the full plastic moment of resistance for grade S275 steel about the major axis.
   (d) Find the reduced plastic moment of resistance about the major axis when $F = 800$ kN using grade S275 steel.

231

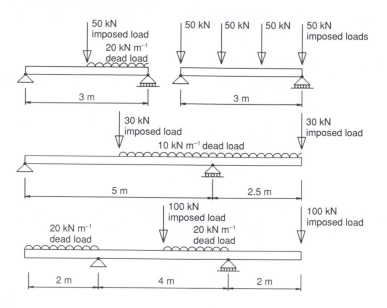

Fig. 4.60

(e) When $M_x = 500$ kN, check the local and overall capacity of the column assuming the base is pinned.

10. Select a suitable short column section in grade S275 steel to support a factored axial concentric load of 1000 kN and factored bending moments of 400 kN m about the major axis, and 100 kN m about the minor axis.

11. Select a suitable column section in grade S275 steel to support a factored axial concentric load of 1000 kN and factored bending moments of 400 kN m about the major axis, and 100 kN m about the minor axis, both applied at each end. The column is 10 m long and is fully fixed against rotation at top and bottom, and the floors it supports are braced against sway.

12. Design a splice connection for a 686 × 254 × 140 UB section in grade S275 steel to cater for half bending strength and half the shear capacity of the section.

13. (a) Explain with the aid of sketches the following terms used in welded connections:
    (i)   fillet welds
    (ii) butt welds
    (iii) throat size.
   (b) A bracket made from a 406 × 178 × 60 universal beam is welded to a steel column as shown below. The bracket is designed to support an ultimate load of 500 kN. Show the proposed welding scheme for this connection is adequate. Assume the steel is of grade S275 and the electrode strength is 42.

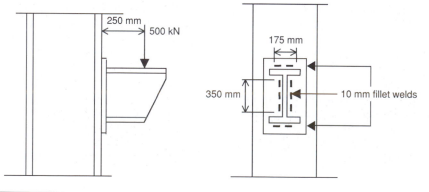

Chapter 5

Design in unreinforced masonry to BS 5628

This chapter is concerned with the design of unreinforced masonry walls to BS 5628: Part 1. The chapter describes the composition and properties of the three main materials used in masonry construction: bricks, blocks and mortars. Masonry is primarily used nowadays in the construction of loadbearing and non-loadbearing walls. The primary aim of this chapter is to give guidance on the design of single leaf and cavity walls, with and without stiffening piers, subject to either vertical or lateral loading.

5.1 Introduction

Structural masonry was traditionally very widely used in civil and structural works including tunnels, bridges, retaining walls and sewerage systems (*Fig. 5.1*). However, the introduction of steel and concrete with their superior strength and cost characteristics led to a sharp decline in the use of masonry for these applications.

Over the past two decades or so, masonry has recaptured some of the market lost to steel and con-

crete due largely to the research and marketing work sponsored in particular by the Brick Development Association. For instance, everybody now knows that 'brick is beautiful'. Less well appreciated,

Fig. 5.1 *Traditional application of masonry in construction: (a) brick bridge; (b) brick sewer; (c) brick retaining wall.*

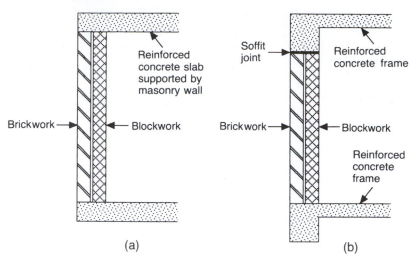

Fig. 5.2 (a) Loadbearing and (b) non-loadbearing masonry walls.

perhaps, is the fact that masonry has excellent structural, thermal and acoustic properties. Furthermore, it displays good resistance to fire and the weather. On this basis it has been argued that it is faster and cheaper to build certain types of buildings using only masonry rather than using a combination of materials to provide these properties.

Brick manufacturers have also pointed out that, unlike steelwork, masonry does not require regular maintenance nor, indeed, suffers from the durability problems which have plagued concrete. The application of reinforcing and prestressing techniques to masonry design will, it is believed, considerably improve its structural properties and hence its appeal to designers.

Despite the above, masonry is primarily used nowadays for the construction of loadbearing and non-loadbearing walls (*Fig. 5.2*). These structures are principally designed to resist lateral and vertical loading. The lateral loading arises mainly from the wind pressure acting on the wall. The vertical loading is attributable to dead plus imposed loading from any supported floors, roofs, etc. and/or self-weight of the wall.

Design of masonry structures in the UK is carried out in accordance with the recommendations given in BS 5628: *Code of Practice for Use of Masonry*. This code is divided into three parts:

Part 1: *Structural Use of Unreinforced Masonry.*
Part 2: *Structural Use of Reinforced and Prestressed Masonry.*
Part 3: *Materials and Components, Design and Workmanship.*

Part 1 was originally published in 1978 and Parts 2 and 3 were issued in 1985. This book deals only with the design of unreinforced masonry walls, subject to vertical or lateral loading. It should, therefore, be assumed that all references to BS 5628 refer to Part 1, unless otherwise noted.

Before looking in detail at the design of masonry walls, the composition and properties of the component materials are considered in the following sections.

5.2 Materials

Structural masonry basically consists of bricks or blocks bonded together using mortar or grout. In cavity walls, wall ties complying with BS 1243 or DD 140 are also used to tie together the two skins of masonry (*Fig. 5.3*). In external walls, damp-proof courses are also necessary to prevent moisture ingress to the building fabric (*Fig. 5.4*).

The following discussion will concentrate on the composition and properties of the three main components of structural masonry, namely (i) bricks, (ii) blocks and (iii) mortar. The reader is referred to BS 5628: Parts 1 and 3 and BRE Digest 380 for guidance on the design and specification of wall ties and damp-proof systems for normal applications.

5.2.1 BRICKS
Bricks are manufactured from a variety of materials such as clay, lime and sand/flint, concrete and natural stone. Of these, clay bricks are by far the most commonly used variety in the UK.

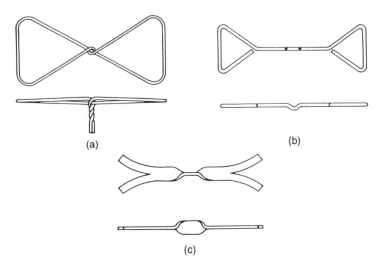

Fig. 5.3 *Wall ties to BS 1243: (a) butterfly tie; (b) double triangle tie; (c) vertical twist tie.*

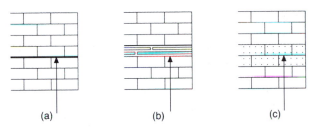

Fig. 5.4 *Damp-proof courses (a) bitumen, pitch and bitumen, polymer, lead, copper, polyethylene d.p.c.; (b) two courses of bonded slate d.p.c.; (c) brick d.p.c. (usually two courses) (based on Table 12, BS 5628: Part 3).*

Clay bricks are manufactured by shaping suitable clays to units of standard size, normally taken to be 215 × 102.5 × 65 mm (*Fig. 5.5*). Sand facings and face textures may then be applied to the 'green' clay. Alternatively, the clay units may be perforated or frogged in order to reduce the self-weight of the unit. Thereafter, the clay units are fired in kilns to a temperature in the range 900–1500 °C in order to produce a brick suitable for structural use. It is because of the firing process that bricks have excellent fire-resistant properties.

In design it is normal to refer to the coordinating size of bricks. This is usually taken to be 225 × 112.5 × 75 mm and is based on the actual or work size of the brick, i.e. 215 × 102.5 × 65 mm, plus an allowance of 10 mm for the mortar joint (*Fig. 5.6*). Clay bricks are also manufactured in metric modular format having a coordinating size of 200 × 100 × 75 mm. Other cuboid and special shapes are also available (BS 4729).

Bricks are normally available in three categories, namely common, facing and engineering. Common bricks are those suitable for general construction

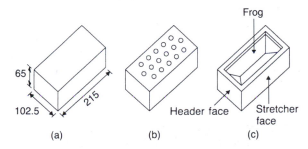

Fig. 5.5 *Typical bricks: (a) solid; (b) perforated; (c) frogged.*

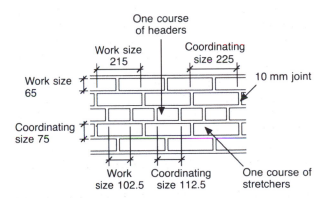

Fig. 5.6 *Coordinating and work size of bricks.*

235

Table 5.1 Classification of bricks by compressive strength and water absorption (Table 4, BS 3921)

Type	Average compressive strength ($N\ mm^{-2}$) not less than	Average absorption (% by weight) not greater than
Engineering A	70	4.5
Engineering B	50	7.0
Damp-proof course 1	5	4.5
Damp-proof course 2	5	7.0
All others	5	No limit

Table 5.2 Frost resistance and soluble salt content designations for clay bricks (BS 3921)

Designation	Frost resistance	Soluble salt content
FL	Frost resistant (F)	Low (L)
FN	Frost resistant (F)	Normal (N)
ML	Moderately frost resistant (M)	Low (L)
MN	Moderately frost resistant (M)	Normal (N)
OL	Not frost resistant (O)	Low (L)
ON	Not frost resistant (O)	Normal (N)

work, with no special claim to give an attractive appearance. Facing bricks are specially made or selected to give an attractive appearance on the basis of colour and texture. Engineering bricks tend to be dense and strong and are designated class A and class B on the basis of strength and water absorption (*Table 5.1*).

BS 3921 also classifies clay bricks with respect to their resistance to frost and the maximum soluble salt content, using the designations shown in *Table 5.2*. The reader is referred to BS 3921 for definitions of these designations. Guidance on the selection of clay bricks most appropriate for particular situations as regards frost resistance and soluble salt content can be found in Table 13 of BS 5628: Part 3.

5.2.2 BLOCKS

Blocks are walling units but, unlike bricks, are normally made from concrete. They are available in two basic types: aerated concrete and aggregate concrete. The aerated blocks are made from a mixture of sand, pulverized fuel ash, cement and aluminium powder. The aggregate blocks have a composition similar to that of normal concrete, consisting chiefly of sand, coarse and fine aggregate and cement plus extenders. Aerated blocks tend to

have lower densities than aggregate blocks, which accounts for the former's superior thermal properties, lower unit weights and lower strengths. Generally, aerated blocks are more expensive than aggregate blocks.

Blocks are manufactured in three basic forms: solid, cellular and hollow (*Fig. 5.7*). Solid blocks have no formed holes or cavities other than those inherent in the material. Cellular blocks have one or more formed holes or cavities which do not pass through the block. Hollow blocks are similar to cellular blocks except that the holes or cavities pass through the block.

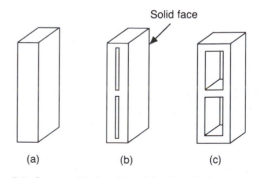

Fig. 5.7 Concrete blocks: (a) solid; (b) cellular; (c) hollow.

Table 5.3 Work sizes of concrete blocks (Table 1, BS 6073: Part 2)

Length (mm)	Height (mm)	Thickness (mm)														
		60	75	90	100	115	125	140	150	175	190	200	215	220	225	250
390	190	x	x	x	x	x		x	x		x	x				
440	140	x	x	x	x			x	x		x	x			x	
440	190	x	x	x	x			x	x		x		x	x		
440	215	x	x	x	x	x	x	x	x	x	x	x	x	x	x	x
440	290	x	x	x	x			x	x		x	x	x			
590	140		x	x	x			x	x		x	x	x			
590	190		x	x	x			x	x		x	x	x			
590	215		x	x	x		x	x	x	x		x	x		x	x

Table 5.4 Compressive strength of concrete blocks (Clause 5.3, BS 6073: Part 2)

Compressive strengths ($N\ mm^{-2}$)
2.8
3.5
5.0
7.0
10.0
15.0
20.0
35.0

For structural design, the two most important properties of blocks are their size and compressive strength. *Tables 5.3* and *5.4* give the most commonly available sizes and compressive strengths of concrete blocks. The most frequently used block has a work face of 440×215 mm, thickness 100 mm and compressive strength 3.5 N mm^{-2}; 2.8 N mm^{-2} is a popular strength for aerated concrete blocks, and 7.0 N mm^{-2} for aggregate concrete blocks as it can be used below ground. Guidance on the selection and specification of concrete blocks in masonry construction can be found in BS 5628: Part 3 and BS 6073 respectively.

5.2.3 MORTARS

The primary function of the mortar is to bind together the individual brick or block units, thereby allowing the transfer of compression, shear and tensile stresses between adjacent units. However, there are several other properties which the mortar must possess for ease of construction and maintenance.

For instance, the mortar should be easy to spread and remain plastic for a sufficient length of time in order that the units can be accurately positioned before setting occurs. On the other hand, the setting time should not be too excessive otherwise the mortar may be squeezed out as successive courses of units are laid. Additionally, the mortar should be able to resist water uptake by the absorbent bricks, otherwise hydration and hence full development of the mortar strength may be prevented.

In its most basic form, mortar simply consists of a mixture of sand and ordinary Portland cement (OPC). However, such a mix is generally unsuitable for use in masonry since it will tend to be too strong in comparison with the strength of the bricks/blocks. It is generally desirable to provide the lowest grade of mortar possible, taking into account the strength and durability of the bricks/blocks to be used. This is to ensure that any cracking which occurs in the masonry due to settlement of the foundations, caused by thermal/moisture movements, etc., will occur in the mortar rather than the masonry units themselves. Cracks in the mortar tend to be smaller because the mortar is more flexible than the masonry units and, hence, are easier to repair.

In order to produce a more practicable mix, it is normal to add lime to the cement mortar. This has the effect of reducing the strength of the mix, but at the same time increasing its workability and bonding properties. For most forms of masonry construction a 1 : 1 : 6 cement : lime : sand mortar is suitable. Similar properties can also be imparted to the mortar by adding a plasticizer to the mix. Alternatively, the OPC can be substituted with masonry cement which is a mixture of approximately 75% OPC, an inert filler and a plasticizer. Such

Table 5.5   Types of mortars (Table 1, BS 5628)

	Mortar designation	Type of mortar (proportion by volume)			Mean compressive strength at 28 days (N mm^{-2})	
		Cement : lime : sand	Masonry cement : sand	Cement : sand with plasticizer	Preliminary (laboratory) tests	Site tests
Increasing strength / Increasing ability to accommodate movement, (i)	(i)	$1 : 0$ to $\frac{1}{4} : 3$	–	–	16.0	11.0
e.g. due to settlement,	(ii)	$1 : \frac{1}{2} : 4$ to $4\frac{1}{2}$	$1 : 2\frac{1}{2}$ to $3\frac{1}{2}$	$1 : 3$ to 4	6.5	4.5
temperature and moisture changes	(iii)	$1 : 1 : 5$ to 6	$1 : 4$ to 5	$1 : 5$ to 6	3.6	2.5
	(iv)	$1 : 2 : 8$ to 9	$1 : 5\frac{1}{2}$ to $6\frac{1}{2}$	$1 : 7$ to 8	1.5	1.0

Direction of change in properties is shown by the arrows

Increasing resistance to frost attack during construction →

← Improvement in bond and consequent resistance to rain penetration

a mix will have intermediate properties to those displayed by the cement, lime, sand mortar and cement, sand, plasticizer mortar.

By varying the proportions of the constituents referred to above, mortars of differing compressive and bond strengths, plasticity and frost susceptibility can be produced as shown in *Table 5.5*. As will be noted from the table there are three basic types of mortars and four mortar strengths associated with each mortar type.

Table 5.6, which is based on Table 13 of BS 5628: Part 3, gives guidance on the selection of masonry units and mortars for particular applications as regards durability.

5.3 Masonry design

The foregoing has summarized some of the more important properties of the component materials which are used in masonry construction. As noted at the beginning of this chapter, masonry is mostly used these days in the construction of loadbearing and panel walls. Loadbearing walls primarily resist vertical loading while panel walls primarily resist lateral loading. *Figure 5.8* shows some commonly used wall types.

The remainder of this chapter considers the design of (1) loadbearing walls, with and without stiffening piers, resisting vertical compression loading (*section 5.5*) and (2) panel walls resisting lateral loading (*section 5.6*).

5.4 Symbols

For the purposes of this chapter, the following symbols have been used. These have largely been taken from BS 5628.

GEOMETRIC PROPERTIES

t	actual thickness of wall or leaf
h	height of panel between restraints
L	length of wall between restraints
A	cross-sectional area of loaded wall
Z	sectional modulus
t_{ef}	effective thickness of wall
h_{ef}	effective height of wall
K	effective thickness coefficient
e_x	eccentricity of loading at top of wall
λ	slenderness ratio
β	capacity reduction factor

Table 5.6 Desirable minimum quality of brick, block and mortar for durability (based on Table 13, BS 5628: Part 3)

Location	Clay bricks	Concrete blocks (type)[a]	Mortar designation	
			NF[b]	PF[b]
Internal walls, inner leaf of external cavity walls, backing (inner face) of external solid walls	NF[b] (O) PF[b] (M)	T	(iv)	(iii)
External walls, outer leaf of external cavity walls	(1)[c] (M) (2) (M) (3) (F)	T R R[g]	(iv)[d] (iii)[e] (ii)[e]	(iii)[d] (iii)[e,f] (ii)[e,f]
Facing (outer face) of external solid construction	(F) —	— (see External walls)	(ii)[h] (see External walls)	(ii)[h]
Damp-proof course (d.p.c)	d.p.c.	(Not applicable)	(i)	(i)
External freestanding walls (with coping)	(F)	S	(iii)	(iii)
Earth retaining walls (back filled with free-draining material)	(F)	R[g]	(ii)[e,i]	(ii)[e,i]
Sewerage work (foul water)	A	—	(ii)[h,i]	(ii)[h,i]

Notes. [a] R: minimum density 1500 kg m^{-3} *and* average compressive strength $\geqslant$ 7 N mm^{-2}. S: minimum density 1500 kg m^{-3} *and* average compressive strength $\geqslant$ 3.5 N mm^{-2} or average compressive strength $\geqslant$ 7 N mm^{-2}. T: all types except that blocks less than 75 mm thick generally only suitable for internal partitions.
[b] NF, no risk of frost during construction; PF, possibility of frost during construction; (O) not frost resistant; (F) frost resistant; (M) moderately frost resistant.
[c] (1) Above d.p.c. (2) Below d.p.c. but not closer than 150 mm to finished ground level. (3) Below 150 mm above finished ground level.
[d] For clay brickwork, not less than mortar designation (iii) should be used; if this brickwork is to be rendered, see footnote f.
[e] Where sulphates are present in the soil or ground water, see footnote h.
[f] Where clay bricks inadvertently become wet, the use of plasticized mortars is desirable.
[g] Concrete blocks are generally not suitable for use in contact with ground from which there is any danger of sulphate attack.
[h] Cement mortars with sulphate-resisting cement in place of ordinary Portland cement should be used.
[i] For clay bricks, mortar designation (i) should be used.

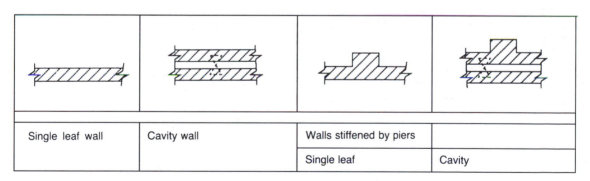

Single leaf wall	Cavity wall	Walls stiffened by piers	
		Single leaf	Cavity

Fig. 5.8 Masonry walls.

COMPRESSION

G_k	characteristic dead load
Q_k	characteristic imposed load
W_k	characteristic wind load
γ_f	partial safety factor for load
γ_m	partial safety factor for materials
f_k	characteristic compressive strength of masonry
N	ultimate design vertical load
N_R	ultimate design vertical load resistance of wall

FLEXURE

$f_{kx\ par}$ characteristic flexural strength of masonry with plane of failure parallel to bed joint

$f_{kx\ perp}$ characteristic flexural strength of masonry with plane of failure perpendicular to bed joint

α bending moment coefficient

μ orthogonal ratio

M ultimate design moment

M_R design moment of resistance

M_{par} design moment with plane of failure parallel to bed joint

M_{perp} design moment with plane of failure perpendicular to bed joint

$M_{k\ par}$ design moment of resistance with plane of failure parallel to bed joint

$M_{k\ perp}$ design moment of resistance with plane of failure perpendicular to bed joint

5.5 Design of vertically loaded masonry walls

In common with most modern codes of practice dealing with structural design, BS 5628: *Code of Practice for Use of Masonry* is based on the limit state philosophy (*Chapter 1*). This code states that the primary aim of design is to ensure an adequate margin of safety against the ultimate limit state being reached. In the case of vertically loaded walls

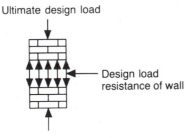

Ultimate design load

Design load resistance of wall

this is achieved by ensuring that the ultimate design load (N) does not exceed the design load resistance of the wall (N_R):

$$N \leq N_R \qquad (5.1)$$

The ultimate design load is a function of the actual loads bearing down on the wall. The design load resistance is related to the design strength of the masonry wall. The following subsections discuss the procedures for estimating the:

1. ultimate design load
2. design strength of masonry walls and
3. design load resistance of masonry walls.

5.5.1 ULTIMATE DESIGN LOADS, N

As discussed in *section 2.3*, the loads acting on a structure are divided into three basic types, namely dead loads, imposed (or live) loads and wind loads. Generally, the ultimate design load is obtained by multiplying the characteristic (dead/imposed/wind) loads (F_k) by the appropriate partial safety factor for loads (γ_f)

$$N = \gamma_f F_k \qquad (5.2)$$

5.5.1.1 Characteristic loads (clause 21, BS 5628)

The characteristic dead loads (G_k) and imposed loads (Q_k) are obtained from the following: (i) BS 648: *Schedule of Weights of Building Materials* and (ii) BS 6399: *Design Loadings for Buildings*, Part 1: *Code of Practice for Dead and Imposed Loads*. The characteristic wind load (W_k) is calculated in accordance with CP 3: Chapter V: Part 2: *Wind Loads* or Part 2 of BS 6399: *Code of Practice for Wind Loads*.

5.5.1.2 Partial safety factors for loads (γ_f) (clause 22, BS 5628)

As discussed in *section 2.3.2*, the applied loads may be greater than anticipated for a number of reasons. Therefore, it is normal practice to factor up the characteristic loads. *Table 5.7* shows the partial

Table 5.7 Partial factors of safety on loading, γ_f, for various load combinations

Load combinations	Ultimate limit state		
	Dead	Imposed	Wind
Dead and imposed	$1.4G_k$ or $0.9G_k$	$1.6Q_k$	
Dead and wind	$1.4G_k$ or $0.9G_k$		The larger of $1.4W_k$ or $0.015G_k$
Dead, imposed and wind	$1.2G_k$	$1.2Q_k$	The larger of $1.2W_k$ or $0.015G_k$

safety factors on loading for various load combinations. Thus, with structures subject to only dead and imposed loads the partial safety factors for the ultimate limit state are usually taken to be 1.4 and 1.6 respectively. The ultimate design load for this load combination is given by

$$N = 1.4G_k + 1.6Q_k \qquad (5.3)$$

In assessing the effect of dead, imposed and wind loads for the ultimate limit state, the partial safety factor is generally taken to be 1.2 for all the load types. Hence

$$N = 1.2(G_k + Q_k + W_k) \qquad (5.4)$$

5.5.2 DESIGN STRENGTH
The design strength of masonry is given by

$$\text{Design strength} = \frac{\beta f_k}{\gamma_m} \qquad (5.5)$$

where f_k is the characteristic compressive strength of masonry, γ_m the partial safety factor for materials and β the capacity reduction factor. Each of these factors are discussed in the following sections.

5.5.2.1 Characteristic compressive strength of masonry, f_k (clause 23.1, BS 5628)
The basic characteristic compressive strengths of normally bonded masonry constructed under laboratory conditions and tested at an age of 28 days are given in *Table 5.8*. As will be noted from the table, the compressive strength is a function of several factors including the type and size of masonry unit, compressive strength of the unit and quality of the mortar used. The following discusses how the characteristic strength of masonry constructed from (a) brickwork and (b) blockwork is determined using *Table 5.8*.

Brickwork. The basic characteristic strengths of masonry built with standard format bricks and mortar designations (i)–(iv) (*Table 5.5*) are obtained from *Table 5.8(a)*. The values in the table may be modified where the following conditions, among others, apply:

1. If the horizontal cross-sectional area of the loaded wall (A) is less than 0.2 m², the basic compressive strength should be multiplied by the factor $(0.7 + 1.5A)$ (clause 23.1.1, BS 5628).

2. When brick walls are constructed so that the thickness of the wall or loaded inner leaf of a cavity wall is equal to the width of a standard format brick, the basic compressive strength may be multiplied by 1.15 (clause 23.1.2, BS 5628).

Blockwork. *Tables 5.8(b)*, *(c)* and *(d)* are used, singly or in combination, to determine the basic characteristic compressive strength of blockwork masonry. As in the case of brick masonry, the values given in these tables may be modified where the cross-sectional area of loaded wall is less than 0.2 m² (see above).

Table 5.8(b) applies to masonry constructed using (solid or hollow) blocks having a shape factor of 0.6. The shape factor is the ratio of the height to least horizontal length of the block (*Fig. 5.9*). *Table 5.8(c)* applies to masonry constructed using hollow blocks having a shape factor of between 2.0 and 4.0. *Table 5.8(d)* applies to masonry constructed using solid blocks having a shape factor of between 2.0 and 4.0.

When walls are constructed using hollow blocks having a shape factor of between 0.6 and 2.0, the characteristic strengths should be determined by interpolating between the values in *Tables 5.8(b)* and *(c)*. When walls are constructed using solid blocks having a shape factor of between 0.6 and 2.0, the characteristic strengths should be determined by interpolating between the values in *Tables 5.8(b)* and *(d)*.

5.5.2.2 Partial safety factor for materials (γ_m)
The quality control exercised both during the manufacture of individual structural units and during construction influences the design strength of the masonry element. Two degrees of quality control are recognized: normal and special. *Table 5.9* shows the values of the partial safety factor for material strengths for various combinations of categories of manufacturing control of structural units and construction control.

The degree of quality control for manufacture and construction will usually be assumed to be 'normal' unless the provisions outlined in clauses 27.2.1.2 and 27.2.2.2 (BS 5628) relating to the 'special' category can clearly be met.

5.5.2.3 Capacity reduction factor (β)
Masonry walls which are tall and slender are likely to be less stable under compressive loading than

Table 5.8 Characteristic compressive strength of masonry, f_k (Table 2, BS 5628)

(a) Constructed with standard format bricks

Mortar designation	Compressive strength of unit (N mm^{-2})								
	5	10	15	20	27.5	35	50	70	100
(i)	2.5	4.4	6.0	7.4	9.2	11.4	15.0	19.2	24.0
(ii)	2.5	4.2	5.3	6.4	7.9	9.4	12.2	15.1	18.2
(iii)	2.5	4.1	5.0	5.8	7.1	8.5	10.6	13.1	15.5
(iv)	2.2	3.5	4.4	5.2	6.2	7.3	9.0	10.8	12.7

(b) Constructed with blocks having a shape factor of 0.6

Mortar designation	Compressive strength of unit (N mm^{-2})							
	2.8	3.5	5.0	7.0	10	15	20	35 or greater
(i)	1.4	1.7	2.5	3.4	4.4	6.0	7.4	11.4
(ii)	1.4	1.7	2.5	3.2	4.2	5.3	6.4	9.4
(iii)	1.4	1.7	2.5	3.2	4.1	5.0	5.8	8.5
(iv)	1.4	1.7	2.2	2.8	3.5	4.4	5.2	7.3

(c) Constructed with hollow blocks having a shape factor of between 2 and 4

Mortar designation	Compressive strength of unit (N mm^{-2})							
	2.8	3.5	5.0	7.0	10	15	20	35 or greater
(i)	2.8	3.5	5.0	5.7	6.1	6.8	7.5	11.4
(ii)	2.8	3.5	5.0	5.5	5.7	6.1	6.5	9.4
(iii)	2.8	3.5	5.0	5.4	5.5	5.7	5.9	8.5
(iv)	2.8	3.5	4.4	4.8	4.9	5.1	5.3	7.3

(d) Constructed with solid concrete blocks having a shape factor of between 2 and 4

Mortar designation	Compressive strength of unit (N mm^{-2})							
	2.8	3.5	5.0	7.0	10	15	20	35 or greater
(i)	2.8	3.5	5.0	6.8	8.8	12.0	14.8	22.8
(ii)	2.8	3.5	5.0	6.4	8.4	10.6	12.8	18.8
(iii)	2.8	3.5	5.0	6.4	8.2	10.0	11.6	17.0
(iv)	2.8	3.5	4.4	5.6	7.0	8.8	10.4	14.6

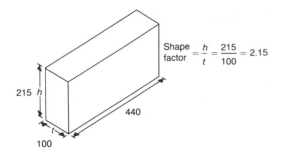

Shape factor $= \dfrac{h}{t} = \dfrac{215}{100} = 2.15$

Fig. 5.9 *Definition of shape factor.*

Table 5.9 Partial safety factors for material strengths, γ_m (Table 4, BS 5628)

		Category of construction control	
		Special	*Normal*
Category of manufacturing control of structural units	Special	2.5	3.1
	Normal	2.8	3.5

Table 5.10 Capacity reduction factor, β (Table 7, BS 5628)

Slenderness	Eccentricity at top of wall, e_x, ratio			
h_{ef}/t_{ef}	*0.05t*	*0.1t*	*0.2t*	*0.3t*
0	1.00	0.88	0.66	0.44
6	1.00	0.88	0.66	0.44
8	1.00	0.88	0.66	0.44
10	0.97	0.88	0.66	0.44
12	0.93	0.87	0.66	0.44
14	0.89	0.83	0.66	0.44
16	0.83	0.77	0.64	0.44
18	0.77	0.70	0.57	0.44
20	0.70	0.64	0.51	0.37
22	0.62	0.56	0.43	0.30
24	0.53	0.47	0.34	
26	0.45	0.38		
27	0.40	0.33		

walls which are short and stocky. Similarly, eccentric loading will reduce the compressive load capacity of the wall. In both cases, the reduction in load capacity reflects the increased risk of failure due to instability rather than crushing of the materials. These two effects are taken into account by means of the capacity reduction factor, β (*Table 5.10*). This factor generally reduces the design strength of the member, in some cases by as much as 70%, and is a function of the following factors which are discussed below: (a) slenderness ratio and (b) eccentricity of loading.

Slenderness ratio (SR). The SR indicates the type of failure which may arise when a member is subject to compressive loading. Thus, walls which are short and stocky will have a low SR and tend to fail by crushing. Walls which are tall and slender will have higher SRs and will also fail by crushing. However, in this case, failure will arise as a result of excessive bending of the wall rather than direct crushing of the materials. Similarly, walls which are rigidly fixed at their ends will tend to have lower SRs and hence higher

design strengths than walls which are partially fixed or unrestrained.

As can be appreciated from the above, the SR depends upon the height of the member, the cross-sectional area of loaded wall and the restraints at the member ends. BS 5628 defines SR as

$$\text{SR} = \frac{\text{effective height}}{\text{effective thickness}} = \frac{h_{ef}}{t_{ef}} \quad (5.6)$$

The following discusses how the (i) effective height and (ii) effective thickness of masonry walls are determined.

EFFECTIVE HEIGHT (h_{ef}). The effective height of a loadbearing wall depends on the actual height of the member and the type of restraint at the wall ends. Such restraint that exists is normally provided by any supported floors, roofs, etc. and may be designated 'simple' (i.e. pin-ended) or 'enhanced' (i.e. partially fixed), depending on the actual construction details used. *Figures 5.10* and *5.11* show typical examples of horizontal restraints which provide simple and enhanced resistance respectively.

According to clause 28.3.2 of BS 5628 the effective height of a wall may be taken as:

1. 0.75 times the clear distance between lateral supports which provide enhanced resistance to lateral movements; or

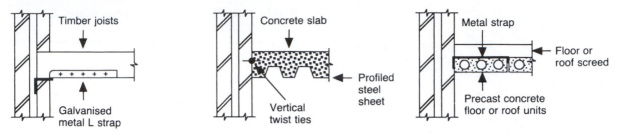

Fig. 5.10 *Details of horizontal supports providing simple resistance.*

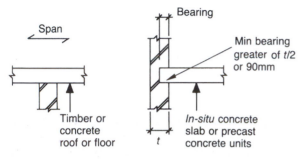

Fig. 5.11 *Details of horizontal supports providing enhanced resistance.*

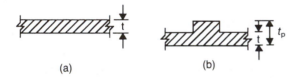

Fig. 5.12 *Effective thickness: (a) single leaf wall; (b) single leaf wall stiffened with piers (based on Fig. 3 of BS 5628).*

Table 5.11   Stiffness coefficient for walls stiffened by piers (Table 5, BS 5628)

Ratio of pier spacing (centre to centre) to pier width	Ratio t_p/t of pier thickness to actual thickness of wall to which it is bonded		
	1	2	3
6	1.0	1.4	2.0
10	1.0	1.2	1.4
20	1.0	1.0	1.0

Note. Linear interpolation between the values given in the table is permissible, but not extrapolation outside the limits given.

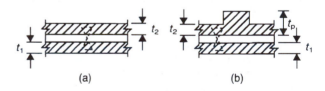

Fig. 5.13 *Effective thickness: (a) cavity wall; (b) cavity wall stiffened with piers (based on* Fig. 3 of BS 5628*).*

2. the clear distance between lateral supports which provide simple resistance to lateral movement.

EFFECTIVE THICKNESS (t_{ef}). For single leaf walls the effective thickness is taken as the actual thickness (t) of the wall (*Fig. 5.12(a)*):

$$t_{ef} = t \quad \text{(single leaf wall)}$$

If the wall was stiffened with piers (*Fig. 5.12(b)*), in order to increase its load capacity, the effective thickness is given by $t_{ef} = tK$ (single leaf wall stiffened with piers) where K is obtained from *Table 5.11*.

For cavity walls the effective thickness should be taken as the greater of (i) two-thirds the sum of the actual thickness of the two leaves, i.e. $^2/_3(t_1 + t_2)$ or

(ii) the actual thickness of the thicker leaf, i.e. t_1 or t_2 (*Fig. 5.13(a)*). Where the cavity wall is stiffened by piers, the effective thickness of the wall is taken as the greatest of (i) $^2/_3(t_1 + Kt_2)$ or (ii) t_1 or (iii) Kt_2 (*Fig. 5.13(b)*).

Eccentricity of vertical loading. In addition to the slenderness ratio, the capacity reduction factor is also a function of the eccentricity of loading. When considering the design of walls it is not

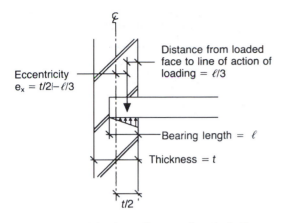

Fig. 5.14 *Eccentricity for wall supporting single floor.*

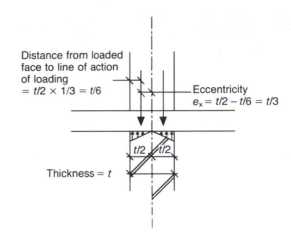

Fig. 5.15 *Eccentricity for wall supporting continuous floor.*

realistic to assume that the load will be applied truly axially, but rather that it will occur at some eccentricity to the centroid of the wall. This eccentricity is normally expressed as a fraction of the wall thickness.

Clause 31 of BS 5628 recommends that the loads transmitted to a wall by a single floor or roof act at one-third of the length of the bearing surface from the loaded edge (*Fig. 5.14*). This is based on the assumption that the stress distribution under the bearing surface is triangular in shape as shown in *Fig. 5.14*. Where a uniform floor is continuous over a wall, each span of the floor should be taken as being supported individually on half the total bearing area (*Fig. 5.15*).

5.5.3 DESIGN VERTICAL LOAD RESISTANCE OF WALLS (N_R)

The foregoing has discussed how to evaluate the design strength of a masonry wall being equal to the characteristic strength (f_k) multiplied by the capacity reduction factor (β) and divided by the appropriate safety factor for materials (γ_m). The characteristic strengths and safety factors for materials are obtained from *Tables 5.8* and *5.9*

respectively. The effective thickness and effective height of the member are used to determine the slenderness ratio and thence, together with the eccentricity of loading, the capacity reduction factor via *Table 5.10*.

The design strength is used to estimate the vertical load resistance of a wall, N_R, which is given by

$$N_R = \text{stress} \times \text{area} = \frac{\beta f_k}{\gamma_m} \times t \times 1 = \frac{\beta t f_k}{\gamma_m} \quad (5.7)$$

As stated earlier, the primary aim of design is to ensure that the ultimate design load does not exceed the design load resistance of the wall. Thus from equations 5.1, 5.2 and 5.7

$$N \leqslant N_R$$

$$\gamma_f(G_k, Q_k, \text{ or } W_k) \leqslant \frac{\beta t f_k}{\gamma_m} \quad (5.8)$$

Equation 5.8 provides the basis for the design of vertically loaded walls. The full design procedure is summarized in *Fig. 5.16*.

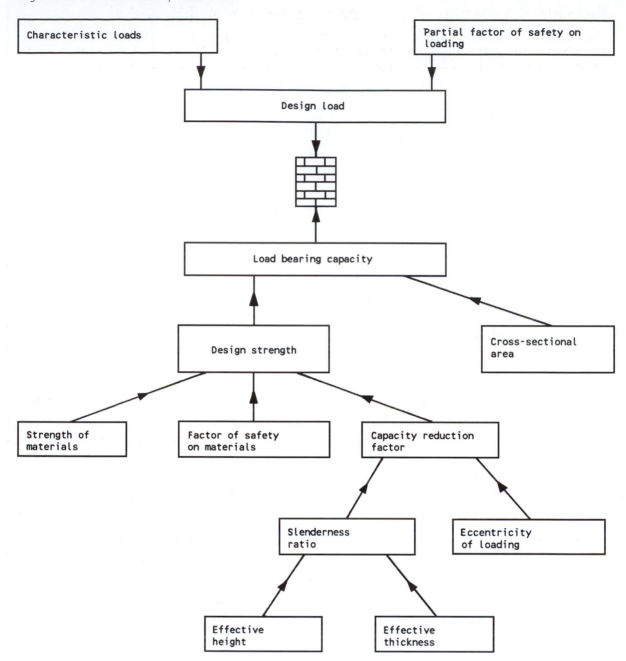

Fig. 5.16 *Design procedure for vertically loaded walls.*

Example 5.1 Design of a loadbearing brick wall (BS 5628)

The internal loadbearing brick wall shown in *Fig. 5.17* supports an ultimate axial load of 140 kN m^{-1} run including self-weight of the wall. The wall is 102.5 mm thick and 4 m long. Assuming that normal manufacturing and construction controls apply, design the wall.

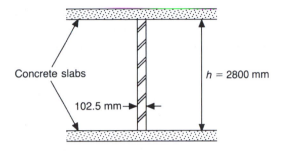

Fig. 5.17

LOADING
Ultimate design load, $N = 140$ kN m^{-1} = 140 N mm^{-1}

DESIGN VERTICAL LOAD RESISTANCE OF WALL

Characteristic compressive strength

$$\text{Basic value} = f_k$$

Check modification factor

SMALL PLAN AREA. Modification factor does not apply since horizontal cross-sectional area of wall, $A = 0.1025 \times 4.0 = 0.41$ m^2 ≮ 0.2 m^2.

NARROW BRICK WALL. Modification factor is 1.15 since wall is one brick thick. Hence modified characteristic compressive strength is $1.15f_k$.

Safety factor for materials (γ_m)
Manufacturing and construction controls are 'normal'. Hence from *Table 5.9*, $\gamma_m = 3.5$.

Capacity reduction factor (β)

Eccentricity. Since wall is axially loaded assume eccentricity of loading, $e_x < 0.05t$.

Slenderness ratio (SR). Concrete slab provides 'enhanced' resistance to wall:

$$h_{ef} = 0.75 \times \text{height} = 0.75 \times 2800 = 2100 \text{ mm}$$
$$t_{ef} = \text{actual thickness (single leaf)} = 102.5 \text{ mm}$$
$$\text{SR} = h_{ef}/t_{ef} = 2100/102.5 = 20.5 < \text{permissible} = 27$$

Hence, from *Table 5.10*, $\beta = 0.68$.

Design vertical load resistance of wall (N_R)

$$N_R = \frac{\beta \times \text{modified characteristic strength} \times t}{\gamma_m} = \frac{0.68 \times (1.15f_k)102.5}{3.5}$$

DETERMINATION OF f_K
For structural stability

$$N_R \geqslant N$$
$$\frac{0.68 \times 1.15f_k \times 102.5}{3.5} \geqslant 140$$

$$\text{Hence } f_k \geqslant \frac{140}{22.9} = 6.1 \text{ N mm}^{-2}$$

SELECTION OF BRICK AND MORTAR TYPE

From *Table 5.8(a)*, the following brick/mortar combinations would be appropriate:

Compressive strength of bricks $(N\ mm^{-2})$	Mortar designation	f_k $(N\ mm^{-2})$
27.5	(iii)	7.1
20	(ii)	6.4

The actual brick type that will be specified on the working drawings will depend not only upon the structural requirements but also on aesthetics, durability, buildability and cost considerations. In this particular case the designer may specify the minimum requirements as brick strength: $\geqslant 27.5$ N mm^{-2}, mortar mix: 1 : 1 : 6, frost resistance/soluble salt content: ON. Assuming that the wall will be plastered on both sides, appearance of the brick will not need to be specified.

Example 5.2 Design of a brick wall with a 'small' plan area (BS 5628)

Redesign the wall in *Example 5.1* assuming that it is only 1.5 m long.

The calculations for this case are essentially the same as for the 4 m long wall except for the fact that the plan area of the wall, A, is now less than 0.2 m^2, being equal to $0.1025 \times 1.5 = 0.1538$ m^2. Hence, the 'plan area' modification factor is equal to

$$(0.70 + 1.5A) = 0.7 + 1.5 \times 0.1538 = 0.93$$

The 1.15 factor applicable to narrow brick walls still applies, and therefore the modified characteristic compressive strength of masonry is given by

$$0.93 \times 1.15 f_k = 1.07 f_k$$

From the above $\gamma_m = 3.5$ and $\beta = 0.68$. Hence, the required characteristic compressive strength of masonry, f_k, is given by

$$\frac{0.68 \times 1.07 f_k \times 102.5}{3.5} \geqslant 140 \text{ kN m}^{-1} \Rightarrow f_k \geqslant \frac{140}{21.3} = 6.6 \text{ N mm}^{-2}$$

From *Table 5.8(a)*, any of the following brick/mortar combinations would be appropriate:

Compressive strength of bricks $(N\ mm^{-2})$	Mortar designation	f_k $(N\ mm^{-2})$
35	(iv)	7.3
27.5	(iii)	7.1
20	(i)	7.4

Hence it can be immediately seen that walls having similar construction details but a plan area of < 0.2 m^2 will have lower load-carrying capacities.

Example 5.3 Analysis of brick walls stiffened with piers (BS 5628)

A 3.5 m high wall shown in cross-section in *Fig. 5.18* is constructed from clay bricks having a compressive strength of 20.5 N mm^{-2} (i.e. Class 3) laid in a 1 : 1 : 6 mortar. Calculate the ultimate loadbearing capacity of the wall assuming the partial safety factor for materials is 3.5 and the resistance to lateral loading is (a) 'enhanced' and (b) 'simple'.

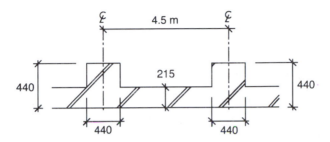

Fig. 5.18

ASSUMING 'ENHANCED' RESISTANCE

Characteristic compressive strength (f_k). Assume compressive strength of bricks is 20 N mm^{-2} bricks. From *Table 5.5*, 1 : 1 : 6 mix corresponds to a grade (iii) mortar. This implies that $f_k = 5.8$ N mm^{-2} (*Table 5.8*).

Safety factor for materials (γ_m)

$$\gamma_m = 3.5$$

Capacity reduction factor (β)

ECCENTRICITY. Assume wall is axially loaded. Hence $e_x < 0.05t$.

SLENDERNESS RATIO (SR). With 'enhanced' resistance

$$h_{ef} = 0.75 \times \text{height} = 0.75 \times 3500 = 2625 \text{ mm}$$

$$\frac{\text{Pier spacing}}{\text{Pier width}} = \frac{4500}{440} = 10.2$$

$$\frac{\text{Pier thickness}}{\text{Thickness of wall}} = \frac{440}{215} = 2.0$$

Hence from *Table 5.11*, $K = 1.2$. The effective thickness of the wall, t_{ef}, is equal to

$$t_{ef} = tK = 215 \times 1.2 = 258 \text{ mm}$$

$$\text{SR} = \frac{h_{ef}}{t_{ef}} = \frac{2625}{258} = 10.2 < \text{permissible} = 27$$

Hence, from *Table 5.10*, $\beta = 0.96$.

Design vertical load resistance of wall (N_R)

$$N_R = \frac{\beta f_k t}{\gamma_m} = \frac{0.96 \times 5.8 \times 215}{3.5}$$

$$= 342 \text{ N mm}^{-1} \text{ run of wall}$$

$$= 342 \text{ kN m}^{-1} \text{ run of wall}$$

Hence the ultimate load capacity of wall, assuming enhanced resistance, is 342 kN m^{-1} run of wall.

ASSUMING 'SIMPLE' RESISTANCE

The calculation for this case is essentially the same as for 'enhanced' resistance except

$$h_{ef} = \text{actual height} = 3500 \text{ mm}$$

$$t_{ef} = 258 \text{ mm} \quad \text{(as above)}$$

Hence

$$SR = \frac{h_{ef}}{t_{ef}} = \frac{3500}{258} = 13.6.$$

Assuming $e_x < 0.05t$ (as above), this implies that $\beta = 0.9$ (*Table 5.10*). The design vertical load resistance of the wall, N_R, is given by

$$N_R = \frac{\beta f_k t}{\gamma_m} = \frac{0.9 \times 5.8 \times 215}{3.5}$$

$$= 320 \text{ N mm}^{-1} \text{ run of wall}$$

$$= 320 \text{ kN m}^{-1} \text{ run of wall}$$

Hence it can be immediately seen that, all other factors being equal, walls having simple resistance have a lower resistance to failure.

Example 5.4 Design of single leaf brick and block walls (BS 5628)

Design the single leaf loadbearing wall shown in *Fig. 5.19* using mortar designation (iii) and either (a) standard format bricks or (b) solid concrete blocks of length 390 mm, height 190 mm and thickness 100 mm. Assume that the manufacturing and construction controls are special and normal respectively.

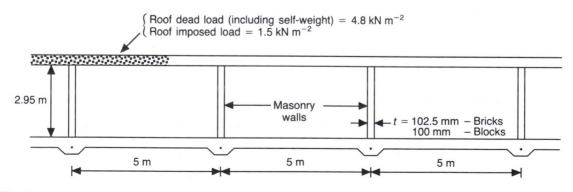

Fig. 5.19

DESIGN FOR STANDARD FORMAT BRICKS

Loading

Characteristic dead load g_k. Assuming that the bricks are made from clay, the density of the brickwork can be assumed to be 55 kg m^{-2} per 25 mm thickness (*Table 2.1*) or $(55 \times (102.5/25)9.81 \times 10^{-3} =)$ 2.2 kN m^{-2}. Therefore self-weight of wall is

$$SW = 2.2 \times 2.95 \times 1 = 6.49 \text{ kN m}^{-1} \text{ run of wall.}$$

$$\text{Dead load due to roof} = \frac{5+5}{2} \times 4.8 = 24 \text{ kN m}^{-1} \text{ run}$$

$$g_k = SW + \text{roof load}$$

$$= 6.49 + 24 = 30.49 \text{ kN m}^{-1} \text{ run of wall}$$

Characteristic imposed load, q_k

$$q_k = \text{roof load}$$

$$= \frac{5+5}{2} \times 1.5 = 7.5 \text{ kN m}^{-1} \text{ run of wall}$$

Ultimate design load, N

$$N = 1.4g_k + 1.6q_k$$

$$= 1.4 \times 30.49 + 1.6 \times 7.5 = 54.7 \text{ kN m}^{-1} \text{ run of wall}$$

$$= 54.7 \text{ N mm}^{-1} \text{ run of wall}$$

Design vertical load resistance of wall

Characteristic compressive strength

$$\text{Basic value} = f_k$$

MODIFICATION FACTORS

1. Small plan area – modification factor will not apply provided loaded plan area $A < 0.2$ m^2, i.e. provided that the wall length exceeds 2 m $\approx$ 0.2 m^2/0.1025 m.
2. Narrow brick wall – since wall is one brick thick, modification factor = 1.15.

Hence, modified characteristic compressive strength = $1.15f_k$.

Safety factor for materials (γ_m). Manufacturing control is special; construction control is normal. Hence from *Table 5.9* $\gamma_m = 3.1$.

Capacity reduction factor (β)

ECCENTRICITY. Assuming the wall is symmetrically loaded, eccentricity of loading, $e_x < 0.05t$.

SLENDERNESS RATIO (SR). Concrete slab provides 'enhanced' resistance to wall:

$$h_{ef} = 0.75 \times \text{height} = 0.75 \times 2950 = 2212.5 \text{ mm}$$

$$t_{ef} = \text{actual thickness (single leaf)} = 102.5 \text{ mm}$$

$$SR = \frac{h_{ef}}{t_{ef}} = \frac{2212.5}{102.5} = 21.6$$

Hence from *Table 5.10* $\beta = 0.63$.

Design vertical load resistance of wall (N_R)

$$N_R = \frac{\beta \times (\text{modified characteristic strength}) \times t}{\gamma_m}$$

$$= \frac{0.63 \times (1.15 f_k) 102.5}{3.1}$$

Determination of f_k
For structural stability

$$N_R \geqslant N$$

$$\frac{0.63 \times 1.15 f_k \times 102.5}{3.1} \geqslant 54.7$$

$$f_k \geqslant \frac{54.7}{23.95} = 2.3 \text{ N mm}^{-2}$$

Selection of brick and mortar type
From *Table 5.8(a)*, any of the following brick/mortar combinations would be appropriate:

Compressive strength of bricks (N mm^{-2})	Mortar designation	f_k (N mm^{-2})
10	(iv)	3.5
5	(iii)	2.5

DESIGN FOR SOLID CONCRETE BLOCKS

Loading
Assuming that the density of the blockwork is the same as that for brickwork, i.e. 55 kg m^{-2} per 25 mm thickness (*Table 2.1*), self-weight of wall is 6.49 kN m^{-1} run of wall.

Dead load due to roof = 24 kN m^{-1} run of wall

Imposed load from roof = 7.5 kN m^{-1} run of wall

Ultimate design load, $N = 1.4(6.49 + 24) + 1.6 \times 7.5 = 54.7$ kN m^{-1} run of wall

Design vertical load resistance of wall

Characteristic compressive strength. Characteristic strength is f_k assuming that plan area of wall is greater than 0.2 m^2. Note that the narrow wall modification factor only applies to brick walls.

Safety factor for materials

$$\gamma_m = 3.1$$

Capacity reduction factor

ECCENTRICITY

$$e_x < 0.05t$$

SLENDERNESS RATIO (SR). Concrete slab provides 'enhanced' resistance to wall:

$$h_{ef} = 0.75 \times \text{height} = 0.75 \times 2950 = 2212.5 \text{ mm}$$

$$t_{ef} = \text{actual thickness (single leaf)} = 100 \text{ mm}$$

$$\text{SR} = \frac{h_{ef}}{t_{ef}} = \frac{2212.5}{100} = 22.1$$

Hence from *Table 5.10*, $\beta = 0.61$.

Design vertical load resistance of wall (N_R)

$$N_R = \frac{\beta \times \text{(modified characteristic strength)} \times t}{\gamma_m}$$

$$= \frac{0.61 \times (f_k)\,100}{3.1}$$

Determination of f_k

For structural stability

$$N_R \geqslant N$$

$$\frac{0.61 \times (f_k)100}{3.1} \geqslant 54.7$$

$$f_k \geqslant \frac{54.7}{19.68} = 2.8 \text{ N mm}^{-2}$$

Selection of block and mortar type

$$\text{Shape factor for block} = \frac{\text{height}}{\text{thickness}} = \frac{190}{100} = 1.9$$

Interpolating between *Tables 5.8(b)* and *(d)* a solid block of compressive strength 3.5 N mm^{-2} used with mortar designation (iv) would produce masonry of compressive strength 3.3 N mm^{-2} which would be appropriate here.

Example 5.5 Design of a cavity wall (BS 5628)

A cavity wall of length 6 m supports the loads shown in *Fig. 5.20*. The inner loadbearing leaf is built using concrete blocks of length 440 mm, height 215 mm, thickness 100 mm and faced with plaster, and the outer leaf from standard format clay bricks. Design the wall assuming that the manufacturing control is special and construction control is normal. The self-weight of the blocks and plaster can be taken to be 2.4 kN m^{-2}.

Clause 29.1 of BS 5628 on cavity walls states that 'where the load is carried by one leaf only, the loadbearing capacity of the wall should be based on the horizontal cross-sectional area of that leaf alone, although the stiffening effect of the other leaf can be taken into account when calculating the slenderness ratio'. Thus, it should be noted that the following calculations relate to the design of the inner loadbearing leaf.

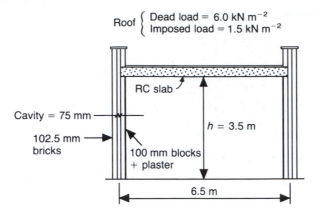

Fig. 5.20

LOADING

Characteristic dead load g_k

$$g_k = \text{roof load} + \text{self-weight of wall}$$

$$= \frac{(6.5 \times 1)6}{2} + (3.5 \times 1)2.4$$

$$= 19.5 + 8.4 = 27.9 \text{ kN m}^{-1} \text{ run of wall}$$

Characteristic imposed load, q_k

$$q_k = \text{roof load}$$

$$= \frac{(6.5 \times 1)1.5}{2} = 4.9 \text{ kN m}^{-1} \text{ run of wall}$$

Ultimate design load, N

$$N = 1.4g_k + 1.6q_k = 1.4 \times 27.9 + 1.6 \times 4.9 = 46.9 \text{ kN m}^{-1} \text{ run of wall}$$

DESIGN VERTICAL LOAD RESISTANCE OF WALL

Characteristic compressive strength

$$\text{Basic value} = f_k$$

Check modification factors

1. Small plan area – modification factor does not apply since loaded plan area, $A > 0.2$ m² where $A = 6 \times 0.1 = 0.6$ m².
2. Narrow brick wall – modification factor does not apply either since wall is of cavity construction.

Safety factor for materials
Manufacturing control is special; construction control is normal. Hence, from *Table 5.9*, $\gamma_m = 3.1$.

Capacity reduction factor

Eccentricity. The load from the concrete roof will be applied eccentrically as shown in the figure. The eccentricity is given by

$$e_x = \frac{t}{2} - \frac{t}{3} = \frac{t}{6} = 0.167t$$

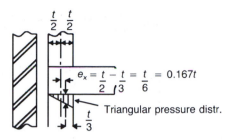

$$e_x = \frac{t}{2} - \frac{t}{3} = \frac{t}{6} = 0.167t$$

Triangular pressure distr.

Slenderness ratio (SR). Concrete slab provides 'enhanced' resistance to wall:

$$h_{\mathrm{ef}} = 0.75 \times \text{height} = 0.75 \times 3500 = 2625 \text{ mm}$$
$$t_{\mathrm{ef}} = 135 \text{ mm}$$

since t_{ef} is the greater of

$$2/3(t_1 + t_2) \ [= 2/3(102.5 + 100) = 135 \text{ mm}]$$

and t_1 [= 102.5 mm] or t_2 [= 100 mm].

$$\mathrm{SR} = \frac{h_{\mathrm{ef}}}{t_{\mathrm{ef}}} = \frac{2625}{135} = 19.4 < \text{permissible} = 27$$

Slenderness ratio $h_{\mathrm{ef}}/t_{\mathrm{ef}}$	Eccentricity at top of wall, e_x		
	0.1t	0.167t	0.2t
18	0.70		0.57
19.4	0.658	0.570	0.528
20	0.64		0.51

Hence by interpolating between the values in *Table 5.10*, β = 0.57.

Design vertical load resistance of wall (N_R)

$$N_R = \frac{\beta f_k t}{\gamma_m} = \frac{0.57 \times f_k \times 100}{3.1}$$

DETERMINATION OF f_k
For structural stability

$$N_R \geqslant N$$

$$\frac{0.57 f_k 100}{3.1} \geqslant 46.9$$

$$f_k \geqslant \frac{46.9}{18.39} = 2.6 \text{ N mm}^{-2}$$

SELECTION OF BRICK AND MORTAR TYPE

$$\text{Shape factor} = \frac{215}{100} = 2.15$$

From *Table 5.8(d)*, a solid concrete block with a compressive strength of 2.8 N mm^{-2} used with mortar type (iv) would be appropriate. The bricks and mortar for the outer leaf would be selected on the basis of appearance and durability.

5.6 Design of laterally loaded wall panels

A non-loadbearing wall which is supported on a number of sides is usually referred to as a panel wall. Panel walls are very common in the UK and are mainly used to clad framed buildings. These walls are primarily designed to resist lateral loading from the wind.

The purpose of this section is to describe the design of laterally loaded panel walls. This requires an understanding of the following factors which are discussed below:

1. characteristic flexural strength
2. orthogonal ratio
3. support conditions
4. limiting dimensions
5. basis of design.

5.6.1 CHARACTERISTIC FLEXURAL STRENGTH OF MASONRY (f_{kx})

The majority of panel walls tend to behave as a two-way spanning plate when subject to lateral loading (*Fig. 5.21*). However, it is more convenient to understand the behaviour of such walls if the bending processes in the horizontal and vertical directions are studied separately.

Bending tests carried out on panel walls show that the flexural strength of masonry is significantly greater when the plane of failure occurs perpendicular to the bed joint rather than when failure occurs parallel to the bed joint (*Fig. 5.22*). This is

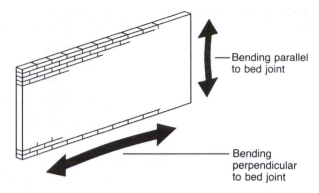

Fig. 5.21 *Panel wall subject to two-way bending.*

because, in the former case, failure is a complex mix of shear, etc. and end bearing, while in the latter case cracks need only form at the mortar/brick interface.

Table 5.12 shows the characteristic flexural strengths of masonry constructed using a range of brick/block types and mortar designations for the two failure modes. Note that for masonry built with clay bricks, the flexural strengths are primarily related to the water-absorption properties of the brick, but for masonry built with concrete blocks the flexural strengths are related to the compressive strengths and thicknesses of the blocks.

5.6.2 ORTHOGONAL RATIO, μ

The orthogonal ratio is the ratio of the characteristic flexural strength of masonry when failure occurs

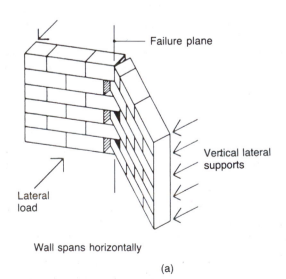

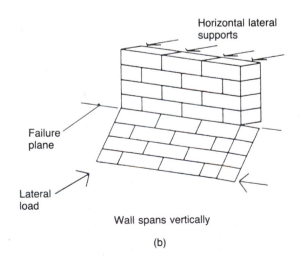

Fig. 5.22 *Failure modes of walls subject to lateral loading: (a) failure perpendicular to bed joint; (b) failure parallel to bed joint.*

Table 5.12 Characteristic flexural strength of masonry, f_{kx} in N mm^{-2} (Table 3, BS 5628)

	Plane of failure parallel to bed joints			Plane of failure perpendicular to bed joints		
Mortar designation	*(i)*	*(ii) and (iii)*	*(iv)*	*(i)*	*(ii) and (iii)*	*(iv)*
Clay bricks having a water absorption						
less than 7%	0.7	0.5	0.4	2.0	1.5	1.2
between 7% and 12%	0.5	0.4	0.35	1.5	1.1	1.0
over 12%	0.4	0.3	0.25	1.1	0.9	0.8
Calcium silicate bricks	0.3	0.3	0.2	0.9	0.9	0.6
Concrete bricks	0.3	0.3	0.2	0.9	0.9	0.6
Concrete blocks of compressive strength in N/mm^2:						
2.8 ⎤ used in walls of					0.40	0.4
3.5 ⎬ thickness★	0.25		0.2		0.45	0.4
7.0 ⎦ up to 100 mm					0.60	0.5
2.8 ⎤ used in walls					0.25	0.2
3.5 ⎬ of thickness★	0.15		0.1		0.25	0.2
7.0 ⎦ 250 mm					0.35	0.3
10.5 ⎤ used in walls					0.75	0.6
14.0 ⎬ of any thickness★	0.25		0.2		0.90†	0.7†
and over}						

Notes. ★ The thickness should be taken to be the thickness of the wall, for a single-leaf wall, or the thickness of the leaf, for a cavity wall.
† When used with flexural strength in parallel direction, assume the orthogonal ratio $\mu = 0.3$.

parallel to the bed joints ($f_{kx\,par}$) to that when failure occurs perpendicular to the bed joint ($f_{kx\,perp}$):

$$\mu = \frac{f_{kx\,par}}{f_{kx\,perp}} \qquad (5.9)$$

As will be discussed later, the orthogonal ratio is used to calculate the size of the bending moment in panel walls.

For masonry constructed using clay, calcium silicate or concrete bricks a value of 0.35 for μ can normally be assumed in design. No such unique value exists for masonry walls built with concrete blocks and must, therefore, be determined for individual block types.

5.6.3 SUPPORT CONDITIONS

In order to assess the lateral resistance of masonry panels it is necessary to take into account the support conditions at the edges. Three edge conditions are possible:

1. a free edge
2. a simply supported edge
3. a restrained edge.

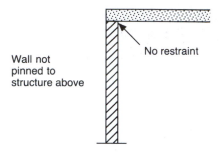

Fig. 5.23 *Detail providing free edge.*

A free edge is one that is unsupported (*Fig. 5.23*). A restrained edge is one that, when the panel is loaded, will fail in flexure before rotating. All other supported edges are assumed to be simply supported. This is despite the fact that some degree of fixity may actually exist in practice. *Figures 5.24* and *5.25* show typical detail providing simple and restrained support conditions respectively.

5.6.4 LIMITING DIMENSIONS
Clause 36.3 of BS 5628 lays down various limits on the dimensions of laterally loaded panel walls depending upon the support conditions. These are chiefly in respect of (i) the area of the panel (ii) the height and length of the panel. Specifically, the code mentions the following limits:

1. *Panel supported on three edges.*
   (a) *two or more sides continuous*: height × length equal to $1500t_{ef}^2$ or less;
   (b) *all other cases*: height × length equal to $1350t_{ef}^2$ or less.
2. *Panel supported on four edges.*
   (a) *three or more sides continuous*: height × length equal to $2250t_{ef}^2$ or less;
   (b) *all other cases*: height × length equal to $2025t_{ef}^2$ or less.
3. *Panel simply supported at top and bottom.* Height equal to $40t_{ef}$ or less.

Figure 5.26 illustrates the above limits on panel sizes. The height and length restrictions referred to in (ii) above only apply to panels which are supported on three or four edges, i.e. (1) and (2). In such cases the code states that no dimension should exceed 50 times the effective thickness of the wall (t_{ef}).

5.6.5 BASIS OF DESIGN (CLAUSE 36.4, BS 5628)
The preceding sections have summarized much of the background material needed to design laterally

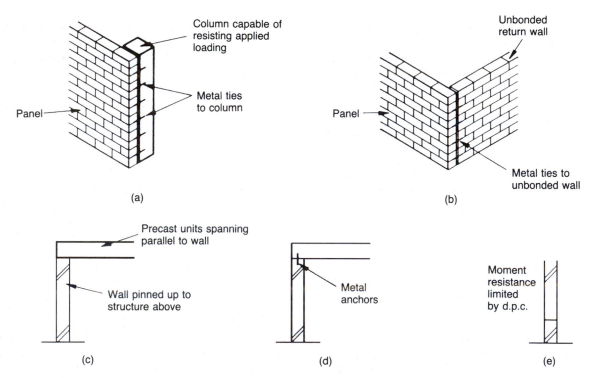

(a)

Column capable of resisting applied loading

Metal ties to column

Panel

(b)

Unbonded return wall

Panel

Metal ties to unbonded wall

(c)

Precast units spanning parallel to wall

Wall pinned up to structure above

(d)

Metal anchors

(e)

Moment resistance limited by d.p.c.

Fig. 5.24 *Details providing simple support conditions: (a) and (b) vertical support; (c)–(e) horizontal support (based on Figs 7 and 8, BS 5628).*

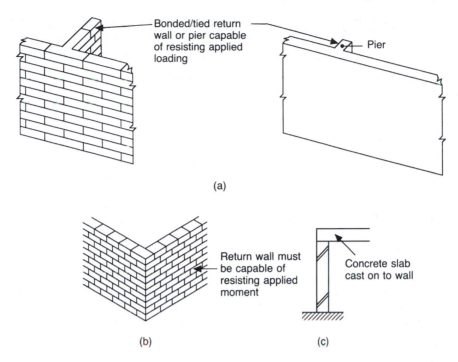

Fig. 5.25 *Details providing restrained support conditions: (a) and (b) vertical support; (c) horizontal support at head of wall (based on Figs 7 and 8, BS 5628).*

loaded panel walls. This section now considers the design procedure in detail.

The principal aim of design is to ensure that the ultimate design moment (M) does not exceed the design moment of resistance of the panel (M_d):

$$M \leq M_d \tag{5.10}$$

Since failure may take place about either axis (*Fig. 5.21*), for a given panel, there will be two design moments and two corresponding moments of resistance. The ultimate design moment per unit height of a panel when the plane of failure is perpendicular to the bed joint, M_{perp} (*Fig. 5.22*), is given by

$$M_{perp} = \alpha W_k \gamma_f L^2 \tag{5.11}$$

The ultimate design moment per unit height of a panel when the plane of failure is parallel to the bed joint, M_{par}, is given by

$$M_{par} = \mu \alpha W_k \gamma_f L^2 \tag{5.12}$$

where μ is the orthogonal ratio, α the bending moment coefficient taken from *Table 5.13*, γ_f the partial safety factor for loads (*Table 5.7*), L the length of the panel between supports and W_k the characteristic wind load per unit area.

The corresponding design moments of resistance when the plane of bending is perpendicular, $M_{k\,perp}$, or parallel, $M_{k\,par}$, to the bed joint is given by equations 5.13 and 5.14 respectively:

$$M_{k\,perp} = \frac{f_{kx\,perp}Z}{\gamma_m} \tag{5.13}$$

$$M_{k\,par} = \frac{f_{kx\,par}Z}{\gamma_m} \tag{5.14}$$

where $f_{kx\,perp}$ is the characteristic flexural strength perpendicular to the plane of bending, $f_{kx\,par}$ the characteristic flexural strength parallel to the plane of bending (*Table 5.12*), Z the section modulus and γ_m the partial safety factor for materials (*Table 5.9*).

Equations 5.10–5.14 form the basis for the design of laterally loaded panel walls. It should be noted, however, since by definition $\mu = f_{kx\,par}/f_{kx\,perp}$, $M_{k\,par} = \mu M_{k\,perp}$ (by dividing equation 5.14 by equation 5.13) and $M_{par} = \mu M_{perp}$ (by dividing equations 5.11 by equation 5.12) that either equations 5.11 and 5.13 or equations 5.12 and 5.14 can be used in design. The full design procedure is summarized in *Fig. 5.27* and illustrated by means of the following examples.

259

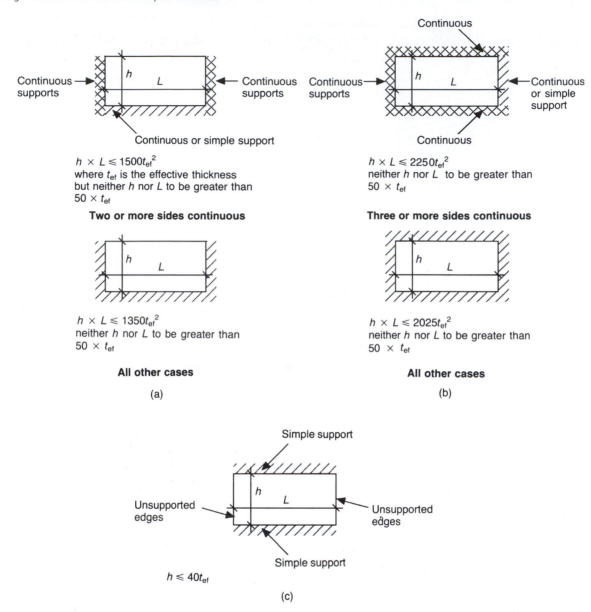

$h \times L \leqslant 1500t_{ef}^2$
where t_{ef} is the effective thickness
but neither h nor L to be greater than
$50 \times t_{ef}$

Two or more sides continuous

$h \times L \leqslant 2250t_{ef}^2$
neither h nor L to be greater than
$50 \times t_{ef}$

Three or more sides continuous

$h \times L \leqslant 1350t_{ef}^2$
neither h nor L to be greater than
$50 \times t_{ef}$

All other cases

(a)

$h \times L \leqslant 2025t_{ef}^2$
neither h nor L to be greater than
$50 \times t_{ef}$

All other cases

(b)

$h \leqslant 40t_{ef}$

(c)

Fig. 5.26 *Panel sizes: (a) panels supported on three sides; (b) panels supported on four edges; (c) panel simply supported top and bottom.*

5.7 Summary

This chapter has explained the basic properties of the materials used in unreinforced masonry design and the design procedures involved in the design of (a) single leaf and cavity walls, with and without stiffening piers, subject to vertical loading and (b) panel walls resisting lateral loading. The design procedures outlined are in accordance with BS 5628: Part 1: *Unreinforced Masonry* which is based on the limit state principles. In the case of vertically loaded walls it was found that the design process simply involves ensuring that the ultimate design load does not exceed the design strength of the wall. Panel walls, on the other hand, are normally subject to two-way bending. This means that, firstly, the designer must evaluate the design moment acting on the panel about one axis and then ensure that this does not exceed the corresponding moment of resistance of the panel.

Example 5.6 Analysis of a one-way spanning wall panel (BS 5628)

Estimate the design wind pressure that the cladding panel shown below can resist given that it is constructed using clay bricks having a water absorption of $< 7\%$ and mortar designation (iii). Assume $\gamma_f = 1.2$ and $\gamma_m = 3.5$.

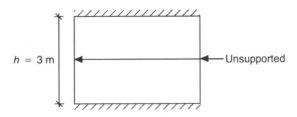

ULTIMATE DESIGN MOMENT (M)

Since the vertical edges are unsupported, the panel must span vertically and, therefore, equations 5.11 and 5.12 cannot be used to determine the design moment here. The critical plane of bending will be parallel to the bed joint and the ultimate design moment at mid-height of the panel, M, is given by (clause 36.4.2 of BS 5628)

$$M = \frac{\text{ultimate load} \times \text{height}}{8}$$

$$\text{Ultimate load on the panel per unit length of wall} = \text{wind pressure} \times \text{area}$$

$$= (\gamma_f W_k)(\text{height} \times \text{length of panel})$$

$$= 1.2 W_k 3000 \times 1000$$

$$= 3.6 W_k 10^6 \text{ N m}^{-1} \text{ length of wall}$$

Hence

$$M = \frac{3.6 W_k \times 10^6 \times 3000}{8} = 1.35 \times 10^9 W_k \text{ m}^{-1} \text{ length of wall}$$

MOMENT OF RESISTANCE (M_d)

Section modulus (Z)

$$Z = \frac{bd^2}{6} = \frac{10^3 \times 102.5^2}{6} = 1.75 \times 10^6 \text{ mm}^3 \text{ m}^{-1} \text{ length of wall}$$

Moment of resistance

The design moment of resistance, M_d, is equal to the moment of resistance when the plane of failure is parallel to the bed joint, $M_{k\,par}$. Hence

$$M_d = M_{k\,par} = \frac{f_{kx\,par} Z}{\gamma_m} = \frac{0.5 \times 1.75 \times 10^6}{3.5} = 0.25 \times 10^6 \text{ N mm m}^{-1}$$

where $f_{kx\,par} = 0.5 \text{ N mm}^{-2}$ from *Table 5.12*, since water absorption of bricks $< 7\%$ and the mortar is designation (iii)

DETERMINATION OF MAXIMUM WIND PRESSURE (W_k)

For structural stability:

$$M \leqslant M_d$$

$$1.35 \times 10^9 W_k \leqslant 0.25 \times 10^6$$

$$W_k \leqslant 0.185 \times 10^{-3} \text{ N mm}^{-2}$$

Hence the maximum wind pressure that the panel can resist is 0.185×10^{-3} N mm^{-2} or 185 N m^{-2}.

Table 5.13 Bending moment coefficients in laterally loaded wall panels (based on Table 9, BS 5628)

Key to support conditions

——— denotes free edge
⊿⊿⊿ simply supported edge
⋊⋊⋊ an edge over which full continuity exists

	μ	Values of α h/L 0.30	0.50	0.75	1.00	1.25	1.50	1.75
A	1.00	0.031	0.045	0.059	0.071	0.079	0.085	0.090
	0.90	0.032	0.047	0.061	0.073	0.081	0.087	0.092
	0.80	0.034	0.049	0.064	0.075	0.083	0.089	0.093
	0.70	0.035	0.051	0.066	0.077	0.085	0.091	0.095
	0.60	0.038	0.053	0.069	0.080	0.088	0.093	0.097
	0.50	0.040	0.056	0.073	0.083	0.090	0.095	0.099
	0.40	0.043	0.061	0.077	0.087	0.093	0.098	0.101
	0.35	0.045	0.064	0.080	0.089	0.095	0.100	0.103
	0.30	0.048	0.067	0.082	0.091	0.097	0.101	0.104
C	1.00	0.020	0.028	0.037	0.042	0.045	0.048	0.050
	0.90	0.021	0.029	0.038	0.043	0.046	0.048	0.050
	0.80	0.022	0.031	0.039	0.043	0.047	0.049	0.051
	0.70	0.023	0.032	0.040	0.044	0.048	0.050	0.051
	0.60	0.024	0.034	0.041	0.046	0.049	0.051	0.052
	0.50	0.025	0.035	0.043	0.047	0.050	0.052	0.053
	0.40	0.027	0.038	0.044	0.048	0.051	0.053	0.054
	0.35	0.029	0.039	0.045	0.049	0.052	0.053	0.054
	0.30	0.030	0.040	0.046	0.050	0.052	0.054	0.054
E	1.00	0.008	0.018	0.030	0.042	0.051	0.059	0.066
	0.90	0.009	0.019	0.032	0.044	0.054	0.062	0.068
	0.80	0.010	0.021	0.035	0.046	0.056	0.064	0.071
	0.70	0.011	0.023	0.037	0.049	0.059	0.067	0.073
	0.60	0.012	0.025	0.040	0.053	0.062	0.070	0.076
	0.50	0.014	0.028	0.044	0.057	0.066	0.074	0.080
	0.40	0.017	0.032	0.049	0.062	0.071	0.078	0.084
	0.35	0.018	0.035	0.052	0.064	0.074	0.081	0.086
	0.30	0.020	0.038	0.055	0.068	0.077	0.083	0.089

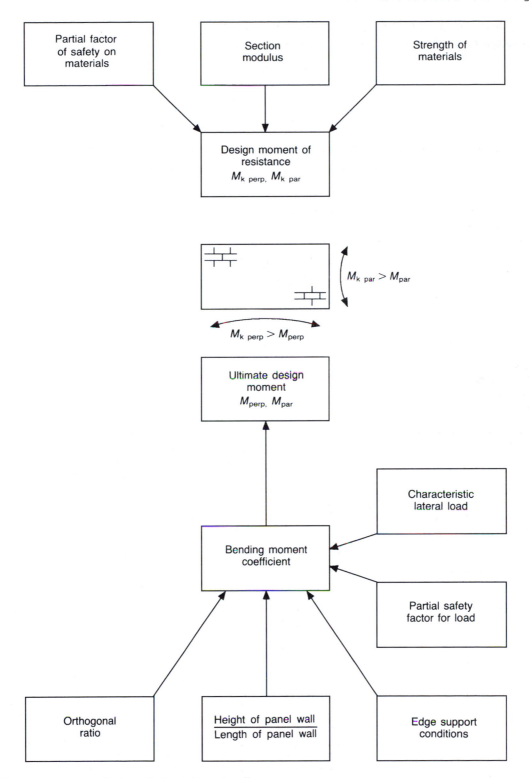

Fig. 5.27 *Design procedure for laterally loaded panel walls.*

Example 5.7 Analysis of a two-way spanning panel wall (BS 5628)

The panel wall shown in *Fig. 5.28* is constructed using clay bricks having a water absorption of greater than 12% and mortar designation (ii). If the manufacturing and construction controls are both normal, calculate the design wind pressure, W_k, the wall can withstand. Assume that the wall is simply supported on all four edges.

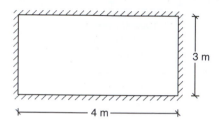

Fig. 5.28

ULTIMATE DESIGN MOMENT (M)

Orthogonal ratio (μ)

$$\mu = f_{kx\,par}/f_{kx\,perp} = 0.3/0.9 = 0.33 \quad (Table\ 5.12)$$

Bending moment coefficient (α)

$$h/L = 3/4 = 0.75$$

Hence, from *Table 5.13(E)*, $\alpha = 0.053$.

Ultimate design moment (M)

$$M = M_{perp} = \alpha\gamma_f W_k L^2 = 0.053 \times 1.2 W_k \times 4^2 \text{ kN m m}^{-1} \text{ run} = 1.0176 \times 10^6 W_k \text{ N mm m}^{-1} \text{ run}$$

MOMENT OF RESISTANCE (M$_d$)

Safety factor for materials (γ$_m$)

$$\gamma_m = 3.5 \quad (Table\ 5.9)$$

Section modulus (Z)

$$Z = \frac{bd^2}{6} = \frac{10^3 \times 102.5^2}{6} = 1.75 \times 10^6 \text{ mm}^3$$

Moment of resistance (M$_d$)

$$M_d = M_{k\,perp} = \frac{f_{kx\,perp}Z}{\gamma_m} = \frac{0.9 \times 1.75 \times 10^6}{3.5} = 0.45 \times 10^6 \text{ N mm m}^{-1} \text{ run}$$

DETERMINATION OF MAXIMUM WIND PRESSURE (W^k)

For structural stability

$$M \leq M_d$$

$$1.0176 \times 10^6 W_k \leq 0.45 \times 10^6$$

$$\Rightarrow W_k \leq 0.44 \text{ kN m}^{-2}$$

Hence the panel is able to resist a wind pressure of 0.44 kN m^{-2}.

Example 5.8 Design of a two-way spanning single-leaf panel wall (BS 5628)

A 102.5 mm brick wall is to be designed to withstand a wind pressure, W_k, of 0.65 kN m^{-2}. Assuming that the edge support conditions are as shown in *Fig. 5.29* and that both the manufacturing and construction controls are normal, determine suitable brick/mortar combination(s) for the wall.

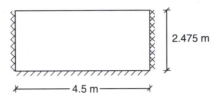

2.475 m

4.5 m

Fig. 5.29

ULTIMATE DESIGN MOMENT (M)

Orthogonal ratio (μ)
Assume $\mu = 0.35$.

Bending moment coefficient (α)

$$\frac{h}{L} = \frac{2475}{4500} = 0.55$$

Hence, from *Table 5.13(C)*, $\alpha = 0.040$.

Ultimate design moment (M)

$$M = M_{\text{perp}} = \alpha \gamma_f W_k L^2 = 0.040 \times 1.2 \times 0.65 \times 4.5^2 \text{ kN m m}^{-1} \text{ run}$$

$$= 0.632 \text{ kN m m}^{-1} \text{ run} = 0.632 \times 10^6 \text{ N mm m}^{-1} \text{ run}$$

MOMENT OF RESISTANCE (M_d)

Safety factor for materials (γ_m)

$$\gamma_m = 3.5 \quad (Table\ 5.9)$$

Section modulus (Z)

$$Z = \frac{bd^2}{6} = \frac{10^3 \times 102.5^2}{6} = 1.75 \times 10^6 \text{ mm}^3$$

Moment of resistance (M_d)

$$M_d = M_{\text{k perp}} = \frac{f_{\text{kx perp}} Z}{\gamma_m} = \frac{1.75 \times 10^6 f_{\text{kx perp}}}{3.5} = 0.5 \times 10^6 f_{\text{kx perp}} \text{ N mm m}^{-1} \text{ run}$$

DETERMINATION OF $f_{\text{kx perp}}$
For structural stability:

$$M \leq M_d$$

$$0.632 \times 10^6 \leq 0.5 \times 10^6 f_{\text{kx perp}}$$

$$\Rightarrow f_{\text{kx perp}} \leq 1.26 \text{ N mm}^{-2}$$

SELECTION OF BRICK AND MORTAR TYPE

From *Table 5.12*, any of the following brick/mortar combinations would be appropriate:

Clay brick having a water absorption:	Mortar designation	$f_{kx\ perp}$ (N mm^{-2})
< 7%	(iii)	1.5
7%–12%	(i)	1.5

Note that the actual value of μ for the brick/mortar combinations above is 0.33 and not 0.35 as assumed. However, this difference is insignificant and will not affect the choice of the brick/mortar combinations shown in the above table.

Example 5.9 Analysis of a two-way spanning cavity panel wall (BS 5628)

Determine the maximum design wind pressure, W_k, which can be resisted by the cavity wall shown in *Fig. 5.30* if the construction details are as follows:

1. *Outer leaf*: clay brick having a water absorption < 7% laid in a 1 : 1 : 6 mortar (i.e. designation (iii)).
2. *Inner leaf*: solid concrete blocks of compressive strength 3.5 N mm^{-2} and length 30 mm, height 190 mm and thickness 100 mm also laid in a 1 : 1 : 6 mortar.

Assume that the top edge of the wall is unsupported, but that the base and vertical edges are simply supported. The manufacturing and construction controls can both be assumed to be normal.

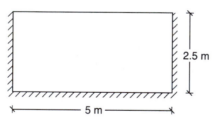

Fig. 5.30

The design wind pressure, W_k, depends upon the available capacities of both leaves of the wall. Therefore, it is normal practice to work the capacities of the outer and inner leaf separately, and to then sum them in order to determine W_k.

OUTER LEAF

Ultimate design moment (M)
Orthogonal ratio (μ)

$$\mu = \frac{f_{kx\ par}}{f_{kx\ perp}} = \frac{0.5}{1.5} = 0.33 \quad (Table\ 5.12)$$

Bending moment coefficient (α)

$$\frac{h}{L} = \frac{2.5}{5.0} = 0.5$$

Hence from *Table 5.13(A)*, $\alpha = 0.065$.

Ultimate design moment (M)

$$M = M_{\text{perp}} = \alpha \gamma_f W_k L^2 = 0.065 \times 1.2 (W_k)_{\text{outer}} \times 5^2 \text{ kN m m}^{-1} \text{ run}$$
$$= 1.95 \times 10^6 (W_k)_{\text{outer}} \text{ N mm m}^{-1} \text{ run}$$

Moment of resistance (M_d)

Safety factor for materials (γ_m)

$$\gamma_m = 3.5 \quad (\textit{Table 5.9})$$

Section modulus (Z)

$$Z = \frac{bd^2}{6} = \frac{10^3 \times 102.5^2}{6} = 1.75 \times 10^6 \text{ mm}^3$$

Moment of resistance (M_d)

$$M_d = M_{k \text{ perp}} = \frac{f_{kx \text{ perp}} Z}{\gamma_m} = \frac{1.5 \times 1.75 \times 10^6}{3.5} = 0.75 \times 10^6 \text{ N mm m}^{-1}$$

Maximum permissible wind pressure on outer leaf ($W_k)_{\text{outer}}$

For structural stability:

$$M \leqslant M_d$$
$$1.95(W_k)_{\text{outer}} \times 10^6 \leqslant 0.75 \times 10^6$$
$$(W_k)_{\text{outer}} \leqslant 0.38 \text{ kN m}^{-2}$$

INNER LEAF

Ultimate design moment (M)

Orthogonal ratio (μ)

$$\mu = \frac{f_{kx \text{ par}}}{f_{kx \text{ perp}}} = \frac{0.25}{0.45} = 0.55 \quad (\textit{Table 5.12})$$

Bending moment coefficient (α)

$$\frac{h}{L} = \frac{2.5}{5.0} = 0.5$$

Hence from *Table 5.13(A)*, $\alpha = 0.055$.

Ultimate design moment

$$M = M_{\text{perp}} = \alpha \gamma_f W_k L^2 = 0.055 \times 1.2 (W_k)_{\text{inner}} \times 5^2 \text{ kN m m}^{-1} \text{ run}$$
$$= 1.65 \times 10^6 (W_k)_{\text{inner}} \text{ N mm m}^{-1} \text{ run}$$

Moment of resistance (M_d)

Safety factor for materials (γ_m)

$$\gamma_m = 3.5 \quad (\textit{Table 5.9})$$

Section modulus (Z)

$$Z = \frac{bd^2}{6} = \frac{10^3 \times 100^2}{6} = 1.67 \times 10^6 \text{ mm}^3$$

Moment of resistance (M_d)

$$M_d = M_{k\,perp} = \frac{f_{kx\,perp}Z}{\gamma_m} = \frac{0.45 \times 1.67 \times 10^6}{3.5} = 0.21 \times 10^6 \text{ N mm m}^{-1}$$

Maximum permissible wind pressure on inner leaf $(W_k)_{inner}$
For structural stability:

$$M \leqslant M_d$$

$$1.65(W_k)_{inner} \times 10^6 \leqslant 0.21 \times 10^6 \text{ N mm m}^{-1} \text{ run}$$

$$(W_k)_{inner} \leqslant 0.13 \text{ kN m}^{-2}$$

Hence maximum wind pressure that the panel can withstand, W_k, is given by

$$W_k = (W_k)_{inner} + (W_k)_{outer} = 0.13 + 0.38 = 0.51 \text{ kN m}^{-2}$$

Questions

1. (a) Discuss why there was a decline and what factors have led to a revival in the use of masonry in construction over recent years.
   (b) The 4.0 m high wall shown adjacent is constructed from clay bricks having a compressive strength of 20 N mm^{-2} laid in a 1 : 1 : 6 mortar. Calculate the design load resistance of the wall assuming the partial safety factor for materials is 3.5 and the resistance to lateral loading is simple.

2. (a) Discuss the factors to be considered in the selection of brick, block and mortar types for particular applications.
   (b) Design the external cavity wall and internal single leaf loadbearing wall for the single-storey building shown below. Assume that the internal

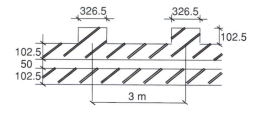

wall is made from solid concrete blocks of length 390 mm, height 190 mm and thickness 100 mm and that the external wall consists of 102.5 mm thick brick outer leaf with 390 × 190 × 100 mm concrete blocks. Assume that the manufacturing and construction controls are special and normal respectively and that the self-weight of the brick and blockwork is 2.2 kN m^{-2}.

Roof dead load (including self-weight) = 5.5 kN m^{-2}
Roof imposed load = 1.5 kN m^{-2}

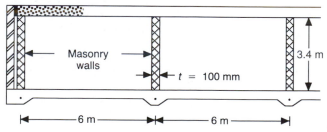

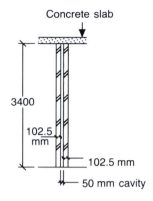

Concrete slab

3400

102.5 mm

102.5 mm

50 mm cavity

3. (a) Explain the difference between simple and enhanced resistance and sketch typical construction details showing examples of both types of restraint.

(b) The brick cavity wall shown above supports an ultimate axial load of 200 kN m^{-1}. Assuming that the load is equally shared by both leaves and the manufacturing and construction controls are normal and special respectively, design the wall.

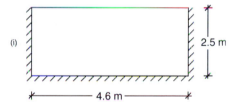

(i)

2.5 m

4.6 m

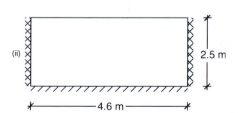

(ii)

2.5 m

4.6 m

4. (a) Explain with the use of sketches the limiting dimensions of laterally loaded panel walls.

(b) A 102.5 mm brick wall is to be designed to withstand a wind pressure of 550 N m^{-2}. For the support conditions shown in (i) and (ii) below (left), determine suitable brick/mortar combinations for the two cases. Assume that both the manufacturing and construction controls are normal.

5. (a) In all the examples on the design of laterally loaded panel wall discussed in this chapter the self-weight of the wall has been ignored. What effect do you think that including the self-weight would have on the design compressive strength of masonry?

(b) Determine the maximum design wind pressure which can be resisted by the cavity wall shown below if the construction details are as follows: (1) outer leaf: clay bricks having a water absorption 7–12% laid in mortar designation (ii); (2) inner leaf: solid concrete blocks of compressive strength 3.5 N mm^{-2} and length 390 mm, height 190 mm and thickness 100 mm also laid in mortar designation (ii). Assume that the manufacturing and construction controls are both special and normal respectively and the vertical edges are unsupported.

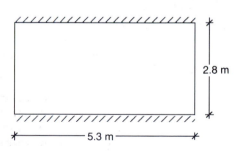

2.8 m

5.3 m

Chapter 6

Design of timber elements to BS 5268

This chapter is concerned with the design of timber elements to British Standard 5268: Part 2, which is based on the permissible stress philosophy. The chapter describes how timber is specified for structural purposes and discusses some of the basic concepts involved such as stress grading, grade stresses and strength classes. The primary aim of this chapter is to give guidance on the design of flexural members, e.g. beams and joists, compression members, e.g. posts and columns and load sharing systems, e.g. stud walling.

6.1 Introduction

Wood is a very versatile raw material and is still widely used in construction, especially in countries such as Canada, Sweden, Finland, Norway and Poland, where there is an abundance of good-quality timber. Timber can be used in a range of structural applications including marine works: construction of wharves, piers, cofferdams; heavy civil works: bridges, piles, shoring, pylons; domestic housing: roofs, floors, partitions; shuttering for precast and *in situ* concrete; falsework for brick or stone construction.

Of all the construction materials which have been discussed in this book, only timber is naturally occurring. This makes it a very difficult material to characterise and partly accounts for the wide variation in the strength of timber, not only between different species but also between timber of the same species and even from the same log. Quite naturally, this led to uneconomical use of timber which was costly for individuals and the nation as a whole. However, this problem has now been largely overcome by specifying stress graded timber (*section 6.2*).

There is an enormous variety of timber species. They are divided into softwoods and hardwoods, a botanical distinction, not on the basis of mechanical strength. Softwoods are derived from trees with needle-shaped leaves and are usually evergreen, e.g. fir, larch, spruce, hemlock, pine. Hardwoods are derived from trees with broad leaves and are usually deciduous, e.g. ash, elm, oak, teak, iroko, ekki, greeheart. Obviously the suitability of a particular timber type for any given purpose will depend upon various factors such as performance, cost, appearance and availability. This makes specification very difficult. The task of the structural engineer has been simplified, however, by grouping timber species into sixteen strength classes for which typical design parameters, e.g. grade stresses and moduli of elasticity, have been produced (*section 6.3*). Most standard design in the UK is with softwoods.

Design of timber elements is normally carried out in accordance with BS 5268: *Structural Use of Timber*. This is divided into the following parts:

Part 2: *Code of Practice for Permissible Stress Design, Materials and Workmanship.*
Part 3: *Code of Practice for Trussed Rafter Roofs.*
Part 4: *Fire Resistance of Timber Structures.*
Part 5: *Code of Practice for the Preservative Treatment of Structural Timber.*
Part 6: *Code of Practice for Timber Frame Walls.*
Part 7: *Recommendations for the Calculation Basis for Span Tables.*

The design principles which will be outlined in this chapter are mostly based on the contents of Part 2 of the code. It should therefore be assumed that all future references to BS 5268 refer exclusively to Part 2. As pointed out in Chapter 1 of this book, BS 5268 is based on permissible stress design rather than limit state design. This means in practice that a partial safety factor is applied only to material properties, i.e. the permissible stresses (*section 6.4*) and not the loading.

Specifically, this chapter gives guidance on the design of timber beams, joists, columns and stud walling. The design of timber formwork is not covered here as it was considered to be rather too specialised a topic and, therefore, inappropriate for a book of this nature. Before discussing the design process in detail, the following sections will expand on the more general aspects mentioned above, namely:

a) stress grading
b) grade stress and strength class
c) permissible stress.

6.2 Stress grading

The strength of timber is a function of several parameters including the moisture content, density, size of specimen and the presence of various strength-reducing characteristics such as knots, slope of grain, fissures and wane. Prior to the introduction of BS 5268 the strength of timber was determined by carrying out short-term loading tests on small timber specimens free from all defects. The data were used to estimate the minimum strength which was taken as the value below which not more than 1% of the test results fell. These strengths were multiplied by a reduction factor to give basic stresses. The reduction factor made an allowance for the reduction in strength due to duration of loading, size of specimen and other effects normally associated with a safety factor such as accidental overload, simplifying assumptions made during design and design inaccuracies, together with poor workmanship. Basic stress was defined as the stress which could safely be permanently sustained by timber free from any strength-reducing characteristics. Basic stress, however, was not directly applicable to structural size timber since structural size timber invariably contains defects, which further reduces its strength. To take account of this, timber was visually classified into one of four grades, namely 75, 65, 50 and 40, which indicated the percentage free from defects. The grade stress for structural size timber was finally obtained by multiplying the grade designations expressed as a percentage (e.g. 75%, 65% etc.) by the basic stress for the timber.

With the introduction of BS 5268 the concept of basic stress was largely abandoned and a revised procedure for assessing the strength of timber adopted. From then on, the first step involved grading structural size timber. Grading was still carried out visually, although it was now common practice to do this mechanically. The latter approach offered the advantage of greater economy in the use of timber since it took into account the density of

timber which significantly influences its strength.

Mechanical stress grading is based on the fact that there is a direct relationship between the modulus of elasticity measured over a relatively short span, i.e. stiffness, and bending strength. The stiffness is assessed non-destructively by feeding individual pieces of timber through a series of rollers on a machine which automatically applies small transverse loads over short successive lengths and measures the deflections. These are compared with permitted deflections appropriate to given stress grades and the machine assesses the grade of the timber over its entire length.

When BS 5268 was published in 1984 the numbered grades (i.e. 75, 65, 50 and 40) were withdrawn and replaced by two visual grades: General Structural (GS) and Special Structural (SS) and four machine grades: MGS, MSS, M75 and M50. The SS grade timber was used as the basis for strength and modulus of elasticity determinations by subjecting a large number of structural sized specimens to short-term load tests. The results were used to obtain the fifth percentile stresses, defined as the value below which not more than 5% of test results fell (*Fig. 6.1*). The fifth percentile values for other grades of the same species were derived using grade relativity factors established from the same series of tests. Finally, the grade stresses were obtained by dividing the fifth percentile stresses by a reduction factor, which included adjustments for a standard depth of specimen of 300 mm, duration of load and a factor of safety. The two visual grades are still referred to in the latest revision of BS 5268 published in 2002. However, machine graded timber is now graded directly to one of sixteen strength classes defined in BS EN 519, principally on the basis of bending stress, mean modulus of elasticity and characteristic density, and marked accordingly.

6.3 Grade stress and strength class

Table 6.1 shows typical timber species/grade combinations and associated grade stresses and moduli of elasticity. This information would enable the designer to determine the size of a timber member given the intensity and distribution of the loads to be carried. However, it would mean that the contractor's choice of material would be limited to one particular species/grade combination, which could be difficult to obtain. It would obviously be better if a range of species/grade combinations could

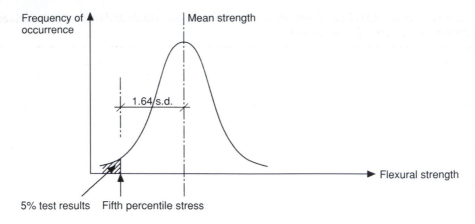

Fig. 6.1 *Frequency distribution curve for flexural strength of timber.*

Table 6.1 Grade stresses for softwoods graded in accordance with BS 4978: for service classes 1 and 2 (Table 10, BS 5268)

Standard name	Grade	Bending parallel to grain[1] N/mm²	Tension parallel to grain[1] N/mm²	Compression		Shear parallel to grain N/mm²	Modulus of elasticity	
				Parallel to grain N/mm²	Perpendicular to grain[2] N/mm²		Mean N/mm²	Minimum N/mm²
Redwood/whitewood	SS	7.5	4.5	7.9	2.1	0.82	10 500	7 000
(imported)	GS	5.3	3.2	6.8	1.8	0.82	9 000	6 000
British larch	SS	7.5	4.5	7.9	2.1	0.82	10 500	7 000
	GS	5.3	3.2	6.8	1.8	0.82	9 000	6 000
British pine	SS	6.8	4.1	7.5	2.1	0.82	10 500	7 000
	GS	4.7	2.9	6.1	1.8	0.82	9 000	6 000
British spruce	SS	5.7	3.4	6.1	1.6	0.64	8 000	5 000
	GS	4.1	2.5	5.2	1.4	0.64	6 500	4 500
Douglas fir	SS	6.2	3.7	6.6	2.4	0.88	11 000	7 000
(British grown)	GS	4.4	2.6	5.2	2.1	0.88	9 500	6 000
Parana pine	SS	9.0	5.4	9.5	2.4	1.03	11 000	7 500
(imported)	GS	6.4	3.8	8.1	2.2	1.03	9 500	6 000
Pitch pine	SS	10.5	6.3	11.0	3.2	1.16	13 500	9 000
(Caribbean)	GS	7.4	4.4	9.4	2.8	1.16	11 000	7 500
Western red cedar	SS	5.7	3.4	6.1	1.7	0.63	8 500	5 500
(imported)	GS	4.1	2.5	5.2	1.6	0.63	7 000	4 500
Douglas fir-larch	SS	7.5	4.5	7.9	2.4	0.85	11 000	7 500
(Canada and USA)	GS	5.3	3.2	6.8	2.2	0.85	10 000	6 500
Hem-fir	SS	7.5	4.5	7.9	1.9	0.68	11 000	7 500
(Canada and USA)	GS	5.3	3.2	6.8	1.7	0.68	9 000	6 000
Spruce-pine-fir	SS	7.5	4.5	7.9	1.8	0.68	10 000	6 500
(Canada and USA)	GS	5.3	3.2	6.8	1.6	0.68	8 500	5 500
Sitka spruce	SS	6.6	4.0	7.0	1.7	0.66	10 000	6 500
(Canada)	GS	4.7	2.8	6.0	1.5	0.66	8 000	5 500
Western whitewoods	SS	6.6	4.0	7.0	1.7	0.66	9 000	6 000
(USA)	GS	4.7	2.8	6.0	1.5	0.66	7 500	5 000
Southern pine	SS	9.6	5.8	10.2	2.5	0.98	12 500	8 500
(USA)	GS	6.8	4.1	8.7	2.2	0.98	10 500	7 000

Notes. [1] Stresses applicable to timber 300 mm deep (or wide): for other section sizes see 2.10.6 and 2.12.2 of BS 5268.
[2] When the specifications specifically prohibit wane at bearing areas, the SS grade compression perpendicular to grain stress may be multiplied by 1.33 and used for all grades.

Table 6.2 Softwood combinations of species and visual grades which satisfy the requirements for various strength classes (Table 2, BS 5268)

Standard name	Strength classes						
	C14	C16	C18	C22	C24	C27	C30
Imported							
Parana pine		GS			SS		
Caribbean pitch pine			GS			SS	
Redwood		GS			SS		
Whitewood		GS			SS		
Western red cedar	GS		SS				
Douglas fir-larch (Canada and USA)		GS			SS		
Hem-fir (Canada and USA)		GS			SS		
Spruce-pine-fir (Canada and USA)		GS			SS		
Sitka spruce (Canada)	GS		SS				
Western whitewoods (USA)	GS		SS				
Southern pine (USA)			GS		SS		
British grown							
Douglas fir	GS		SS				
Larch		GS			SS		
British pine	GS			SS			
British spruce	GS		SS				

be specified and the contractor could then select the most economical one. Such an approach forms the basis of grouping timber species/grade combinations with similar strength characteristics into strength classes (*Table 6.2*).

In all there are sixteen strength classes, C14, C16, C18, C22, C24, TR26, C27, C30, C35, C40, D30, D35, D40, D50, D60 and D70, with C14 having the lowest strength characteristics. The strength class designations indicate the bending strength of the timber. Strength classes C14 to C40 and TR26 are for softwoods and D30 to D70 are for hardwoods. Strength class TR26 is intended for use in the design of trussed rafters. The grade stresses and moduli of elasticity associated with each strength class are shown in *Table 6.3*. In the UK structural timber design is normally based on strength classes C16 to C27. These classes cover a wide range of softwoods which display good structural properties and are both plentiful and cheap.

6.4 Permissible stresses

The grade stresses given in *Tables 6.1* and *6.3* were derived assuming particular conditions of service

and loading. In order to take account of the actual conditions that individual members will be subject to during their design life, the grade stresses are multiplied by modification factors known as K-factors. The modified stresses are termed permissible stresses.

BS 5268 lists over 80 K-factors. However, the following subsections consider only those modification factors relevant to the design of simple flexural and compression members, namely:

K_2: Moisture content factor
K_3: Duration of loading factor
K_5: Notched ends factor
K_7: Depth factor
K_8: Load-sharing systems factor
K_{12}: Compression member stress factor.

6.4.1 MOISTURE CONTENT, K_2

The strength and stiffness of timber decreases with increasing moisture content. This effect is taken into account by assigning timber used for structural work to a service class. BS 5628 recognizes three service classes as follows:

Service class 1 is characterized by a moisture content in the material corresponding to a temperature

Table 6.3 Grade stresses and moduli of elasticity for various strength classes: for service classes 1 and 2 (based on Table 8, BS 5268)

Strength class	Bending parallel to grain $(\sigma_{m,g,||})$ N/mm^2	Tension parallel to grain N/mm^2	Compression parallel to grain $(\sigma_{c,g,||})$ N/mm^2	Compression perpendicular to grain[1] $(\sigma_{c,g,\perp})$ N/mm^2	N/mm^2	Shear parallel to grain (τ_g) N/mm^2	Modulus of elasticity		Density[2] ρ_k kg/m^3	Density[2] ρ_{mean} kg/m^3
							E_{mean} N/mm^2	E_{min} N/mm^2		
C14	4.1	2.5	5.2	2.1	1.6	0.60	6 800	4 600	290	350
C16	5.3	3.2	6.8	2.2	1.7	0.67	8 800	5 800	310	370
C18	5.8	3.5	7.1	2.2	1.7	0.67	9 100	6 000	320	380
C22	6.8	4.1	7.5	2.3	1.7	0.71	9 700	6 500	340	410
C24	7.5	4.5	7.9	2.4	1.9	0.71	10 800	7 200	350	420
TR26	10.0	6.0	8.2	2.5	2.0	1.10	11 000	7 400	370	450
C27	10.0	6.0	8.2	2.5	2.0	1.10	12 300	8 200	370	450
C30	11.0	6.6	8.6	2.7	2.2	1.20	12 300	8 200	380	460
C35	12.0	7.2	8.7	2.9	2.4	1.30	13 400	9 000	400	480
C40	13.0	7.8	8.7	3.0	2.6	1.40	14 500	10 000	420	500
D30	9.0	5.4	8.1	2.8	2.2	1.40	9 500	6 000	530	640
D35	11.0	6.6	8.6	3.4	2.6	1.70	10 000	6 500	560	670
D40	12.5	7.5	12.6	3.9	3.0	2.00	10 800	7 500	590	700
D50	16.0	9.6	15.2	4.5	3.5	2.20	15 000	12 600	650	780
D60	18.0	10.8	18.0	5.2	4.0	2.40	18 500	15 600	700	840
D70	23.0	13.8	23.0	6.0	4.6	2.60	21 000	18 000	900	1 080

[1] When the specification specifically prohibits wane at bearing areas, the higher values may be used.
[2] For the calculation of dead load, the average density should be used.

of 20°C and the relative humidity of the surrounding air only exceeding 65% for a few weeks per year. Timbers used internally in a continuously heated building normally experience this environment. In such environments most timbers will attain an average moisture content not exceeding 12%.

Service class 2 is characterized by a moisture content in the material corresponding to a temperature of 20°C and the relative humidity of the surrounding air only exceeding 85% for a few weeks per year. Timbers used in covered buildings will normally experience this environment. In such environments most timbers will attain an average moisture content not exceeding 20%.

Service class 3, due to climatic conditions, is characterized by higher moisture contents than service class 2 and is applicable to timbers used externally and fully exposed.

The grade stresses and moduli of elasticity quoted in *Tables 6.1* and *6.3* are applicable to timber exposed to service classes 1 and 2. According to clause 2.6.2 of BS 5268, grade stress values for timber exposed to service class 3 should be obtained by multiplying the values in *Tables 6.1* and *6.3* by the modification factor K_2 from Table 16 of BS 5268,

Table 6.4 Modification factor K_2 by which stresses and moduli for service classes 1 and 2 should be multiplied to obtain stresses and moduli applicable to service class 3 (Table 16, BS 5268)

Property	K_2
Bending parallel to grain	0.8
Tension parallel to grain	0.8
Compression parallel to grain	0.6
Compression perpendicular to grain	0.6
Shear parallel to grain	0.9
Mean and minimum modulus of elasticity	0.8

reproduced here as *Table 6.4*. Clause 2.6.1 also notes that because it is difficult to dry thick timber, service class 3 stresses and moduli should be used for solid timber members more than 100 mm thick, unless they have been specially dried.

6.4.2 DURATION OF LOADING, K_3
The stresses given in *Tables 6.1* and *6.3* apply to long term loading. Where the applied loads will act for shorter durations e.g. snow and wind, the grade

Table 6.5 Modification factor K_3 for duration of loading (Table 17, BS 5268)

Duration of loading	Value of K_3
Long term (e.g. dead + permanent imposed[a])	1.00
Medium term (e.g. dead + snow, dead + temporary imposed)	1.25
Short term (e.g. dead + imposed + wind[b], dead + imposed + snow + wind[b])	1.50
Very short term (e.g. dead + imposed + wind[c])	1.75

Notes. [a] For uniformly distributed imposed floor loads $K_3 = 1.00$ except for types 2 and 3 buildings (Table 1, BS 6399: Part 1: 1996) where for foot traffic on corridors, hallways, landings and stairways only, K_3 may be assumed to be 1.5.
[b] For wind, short term category applies to class C (15 s gust) as defined in CP3 : Chapter V : Part 2 or, where the largest diagonal dimension of the loaded area a, as defined in BS 6399: Part 2, exceeds 50 m.
[c] For wind, very short-term category applies to classes A and B (3 s or 5 s gust) as defined in CP3 : Chapter V : Part 2 or, where the largest diagonal dimension of the loaded area a, as defined in BS 6399: Part 2, does not exceed 50 m.

stresses can be increased. Table 17 of BS 5268, reproduced here as *Table 6.5*, gives the modification factor K_3 by which these values should be multiplied for various load combinations.

6.4.3 NOTCHED ENDS, K_5
Notches at the ends of flexural members will result in high shear concentrations which may cause structural failure and must, therefore, be taken into account during design (*Fig. 6.2*).

In notched members the grade shear stresses parallel to the grain (*Tables 6.1* and *6.3*) are multiplied by a modification factor K_5 calculated as follows:

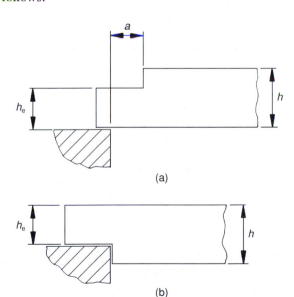

(a)

(b)

Fig. 6.2 *Notched beams: (a) beam with notch on top edge; (b) beam with notch on underside (Fig. 2, BS 5268).*

1. For a notch on the top edge (*Fig. 6.2(a)*):

$$K_5 = \frac{h(h_e - a) + ah_e}{h_e^2} \quad \text{for } a \leqslant h_e \quad (6.1)$$

$$K_5 = 1.0 \quad \text{for } a > h_e \quad (6.2)$$

2. For a notch on the underside (*Fig. 6.2(b)*):

$$K_5 = \frac{h_e}{h} \quad (6.3)$$

Clause 2.10.4 of BS 5268 also notes that the effective depth, h_e, should not be less than 0.5 h, i.e. $K_5 \geqslant 0.5$.

6.4.4 DEPTH FACTOR, K_7
The grade bending stresses given in *Table 6.3* only apply to timber sections having a depth h of 300 mm. For other depths of beams, the grade bending stresses are multiplied by the depth factor K_7, defined in clause 2.10.6 of BS 5268 as follows:

$$K_7 = 1.17 \text{ for solid beams having a depth} \leqslant 72 \text{ mm}$$

$$K_7 = \left(\frac{300}{h}\right)^{0.11} \text{ for solid beams with}$$

$$72 \text{ mm} < h < 300 \text{ mm} \quad (6.4)$$

$$K_7 = \frac{0.81(h^2 + 92\,300)}{(h^2 + 56\,800)} \text{ for solid beams}$$

with $h > 300$ mm

6.4.5 LOAD-SHARING SYSTEMS, K_8
The grade stresses given in *Tables 6.1* and *6.3* apply to individual members, e.g. isolated beams and columns, rather than assemblies. When four

275

or more members such as rafters, joists or wall studs, spaced a maximum of 610 mm centre to centre act together to resist a common load, the grade stress should be multiplied by a load-sharing factor K_8 which has a value of 1.1 (clause 2.9, BS 5268).

6.4.6 COMPRESSION MEMBERS, K_{12}

The grade compression stresses parallel to the grain given in *Tables 6.1* and *6.3* are used to design struts and columns. These values apply to compression members with slenderness ratios less than 5 which would fail by crushing. Where the slenderness ratio of the member is equal to or greater than 5 the grade stresses should be multiplied by the modification factor K_{12} given in Table 22 of BS 5268, reproduced here as *Table 6.6*. Alternatively Appendix B of BS 5268 gives a formula for K_{12} which could be used. This is based on the Perry–Robertson equation which has also been used to develop the steel column design Tables 24(a)–(d) in BS 5950 (section 4.9). The factor K_{12} takes into account the tendency of the member to fail by buckling and allows for imperfections such as out of straightness and accidental load eccentricities.

The factor K_{12} is based on the minimum modulus of elasticity, E_{min}, irrespective of whether the compression member acts alone or forms part of a load-sharing system and the compression stress, $\sigma_{c,||}$, is given by:

$$\sigma_{c,||} = \sigma_{c,g,||}K_3 \qquad (6.5)$$

6.5 Timber design

Having discussed some of the more general aspects, the following sections will consider in detail the design of:

1. flexural members
2. compression members
3. stud walling.

6.6 Symbols

For the purposes of this chapter, the following symbols have been used. These have largely been taken from BS 5268.

GEOMETRICAL PROPERTIES

b	breadth of beam
h	depth of beam
A	total cross-sectional area

i	radius of gyration
I	second moment of area
Z	section modulus

BENDING

L	effective span
M	design moment
M_R	moment of resistance
$\sigma_{m,a,||}$	applied bending stress parallel to grain
$\sigma_{m,g,||}$	grade bending stress parallel to grain
$\sigma_{m,adm,||}$	permissible bending stress parallel to grain

DEFLECTION

δ_t	total deflection
δ_m	bending deflection
δ_v	shear deflection
δ_p	permissible deflection
E	modulus of elasticity
E_{mean}	mean modulus of elasticity
E_{min}	minimum modulus of elasticity
G	shear modulus

SHEAR

F_v	design shear force
τ_a	applied shear stress parallel to grain
τ_g	grade shear stress parallel to grain
τ_{adm}	permissible shear stress parallel to grain

BEARING

F	bearing force
l_b	length of bearing
$\sigma_{c,a,\perp}$	applied compression stress perpendicular to grain
$\sigma_{c,g,\perp}$	grade compression stress perpendicular to grain
$\sigma_{c,adm,\perp}$	permissible bending stress perpendicular to grain

COMPRESSION

L_e	effective length of a column
λ	slenderness ratio
N	axial load
$\sigma_{c,a,||}$	applied compression stress parallel to grain
$\sigma_{c,g,||}$	grade compression stress parallel to grain
$\sigma_{c,adm,||}$	permissible compression stress parallel to grain
$\sigma_{c,||}$	compression stress = $\sigma_{c,g,||}K_3$
σ_e	Euler critical stress

Table 6.6 Modification factor K_{12} for compression members (Table 22, BS 5268)

$E/\sigma_{c,\parallel}$	Values of slenderness ratio λ $(= L_e/i)$																			
	<5	5	10	20	30	40	50	60	70	80	90	100	120	140	160	180	200	220	240	250
	Equivalent L_e/b (for rectangular sections)																			
	<1.4	1.4	2.9	5.8	8.7	11.6	14.5	17.3	20.2	23.1	26.0	28.9	34.7	40.5	46.2	52.0	57.8	63.6	69.4	72.3
	Value of K_{12}																			
400	1.000	0.975	0.951	0.896	0.827	0.735	0.621	0.506	0.408	0.330	0.271	0.225	0.162	0.121	0.094	0.075	0.061	0.051	0.043	0.040
500	1.000	0.975	0.951	0.899	0.837	0.759	0.664	0.562	0.466	0.385	0.320	0.269	0.195	0.148	0.115	0.092	0.076	0.063	0.053	0.049
600	1.000	0.975	0.951	0.901	0.843	0.774	0.692	0.601	0.511	0.430	0.363	0.307	0.226	0.172	0.135	0.109	0.089	0.074	0.063	0.058
700	1.000	0.975	0.951	0.902	0.848	0.784	0.711	0.629	0.545	0.467	0.399	0.341	0.254	0.195	0.154	0.124	0.102	0.085	0.072	0.067
800	1.000	0.975	0.952	0.903	0.851	0.792	0.724	0.649	0.572	0.497	0.430	0.371	0.280	0.217	0.172	0.139	0.115	0.096	0.082	0.076
900	1.000	0.976	0.952	0.904	0.853	0.797	0.734	0.665	0.593	0.522	0.456	0.397	0.304	0.237	0.188	0.153	0.127	0.106	0.091	0.084
1000	1.000	0.976	0.952	0.904	0.855	0.801	0.742	0.677	0.609	0.542	0.478	0.420	0.325	0.255	0.204	0.167	0.138	0.116	0.099	0.092
1100	1.000	0.976	0.952	0.905	0.856	0.804	0.748	0.687	0.623	0.559	0.497	0.440	0.344	0.272	0.219	0.179	0.149	0.126	0.107	0.100
1200	1.000	0.976	0.952	0.905	0.857	0.807	0.753	0.695	0.634	0.573	0.513	0.457	0.362	0.288	0.233	0.192	0.160	0.135	0.116	0.107
1300	1.000	0.976	0.952	0.905	0.858	0.809	0.757	0.701	0.643	0.584	0.527	0.472	0.378	0.303	0.247	0.203	0.170	0.144	0.123	0.115
1400	1.000	0.976	0.952	0.906	0.859	0.811	0.760	0.707	0.651	0.595	0.539	0.486	0.392	0.317	0.259	0.214	0.180	0.153	0.131	0.122
1500	1.000	0.976	0.952	0.906	0.860	0.813	0.763	0.712	0.658	0.603	0.550	0.498	0.405	0.330	0.271	0.225	0.189	0.161	0.138	0.129
1600	1.000	0.976	0.952	0.906	0.861	0.814	0.766	0.716	0.664	0.611	0.559	0.508	0.417	0.342	0.282	0.235	0.198	0.169	0.145	0.135
1700	1.000	0.976	0.952	0.906	0.861	0.815	0.768	0.719	0.669	0.618	0.567	0.518	0.428	0.353	0.292	0.245	0.207	0.177	0.152	0.142
1800	1.000	0.976	0.952	0.906	0.862	0.816	0.770	0.722	0.673	0.624	0.574	0.526	0.438	0.363	0.302	0.254	0.215	0.184	0.159	0.148
1900	1.000	0.976	0.952	0.907	0.862	0.817	0.772	0.725	0.677	0.629	0.581	0.534	0.447	0.373	0.312	0.262	0.223	0.191	0.165	0.154
2000	1.000	0.976	0.952	0.907	0.863	0.818	0.773	0.728	0.681	0.634	0.587	0.541	0.455	0.382	0.320	0.271	0.230	0.198	0.172	0.160

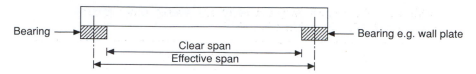

Fig. 6.3 *Effective span of simply supported beams.*

6.7 Flexural members

Beams, rafters and joists are examples of flexural members. All calculations relating to their design are based on the effective span and principally involves consideration of the following aspects which are discussed below:

1. bending
2. deflection
3. lateral buckling
4. shear
5. bearing.

Generally, for medium-span beams the design process follows the sequence indicated above. However, deflection is usually critical for long-span beams and shear for heavily loaded short-span beams.

6.7.1 EFFECTIVE SPAN
According to clause 2.10.3 of BS 5268, for simply supported beams the effective span is normally taken as the distance between the centres of bearings (*Fig. 6.3*).

6.7.2 BENDING
If flexural members are not to fail in bending, the design moment, M, must not exceed the moment of resistance, M_R

$$M \leqslant M_R \qquad (6.6)$$

The design moment is a function of the applied loads. The moment of resistance for a beam can be derived from the theory of bending (equation 2.5, *Chapter 2*) and is given by

$$M_R = \sigma_{m,adm,||} Z_{xx} \qquad (6.7)$$

where

$\sigma_{m,adm,||}$ permissible bending stress parallel to grain
Z_{xx} section modulus

For rectangular sections $Z_{xx} = \dfrac{bd^2}{6}$ (*Fig. 6.4*)

where
b breadth of section
d depth of section

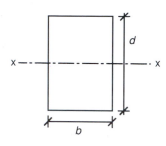

Fig. 6.4 *Section modulus.*

The permissible bending stress is calculated by multiplying the grade bending stress, $\sigma_{m,g,||}$, by any relevant K-factors:

$$\sigma_{m,adm,||} = \sigma_{m,g,||} K_2 K_3 K_7 K_8 \quad \text{(as appropriate)} \qquad (6.8)$$

For a given design moment the minimum required section modulus, Z_{xx} req, can be calculated using equation 6.9, obtained by combining equations 6.6 and 6.7:

$$Z_{xx} \text{ req} \geqslant \frac{M}{\sigma_{m,adm,||}} \qquad (6.9)$$

A suitable timber section can then be selected from Tables NA.2, NA.3 and NA.4 of BS ENV 336: *Structural timber. Coniferous and Poplar. Sizes. Permissible deviations.* These tables give the customary target sizes of, respectively, sawn timber, timber machined on the width and timber machined on all four sides. Table NA.2 is reproduced here as *Table 6.7. Table 6.8* is an expanded version which includes a number of useful section properties to aid design. Finally, the chosen section should be checked for deflection, lateral buckling, shear and bearing to assess its suitability as discussed below.

6.7.3 DEFLECTION
Excessive deflection of flexural members may result in damage to surfacing materials, ceilings, partitions and finishes, and to the functional needs as well as aesthetic requirements.

Clause 2.10.7 of BS 5268 recommends that generally such damage can be avoided if the total

Table 6.7 Customary target sizes of sawn structural timber (based on Table NA.2, BS ENV 336)

Thickness (mm)	Width (mm)									
	75	100	125	150	175	200	225	250	275	300
22		X	X	X	X	X	X			
25	X	X	X	X	X	X	X			
38	X	X	X	X	X	X	X			
47	X	X	X	X	X	X	X	X		X
63		X	X	X	X	X	X			
75		X	X	X	X	X	X	X	X	X
100		X		X		X	X	X		X
150				X		X				X
250										
300								X		X

Table 6.8 Geometrical properties of sawn softwoods

Customary target size*	Area	Section Modulus		Second of moment area		Radius of gyration	
		About x–x	About y–y	About x–x	About y–y	About x–x	About y–y
mm	$10^3 mm^2$	$10^3 mm^3$	$10^3 mm^3$	$10^6 mm^4$	$10^6 mm^4$	mm	mm
22 × 100	2.20	36.6	8.1	1.83	0.089	28.9	6.35
22 × 125	2.75	57.3	10.1	3.58	0.111	36.1	6.35
22 × 150	3.30	82.5	12.1	6.19	0.133	43.3	6.35
22 × 175	3.85	112.3	14.1	9.83	0.155	50.5	6.35
22 × 200	4.40	146.7	16.1	14.7	0.178	57.7	6.35
22 × 225	4.95	185.6	18.2	20.9	0.200	65.0	6.35
25 × 75	1.875	23.4	7.8	0.88	0.098	21.7	7.22
25 × 100	2.50	41.7	10.4	2.08	0.130	28.9	7.22
25 × 125	3.125	65.1	13.0	4.07	0.163	36.1	7.22
25 × 150	3.75	93.8	15.6	7.03	0.195	43.3	7.22
25 × 175	4.375	128	18.2	11.2	0.228	50.5	7.22
25 × 200	5.00	167	20.8	16.7	0.260	57.7	7.22
25 × 225	5.625	211	23.4	23.7	0.293	65.0	7.22
38 × 75	2.85	35.6	18.1	1.34	0.343	21.7	11.0
38 × 100	3.80	63.3	24.1	3.17	0.457	28.9	11.0
38 × 125	4.75	99.0	30.1	6.18	0.572	36.1	11.0
38 × 150	5.70	143	36.1	10.7	0.686	43.3	11.0
38 × 175	6.54	194	42.1	17.0	0.800	50.5	11.0
38 × 200	7.60	253	48.1	25.3	0.915	57.7	11.0
38 × 225	8.55	321	54.2	36.1	1.03	65.0	11.0
47 × 75	3.53	44.1	27.6	1.65	0.649	21.7	13.6
47 × 100	4.70	78.3	36.8	3.92	0.865	28.9	13.6
47 × 125	5.88	122	46.0	7.65	1.08	36.1	13.6
47 × 150	7.05	176	55.2	13.2	1.30	43.3	13.6
47 × 175	8.23	240	64.4	21.0	1.51	50.5	13.6

Table 6.8 *(cont'd)*

Customary target size*	Area	Section Modulus		Second moment of area		Radius of gyration	
		About x–x	*About y–y*	*About x–x*	*About y–y*	*About x–x*	*About y–y*
mm	$10^3 mm^2$	$10^3 mm^3$	$10^3 mm^3$	$10^6 mm^4$	$10^6 mm^4$	*mm*	*mm*
47 × 200	9.40	313	73.6	31.3	1.73	57.7	13.6
47 × 225	10.6	397	82.8	44.6	1.95	65.0	13.6
47 × 250	11.8	490	92.0	61.2	2.16	72.2	13.6
47 × 300	14.1	705	110	106	2.60	86.6	13.6
63 × 100	6.30	105	66.2	5.25	2.08	28.9	18.2
63 × 125	7.88	164	82.7	10.3	2.60	36.1	18.2
63 × 163	9.45	236	99.2	17.7	3.13	43.3	18.2
63 × 175	11.0	322	116	28.1	3.65	50.5	18.2
63 × 200	12.6	420	132	42.0	4.17	57.7	18.2
63 × 225	14.2	532	149	59.8	4.69	65.0	18.2
75 × 100	7.50	125	93.8	6.25	3.52	28.9	21.7
75 × 125	9.38	195	117	12.2	4.39	36.1	21.7
75 × 150	11.3	281	141	21.1	5.27	43.3	21.7
75 × 175	13.1	383	164	33.5	6.15	50.5	21.7
75 × 200	15.0	500	188	50.0	7.03	57.7	21.7
75 × 225	16.9	633	211	71.2	7.91	65.0	21.7
75 × 250	18.8	781	234	97.7	8.79	72.2	21.7
75 × 275	20.6	945	258	130	9.67	79.4	21.7
75 × 300	22.5	1130	281	169	10.5	86.6	21.7
100 × 100	10.0	167	167	8.33	8.33	28.9	28.9
100 × 150	15.0	375	250	28.1	12.5	43.3	28.9
100 × 200	20.0	667	333	66.7	16.7	57.7	28.9
100 × 225	22.5	844	375	94.9	18.8	65.0	28.9
100 × 250	25.0	1010	417	130	20.8	72.2	28.9
100 × 300	30.0	1500	500	225	25.0	86.6	28.9
150 × 150	20.0	563	563	42.2	42.2	43.3	43.3
150 × 200	25.5	1000	750	100	56.3	57.7	43.3
150 × 300	30.0	2250	1130	338	84.4	86.6	43.3
250 × 250	62.5	2600	2600	326	326	72.2	72.2
300 × 300	90.0	4500	4500	675	675	86.6	86.6

Note. * Desired size of timber measured at 20% moisture content

deflection, δ_t, of the member when fully loaded does not exceed the permissible deflection, δ_p:

$$\delta_t \leq \delta_p \qquad (6.10)$$

The permissible deflection is generally given by

$$\delta_p = 0.003 \times span \qquad (6.11)$$

but for longer-span domestic floor joists, i.e. spans over 4.67 m, should not exceed 14 mm:

$$\delta_p \leq 14 \text{ mm} \qquad (6.12)$$

The total deflection, δ_t, is the summation of the bending deflection, δ_m, plus the shear deflection, δ_v:

$$\delta_t = \delta_m + \delta_v \qquad (6.13)$$

Table 6.9 gives the bending and shear deflection formulae for some common loading cases for beams of rectangular cross-section. The formulae have been derived by assuming that the shear modulus is equal to one-sixteenth of the permissible modulus of elasticity in accordance with clause 2.7 of BS 5268.

Table 6.9 Bending and shear deflections assuming $G = E/16$

Load distribution and supports	Deflection at Centre C or end E	
	Bending	Shear
![w C, L, simply supported UDL]	$\dfrac{5}{384} \times \dfrac{wL^4}{EI}$	$\dfrac{12}{5} \times \dfrac{wL^2}{EA}$
![W, L/2 C L/2, simply supported point load]	$\dfrac{WL^3}{48EI}$	$\dfrac{24}{5} \times \dfrac{WL}{EA}$
![W W, a C a, L, two point loads]	$\dfrac{Wa}{EI}\left[\dfrac{L^2}{8} - \dfrac{a^2}{6}\right]$	$\dfrac{96}{5} \times \dfrac{Wa}{EA}$
![w C, L, fixed ends UDL]	$\dfrac{wL^4}{384EI}$	$\dfrac{12}{5} \times \dfrac{wL^2}{EA}$
![W, L/2 C L/2, fixed ends point load]	$\dfrac{WL^3}{192EI}$	$\dfrac{24}{5} \times \dfrac{WL}{EA}$
![w, L E, cantilever UDL]	$\dfrac{wL^4}{8EI}$	$\dfrac{48}{5} \times \dfrac{wL^2}{EA}$
![W, L E, cantilever point load]	$\dfrac{WL^3}{3EI}$	$\dfrac{96}{5} \times \dfrac{WL}{EA}$

For solid timber members acting alone the deflections should be calculated using the minimum modulus of elasticity, but for load-sharing systems the deflections should be based on the mean modulus of elasticity.

6.7.4 LATERAL BUCKLING

If flexural members are not effectively laterally restrained, it is possible for the member to twist sideways before developing its full flexural strength (*Fig. 6.5*), thereby causing it to fail in bending, shear or deflection. This phenomenon is called lateral buckling and can be avoided by ensuring that the depth to breadth ratios given in *Table 6.10* are complied with.

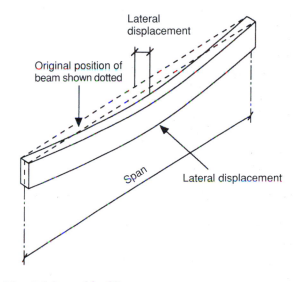

Fig. 6.5 *Lateral buckling.*

Table 6.10 Maximum depth to breadth ratios (Table 19, BS 5268)

Degree of lateral support	Maximum depth to breadth ratio
No lateral support	2
Ends held in position	3
Ends held in position and member held in line, as by purlins or tie-rods at centres not more than 30 times the breadth of the member	4
Ends held in position and compression edge held in line, as by direct connection of sheathing, deck or joists	5
Ends held in position and compression edge held in line, as by direct connection of sheathing, deck or joists, together with adequate bridging or blocking spaced at intervals not exceeding six times the depth	6
Ends held in position and both edges held firmly in line	7

6.7.5 SHEAR

If flexural members are not to fail in shear, the applied shear stress parallel to the grain, τ_a, should not exceed the permissible shear stress, τ_{adm}:

$$\tau_a \leq \tau_{adm} \qquad (6.14)$$

For a beam with a rectangular cross-section, the maximum applied shear stress occurs at the neutral axis and is given by:

$$\tau_a = \frac{3F_v}{2A} \qquad (6.15)$$

where
F_v applied maximum vertical shear force
A cross-sectional area

The permissible shear stress is given by

$$\tau_{adm} = \tau_g K_2 K_3 K_5 K_8 \quad \text{(as appropriate)} \qquad (6.16)$$

where τ_g is the grade shear stress parallel to the grain (*Tables 6.1* and *6.3*).

6.7.6 BEARING PERPENDICULAR TO GRAIN

Bearing failure may arise in flexural members which are supported at their ends on narrow beams or wall plates. Such failures can be avoided by ensuring that the applied bearing stress, $\sigma_{c,a,\perp}$, never exceeds the permissible compression stress perpendicular to the grain, $\sigma_{c,adm,\perp}$:

$$\sigma_{c,a,\perp} \leq \sigma_{c,adm,\perp} \qquad (6.17)$$

The applied bearing stress is given by

$$\sigma_{c,a,\perp} = \frac{F}{bl_b} \qquad (6.18)$$

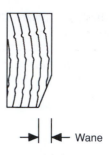

Fig. 6.6 *Wane.*

where
F bearing force (usually maximum reaction)
b breadth of section
l_b bearing length.

The permissible compression stress is obtained by multiplying the grade compression stress perpendicular to the grain, $\sigma_{c,g,\perp}$ by the K-factors for moisture content (K_2), load duration (K_3) and load sharing (K_8) as appropriate:

$$\sigma_{c,adm,\perp} = \sigma_{c,g,\perp} K_2 K_3 K_8 \qquad (6.19)$$

It should be noted that the grade compression stresses perpendicular to the grain given in *Tables 6.1* and *6.3* apply to (i) bearings of any length at the ends of members and (ii) bearings 150 mm or more in length at any position. Moreover, two values for the grade compression stress perpendicular to the grain are given for each strength class (*Table 6.3*). The lower value takes into account the amount of wane which is permitted within each stress grade (*Fig. 6.6*). If, however, the specification prohibits wane from occurring at bearing areas the higher value may be used.

Example 6.1 Design of a timber beam (BS 5268)

A timber beam with a clear span of 2.85 m supports a uniformly distributed load of 10 kN including self-weight of beam. Determine a suitable section for the beam using timber of strength class C16 under service class 1. Assume that the bearing length is 150 mm and that the ends of the beam are held in position and compression edge held in line.

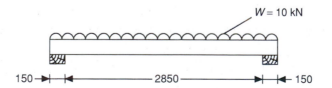

1. EFFECTIVE SPAN

Distance between centres of bearing (l) = 3000 mm

2. GRADE STRESS AND MODULUS OF ELASTICITY FOR C16

Values in N/mm^2 are as follows

Bending parallel to grain $\sigma_{m,g,\|}$	Shear parallel to grain τ_g	Compression perpendicular to grain $\sigma_{c,g,\perp}$	Modulus of elasticity E_{min}
5.3	0.67	1.7	5800

3. MODIFICATION FACTORS

K_2, moisture content factor does not apply since the beam is subject to service class 1

K_3, duration of loading factor = 1.0

K_8, load sharing factor, does not apply since there is only a single beam

K_7, depth factor = $\left(\dfrac{300}{h}\right)^{0.11}$

Assuming $h = 250$ implies that $K_7 = 1.020$

4. BENDING

$$M = \frac{Wl}{8} = \frac{10 \times 3}{8} = 3.75 \text{ kN m}$$

$$\sigma_{m,adm,\|} (\text{assuming } h = 250) = \sigma_{m,g,\|}K_3K_7 = 5.3 \times 1.0 \times 1.020 = 5.406 \text{ N/mm}^2$$

$$Z_{xx}\text{req} \geqslant \frac{M}{\sigma_{m,adm,\|}} = \frac{3.75 \times 10^6}{5.406}$$

$$= 694 \times 10^3 \text{ mm}^3$$

5. DEFLECTION

Permissible deflection $(\delta_p) = 0.003 \times$ span

The deflection due to shear (δ_s) is likely to be insignificant in comparison to the bending deflection (δ_b) and may be ignored in order to make a first estimate of the total deflection (δ_t):

$$\delta_t(\text{ignoring shear deflection}) = \frac{5Wl^3}{384E_{min}I_{xx}} \quad (Table\ 6.9)$$

$$= \frac{5 \times 10^4 \times 3000^3}{384 \times 5800 \times I_{xx}}$$

Since $\delta_p \geqslant \delta_t$

$$0.003 \times 3000 \geqslant \frac{5 \times 10^4 \times 3000^3}{384 \times 5800 \times I_{xx}}$$

$$I_{xx}\text{req} \geqslant 67.3 \times 10^6 \text{ mm}^4$$

From *Table 6.8*, section 75 × 250 provides

$$Z_{xx} = 781 \times 10^3 \text{ mm}^3 \qquad I_{xx} = 97.7 \times 10^6 \text{ mm}^4 \qquad A = 18.8 \times 10^3 \text{ mm}^2$$

The total deflection including shear deflection can now be calculated and is given by

$$\frac{5Wl^3}{384E_{min}I_{xx}} + \frac{12Wl}{5E_{min}A} = \frac{5 \times 10^4 \times 3000^3}{384 \times 5800 \times 97.7 \times 10^6} + \frac{12 \times 10^4 \times 3000}{5 \times 5800 \times 18.8 \times 10^3}$$

$$= 6.2 \text{ mm} + 0.7 \text{ mm}$$

$$= 6.9 \text{ mm} \leqslant \delta_p = 0.003 \times 3000 = 9 \text{ mm}$$

Therefore a beam with a 75 × 250 section is adequate for bending and deflection.

6. LATERAL BUCKLING

Permissible $\dfrac{d}{b} = 5$ (*Table 6.10*)

Actual $\dfrac{d}{b} = \dfrac{250}{75} = 3.3 <$ permissible

Hence the section is adequate for lateral buckling.

7. SHEAR

Permissible shear stress is

$$\tau_{adm} = \tau_g K_3 = 0.67 \times 1.0 = 0.67 \text{ N/mm}^2$$

Maximum shear force is

$$F_v = \frac{W}{2} = \frac{10 \times 10^3}{2} = 5 \times 10^3 \text{ N}$$

Maximum shear stress at neutral axis is

$$\tau_a = \frac{3}{2}\frac{F_v}{A} = \frac{3}{2} \times \frac{5 \times 10^3}{18.8 \times 10^3} = 0.4 \text{ N/mm}^2 < \text{permissible}$$

Therefore the section is adequate in shear.

8. BEARING

Permissible bearing stress is

$$\sigma_{c,adm,\perp} = \sigma_{c,g,\perp} K_3 = 1.7 \times 1.0 = 1.7 \text{ N/mm}^2$$

End reaction, F, is

$$\frac{W}{2} = \frac{10 \times 10^3}{2} = 5 \times 10^3 \text{ N}$$

$$\sigma_{c,a,\perp} = \frac{F}{bl_b} = \frac{5 \times 10^3}{75 \times 150} = 0.44 \text{ N/mm}^2 < \text{permissible}$$

Therefore the section is adequate in bearing. Since all the checks are satisfactory, use 75 mm × 250 mm sawn C16 beam.

Example 6.2 Design of timber floor joists (BS 5268)

Design the timber floor joist for a domestic dwelling using timber of strength class C18 given that:

a) the joists are spaced at 400 mm centres;
b) the floor has an effective span of 3.8 m;
c) the flooring is tongue and groove boarding with a self-weight of 0.1 kN/m^2;
d) the ceiling is of plasterboard with a self weight of 0.2 kN/m^2.

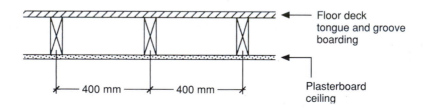

Floor deck
tongue and groove
boarding

Plasterboard
ceiling

|— 400 mm —|— 400 mm —|

1. DESIGN LOADING

Tongue and groove boarding	$= 0.10 \text{ kN/m}^2$
Ceiling	$= 0.20 \text{ kN/m}^2$
Joists (say)	$= 0.10 \text{ kN/m}^2$
Imposed floor load for domestic dwelling (*Table 2.2*)	$= 1.50 \text{ kN/m}^2$
Total load	$= 1.90 \text{ kN/m}^2$

Uniformly distributed load/joist (W) is

$$W = \text{joist spacing} \times \text{effective span} \times \text{load}$$

$$= 0.4 \times 3.8 \times 1.9 = 2.9 \text{ kN}$$

2. GRADE STRESSES AND MODULUS OF ELASTICITY FOR C18

Values in N/mm^2 are as follows

Bending parallel to grain $\sigma_{m,g,\parallel}$	Compression perpendicular to grain $\sigma_{c,g,\perp}$	Shear parallel to grain τ_g	Modulus of elasticity E_{mean}
5.8	1.7	0.67	9100

3. MODIFICATION FACTORS

K_2, moisture content factor does not apply since joists are exposed to service class 2

K_3, duration of loading $= 1.0$

K_8, load-sharing system $= 1.1$

K_7, depth factor $\quad = \left(\dfrac{300}{h}\right)^{0.11}$

where

$h = 225,\ K_7 = 1.032$

$h = 200,\ K_7 = 1.046$

$h = 175,\ K_7 = 1.061$

4. BENDING

Bending moment (M) $\dfrac{Wl}{8} = \dfrac{2.9 \times 3.8}{8} = 1.4 \text{ kN m}$

$$\sigma_{m,adm,\parallel} (\text{ignoring } K_7) = \sigma_{m,g,\parallel} K_3 K_8 = 5.8 \times 1.0 \times 1.1 = 6.38 \text{ N/mm}^2$$

$$Z_{xx}\text{req} \geqslant \frac{M}{\sigma_{m,adm,\parallel}} = \frac{1.4 \times 10^6}{6.38}$$

$$= 219 \times 10^3 \text{ mm}^3$$

From *Table 6.8* a 47×200 mm joist would be suitable ($Z_{xx} = 313 \times 10^3 \text{ mm}^3$, $I_{xx} = 31.3 \times 10^6 \text{ mm}^4$, $A = 9.4 \times 10^3 \text{ mm}^2$)

Hence $K_7 = 1.046$. Therefore

$$Z_{xx}\text{req} = \frac{219 \times 10^3}{1.046} = 209 \times 10^3 \text{ mm}^3 < \text{provided} \quad \text{OK}$$

5. DEFLECTION

Permissible deflection = $0.003 \times$ span

$$= 0.003 \times 3800 = 11.4 \text{ mm}$$

Total deflection $(\delta_t) = $ bending deflection $(\delta_m) +$ shear deflection (δ_v)

$$= \frac{5Wl^3}{384E_{\text{mean}}I_{\text{xx}}} + \frac{12Wl}{5E_{\text{mean}}A}$$

$$= \frac{5 \times 2.9 \times 10^3 \times \left(3.8 \times 10^3\right)^3}{384 \times 9.1 \times 10^3 \times 31.3 \times 10^6} + \frac{12 \times 2.9 \times 10^3 \times 3.8 \times 10^3}{5 \times 9.1 \times 10^3 \times 9.4 \times 10^3}$$

$$= 7.3 \text{ mm} + 0.3 \text{ mm} = 7.6 \text{ mm} < \text{permissible}$$

Therefore 47 mm $\times$ 200 mm joist is adequate in bending and deflection.

6. LATERAL BUCKLING

Permissible $\dfrac{d}{b} = 5$ (*Table 6.10*)

Actual $\quad \dfrac{d}{b} = \dfrac{200}{47} = 4.3 < \text{permissible}$

Hence the section is adequate for lateral buckling.

7. SHEAR

Permissible shear stress is

$$\tau_{\text{adm}} = \tau_{\text{g}}K_3K_8 = 0.67 \times 1.0 \times 1.1 = 0.737 \text{ N/mm}^2$$

Maximum shear force is

$$F_{\text{v}} = \frac{W}{2} = \frac{2.9 \times 10^3}{2} = 1.45 \times 10^3 \text{ N}$$

Maximum shear stress at neutral axis is

$$\tau_{\text{a}} = \frac{3}{2} \frac{F_{\text{v}}}{A} = \frac{3}{2} \times \frac{1.45 \times 10^3}{9.4 \times 10^3} = 0.23 \text{ N/mm}^2 < \text{permissible}$$

Therefore the section is adequate in shear.

8. BEARING

Permissible compression stress perpendicular to grain is

$$\sigma_{\text{c,adm},\perp} = \sigma_{\text{c,g},\perp}K_3K_8 = 1.7 \times 1.0 \times 1.1 = 1.87 \text{ N/mm}^2$$

Maximum end reaction is

$$F = \frac{W}{2} = \frac{2.9 \times 10^3}{2} = 1.45 \times 10^3 \text{ N}$$

Assuming that the floor joists span on to 100 mm wide wall plates the bearing stress is given by

$$\sigma_{\text{c,a},\perp} = \frac{F}{bl_{\text{b}}} = \frac{1.45 \times 10^3}{47 \times 100} = 0.31 \text{ N/mm}^2 < \text{permissible}$$

Therefore the section is adequate in bearing.

9. CHECK ASSUMED SELF-WEIGHT OF JOISTS

From *Table 6.3*, the average density of timber of strength class C18 is 380 kg/m³. Hence, self-weight of the joists is

$$\frac{\left(47 \times 200 \times 10^{-3}\right) \times 380 \text{ kg/m}^{-3} \times 9.81 \times 10^{-3}}{0.4} = 0.088 \text{ kN/m}^2 < 0.10 \text{ kN/m}^2 \quad \text{(assumed)}$$

Since all the checks are satisfactory use 47 mm × 200 mm C18 sawn floor joists.

Example 6.3 Design of a notched floor joist (BS 5268)

The joists in *Example 6.2* are to be notched at the bearings with a 75 m deep notch as shown below. Check that the notched section is still adequate.

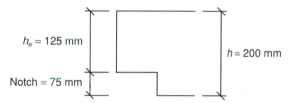

The presence of the notch affects only the shear stresses in the joists. For a notched member the permissible shear stress is given by

$$\tau_{\text{adm}} = \tau_g K_3 K_5 K_8$$

where

$$K_5 = \frac{h_e}{h} = \frac{125}{200} = 0.625 > \text{min. } (= 0.5)$$

Hence

$$\tau_{\text{adm}} = 0.67 \times 1.0 \times 0.625 \times 1.1 = 0.46 \text{ N/mm}^2$$

Applied shear parallel to grain, τ_a (from above) is

$$0.23 \text{ N/mm}^2 < \text{permissible}$$

Therefore the 47 mm × 200 mm sawn joists are also adequate when notched with a 75 mm deep bottom edge notch at the bearing.

Example 6.4 Analysis of a timber roof (BS 5268)

A flat roof spanning 4.5 m is constructed using timber joists of grade GS whitewood with a section size of 47 mm × 225 mm and spaced at 450 mm centres. The total dead load due to the roof covering and ceiling including the self-weight of the joists is 1 kN/m². Calculate the maximum imposed load the roof can carry assuming that the duration of loading is (a) long term (b) medium term.

1. DESIGN LOADING

$$\text{Dead load} = 1 \text{ kN/m}^2$$
$$\text{Live load} = q \text{ kN/m}^2$$

Uniformly distributed load/joist, W, is

$$W = \text{joist spacing} \times \text{effective span} \times (\text{dead} + \text{live}) = 0.45 \times 4.5 \ (1 + q)$$

2. GRADE STRESSES AND MODULUS OF ELASTICITY

Grade GS whitewood timber belongs to strength class C16 (*Table 6.2*). Values in N/mm^2 are as follows:

Bending parallel to grain $\sigma_{m,g,\|\|}$	Compression perpendicular to grain $\sigma_{c,g,\perp}$	Shear parallel to grain τ_g	Modulus of elasticity E_{mean}
5.3	1.7	0.67	8800

3. MODIFICATION FACTORS

K_3, duration of loading (*Table 6.5*) = 1.0 (long term) = 1.25 (medium term)

K_8, load-sharing system = 1.1

K_7, depth factor $= \left(\dfrac{300}{h}\right)^{0.11}$

where $h = 225$, $K_7 = 1.032$

4. GEOMETRICAL PROPERTIES

From *Table 6.8*, 47 × 225 section provides:

$$\begin{aligned}
\text{Cross-sectional area, } A &= 10.6 \times 10^3 \text{ mm}^2 \\
\text{Section modulus about x–x, } Z_{xx} &= 397 \times 10^3 \text{ mm}^3 \\
\text{Second moment of area about x–x, } I_{xx} &= 44.6 \times 10^6 \text{ mm}^4
\end{aligned}$$

5. BENDING

(a) Long term

Permissible bending stress parallel to grain is

$$\sigma_{m,adm,\|\|} = \sigma_{m,g,\|\|}K_3K_7K_8 = 5.3 \times 1.0 \times 1.032 \times 1.1 = 6.02 \text{ N/mm}^2$$

Moment of resistance, M_R, is

$$M_R = \sigma_{m,adm,\|\|}Z_{xx} = 6.02 \times 397 \times 10^3 \times 10^{-6} = 2.39 \text{ kN m}$$

Design moment, $M = \dfrac{Wl}{8} = 0.45 \times 4.5(1 + q)\dfrac{4.5}{8} = 1.139(1 + q)$

Equating $M_R = M$,

$$2.39 = 1.139 \ (1 + q) \Rightarrow q = 1.09 \text{ kN/m}^2$$

(b) Medium term

From above

$$\sigma_{m,adm,\|\|} = 6.02 K_3 (\text{medium term}) = 6.02 \times 1.25 = 7.52 \text{ N/mm}^2$$

$$M_R = 7.52 \times 397 \times 10^3 \times 10^{-6} = 2.98 \text{ kNm}$$

Equating $M_R = M$,

$$2.98 = 1.139(1 + q) \Rightarrow q = 1.62 \text{ kN/m}^2$$

6. DEFLECTION

Maximum total deflection = bending deflection (δ_m) + shear deflection (δ_v)

$$0.003L = \frac{5WL^3}{384E_{mean}I_{xx}} + \frac{12WL}{5E_{mean}A}$$

$$0.003 = W \left[\frac{5 \times (4.5 \times 10^3)^2}{384 \times 8800 \times 44.6 \times 10^6} + \frac{12}{5 \times 8800 \times 10.6 \times 10^3} \right]$$

$$= 6.975 \times 10^{-7} \; W$$

$$W = 4300 \; \text{N per joist}$$

Load per unit area is

$$\frac{W}{\text{joist spacing} \times \text{span}} = \frac{4.3}{0.45 \times 4.5}$$

Hence

$$q = 2.12 - \text{dead load} = 2.12 - 1 = 1.12 \; \text{kN/m}^2$$

7. SHEAR

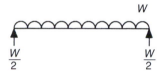

Permissible shear parallel to grain is

$$\tau_{adm} = \tau_g K_3 K_8 = 0.67 \times 1.0 \times 1.1 = 0.737 \; \text{N/mm}^2$$

Maximum shear force $F_v = \dfrac{2}{3} \tau_{adm} A$ (equation 6.15)

$$= \left(\frac{2}{3} \times 0.737 \times 10.6 \times 10^3 \right) \times 10^{-3} = 5.2 \; \text{kN}$$

Total load per joist $= 2F_v = 10.4 \; \text{kN}$

Load per unit area $= \dfrac{10.4}{0.45 \times 4.5} = 5.13 \; \text{kN/m}^2$

Hence

$$q = 5.13 - 1 = 4.13 \; \text{kN/m}^2 \quad \text{(long term)}$$

and

$$q = 5.43 \; \text{kN/mm}^2 \quad \text{(medium term, } K_3 = 1.25)$$

Hence the safe long-term imposed load that the roof can support is 1.09 kN/m^2 (bending critical) and the safe medium-term imposed load is 1.12 kN/m^2 (deflection critical).

6.8 Design of compression members

Struts and columns are examples of compression members. For design purposes BS 5268 divides compression members into two categories (1) members subject to axial compression only and (2) members subject to combined bending and axial compression.

The principal considerations in the design of compression members are:

1. slenderness ratio
2. axial compressive stress
3. permissible compressive stress.

The following subsections consider these more general aspects before describing in detail the design of the above two categories of compression members.

6.8.1 SLENDERNESS RATIO

The load-carrying capacity of compression members is a function of the slenderness ratio, λ, which is given by

Table 6.11 Effective length of compression members (Table 21, BS 5268)

End conditions	Effective length / Actual length (L_e/L)
(a) Restrained at both ends in position and in direction	0.7
(b) Restrained at both ends in position and one end in direction	0.85
(c) Restrained at both ends in position but not in direction	1.0
(d) Restrained at one end in position and in direction and at the other end in direction but not in position	1.5
(e) Restrained at one end in position and in direction and free at the other end	2.0

$$\lambda = \frac{L_e}{i} \quad (6.20)$$

where
L_e effective length
i radius of gyration

According to clause 2.11.4 of BS 5268, the slenderness ratio should not exceed 180 for compression members carrying dead and imposed loads other than loads resulting from wind in which case a slenderness ratio of 250 may be acceptable.

The radius of gyration, i, is given by

$$i = \sqrt{I/A} \quad (6.21)$$

where
I moment of inertia
A cross-section area.

For rectangular sections

$$i = b/\sqrt{12} \quad (6.22)$$

where b is the least lateral dimension.

The effective length, L_e, of a column is obtained by multiplying the actual length, L, by a coefficient taken from *Table 6.11* which is a function of the fixity at the column ends.

$$L_e = L \times \text{coefficient} \quad (6.23)$$

In *Table 6.11* end condition (a) models the case of a column with both ends fully fixed and no relative horizontal movement possible between the column ends. End condition (c) models the case of a pin-ended column with no relative horizontal movement possible between column ends. End condition (e) models the case of a column with one end fully fixed and the other end free. *Figure 6.7* illustrates all five combinations of end conditions.

6.8.2 AXIAL COMPRESSIVE STRESS

The axial compressive stress is given by

$$\sigma_{c,a,||} = \frac{F}{A} \quad (6.24)$$

where
F axial load
A cross-sectional area.

6.8.3 PERMISSIBLE COMPRESSIVE STRESS

According to clause 2.11.5 of BS 5268, for compression members with slenderness ratios of less than 5, the permissible compressive stress should be taken as the grade compression stress parallel to the grain, $\sigma_{c,g,||}$, modified as appropriate for moisture content, duration of loading and load sharing:

$$\sigma_{c,adm,||} = \sigma_{c,g,||}K_2K_3K_8 \quad \text{for } \lambda < 5 \quad (6.25)$$

For compression members with slenderness ratios equal to or greater than 5, the permissible compressive stress is obtained in the same way but should additionally be modified by the factor K_{12}

$$\sigma_{c,adm,||} = \sigma_{c,g,||}K_2K_3K_8K_{12} \quad \text{for } \lambda \geqslant 5 \quad (6.26)$$

6.8.4 MEMBER DESIGN

Having discussed these common aspects it is now possible to describe in detail the design of compression members. As pointed out earlier, BS 5268 distinguishes between two types of members, that is, those subject to (a) axial compression only and (b) axial compression and bending.

6.8.4.1 Members subject to axial compression only

This category of compression member is designed so that the applied compressive stress, $\sigma_{c,a,||}$, does not exceed the permissible compressive stress parallel to the grain, $\sigma_{c,adm,||}$:

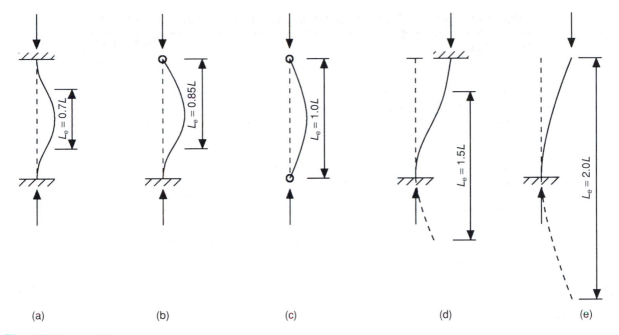

Fig. 6.7 End conditions.

$$\sigma_{c,a,||} \leqslant \sigma_{c,adm,||} \qquad (6.27)$$

The applied compressive stress is calculated using equation 6.24 and the permissible compressive stress is given by equations 6.25 or 6.26 depending upon the slenderness ratio.

6.8.4.2 Members subject to axial compression and bending

This category includes compression members subject to eccentric loading which can be equated to an axial compression force and bending moment. According to clause 2.11.6 of BS 5268, members which are restrained at both ends in position but not direction, which covers most real situations, should be so proportioned that

$$\frac{\sigma_{m,a,||}}{\sigma_{m,adm,||}\left(1 - \dfrac{1.5\sigma_{c,a,||}}{\sigma_e}K_{12}\right)} + \frac{\sigma_{c,a,||}}{\sigma_{c,adm,||}} \leqslant 1 \qquad (6.28)$$

where

$\sigma_{m,a,||}$	applied bending stress
$\sigma_{m,adm,||}$	permissible bending stress
$\sigma_{c,a,||}$	applied compression stress
$\sigma_{c,adm,||}$	permissible compression stress (including K_{12})
σ_e	Euler critical stress $= \pi^2 E_{min}/(L_e/i)^2$

Equation 6.28 is the normal interaction formula used to ensure that lateral instability does not arise in compression members subject to axial force and bending. Thus if the column was subject to compressive loading only, i.e. $M = 0$ and $\sigma_{m,a,||} = 0$, the designer would simply have to ensure that $\sigma_{c,a,||}/\sigma_{c,adm,||} \leqslant 1$. Alternatively, if the column was subject to bending only, i.e. $F = \sigma_{c,a,||} = 0$, the designer should ensure that $\sigma_{m,a,||}/\sigma_{m,adm,||} \leqslant 1$. However, if the column was subject to combined bending and axial compression, then the deflection as a result of the moment M would lead to additional bending due to the eccentricity of the force F as illustrated in Fig. 6.8. This is allowed for by the factor

$$\frac{1}{[1 - (1.5\sigma_{c,a,||}K_{12})/\sigma_e]}$$

in the above expression.

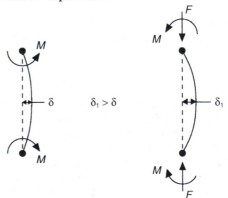

Fig. 6.8 Bending in timber columns.

Example 6.5 Timber column resisting an axial load (BS 5268)

A timber column of redwood GS grade consists of a 100 mm square section which is restrained at both ends in position but not in direction. Assuming that the actual height of the column is 3.75 m, calculate the maximum axial long-term load that the column can support.

1. SLENDERNESS RATIO

$$\lambda = L_e/i \Rightarrow L_e = 1.0 \times h = 1.0 \times 3750 = 3750 \text{ mm}$$

$$i = \sqrt{\frac{I}{A}} = \sqrt{\frac{db^3/12}{db}} = \sqrt{\frac{b^2}{12}} = \frac{100}{\sqrt{12}} = 28.867$$

$$\lambda = \frac{3750}{28.867} = 129.9 < 180 \quad \text{OK}$$

2. GRADE STRESSES AND MODULUS OF ELASTICITY

Grade GS redwood belongs to strength class C16 (*Table 6.2*). Values in N/mm² are as follows

Compression parallel to grain $\sigma_{c,g,\parallel}$	Modulus of elasticity E_{min}
6.8	5800

3. MODIFICATION FACTOR

K_3, duration of loading is 1.0

$$\frac{E_{min}}{\sigma_{c,\parallel}} = \frac{5800}{6.8 \times 1.0} = 852.9 \quad \text{and} \quad \lambda = 129.9$$

From *Table 6.7* by interpolation K_{12} is found to be 0.261.

$\dfrac{E_{min}}{\sigma_{c,\parallel}}$	λ		
	120	129.9	140
800	0.280		0.217
852.9	0.293	0.261	0.228
900	0.304		0.237

4. AXIAL LOAD CAPACITY

Permissible compression stress parallel to grain is

$$\sigma_{c,adm,\parallel} = \sigma_{c,g,\parallel}K_3K_{12} = 6.8 \times 1.0 \times 0.261 = 1.77 \text{ N/mm}^2$$

Hence the long-term axial load capacity of column is

$$\sigma_{c,adm,\parallel}A = 1.77 \times 10^4 \times 10^{-3} = 17.7 \text{ kN}$$

Example 6.6 Timber column resisting an axial load and moment (BS 5268)

Check the adequacy of the column in Example 6.5 to resist a long-term axial load of 10 kN and a bending moment of 350 kN mm.

1. SLENDERNESS RATIO

$$\lambda = L_e/i = 129.9 < 180 \quad (Example\ 6.5)$$

2. GRADE STRESSES AND MODULUS OF ELASTICITY

Values in N/mm^2 for timber of strength class C16 are as follows

Bending parallel to grain $\sigma_{mg,\|\|}$	Compression parallel to grain $\sigma_{c,g,\|\|}$	Modulus of elasticity E_{min}
5.3	6.8	5800

3. MODIFICATION FACTORS

$K_3 = 1.0$

$$K_7 = \left(\frac{300}{h}\right)^{0.11} = \left(\frac{300}{100}\right)^{0.11} = 1.128$$

$K_{12} = 0.261$ (see *Example 6.5*)

4. COMPRESSION AND BENDING STRESSES

Permissible compression stress is

$$\sigma_{c,adm,\|\|} = \sigma_{c,g,\|\|}K_3K_{12} = 6.8 \times 1.0 \times 0.261 = 1.77\ \text{N/mm}^2$$

Applied compression stress is

$$\sigma_{c,a,\|\|} = \frac{\text{axial load}}{A} = \frac{10 \times 10^3}{10^4} = 1\ \text{N/mm}^2$$

Permissible bending stress is

$$\sigma_{m,adm,\|\|} = \sigma_{m,g,\|\|}K_3K_7 = 5.3 \times 1.0 \times 1.128 = 5.98\ \text{N/mm}^2$$

Applied bending stress is

$$\sigma_{m,a,\|\|} = \frac{M}{Z} = \frac{350 \times 10^3}{167 \times 10^3} = 2.10\ \text{N/mm}^2$$

Euler critical stress is

$$\sigma_e = \frac{\pi^2 E_{min}}{(L_e/i)^2} = \frac{\pi^2 \times 5800}{(129.9)^2} = 3.39\ \text{N/mm}^2$$

Since column is restrained at both ends, in position but not in direction, check that the column is so proportioned that

$$\frac{\sigma_{m,a,\|\|}}{\sigma_{m,adm,\|\|}\left(1 - \frac{1.5\sigma_{c,a,\|\|}}{\sigma_e}K_{12}\right)} + \frac{\sigma_{c,a,\|\|}}{\sigma_{c,adm,\|\|}} \leq 1$$

Substituting

$$\frac{2.10}{5.98\left(1 - \dfrac{1.5 \times 1}{3.39} \times 0.261\right)} + \frac{1}{1.77}$$

$$= 0.397 + 0.565 = 0.962 < 1$$

Therefore a 100×100 column is adequate to resist a long-term axial load of 10 kN and a bending moment of 350 kN mm.

6.9 Design of stud walls

In timber frame housing the loadbearing walls are normally constructed using stud walls (*Fig. 6.9*). These walls can be designed to resist not only the vertical loading but also loads normal to the wall due to wind, for example. Stud walls are normally designed in accordance with the requirements of BS 5268: Part 6: *Code of Practice for Timber Frame Walls; Section 6.1: Dwellings not exceeding four storeys*. They basically consist of vertical timber members, commonly referred to as studs, which are held in position by nailing them to timber rails or plates, located along the top and bottom of the studs. The most common stud sizes are 100×50, 47, 38 mm and 75×50, 47, 38 mm. The studs are usually placed at 400 or 600 mm centres depending upon preference, or on the loads they are required to transmit.

The frame is usually covered by a cladding material such as plasterboard which may be required for aesthetic reasons, but will also provide lateral restraint to the studs about the y–y axis. If the wall is not surfaced or only partially surfaced, the studs may be braced along their lengths by internal noggings. Bending about the x–x axis of the stud is assumed to be unaffected by the presence of the cladding material.

Since the centre-to-centre spacing of the stud is normally less than 610 mm, the load-sharing factor K_8 will apply to the design of stud walls. The design of stud walling is illustrated in the following example.

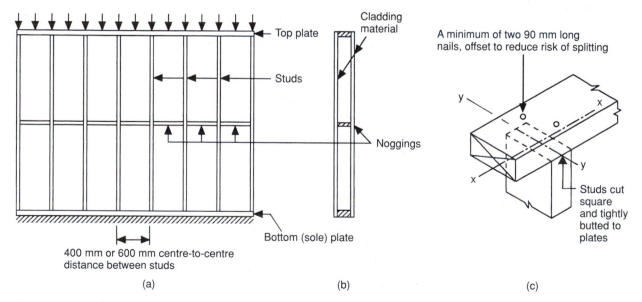

Fig. 6.9 Details of a typical stud wall: (a) elevation; (b) section; (c) typical fixing of top and bottom plates to studs.

Example 6.7 Analysis of a stud wall (BS 5268)

A stud wall panel has an overall height of 3.75 m including top and bottom rails and vertical studs at 600 mm centres with nogging pieces at mid-height. Assuming that the studs, rail framing and nogging pieces comprises 44×100 mm section of strength class C22, calculate the maximum uniformly distributed long term total load the panel is able to support.

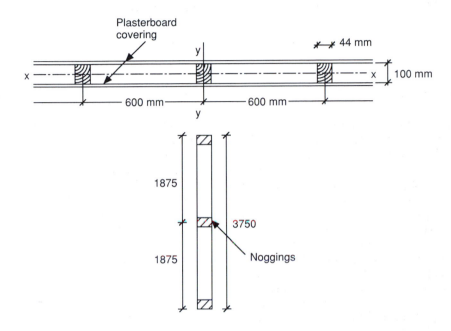

1. SLENDERNESS RATIO

(a) Effective height

$$L_{ex} = \text{coefficient} \times L = 1.0 \times 3750 = 3750 \text{ mm}$$

$$L_{ey} = \text{coefficient} \times L/2 = 1.0 \times 3750/2 = 1875 \text{ mm}$$

(b) Radius of gyration

$$i_{xx} = \sqrt{\frac{I_{xx}}{A}} = \sqrt{\frac{(1/12) \times 44 \times 100^3}{44 \times 100}} = \frac{100}{\sqrt{12}}$$

$$i_{yy} = \sqrt{\frac{I_{yy}}{A}} = \sqrt{\frac{(1/12) \times 100 \times 44^3}{44 \times 100}} = \frac{44}{\sqrt{12}}$$

(c) Slenderness ratio

$$\lambda_{xx} = \frac{L_{ex}}{i_{xx}} = \frac{3750}{100/\sqrt{12}} = 129.9 < 180$$

$$\lambda_{yy} = \frac{L_{ex}}{i_{yy}} = \frac{1875}{44/\sqrt{12}} = 147.6 < 180 \quad \text{(critical)}$$

Note that where two values of λ are possible the larger value must always be used to find $\sigma_{c,adm,||}$.

2. GRADE STRESSES AND MODULUS OF ELASTICITY

For timber of strength class C22, values in N/mm^2 are as follows:

Compression parallel to grain $\sigma_{c,g,||}$	Modulus of elasticity E_{min}
7.5	6500

3. MODIFICATION FACTORS

$$K_3 = 1.0 \qquad K_8 = 1.1$$

$$\frac{E_{min}}{\sigma_{c,||}} = \frac{E_{min}}{\sigma_{c,g,||}K_3} = \frac{6500}{7.5 \times 1.0} = 866.7 \quad \text{and} \quad \lambda = 147.6$$

From *Table 6.6* $K_{12} = 0.212$ by interpolation.

$\dfrac{E_{min}}{\sigma_{c,||}}$	λ		
	140	147.6	160
800	0.217		0.172
866.7	0.230	0.212	0.183
900	0.237		0.188

4. AXIAL STRESSES

Permissible compression stress parallel to grain

$$\sigma_{c,adm,||} = \sigma_{c,g,||}K_3K_8K_{12} = 7.5 \times 1.0 \times 1.1 \times 0.212 = 1.75 \; N/mm^2$$

Axial load capacity of stud is

$$= \sigma_{c,adm,||}A = 1.75 \times 44 \times 100 \times 10^{-3} = 7.7 \; kN$$

Hence uniformly distributed load capacity of stud wall panel is

$$7.7/0.6 = 12.8 \; kN/m$$

Note that the header spans 0.6 m and that this should also be checked as a beam in order to make sure that it is capable of supporting the above load.

6.10 Summary

This chapter has attempted to explain the concepts of stress grading and strength classes and the advantages that they offer to designers and contractors alike involved in specifying timber for structural purposes. The chapter has described the design of flexural and compression members and stud walling, to BS 5628: Part 2: *Structural Use of Timber*, which is based on permissible stress principles. Bending, shear and deflection are found to be the critical factors determining the design of flexural members e.g. beams, rafters and joists. The slenderness ratio influences the load carrying capacity of compression members e.g. struts and columns. Stud walls are normally designed on the assumption that the compression members act together to support a common load.

Questions

1. (a) Discuss the factors which influence the strength of timber and explain how the strength of timber is assessed in practice.

   (b) Simply supported timber roof beams spanning 5 m support a total uniformly distributed load of 11 kN. Determine a suitable section for the beam using timber of strength class C16. Assume that the bearing length is 125 mm and that the compression edge is held in position.

2. (a) Give typical applications of timber in the construction industry and for each case discuss possible desirable properties.

   (b) Redesign the timber joists in *Example 6.2* using timber of strength class C22.

3. (a) Distinguish between softwood and hardwood and grade stress and permissible stress.

   (b) Calculate the maximum long term imposed load that a flat roof can support assuming the following construction details:
   - roof joists are 50 mm × 225 mm of strength class C16 at 600 mm centres
   - effective span is 4.2 m
   - unit weight of woodwool (50 mm thick) is 0.3 kN/m^2

   - unit weight of boarding, bitumen and roofing felt is 0.45 kN/m^2
   - unit weight of plasterboard and skim is 0.22 kN/m^2

4. (a) Discuss the factors accounted for by the modification factor K_{12} in the design of timber compression members.

   (b) Design a timber column of effective length 2.8 m, capable of resisting the following loading:
   (i) medium term axial load of 37.5 kN
   (ii) long term axial load of 30 kN and a bending moment of 300 kN mm.

5. (a) Explain with the aid of sketches connection details which will give rise to the following end conditions:
   (i) restrained in position and direction
   (ii) restrained in position but not in direction
   (iii) unrestrained in position and direction.

   (b) Design a stud wall of length 4.2 m and height 3.8 m, using timber of strength class C16 to support a long-term uniformly distributed load of 14 kN/m.

PART THREE

STRUCTURAL DESIGN TO THE EUROCODES

Part Two of this book has described the design of a number of structural elements in the four media: concrete, steel, masonry and timber to BS 8110, BS 5950, BS 5628 and BS 5268 respectively. The principal aim of this part of the book is to describe the salient features of the structural Eurocodes for these media, Eurocodes 2, 3, 6 and 5 respectively, and highlight the significant differences between the British Standard and the corresponding Eurocode.

The subject-matter has been divided into five chapters as follows:

1. *Chapter 7* introduces the Eurocodes and provides answers to some general questions regarding their nature, role, method of production and layout.

2. *Chapter 8* describes the contents of Part 1.1 of Eurocode 2 for the design of concrete buildings and illustrates the new procedures for designing beams, slabs, pad foundations and columns.

3. *Chapter 9* describes the contents of Part 1.1 of Eurocode 3 for the design of steel buildings and illustrates the new procedures for designing beams, columns and connections.

4. *Chapter 10* describes the contents of Eurocode 6 for the design of masonry structures and high-lights the major difficulties that have delayed its progress.

5. *Chapter 11* describes the contents of Part 1.1 of Eurocode 5 for the design of timber structures and illustrates the new procedures for designing flexural and compression members.

Chapter 7

The structural Eurocodes: An introduction

This chapter describes the nature, objectives and mechanics of producing the new Eurocodes for structural design. The chapter highlights some of the difficulties associated with drafting these standards together with details of how these difficulties were resolved.

7.1 What are Eurocodes?

Eurocodes are the European standards for structural design. *Table 7.1* shows the proposed range of Eurocodes currently under preparation. Like the present UK codes of practice, Eurocodes will come in a number of parts, covering a range of applications.

Eurocodes will have the same legal standing as eventually held by the national equivalent design standards or codes of practice. They will be published first as preliminary standards, designated by ENV (*Norme Vornorme Européenne*). This is equivalent to BSI's *Draft for Development* and must be used in conjunction with a national application

document (NAD), containing supplementary information specific to each member state. The NAD takes precedence over corresponding provisions in the ENV. The complete suit of Eurocodes has been available as ENVs for some years.

The ENV Eurocodes will eventually be revised and reissued as European Standards, designated by EN (*Norme Européenne*). National annexes will replace national application documents in the EN Eurocodes. Following a few years of coexistence, the EN Eurocodes will become mandatory in the sense that all conflicting national standards must be withdrawn. At the time of writing only the lead Eurocode: *Basis of Structural Design* and the first section of the first part of *Eurocode 1: Actions on Structures* have been converted to full EN status. Latest estimates suggest that full implementation of the Eurocodes in every member state will occur by the end of the decade.

7.2 Why are Eurocodes necessary?

The establishment of international standards for structural design is not a new idea. Indeed, the first draft of Eurocode 2 for concrete structures was based on the CEB (Comité Européen du Béton) Model Code of 1978 drawn up by a number of experts from various European countries. Similarly Eurocode 3 was based on the 1977 *Recommendations for Design of Steel Structures* published by ECCS (European Convention for Constructional Steelwork) which was the work of several expert committees drawn from various countries in Europe and beyond. This process of drawing up of European design standards has been given a fresh impetus with the drive towards the political and economic unification of the EC.

There are several advantages to be gained from having design standards which are accepted by all

Table 7.1 Structural Eurocodes currently under preparation

Eurocode	Subject
Eurocode	Basis of Structural Design
1	Actions on Structures
2	Design of Concrete Structures
3	Design of Steel Structures
4	Design of Composite Steel and Concrete Structures
5	Design of Timber Structures
6	Design of Masonry Structures
7	Geotechnical Design
8	Design of Structures for Earthquake Resistance
9	Design of Aluminium Structures

member states. The first and foremost reason is that the provision of Eurocodes and the associated European Standards for construction products will help lower trade barriers between the member states. This will allow contractors and consultants from all member states to compete fairly for work within Europe. Hopefully this will lead to a pooling of resources and the sharing of expertise, thereby lowering production costs. It is further believed that such standards will boost the international standing of European engineers which should help in increasing their chances of winning contracts abroad. A further benefit of having Eurocodes is that they will make it easier for engineers to practise within all EC countries.

7.3 Who produced the structural Eurocodes?

Generally, each Eurocode has been drafted by a small group of experts from various member states. These groups were formerly under contract to the EC Commission but are now under the direct control of CEN (Comité Européen de Normalisation), the European Standards Organization. In addition, a liaison engineer from each member state has been involved in evaluating the final document and discussing with the drafting group the acceptability of the Eurocode in relation to the national standard from the country which they represent.

7.4 Problems associated with the drafting of Eurocodes

The main problems faced by the drafting panels for Eurocodes 2 and 3 included agreeing a common terminology acceptable to all the member states, resolving differing opinions on technical issues, taking into account national differences in materials and design and construction practices, and regional differences in climatic conditions. In addition, it was also considered essential that all Eurocodes should be comprehensive but concise. The following subsections discuss how some of these issues were resolved without compromising the clarity or, indeed, simplicity of the codes.

7.4.1 TERMINOLOGY
At the outset of this work it was necessary to standardize the terminology used in the Eurocodes. Generally, this is similar to that already used in

P(2) In general, a minimum amount of shear reinforcement shall be provided, even where calculation shows that shear reinforcement is unnecessary. This minimum may be omitted in elements such as slabs, (solid, ribbed, hollow), having adequate provision for the transverse distribution of loads, where these are not subjected to significant tensile forces.
Minimum shear reinforcement may also be omitted in members of minor importance which do not contribute significantly to the overall strength and stability of the structure.

(3) Rules for minimum shear reinforcement are given in 5.4. An example of a member of minor importance would be a lintel of less than 2 m span.

Fig. 7.1 Example of principles and application rules (Page 4–36, Oct. 1991 final text of EC2).

the equivalent UK documents. However, there are some minor differences; for example, loads are now called actions while dead and imposed loads are now termed permanent and variable actions respectively. Similarly, bending moments and axial loads are now called internal moments and internal forces respectively. These changes are so minor that they are unlikely to present any major problems to UK engineers.

7.4.2 PRINCIPLES AND APPLICATION RULES
In order to produce a document which is (a) concise, (b) describes the overall aims of design and (c) gives specific guidance as to how these aims can be achieved in practice, the material in the Eurocodes was divided into 'principles' and 'application rules'.

Principles are identified in Eurocode 2 by the letter P and are general statements, definitions, analytical methods, etc. for which no alternative is permitted (*Fig. 7.1*). The application rules are offset to the right of the page in Eurocode 2 and are generally recognized rules which follow the principles and satisfy their requirements (*Fig. 7.1*). In Eurocode 3 different typefaces are used for principles and application rules (*Chapter 9*). The latter approach is also expected to be followed in Eurocode 4.

Application rules each contain one suggested method for satisfying the corresponding principle. It is permissible to use alternative design rules provided that it can be shown that they satisfy the relevant principles and do not negate the other aspects, e.g. serviceability, durability, of the structure.

		Exposure class, according to Table 4.1 of EC2								
		1	2a	2b	3	4a	4b	5a	5b	(3) 5c
(2) Minimum cover (mm)	Reinforcement	15	20	25	40	40	40	25	30	40
	Prestressing steel	25	30	35	50	50	50	35	40	50

Fig. 7.2 *Minimum cover requirements (Table 4.2, Oct. 1991 final text of EC2).*

7.4.3 BOXED VALUES

In Eurocode 2 a number of numerical values, e.g. partial safety factors, minimum concrete covers, coefficients in equations, etc. appear within boxes (*Fig. 7.2*). This signifies that these values are meant to be for guidance only and that other values may be used by individual member states, for the time being. In Eurocode 3 only safety elements, e.g. partial safety factors have been boxed. The actual values to be used in each country are given in the NAD.

This system of identifying certain parameters was introduced in order to account for national differences in material properties, design and construction practices, climatic conditions and so on. Unification of manufacturing and construction practices throughout the EC should see the gradual disappearance of most of these boxed values from the Eurocodes.

7.4.4 APPENDICES/ANNEXES

Some procedures which are not used in everyday design have been included in appendices in Eurocode 2 and annexes in Eurocode 3. Some of the annexes are labelled 'normative' while others are labelled 'informative'. The material which appears in the appendices and the 'normative' annexes has the same status as the rest of the Eurocode but appears there rather than in the body of the code in order to make the document as 'user-friendly' as possible. The material in the 'informative' annexes, however, does not have any status but has been included merely for information.

7.5 What are the differences between Eurocodes and British Standards?

Inevitably, there are many differences between Eurocodes and the national codes. Happily for UK engineers these changes are fairly minor, thanks largely to the work put in by the UK members of the various drafting panels. Consequently it is envisaged that it will not take very much time for engineers in the UK to become familiar with the contents of the Eurocodes and subsequently to use them. It should be borne in mind, however, that many of the European codes and product standards which are needed to support the Eurocodes are currently in preparation. But the strategy devised when the structural Eurocodes were initiated was to produce design standards first and allow these to generate the demand for relevant supporting standards. The date of publication for these documents cannot be predicted with certainty at this stage but it is hoped that they will follow closely after publication of the design standards.

Having discussed these more general aspects, the following chapters will describe the contents of Eurocodes 2, 3, 5 and 6 for concrete, steel, timber and masonry design respectively, and the significant differences between the Eurocode and the corresponding British Standard.

Chapter 8

Eurocode 2: Design of concrete structures

This chapter describes the contents of Part 1.1 of Eurocode 2, the new European standard for the design of buildings in concrete, which is expected to replace BS 8110 by about 2008. The chapter highlights the differences between ENV Eurocode 2: Part 1.1 and BS 8110 and illustrates the new design procedures by means of a number of worked examples on beams, slabs, pad foundation and columns. To help comparison but primarily to ease understanding of the Eurocode the material here has been presented in a similar order to that in Chapter 3 of this book on BS 8110, rather than strictly adhering to the sequence of chapters and clauses adopted in the Eurocode.

8.1 Introduction

Eurocode 2 applies to the design of buildings and civil engineering works in plain, reinforced and prestressed concrete. It is based on limit state principles and comes in several parts as shown in *Table 8.1*.

Part 1.1 of Eurocode 2 gives a general basis for the design of buildings and civil engineering works

Table 8.1 Overall scope of Eurocode 2

Part	Subject
1.1	Reinforced and prestressed concrete for ordinary buildings
1A	Plain concrete
1B	Precast concrete
1C	Lightweight aggregate concrete
1D	Unbonded and external tendons
1E	Fatigue
2	Bridges
3	Foundations
4	Liquid-retaining structures
5	Marine and maritime structures
6	Agriculture structures
7	Massive structures

in reinforced and prestressed concrete made with normal weight aggregates. In addition, it gives some detailing rules which are mainly applicable to ordinary buildings. It is largely similar in scope to Part 1 of BS 8110 which it will replace by about 2008. Part 1.1 of *ENV* Eurocode 2, hereafter referred to as EC2, was issued as a preliminary standard in 1992, ref. no. DD ENV 1992–1–1: 1992. Note that the letters DD signify that the document is a draft for development, the first 1992 is part of the document number and the second 1992 indicates the date it was issued.

The following subjects are covered in Part 1.1:

Chapter 1: Introduction
Chapter 2: Basis of design
Chapter 3: Material properties
Chapter 4: Section and member design
Chapter 5: Detailing provisions
Chapter 6: Construction and workmanship
Chapter 7: Quality control
Appendix 1: Time-dependent effects
Appendix 2: Non-linear analysis
Appendix 3: Additional design procedures for buckling
Appendix 4: Checking deflections by calculation.

The purpose of this chapter is to describe the contents of EC2 and to highlight the principal differences between it and BS 8110. A number of examples covering the design of beams, slabs, pad foundation and columns have also been included to illustrate the new design procedures.

8.2 Structure of EC2

Although the ultimate aim of EC2 and BS 8110 is the same, namely to give guidance on the design of reinforced and prestressed concrete structures, the organization of material in the two documents is rather different. For example, BS 8110 contains separate sections on the design of beams, slabs, columns, bases, etc. However, EC2 divides the material on the basis of structural action, i.e. bending, shear, deflection, torsion, which apply to any element. Furthermore, prestressed concrete is not dealt with separately in EC2 as in BS 8110, but each section on bending, shear, deflection, etc. contains rules relevant to the design of prestressed members.

There is a slight departure from this principle in Chapters 2 and 5 of EC2 which give guidance on the analysis and detailing respectively of specific member types.

8.3 Symbols

The following symbols which have largely been taken from EC2 have been used in this chapter.

GEOMETRIC PROPERTIES

b	width of section
d	effective depth of the tension reinforcement
h	overall depth of section
x	depth to neutral axis
z	lever arm
L_{eff}	effective span of beams and slabs
l_n	clear distance between the forces on the supports
a_1, a_2	distances between the faces of the support to the centre of the effective support at the two ends of the member
c	nominal cover to reinforcement
d'	depth to compression reinforcement

BENDING

g_k, G_k	characteristic permanent action
q_k, Q_k	characteristic variable action
w_k, W_k	characteristic wind load
F_k	characteristic action
F_d	design action
f_{ck}	characteristic compressive cylinder strength of concrete
f_{yk}	characteristic strength of reinforcement
f_{cd}	design concrete strength $= f_{ck}/1.5$
f_{yd}	design steel strength $= f_{yk}/1.15$
X_k	characteristic strength
X_d	design strength
γ_c	partial safety factor for concrete
γ_f	partial safety factor for actions
γ_s	partial safety factor for steel
γ_G	partial safety factor for permanent actions

γ_Q	partial safety factor for variable actions
γ_m	partial safety factor for material
α	reduction factor for sustained compression
C	concrete class
K_o	coefficient given by $M/f_{ck}bd^2$
K_o'	coefficient given by $M_u/f_{ck}bd^2 = 0.167$
M	design ultimate moment
M_u	design ultimate moment of resistance
A_{s1}	area of tension reinforcement
A_{s2}	area of compression reinforcement

SHEAR

S	steel class
τ_{Rd}	basic design shear strength
k	a constant relating to section depth and curtailment
ρ_1	reinforcement ratio corresponding to A_{s1}
σ_{cp}	average stress in concrete due to axial force
f_{ywd}	design yield strength of shear reinforcement
s	spacing of stirrups
v	efficiency factor
V_{Sd}	design shear force due to ultimate loads
V_{Rd1}	design shear resistance of the concrete alone
V_{Rd2}	maximum design shear force that can be carried without crushing of the concrete
V_{Rd3}	design shear force of concrete and shear reinforcement
A_{sw}	cross-sectional area of the shear reinforcement
ρ_w	minimum shear reinforcement area

COMPRESSION

b	width of column
h	depth of section
l_{col}	clear height between centres of restraint
l_0	effective height
β	l_0/l_{col}
I_b	moment of inertia (gross section) of a beam
I_{col}	moment of inertia (gross section) of a column
k_A, k_B	coefficient describing the rigidity of restraint at column ends
α	factor taking into account the conditions of restraint of the beam at the opposite end
i	radius of gyration
λ_{crit}	critical slenderness ratio
υ	angle of inclination of the structure

Table 8.2 Concrete strength classes and characteristic compressive cylinder strengths, f_{ck} (N mm^{-2}) (Table 3.1, EC2)

Strength class of concrete	C12/15	C16/20	C20/25	C25/30	C30/37	C35/45	C40/50	C45/55	C50/60
f_{ck}	12	16	20	25	30	35	40	45	50

v_u	longitudinal force coefficient for an element
e_0	first-order eccentricity = M_{Sd1}/N_{Sd}
e_{01}, e_{02}	respectively the lower and higher values of the first-order eccentricity of the axial load at the ends of the member $\|e_{01}\| \leqslant \|e_{02}\|$
e_{min}	minimum eccentricity = $h/20$
e_e	equivalent eccentricity
e_a	eccentricity covering the effects of geometrical imperfections
e_2	second-order eccentricity
e_{tot}	total eccentricity
N_{Sd}	design axial load
N_{ud}	design ultimate capacity of column subject to axial load only
N_{bal}	design load capacity of balanced section
M_{Rd}	minimum design resisting moment
M_{Sd1}	first-order applied moment at end 1
M_{Sd2}	first-order applied moment at end 2
A_c	cross-section of the concrete
A_s	area of longitudinal reinforcement

8.4 Material properties

8.4.1 CHARACTERISTIC STRENGTHS

8.4.1.1 Concrete (clause 3.1, EC2)

The design rules in EC2 are based on the characteristic 28-day strength of cylinders (f_{ck}). Equivalent cube strengths (f_{cu}) have also been included, but they are only regarded as an alternative method to prove compliance. *Table 8.2* shows the characteristic cylinder strengths of various classes of concrete recommended for use in reinforced and prestressed concrete design. Note that strength class C20/25, for example, refers to cylinder/cube strengths of 20 and 25 N mm^{-2} respectively.

8.4.1.2 Reinforcing steel (clause 3.2, EC2)

Unlike BS 8110, EC2 does not contain details of the reinforcement type to be used in design. For this information the designer must turn to the European standard, which is only available in draft form as provisional standard prEN10080 at this stage.

Therefore, until such time that this standard comes into force, mild steel and high yield steel reinforcing bars of characteristic yield strength, f_{yk}, of 250 and 460 N mm^{-2} respectively will continue to be used in the UK. It should be noted that prEN10080 suggests that grade 460 steel will be replaced by a grade 500 steel and that grade 250 steel will not be included in the standard (see Table 5 of NAD for further information).

8.4.2 DESIGN STRENGTHS (CLAUSE 2.2.3.2, EC2)

The design strengths (X_d) are obtained by dividing the characteristic strengths (X_k) by the appropriate partial safety factor for materials (γ_m):

$$X_d = \frac{X_k}{\gamma_m} \tag{8.1}$$

The partial safety factors for concrete and steel reinforcement are shown in *Table 8.3*. Note that in EC2 the partial safety factor for concrete has a single value of 1.5. This is different from BS 8110 where the partial safety factor varies depending upon the stress type under consideration (*Table 3.3*).

8.5 Actions (clause 2.2.2, EC2)

Actions is the Eurocode terminology for loads and imposed deformations. EC2 defines an action (F) as a force or load applied to a structure. Actions may be 'permanent' (G), e.g. self-weight of structure, fittings and fixed equipment, or 'variable' (Q), e.g. imposed, wind and snow loads.

Table 8.3 Partial safety factors for materials (Table 2.3, EC2)

Combination	Concrete γ_c	Steel reinforcement or prestressing tendons γ_s
Fundamental	1.5	1.15
Accidental (except earthquakes)	1.3	1.0

Table 8.4 Partial safety factors for actions in building structures for persistent and transient design situations (Table 2.2, EC2)

	Permanent actions (γ_G)	Variable actions (γ_Q)	
		One with its characteristic value	Others with their combination values
Favourable effect	1.0	0	0
Unfavourable effect	1.35	1.5	1.5

8.5.1 CHARACTERISTIC ACTIONS (CLAUSE 2.2.2.2, EC2)

The characteristic values of actions (F_k), are specified in Eurocode 1: *Basis of Design and Actions on Structures*, was published as ENV 1993–1–1. Therefore, in the mean time, the NAD requires the designer to refer to the relevant UK loading codes, e.g.

1. BS 648: *Schedule of Weights of Building Materials*;
2. BS 6399: Part 1: 1984 *Code of Practice for Dead and Imposed Loads*;
3. CP 3: Chapter V: Part 2: 1972 *Wind Loads*.

Clause 4 of the NAD notes certain modifications in using the above documents with EC2. However, these are not pertinent to the discussion.

8.5.2 DESIGN ACTIONS (CLAUSE 2.2.2.4, EC2)

The design actions (F_d) are obtained by multiplying the characteristic actions (F_k) by the appropriate partial safety factor for actions (γ_F):

$$F_d = \gamma_F F_k \qquad (8.2)$$

For single beams, the partial safety factors for permanent, γ_G, and variable actions, γ_Q, will normally be 1.35 and 1.5 respectively (*Table 8.4*). The corresponding values in BS 8110 are 1.4 and 1.6 respectively (*Table 3.4*).

Note that where wind acts in combination with permanent and variable actions, all characteristic actions should be multiplied by 1.35. The corresponding value in BS 8110 is 1.2 (*Table 3.4*).

For continuous beams, clause 2.5.1.2 of EC2 recommends that the following load cases will generally be sufficient:

1. alternate spans carrying maximum loads with the others carrying minimum load
2. any two adjacent spans carrying maximum load with the remainder carrying minimum load.

These are different to the load cases suggested in BS 8110 (section 3.6.2). The methods of analysis permitted in EC2, however, are basically the same ones referred to in BS 8110.

8.6 Stress-strain diagrams

8.6.1 CONCRETE (CLAUSE 4.2.1.3.3, EC2)

Figure 8.1 shows the idealized and design stress-strain diagrams for concrete. The basic shape of the two curves, i.e. rectangular – parabolic, is similar to that adopted in BS 8110. Furthermore, the ultimate strain ($\varepsilon_{cu\ max}$) of the concrete in compression is taken to be 0.0035 in EC2 as in BS 8110.

The design concrete strength, f_{cd}, is obtained by dividing the characteristic strength, f_{ck}, of concrete by the partial safety factor for concrete, γ_c (*Table 8.3*):

$$f_{cd} = \frac{f_{ck}}{\gamma_c} \qquad (8.3)$$

However, the design concrete stress is obtained by applying a factor, α, to the design strength:

$$\text{Design stress} = \alpha f_{cd} \qquad (8.4)$$

where α is the reduction factor for sustained compression, generally taken as 0.85 (see Table 3 of NAD).

8.6.2 REINFORCING STEEL (CLAUSE 4.2.2.3.2, EC2)

The design steel stresses (f_{yd}) are derived from the idealized (characteristic) stresses (f_{yk}) by dividing by the partial safety factor for steel, γ_s:

$$f_{yd} = \frac{f_{yk}}{\gamma_s} \qquad (8.5)$$

Figure 8.2 shows the idealized and design stress-strain diagrams for reinforcing steel. However, it will not be possible to use this for design purposes until the Eurocode for reinforcement (EN10080) is published since there are no values of f_{tk} or ε_{uk}.

The stress-strain diagram may be modified with a flatter or horizontal top branch. If a horizontal

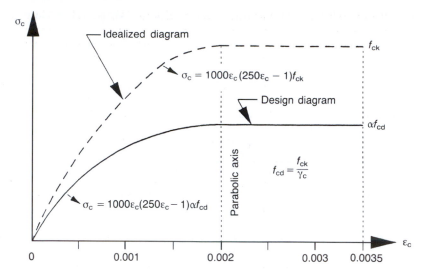

Fig. 8.1 *Parabolic–rectangular stress–strain diagram for concrete in compression (Fig. 4.2, EC2).*

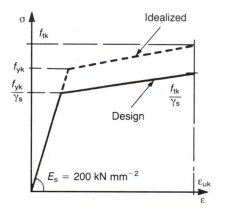

Fig. 8.2 *Design stress-strain diagram for reinforcing steel (Fig. 4.5, EC2).*

top branch is assumed, the design stresses should not exceed f_{yk}/γ_s, although there is no limit to the steel strain. The design equations which have been developed later in this chapter assume that the stress-strain diagram for reinforcement has a horizontal top branch.

8.7 Fire resistance and durability

8.7.1 FIRE RESISTANCE

The European regulations regarding fire resistance of concrete structures are currently being drafted and will, once completed, be published either in an appendix to EC2 or separately as a further part of EC2. In the interim, the NAD requires UK engineers to continue to follow the recommendations contained in Part 2 of BS 8110: 1985.

8.7.2 DURABILITY

In EC2, like BS 8110, the durability of concrete structures is related to

1. environmental conditions
2. cover to reinforcement
3. concrete quality
4. maximum crack width.

With respect to the environment, EC2 identifies nine exposure classes to which a structure could be subject during its design life (*Table 8.5*). They correspond to the first five exposure conditions listed in BS 8110, namely mild, moderate, severe, very severe and most severe (*Table 3.5*). However, there is no class in EC2 which corresponds to the 'abrasive' exposure condition of BS 8110.

Like BS 8110, EC2 allows the designer to reduce the thickness of the concrete cover to the reinforcing bars by specifying better-quality concrete. *Table 8.6* shows the nominal concrete covers to all reinforcement, including links, as a function of concrete quality and exposure class in accordance with EC2 and ENV 206: *Concrete – Performance, Production, Placing and Compliance Criteria: 1992.* (Note that EN206 has now been published.) Both BS 8110 and EC2 recommend that the maximum design crack width should not generally exceed 0.3 mm. The limiting crack width is achieved in practice by (a) providing a minimum amount of reinforcement and (b) limiting the maximum bar spacing or bar

Table 8.5 Exposure classes related to environmental conditions (Table 4.1, EC2)

Exposure class		Examples of environmental conditions
1 Dry environment		Interior of buildings for normal habitation or offices
2 Humid environment	a Without frost	Interior of buildings where humidity is high (e.g. laundries) Exterior components Components in non-aggressive soil and/or water
	b With frost	Exterior components exposed to frost Components in non-aggressive soil and/or water and exposed to frost Interior components when the humidity is high and exposed to frost
3 Humid environment with frost and de-icing salts		Interior and exterior components exposed to frost and de-icing agents
4 Seawater environment	a Without frost	Components completely or partially submerged in seawater, or in the splash zone Components in saturated salt air (coastal area)
	b With frost	Components partially submerged in seawater or in the splash zone and exposed to frost Components in saturated salt air and exposed to frost

The following classes may occur alone or in combination with the above classes:

Exposure class		Examples of environmental conditions
5 Aggressive chemical environment	a	Slightly aggressive chemical environment (gas, liquid or solid) Aggressive industrial atmosphere
	b	Moderately aggressive chemical environment (gas, liquid or solid)
	c	Highly aggressive chemical environment (gas, liquid or solid)

Table 8.6 Nominal cover to reinforcement and concrete quality for durability (based on Tables 6 and NA.1 of EC2 and ENV 206 respectively and the Concise Eurocode for the Design of Concrete Buildings)

Exposure class	Nominal cover (mm)				
1	20	20	20	20	20
2a	–	35	35	30	30
2b	–	–	35	30	30
3	–	–	40	35	35
4a	–	–	40	35	35
4b	–	–	40	35	35
5a	–	–	35	30	30
5b	–	–	–	30	30
5c	–	–	–	–	45
Max. free W/C ratio	0.65	0.60	0.55	0.50	0.45
Min. cement content (kg m^{-3})	260	280	300	300	300
Lowest concrete grade	C25/30	C30/37	C35/45	C40/50	C45/55

size. These requirements will be discussed individually for beams, slabs and columns in *sections 8.8.4, 8.9.2* and *8.11.6* respectively.

8.8 Design of singly and doubly reinforced rectangular beams

8.8.1 BENDING (CLAUSE 4.3.1, EC2)

In analysing a cross-section to determine its ultimate moment of resistance, clause 4.3.1.2 of EC2 recommends that the following assumptions can be made:

1. Plane sections remain plane.
2. The strain in bonded reinforcement is the same as that in the surrounding concrete.
3. The tensile strength of concrete is ignored.
4. The compressive stresses in the concrete may be derived from the design curve in *Fig. 8.1*. (Note that clause 4.2.1.3.3(10) of EC2 states that other idealized stress diagrams may be used, provided they are effectively equivalent to *Fig. 8.1*, e.g. the rectangular stress distribution shown in *Fig. 8.3(d)*.)
5. The stresses in the reinforcement may be derived from *Fig. 8.2*.

These assumptions are similar to those listed in clause 3.4.4.1 of BS 8110 except that in EC2 there is no limit on the lever arm depth and $\gamma_s = 1.15$ rather than 1.05.

Figure 8.3 shows the simplified stress blocks which are used in BS 8110 and EC2 to develop the design equations in bending. In EC2 the maximum concrete compressive stress is taken as $0.85f_{ck}/1.5$ (*Fig. 8.1*) which compares with $0.67f_{cu}/1.5$ in BS 8110 (*Fig. 3.6*). Furthermore, the depth of the compression block is taken as $0.8x$ in EC2 rather than $0.9x$ as in BS 8110, where x is the depth of the neutral axis.

The following subsections derive the equations relevant to the design of singly and doubly reinforced beams according to EC2. The corresponding equations in BS 8110 were derived in *section 3.9.1.1* to which the reader should refer for detailed explanations and the notation used.

8.8.1.1 Singly reinforced beams

Ultimate moment of resistance (M_u)

$$M_u = F_c z \tag{8.6}$$

$$F_c = \frac{(0.85f_{ck})0.8 \times b}{1.5} \tag{8.7}$$

$$z = d - 0.4x \tag{8.8}$$

Clause 2.5.3.4.2 of EC2 limits the depth of the neutral axis (x) to $0.45d$ for concrete grades C12/15 to C35/45 (and $0.35d$ for concrete grades C40/50 and greater) in order to provide a ductile, i.e. under-reinforced, section. Thus

$$x = 0.45d \tag{8.9}$$

Note that the corresponding value for x in BS 8110 is $0.5d$.

Combining equations (8.6)–(8.9) gives

$$M_u = 0.167f_{ck}bd^2 \tag{8.10}$$

Compare $M_u = 0.156f_{cu}bd^2$ (BS 8110).

Area of tensile steel (A_{s1})

$$M = F_s z \tag{8.11}$$

$$F_s = \frac{f_{yk}A_{s1}}{1.15} \tag{8.12}$$

$$A_{s1} = \frac{M}{0.87f_{yk}z} \tag{8.13}$$

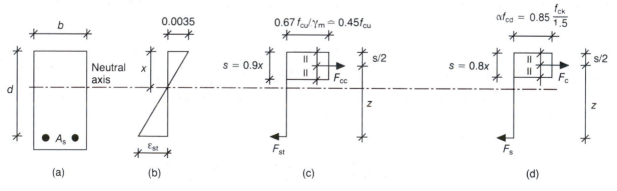

Fig. 8.3 *Singly reinforced section with rectangular stress block: (a) section; (b) strains; (c) stress block (BS 8110); (d) stress block (EC2).*

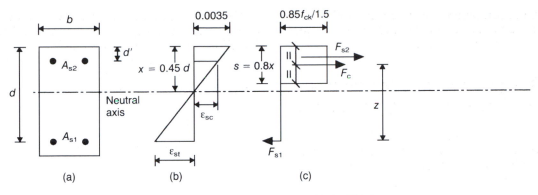

Fig. 8.4 *Doubly reinforced section: (a) section; (b) strains; (c) stress block (EC2).*

Compare $A_s = M/0.95f_y z$ (BS 8110). Equation 8.13 can be used to calculate the area of tension reinforcement provided $M \leqslant M_u$.

Lever arm (z)

$$M = F_c z = \left(\frac{0.85f_{ck}}{1.5}\right)0.8bxz \quad \text{(from equation 8.7)}$$

$$= \frac{3.4}{3}f_{ck}b(d-z)z \quad \text{(from equation 8.8)}$$

Solving for z gives

$$z = d\left[0.5 + \sqrt{(0.25 - 3K_O/3.4)}\right] \quad (8.14)$$

where $K_O = \dfrac{M}{f_{ck}bd^2}$. Compare

$$z = d\left[0.5 + \sqrt{(0.25 - K/0.9)}\right] \quad \text{(BS 8110)}$$

where $K = \dfrac{M}{f_{cu}bd^2}$.

8.8.1.2 Doubly reinforced beams
If the design moment is greater than the ultimate moment of resistance, i.e. $M > M_u$, then compression reinforcement is required. Provided that

$d'/x \ngtr 0.43$ (i.e. compression steel has yielded)

where d' is the depth of the compression steel from the compression face and $x = (d - z)/0.4$, the area of compression reinforcement, A_{s2}, is given by

$$A_{s2} = \frac{M - M_u}{0.87f_{yk}(d - d')} \quad (8.15)$$

and the area of tension reinforcement, A_{s1}, is given by

$$A_{s1} = \frac{M_u}{0.87f_{yk}z} + A_{s2} \quad (8.16)$$

where $z = d\left[0.5 + \sqrt{(0.25 - 3K_o'/3.4)}\right]$

$K_o' = 0.167$

Equations 8.15 and 8.16 have been derived using the stress block shown in *Fig. 8.4*. This is similar to that used to derive the equations for the design of singly reinforced beams (*Fig. 8.3(d)*) except for the additional compression force in the steel.

Example 8.1 Bending reinforcement for a singly reinforced beam (EC2)
Determine the area of main steel, A_{s1}, required for the beam assuming the following material strengths: $f_{ck} = 25$ N mm^{-2} and $f_{yk} = 460$ N mm^{-2}.

$b = 275$

$d = 450$

$q_k = 8$ kN m^{-1}
$g_k = 12$ kN m^{-1}

7 m

Ultimate load, w, is

$$w = 1.35g_k + 1.5q_k = 1.35 \times 12 + 1.5 \times 8 = 28.2 \text{ kN m}^{-1}$$

Design moment, M, is

$$M = \frac{wl^2}{8} = \frac{28.2 \times 7^2}{8} = 172.7 \text{ kN m}$$

Ultimate moment of resistance, M_u, is

$$M_u = 0.167f_{ck}bd^2 = 0.167 \times 25 \times 275 \times 450^2 \times 10^{-6} = 232.5 \text{ kN m}$$

Since $M_u > M$ design as a singly reinforced beam

$$K_o = \frac{M}{f_{ck}bd^2} = \frac{172.7 \times 10^6}{25 \times 275 \times 450^2} = 0.124$$

$$z = d\left[0.5 + \sqrt{(0.25 - 3K_o/3.4)}\right]$$

$$= 450\left[0.5 + \sqrt{(0.25 - 3 \times 0.124/3.4)}\right] = 393.7 \text{ mm}$$

$$A_{s1} = \frac{M}{0.87f_{yk}z} = \frac{172.7 \times 10^6}{0.87 \times 460 \times 393.7} = 1096 \text{ mm}^2$$

Hence from *Table 3.7*, provide 4T20 ($A_{s1} = 1260$ mm^2).

Example 8.2 Bending reinforcement for a doubly reinforced beam (EC2)

Design the bending reinforcement for the beam assuming the cover to the main steel is 40 mm,

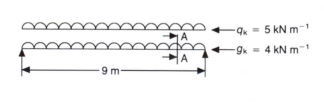

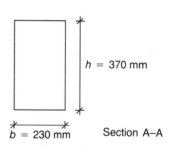

$f_{ck} = 25$ N mm^{-2} and $f_{yk} = 460$ N mm^{-2}.

DESIGN MOMENT (M)
Total ultimate load, W, is

$$W = (1.35g_k + 1.5q_k)\text{span} = (1.35 \times 4 + 1.5 \times 5)9 = 116.1 \text{ kN}$$

Design moment, M, is

$$M = \frac{Wl}{8} = \frac{116.1 \times 9}{8} = 130.6 \text{ kN m}$$

ULTIMATE MOMENT OF RESISTANCE (M_u)
Effective depth
Assume diameter of main bar (Φ) is 25 mm. Effective depth, d, is

$$d = h - c - \Phi/2 = 370 - 40 - 12.5 = 317 \text{ mm}$$

Ultimate moment

Ultimate moment of resistance (M_u) is

$$M_u = 0.167 f_{ck} b d^2 = 0.167 \times 25 \times 230 \times 317^2 \times 10^{-6} = 96.5 \text{ kN m}$$

Since $M_u < M$ design as a doubly reinforced beam.

COMPRESSION REINFORCEMENT (A_{s2})

Assume diameter of compression bars (ϕ) is 16 mm. Effective depth, d', is

$$d' = \text{cover} + \phi/2 = 40 + 16/2 = 48 \text{ mm}$$

$$\frac{d'}{x} = \frac{48}{0.45d} = \frac{48}{0.45 \times 317} = 0.34 \not> 0.43$$

Hence

$$A_{s2} = \frac{M - M_u}{0.87 f_{yk}(d - d')} = \frac{(130.6 - 96.5)10^6}{0.87 \times 460 \times (317 - 48)} = 317 \text{ mm}^2$$

Provide 2T16 ($A_{s2} = 402 \text{ mm}^2$ from *Table 3.7*).

TENSION REINFORCEMENT (A_{s1})

$$z = d\left[0.5 + \sqrt{(0.25 - 3K'_o/3.4)}\right]$$
$$= d\left[0.5 + \sqrt{(0.25 - 3 \times 0.167/3.4)}\right] = 0.82d = 0.82 \times 317 = 260 \text{ mm}$$

$$A_{s1} = \frac{M_u}{0.87 f_{yk}z} + A_{s2} = \frac{96.5 \times 10^6}{0.87 \times 460 \times 260} + 317 = 1245 \text{ mm}^2$$

Provide 3T25 ($A_{s1} = 1470 \text{ mm}^2$ from *Table 3.7*).

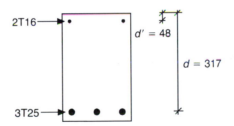

8.8.2 SHEAR (CLAUSE 4.3.2, EC2)

Unlike BS 8110, EC2 compares shear forces rather than shear stresses in order to assess whether shear reinforcement will be required. The notation used in EC2 tends, therefore, to be quite different from that in BS 8110. Furthermore, EC2 generally provides better guidance than BS 8110 regarding the spacing of shear reinforcement.

EC2 identifies four basic shear forces for design purposes, namely V_{Sd}, V_{Rd1}, V_{Rd2} and V_{Rd3} where V_{Sd} is the applied shear force and the remaining three parameters give an indication of the shear resistance of the member: V_{Rd1} is the actual shear resistance of the concrete alone, V_{Rd2} is the max-

imum shear resistance of the concrete and V_{Rd3} the sum of the shear resistances of the concrete and steel and is therefore a function of V_{Rd1}.

Here V_{Rd1} is given by

$$V_{Rd1} = [\tau_{Rd}k(1.2 + 40\rho_1) + 0.15\sigma_{cp}]b_w d \quad (8.17)$$

where τ_{Rd} is the basic shear strength (*Table 8.7*), $k = 1$ for members where $> 50\%$ of the bottom reinforcement is curtailed, otherwise put $k = 1.6 - d \not< 1$ (d in metres), $\rho_1 = A_{s1}/b_w d \not> 0.02$, A_{s1} is the (effective) area of tension reinforcement, b_w the width of section, $\sigma_{cp} = N_{Sd}/A_c$, N_{Sd} is the longitudinal force in section and A_c the cross-sectional area of concrete.

Table 8.7 Values of τ_{Rd} with $\gamma_c = 1.5$ for different concrete strengths (Table 4.8, EC2)

f_{ck}	12	16	20	25	30	35	40
τ_{Rd} (N mm^{-2})	0.18	0.22	0.26	0.30	0.34	0.37	0.41

Here V_{Rd2} is given by

$$V_{Rd2} = \frac{1}{2}\nu f_{cd}b_w 0.9d \tag{8.18}$$

where

$$\nu = 0.7 - \frac{f_{ck}}{200} \not< 0.5 \quad (f_{ck} \text{ in N mm}^{-2}) \tag{8.19}$$

The total shear resistance, V_{Rd3}, is the sum of the shear resistance of the concrete (V_{cd}) and transverse reinforcement (V_{wd}). Here V_{Rd3} can be calculated using one of two methods, namely the **standard method** or the **variable strut inclination method**. Only the former method will be discussed here which is similar in principle to the method contained in BS 8110 (*section 3.9.1.3*). Thus

$$V_{Rd3} = V_{cd} + V_{wd} \tag{8.20}$$

where $V_{cd} = V_{Rd1}$ and

V_{wd} = contribution of the shear reinforcement

$$= \frac{A_{sw}}{s}0.9f_{ywd}d \tag{8.21}$$

where A_{sw} is the cross-sectional area of the shear reinforcement, s the spacing of shear reinforcement (*section 8.8.2.2*) and f_{ywd} the design yield strength of the shear reinforcement.

According to clause 4.3.2.2 of EC2, if $V_{Sd} < V_{Rd1}$, minimum shear reinforcement must be provided (*section 8.8.2.1*) except in members of minor importance where such reinforcement may be omitted. An example of a member of minor importance would be a lintel of less than 2 m span. If $V_{Rd1} < V_{Sd} < V_{Rd2}$, design shear reinforcement must be provided according to equation 8.20, with $V_{Rd3} = V_{Sd}$.

Table 8.8 Minimum values for ρ_w (Table 5.5, EC2)

Concrete classes	Steel classes		
	S220	S400	S500
C12/15 and C20/25	0.0016	0.0009	0.0007
C25/30 to C35/45	0.0024	0.0013	0.0011
C40/50 to C50/60	0.0030	0.0016	0.0013

Table 8.9 Spacing of shear reinforcement

If	$V_{Sd} \leq 1/5\ V_{Rd2}$	$s_{max} = 0.8d \not> 300$ mm
If $1/5\ V_{Rd2} \leq$	$V_{Sd} \leq 2/3\ V_{Rd2}$	$s_{max} = 0.6d \not> 300$ mm
If	$V_{Sd} > 2/3\ V_{Rd2}$	$s_{max} = 0.3d \not> 200$ mm

8.8.2.1 Minimum shear reinforcement areas (ρ_w) (clause 5.4.2.2, EC2)

Minimum values of the shear ratio (ρ_w) are shown in *Table 8.8*, ρ_w being defined as

$$\rho_w = A_{sw}/sb_w \sin \alpha \tag{8.22}$$

where A_{sw} is the area of shear reinforcement within length s, s the spacing of the shear reinforcement, b_w the breadth of the member and $\alpha = 90°$ for vertical stirrups, i.e. $\sin \alpha = 1$. Since the above steel classes do not correspond with those used in the UK, appropriate values of the shear ratio can be obtained by interpolation.

8.8.2.2 Diameter and spacing of shear reinforcement (clause 5.4.2.2, EC2)

EC2 recommends that the diameter of the shear reinforcement should not generally exceed 12 mm. In addition, EC2 recommends that the maximum spacing of shear reinforcement (s_{max}) should lie in the range $0.3d$–$0.8d$ (*Table 8.9*); the actual value will depend upon the ratio of the design shear force to the maximum shear resistance of the concrete (i.e. V_{Sd}/V_{Rd2}). In BS 8110, the only guidance given is that the spacing of the links should not exceed $0.75d$.

Example 8.3 Design of shear reinforcement for a beam (EC2)

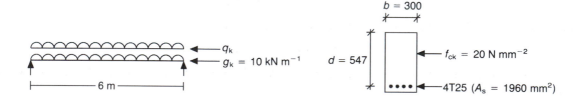

q_k

$g_k = 10$ kN m^{-1}

6 m

$b = 300$

$d = 547$

$f_{ck} = 20$ N mm^{-2}

4T25 ($A_s = 1960$ mm^2)

Design the shear reinforcement for the beam using mild steel ($f_{ywd} = 250$ N mm^{-2}) links for the following load cases: (i) $q_k = 0$, (ii) $q_k = 10$ kN m^{-1}, (iii) $q_k = 27$ kN m^{-1}, (iv) $q_k = 40$ kN m^{-1}.

Shear resistance of concrete alone (V_{Rd1})

$$f_{ck} = 20 \text{ N mm}^{-2} \quad \tau_{Rd} = 0.26 \text{ N mm}^{-2}$$

$$k = 1.6 - d = 1.6 - 0.547 = 1.053$$

$$\rho_1 = \frac{A_{s1}}{b_w d} = \frac{1960}{300 \times 547} = 0.012 \quad \sigma_{cp} = 0$$

$$V_{Rd1} = [\tau_{Rd}k(1.2 + 40\rho_1) + 0.15\sigma_{cp}]b_w d$$

$$= [0.26 \times 1.053(1.2 + 40 \times 0.012)]300 \times 547$$

$$= 75\,478 \text{ N}$$

Maximum shear resistance of concrete (V_{Rd2})

$$\nu = 0.7 - \frac{f_{ck}}{200} = 0.7 - \frac{20}{200} = 0.6$$

$$V_{Rd2} = \frac{1}{2}\nu f_{cd}b_w 0.9d = \frac{1}{2} \times 0.6 \times \frac{20}{1.5} \times 300 \times 0.9 \times 547 = 590\,760 \text{ N}$$

Minimum shear reinforcement (ρ_w)

$$\rho_w = 0.0015 \quad \text{(by interpolation between the values in } Table\ 8.8)$$

Substituting this into $\rho_w = A_{sw}/sb_w \sin \alpha$ gives

$$0.0015 = A_{sw}/s \times 300 \times 1$$

$$\frac{A_{sw}}{s} = 0.45$$

$q_k = 0$

Shear force (V_{Sd})
Total ultimate load, W, is

$$W = (1.35g_k + 1.5q_k) \text{ span} = (1.35 \times 10 \times 10^3)6 = 81\,000 \text{ N}$$

Since beam is symmetrical, maximum shear force $(V_{Sd}) = R_A = R_B = \dfrac{W}{2} = 40\,500$ N

Diameter and spacing of links
Although $V_{Sd} < V_{Rd1}$, member is > 2 m long and cannot therefore be considered to be of minor structural importance (clause 4.3.2.1(2)). Hence provide minimum shear reinforcement, i.e.

$$\frac{A_{sw}}{s} = 0.45 \quad \frac{V_{Sd}}{V_{Rd2}} = \frac{40\,500}{590\,760} = 0.07$$

Hence from *Table 8.9*, maximum spacing of links, s_{max}, is given by

$$s_{max} = 0.8d = 0.8 \times 547 = 437 \text{ mm} \not> 300 \text{ mm}$$

Therefore, provide 30-R8 links at 200 mm centres, $A_{sw}/s = 0.503$, for whole length of beam (*Table 3.10*).

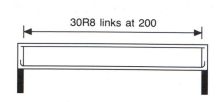
30R8 links at 200

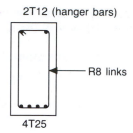
2T12 (hanger bars)

R8 links

4T25

$q_k = 10$ kN m^{-1}

Shear force (V_{Sd})

Total ultimate load, W, is

$$W = (1.35g_k + 1.5q_k)\text{span} = (1.35 \times 10 + 1.5 \times 10)6 = 171 \text{ kN}$$

Since beam is symmetrical, maximum shear force, V_{Sd}, is

$$V_{Sd} = R_A = R_B = \frac{W}{2} = 85\,500 \text{ N}$$

Diameter and spacing of links

Since $V_{Rd1} < V_{Sd} < V_{Rd2}$ provide shear reinforcement according to

$$V_{Rd3} = V_{cd} + V_{wd}$$

where $V_{cd} = V_{Rd1} = 75\,478$ N, $V_{Rd3} = V_{Sd} = 85\,500$ N and

$$V_{wd} = \frac{A_{sw}}{s}0.9f_{ywd}d = \frac{A_{sw}}{s}0.9 \times \frac{250}{1.15} \times 547 = 107\,021\frac{A_{sw}}{s}$$

Hence

$$85\,500 = 75\,478 + 107\,021\frac{A_{sw}}{s}$$

$$\frac{A_{sw}}{s} = \frac{85\,500 - 75\,478}{107\,021} = 0.094 \ll \text{min req } (= 0.45)$$

$$\frac{V_{Sd}}{V_{Rd2}} = \frac{85\,500}{590\,760} = 0.14$$

Hence from *Table 8.9*, maximum spacing of links, s_{max}, is given by

$$s_{max} = 0.8d = 0.8 \times 547 = 437 \text{ mm} \ngtr 300 \text{ mm}$$

Therefore, provide 30-R8 links at 200 mm centres, $A_{sw}/s = 0.503$, for whole length of beam as for case (i), $q_k = 0$, above.

$q_k = 27$ kN m^{-1}

Shear force (V_{Sd})

Total ultimate load, W, is

$$W = (1.35g_k + 1.5q_k) \text{ span} = (1.35 \times 10 + 1.5 \times 27)\,6 = 324 \text{ kN}$$

Since beam is symmetrical, maximum shear force, V_{Sd}, is

$$V_{Sd} = R_A = R_B = \frac{W}{2} = 162\,000 \text{ N}$$

Diameter and spacing of links

Minimum shear reinforcement required where shear force in beam, V_c, is

$$V_c \leqslant V_{cd} + V_{wd}$$

where $V_{cd} = V_{Rd1} = 75\,478$ N and

$$V_{wd} = \frac{A_{sw}}{s}0.9f_{ywd}d = \frac{0.45 \times 0.9 \times 250 \times 547}{1.15} = 48\,160 \text{ N}$$

Hence

$$V_c = 75\,478 + 48\,160 = 123\,638 \text{ N}$$

$$V_c/V_{Rd2} = 123\,638/590\,760$$

$$= 0.21 \Rightarrow s_{max} \not> 300 \text{ mm} \quad (\textit{Table 8.9})$$

Hence from *Table 3.12*, provide R8 links at 200 mm centres ($A_{sw}/s = 0.503$), where $V \leqslant 123\,638$ N, i.e. 2290 mm either side of the mid-span of beam.

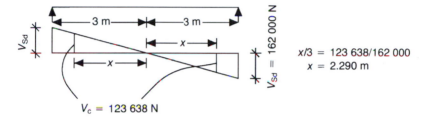

$x/3 = 123\,638/162\,000$
$x = 2.290$ m
$V_c = 123\,638$ N
$V_{Sd} = 162\,000$ N

Where shear force in beam $> V_c$, provide shear reinforcement according to

$$V_{Sd} = V_{cd} + V_{wd}$$

$$162\,000 = 75\,478 + 107\,021 A_{sw}/s$$

$$\frac{A_{sw}}{s} = \frac{162\,000 - 75\,478}{107\,021} = 0.81$$

$$\frac{V_{Sd}}{V_{Rd2}} = \frac{162\,000}{590\,760} = 0.27$$

$$\Rightarrow s_{max} = 0.6d = 0.6 \times 547 = 328 \text{ mm} \not> 300 \text{ mm}$$

Hence from *Table 3.10*, provide R8 links at 100 mm centres ($A_{sw}/s = 1.006$) where shear force in beam exceeds 123 638 N, i.e. 710 mm in from both supports.

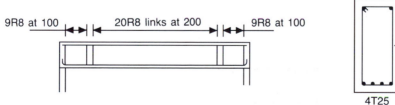

9R8 at 100   20R8 links at 200   9R8 at 100

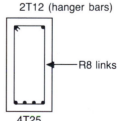

2T12 (hanger bars)
R8 links
4T25

$q_k = 40$ kN m^{-1}

Shear force (V_{Sd})

Total ultimate load, W, is

$$W = (1.35g_k + 1.5q_k)\text{ span} = (1.35 \times 10 + 1.5 \times 40)\,6 = 441 \text{ kN}$$

Since beam is symmetrical, maximum shear force V_{Sd}, is

$$V_{Sd} = R_A = R_B = W/2 = 220\,500 \text{ N}$$

Diameter and spacing of links

As above, minimum shear reinforcement required where shear force in beam, $V_c \leqslant 123\,638$ N. From *Table 3.12*, provide R10 links at 300 mm centres ($A_{sw}/s = 0.523$) where $V \leqslant 123\,638$ N, i.e. 1682 mm either side of the mid-span of beam.

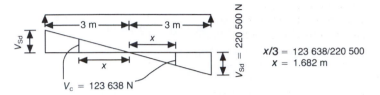

$x/3 = 123\,638/220\,500$
$x = 1.682$ m

$V_c = 123\,638$ N

Where shear force in beam $> V_c$, provide shear reinforcement according to

$$V_{Sd} = V_{cd} + V_{wd}$$

$$220\,500 = 75\,478 + 107\,021\,\frac{A_{sw}}{s}$$

$$\frac{A_{sw}}{s} = \frac{220\,500 - 75\,478}{107\,021} = 1.36$$

$$\frac{V_{Sd}}{V_{Rd2}} = \frac{220\,500}{590\,760} = 0.37$$

$$\Rightarrow s_{max} = 0.6d = 0.6 \times 547$$

$$= 328 \text{ mm} \not> 300 \text{ mm}$$

Hence from *Table 3.10*, provide R10 links at 100 mm centres ($A_{sw}/s = 1.57$) where shear force in beam exceeds 123 638 N, i.e. 1318 mm in from both supports.

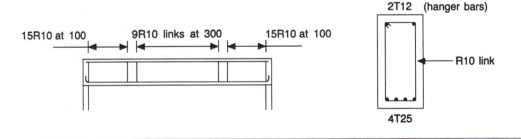

2T12 (hanger bars)

15R10 at 100   9R10 links at 300   15R10 at 100

R10 link

4T25

8.8.3 DEFLECTION (CLAUSE 4.4.3, EC2)

The approach used to check deflection in EC2 is basically the same as in BS 8110 but the presentation has been simplified considerably.

As in BS 8110, EC2 does not require the actual deflections to be calculated explicitly, but recommends that the span/depth ratios shown in *Table 8.10* should not be exceeded. Compliance with these values for the span/depth ratios will normally ensure that the deflection limits in clause 4.4.3.1 of EC2 are satisfied, namely:

1. The calculated sag of a beam, slab or cantilever subjected to the quasi-permanent loads should not exceed span/250.
2. The deflection occurring after construction of the element should not exceed span/500.

The values in *Table 8.10* assume the steel stress at the critical section, σ_s, is 250 N mm^{-2}, corresponding roughly to $f_{yk} = 400$ N mm^{-2}. Where other stress levels are used, the values in the table can be multiplied by $250/\sigma_s$. However, the final span/depth

Table 8.10 Basic ratios of span/effective depth for reinforced concrete members without axial compression (Table 4.14 of EC2 as amended by Table 7 of NAD)

Structural system	Concrete highly stressed	Concrete lightly stressed	Concrete nominally reinforced[a]
1. Simply supported beam, one or two-way spanning simply supported slab	18	25	34
2. End span of continuous beam or one-way continuous slab or two-way spanning slab continuous over one long side	23	32	44
3. Interior span of beam or one-or two-way spanning slab	25	35	48
4. Slab supported on columns without beams (flat slab) (based on longer span)	21	30	41
5. Cantilever	7	10	14

Note. [a] Apply to all stress levels.

ratio should not exceed the basic span/depth ratios for nominally reinforced concrete members given in the table. It will normally be conservative to assume that

$$\sigma_s = \frac{250 f_{yk} A_{s,req}}{400 A_{s,prov}} \tag{8.23}$$

where $A_{s,req}$ is the area of steel required and $A_{s,prov}$ the area of steel provided. Furthermore it should be noted that in *Table 8.10* the actual span/depth ratios for each member type are related to the reinforcement ratio ρ, where $\rho = A_s/bd$. Nominally reinforced values correspond to $\rho = 0.15\%$; lightly stressed members correspond to $\rho = 0.5\%$; highly stressed members correspond to $\rho = 1.5\%$. Values between these cases may be obtained by interpolation.

8.8.4 REINFORCEMENT DETAILS FOR BEAMS
This section outlines EC2 requirements regarding the detailing of beams with respect to:

1. reinforcement percentages
2. spacing of reinforcement
3. anchorage lengths
4. curtailment of reinforcement
5. lap lengths.

8.8.4.1 Reinforcement percentages
(clause 5.4.2.1.1, EC2)
The cross-sectional area of the longitudinal tensile reinforcement, A_{s1}, in beams should be not less than the following:

$$A_{s1} \geqslant \frac{0.6bd}{f_{yk}} \not< 0.0015bd$$

where b is the breadth of section, d the effective depth and f_{yk} the characteristic yield stress of reinforcement (N mm^{-2}). The area of the tension, A_{s1}, and of the compression reinforcement, A_{s2}, should not be greater than the following other than at laps:

$$A_{s1}, A_{s2} \leqslant 0.04A_c$$

where A_c is the cross-sectional area of concrete.

8.8.4.2 Spacing of reinforcement
(clauses 4.4.2.3 and 5.2.1.1, EC2)
The clear horizontal or vertical distance between reinforcing bars should not be less than the following:

1. maximum bar diameter
2. 20 mm
3. $d_g + 5$ mm, if d_g, the maximum aggregate size, exceeds 32 mm.

The maximum spacing between bars is based on the need to ensure that the maximum crack width does not exceed 0.3 mm which may generally be achieved by limiting the bar spacings to the values shown in *Table 8.11*. (clause 4.4.2.3 of EC2 also states that cracks greater than 0.3 mm can be avoided by limiting bar sizes, but this aspect is not discussed here.) The steel stresses in the table can conservatively be estimated using equation 8.23.

8.8.4.3 Anchorage (clause 5.2.2.3, EC2)
EC2 distinguishes between basic and required anchorage lengths. The basic anchorage length, l_b,

Eurocode 2: Design of concrete structures

Table 8.11 Maximum bar spacing for high bond bars (Table 4.12, EC2)

Steel stress (MPa)	Maximum bar spacing (mm)		
	Pure flexure	Pure tension	Prestressed sections (bending)
160	300	200	200
200	250	150	150
240	200	125	100
280	150	75	50
320	100	–	–
360	50	–	–

2. all bars in the lower half of members with an overall depth between 250 mm and 600 mm (*Fig. 8.5(b)*);
3. all bars located at a depth greater than or equal to 300 mm in members with an overall depth of 600 mm or greater (*Fig. 8.5(c)*).

All other conditions will give rise to poor bond conditions.

In conditions of good bond, the design value for the ultimate bond stress, f_{bd}, is given in *Table 8.12*. In all other cases, the values in *Table 8.12* should be multiplied by a coefficient 0.7.

The required anchorage length, $l_{b,net}$, is used to determine the curtailment length of bars in beams as discussed below and is given by

$$l_{b,net} = \alpha_a l_b \frac{A_{s,req}}{A_{s,prov}} \not< l_{b,min} \quad (8.25)$$

where $\alpha_a = 1$ for straight bars and 0.7 for curved bars in tension if the cover perpendicular to the plane of curvature is at least 3ϕ. Also l_b is the basic anchorage length, $A_{s,req}$ the area of reinforcement required by design, $A_{s,prov}$ the area of reinforcement actually provided and $l_{b,min}$ the minimum anchorage length given by

$$l_{b,min} = 0.3 l_b \not< 10\phi \text{ or } \not< 100 \text{ mm} \quad (8.26)$$
(for anchorages in tension)

$$l_{b,min} = 0.6 l_b \not< 10\phi \text{ or } \not< 100 \text{ mm} \quad (8.27)$$
(for anchorages in compression)

where ϕ is the diameter of bar to be anchored.

is the length of bar required to resist the maximum force in the reinforcement, $A_s f_{yd}$, assuming constant bond stress equal to f_{bd} and is given by

$$l_b = (\phi/4)(f_{yd}/f_{bd}) \quad (8.24)$$

where ϕ is the diameter of bar to be anchored, f_{yd} the design strength of reinforcement and f_{bd} the ultimate bond stress (*Table 8.12*). The ultimate bond stress depends upon the quality of the bond between the concrete and steel. EC2 states that bond conditions are considered to be good for:

1. all bars in members with an overall depth (*h*) of less than or equal to 250 mm (*Fig. 8.5(a)*);

Table 8.12 Ultimate bond stress, f_{bd} (N mm^{-2}), assuming good bond conditions (Table 5.3, EC2)

f_{ck}	12	16	20	25	30	35	40	45	50
Plain bars	0.9	1.0	1.1	1.2	1.3	1.4	1.5	1.6	1.7
High bond bars where $\phi \leq 32$ mm or welded mesh fabrics made of ribbed wires	1.6	2.0	2.3	2.7	3.0	3.4	3.7	4.0	4.3

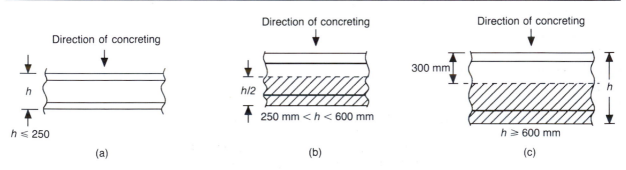

Fig. 8.5 *Definition of bond conditions (Fig. 5.1, EC2).*

Table 8.13 Anchorage lengths, $l_{b,net}$, as multiples of bar size

Steel grade		Concrete class				
		C20/25	C25/30	C30/37	C35/45	C40/45
Plain	Straight bars, compression tension	50	46	42	39	37
$f_{yk} = 250$ N mm^{-2}	Curved bars, tension[a]	35	32	30	28	26
Deformed bars type 2	Straight bars, compression tension	44	37	34	30	27
$f_{yk} = 460$ N mm^{-2}	Curved bars, tension[a]	31	26	24	21	19

Note. [a] In the anchorage region, cover perpendicular to the plane of curvature should be at least 3ϕ.

Table 8.13 shows the anchorage lengths for straight and curved bars as multiples of bar size for plain ($f_{yk} = 250$ N mm^{-2}) and high yield bars ($f_{yk} = 460$ N mm^{-2}) embedded in a range of concrete strengths. Note that the values in the table apply to good bond conditions and to bar sizes less than or equal to 32 mm. Where the bond conditions are poor the values in the table should be divided by 0.7 and where the bar diameter exceeds 32 mm the values should be divided by

$[(132 - \phi)/100]$. The minimum radii to which reinforcement may be bent is shown in *Table 8.14* while the anchorage lengths for straights, hooks, bends and loops are shown in *Fig. 8.6*.

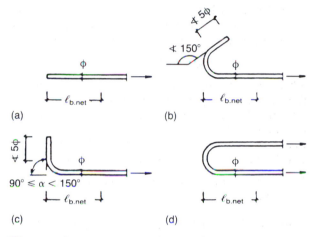

Fig. 8.6 *Required anchorage length (Fig. 5.2, EC2): (a) straight anchor; (b) hook; (c) bend; (d) loop.*

Table 8.14 Minimum diameter of hooks, bends and loops (based on Table 5.1 of EC2 as amended by Table 8 of NAD)

	Bar diameter	
	$\phi < 20$ mm	$\phi \geq 20$ mm
Plain bars S250	4ϕ	4ϕ
High bond bars S460	6ϕ	8ϕ

Example 8.4 Calculation of anchorage lengths (EC2)

Calculate the anchorage lengths for straight and curved bars in tension as multiples of bar size assuming:

1. The bars are high yield of diameter ≤ 32 mm.
2. Concrete strength class is C20/25.
3. Bond conditions are good.

STRAIGHT BARS

High yield bars, f_{yk} = 460 N mm^{-2}
Concrete strength, f_{ck} = 20 N mm^{-2}
Ultimate bond stress, f_{bd} = 2.3 (*Table 8.12*)
Design strength of bars, $f_{yd} = f_{yk}/\gamma_s = 460/1.15$
Coefficient α = 1 (for straight bar)

The basic anchorage length, l_b, is

$$l_b = (\phi/4)(f_{yd}/f_{bd})$$
$$= (\phi/4)(460/1.15) \div 2.3 \approx 44\phi$$

Hence the anchorage length, $l_{b,net}$, is

$$l_{b,net} = \alpha_a l_b = 1 \times 44\phi = 44\phi \quad (Table\ 8.13)$$

CURVED BARS

The calculation is essentially the same for this case except that $\alpha_a = 0.7$ for curved bars and therefore, $l_{b,net}$, is

$$l_{b,net} = \alpha_a l_b = 0.7 \times 44\phi \approx 31\phi$$

8.8.4.4 Curtailment of bars
(clause 5.4.2.1.3, EC2)

The curtailment length of bars in beams is obtained from *Fig. 8.7*. The theoretical cut-off point is based on the design bending moment curve which has been horizontally displaced in the direction of decreasing moment by an amount a_ℓ. The physical (actual) cut-off point occurs at anchorage length, $l_{b,net}$, beyond the 'theoretical' cut-off point.

If the shear resistance is calculated according to the 'standard method' a_ℓ is given by

$$a_\ell = \frac{z}{2}(1 - \cot \alpha) \not< 0 \qquad (8.28)$$

where α is the angle of the shear reinforcement with the longitudinal axis, and z can normally be taken as $0.9d$. With vertical links, for instance, a_ℓ will be equal to

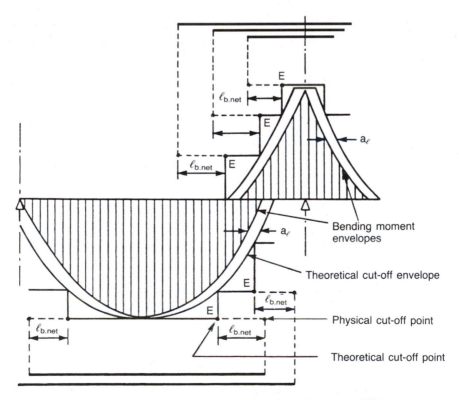

Fig. 8.7 *Envelope line for curtailment of reinforcement in flexural members (Fig. 5.11, EC2).*

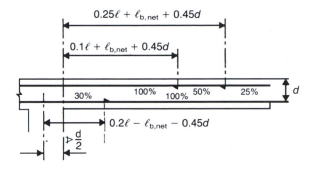

Fig. 8.8 *Curtailment of reinforcement at internal supports of continuous beams.*

$$a_\ell = \frac{0.9d}{2}(1 - \cot 90°) = 0.45d$$

The *Concise Eurocode for the Design of Concrete Buildings* recommends the following simplified rules for the curtailment of reinforcement in beams subjected to predominantly uniformly distributed loads. These are based on the present rules in BS 8110.

Near internal supports in continuous beams of approximately equal spans where $Q_k < G_k$. For curtailment of top reinforcement, at least 25% of the reinforcement required at the supports for the ultimate limit state should be effectively continuous through the spans. All of the reinforcement needed at the supports should extend into the span for a distance from the face of the support of $0.1l + l_{b,net} + 0.45d$. At least 50% of the reinforcement at the support should extend into the span for a distance from the face of the support of $0.25l + l_{b,net} + 0.45d$ (*Fig. 8.8*).

For curtailment of bottom reinforcement, at least 30% of the reinforcement required at mid-span should extend to the support. The remainder should extend to within $0.2l - l_{b,net} - 0.45d$ of the centre line of the support.

Bottom reinforcement near end supports. At least 50% of the reinforcement provided at mid-span should be taken into the support and be anchored as shown in *Fig. 8.9*. The remaining reinforcement should extend to within $0.15l - l_{b,net} - 0.45d$ of the centre line of the support.

8.8.4.5 Lap lengths (clause 5.2.4.13, EC2)
Lap length, l_s, is given by

$$l_s = l_{b,net}\alpha_1 \not< l_{s,min} \qquad (8.29)$$

where $l_{b,net}$ is the anchorage length, $l_{s,min}$ the minimum lap length which should be not less than 15ϕ or 200 mm and α_1 a coefficient which takes the following values:

$\alpha_1 = 1$ for compression laps
$\alpha_1 = 1$ for tension laps where less than 30% of the bars in the section are lapped and where $a \geqslant 6\phi$ and $b \geqslant 2\phi$ (*Fig. 8.10*)
$\alpha_1 = 1.4$ for tension laps where either (i) 30% or more of bars at a section are lapped, or (ii) $a < 6\phi$ or $b < 2\phi$ (*Fig. 8.10*) but not both
$\alpha_1 = 2$ for tension laps if both (i) and (ii) above are satisfied

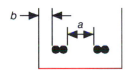

Fig. 8.10 *Evaluation of α_1 (Fig. 5.6, EC2).*

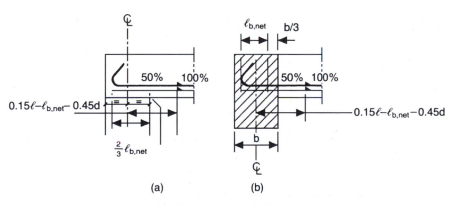

Fig. 8.9 *Curtailments and anchorages of bottom reinforcement at end supports: (a) direct support; (b) indirect support (based on Fig. 5.12, EC2).*

Eurocode 2: Design of concrete structures

Table 8.15 Lap lengths as multiples of bar size

Steel grade		Concrete class				
		C20/25	C25/30	C30/37	C35/45	C40/45
Plain	Laps – compression and tension[a]	50	46	42	39	37
$f_{yk} = 250$ N mm^{-2}	Laps – tension[b]	70	64	60	56	52
	Laps – tension[c]	100	92	84	78	74
Deformed bars	Laps – compression and tension[a]	44	37	34	30	27
type 2	Laps – tension[b]	62	52	48	42	38
$f_{yk} = 460$ N mm^{-2}	Laps – tension[c]	88	74	68	60	54

Notes. [a] $\alpha_1 = 1$.
[b] $\alpha_1 = 1.4$.
[c] $\alpha_1 = 2$.

Table 8.15 shows the lap lengths as multiples of bar size for plain ($f_{yk} = 250$ N mm^{-2}) and high yield bars ($f_{yk} = 460$ N mm^{-2}) embedded in a range of concrete strengths. Note that the values in the table apply to good bond conditions and to bar sizes $\leqslant 32$ mm. Where the bond conditions are poor the values in the table should be divided by 0.7 and where the bar diameter exceeds 32 mm the values should be divided by $[(132 - \phi)/100]$.

Example 8.5 Design of a simply supported beam (EC2)

Design the main steel and shear reinforcement for the beam shown below assuming the following material strengths: $f_{ck} = 25$ N mm^{-2}, $f_{yk} = 460$ N mm^{-2} and $f_{ywd} = 250$ N mm^{-2}. The environmental conditions fall within exposure class 1.

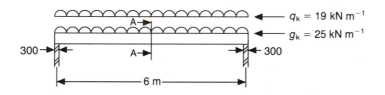
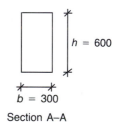

DESIGN MOMENT (M)

Loading

Permanent
Self-weight of beam $= 0.6 \times 0.3 \times 24 = 4.32$ kN m^{-1}
Total permanent load (g_k) $= 25 + 4.32 = 29.32$ kN m^{-1}

Variable
Total variable load (q_k) $= 19$ kN m^{-1}

Ultimate load
Total ultimate load (W) $= (1.35 g_k + 1.5 q_k)$ span
$= (1.35 \times 29.32 + 1.5 \times 19)6$
$= 408.5$ kN

Design moment

Maximum design moment $(M) = \dfrac{Wl}{8} = \dfrac{408.5 \times 6}{8} = 306.4$ kN m

ULTIMATE MOMENT OF RESISTANCE (M_u)

Effective depth

Assume diameter of main bar (ϕ) = 25 mm
Assume diameter of links (ϕ') = 10 mm
Cover for exposure class 1 (c) = 20 mm

Effective depth, d, is

$$d = h - \phi/2 - \phi' - c = 600 - 12.5 - 10 - 20 = 557 \text{ mm}$$

Ultimate moment

Ultimate moment of resistance, M_u is

$$M_u = 0.167 f_{ck} b d^2$$
$$= 0.167 \times 25 \times 300 \times 557^2 \times 10^{-6}$$
$$= 388.6 \text{ kN m}$$

Since $M_u > M$ design as a singly reinforced beam.

MAIN STEEL (A_{s1})

$$K_O = \frac{M}{f_{ck} b d^2} = \frac{306.4 \times 10^6}{25 \times 300 \times 557^2} = 0.132$$

$$z = d\left[0.5 + \sqrt{(0.5 - 3K_O/3.4)}\right]$$
$$= 557\left[0.5 + \sqrt{(0.25 - 3 \times 0.132/3.4)}\right]$$
$$= 482.3 \text{ mm}$$

$$A_{s1} = \frac{M}{0.87 f_{yk} z} = \frac{306.4 \times 10^6}{0.87 \times 460 \times 482.3} = 1588 \text{ mm}^2$$

Therefore, from *Table 3.7*, provide 4T25 (A_{s1} = 1960 mm^2).

SHEAR REINFORCEMENT

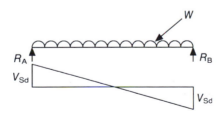

Ultimate load (W) = 408.5 kN

Shear force (V_{Sd})

$$V_{Sd} = \frac{W}{2} = \frac{408\,500}{2} = 204\,250 \text{ N}$$

Shear resistance of concrete alone, V_{Rd1}

$$f_{ck} = 25 \text{ N mm}^{-2} \quad \tau_{Rd} = 0.30 \text{ N mm}^{-2} \quad (Table\ 8.7)$$

$$k = 1.6 - d = 1.6 - 0.557 = 1.043$$

$$\rho_1 = \frac{A_{s1}}{b_w d} = \frac{1960}{300 \times 557} = 0.01173$$

$$\sigma_{cp} = 0$$

$$V_{Rd1} = [\tau_{Rd}k(1.2 + 40\rho_1) + 0.15\sigma_{cp}]b_w d$$

$$V_{Rd1} = [0.30 \times 1.043(1.2 + 40 \times 0.01173)]300 \times 557$$

$$= 87\ 275 \text{ N}$$

Maximum shear resistance of concrete (V_{Rd2})

$$v = 0.7 - \frac{f_{ck}}{200} = 0.7 - \frac{25}{200} = 0.575$$

$$V_{Rd2} = \frac{1}{2}vf_{cd}b_w 0.9d = \frac{1}{2} \times 0.575 \times \frac{25}{1.5} \times 300 \times 0.9 \times 557$$

$$= 720\ 619 \text{ N} > V_{Sd}$$

Minimum shear reinforcement

$\rho_w = 0.0022$ by interpolation between the values in *Table 8.8*, hence

$$\rho_w = A_{sw}/sb_w \sin \alpha$$

$$0.0022 = A_{sw}/s \times 300 \times 1$$

$$\frac{A_{sw}}{s} = 0.66$$

Diameter and spacing of links

Minimum shear reinforcement required where shear force in beam, V_c, is

$$V_c \leq V_{cd} + V_{wd}$$

where $V_{cd} = V_{Rd1} = 87\ 275$ N and

$$V_{wd} = \frac{A_{sw}}{s}0.9f_{ywd}d = \frac{0.66 \times 0.9 \times 250 \times 557}{1.15}$$

$$= 0.66 \times 108\ 978 = 71\ 925 \text{ N}$$

Hence

$$V_c = 87\ 275 + 71\ 925 = 159\ 200 \text{ N}$$

$$V_c/V_{Rd2} = 159\ 200/720\ 619$$

$$= 0.22 \Rightarrow s_{max} \not> 300 \text{ mm} \quad (Table\ 8.9)$$

Hence, from *Table 3.10*, provide R10 links at 200 mm centres ($A_{sw}/s = 0.785$), where $V \leq 159\ 200$ N, i.e. 2338 mm either side of the mid-span of beam.

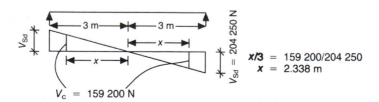

Where shear force in beam $> V_c$, provide shear reinforcement according to

$$V_{Sd} = V_{cd} + V_{wd}$$

$$204\,250 = 87\,275 + 108\,978 A_{sw}/s$$

$$\frac{A_{sw}}{s} = \frac{204\,205 - 87\,275}{108\,978} = 1.07$$

$$\frac{V_{Sd}}{V_{Rd2}} = \frac{204\,250}{720\,619} = 0.28$$

$$\Rightarrow s_{max} = 0.6d = 0.6 \times 557 = 334 \text{ mm} \not> 300 \text{ mm}$$

Hence from *Table 3.10*, provide R10 links at 125 mm centres ($A_{sw}/s = 1.256$) where shear force in beam exceeds 159 200 N, i.e. 662 mm in from both supports.

DEFLECTION
Steel ratio, $\rho = A_s/bd = 1960/557 \times 300 = 0.01173 = 1.173\%$. From *Table 8.10*, basic span/depth ratios for simply supported beams corresponding to $\rho = 0.5$ and 1.5% are 25 and 18 respectively. Interpolating between these values gives a span/depth ratio of 20.3 for $\rho = 1.173\%$.

$$\text{Design service stress, } \sigma_s = \frac{5 f_{yk} A_{req}}{8 A_{prov}} = \frac{5 \times 460 \times 1588}{8 \times 1960}$$

$$= 233 \text{ N mm}^{-2}$$

$$\text{Modification factor} = \frac{250}{\sigma_s} = \frac{250}{233} = 1.07$$

Hence modified span/depth ratio corresponding to $\rho = 1.173\%$ is $20.3 \times 1.07 = 21.7 >$ actual span/depth ratio ($= 6000/557 = 10.7$).

REINFORCEMENT DETAILS
The sketches below show the main reinforcement requirements for the beam. For reasons of buildability the actual reinforcement details may well be slightly different.

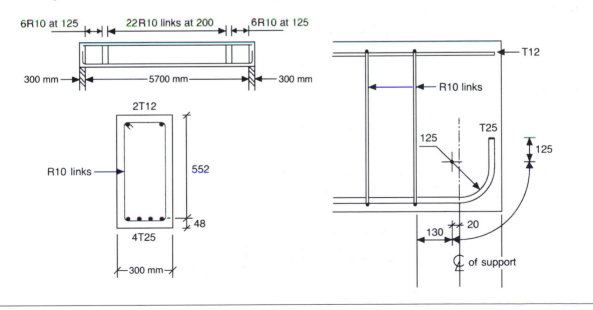

Example 8.6 Analysis of a singly reinforced beam (EC2)

Calculate the maximum variable load that the beam shown below can carry assuming that the load is (i) uniformly distributed and (ii) occurs as a point load at mid-span.

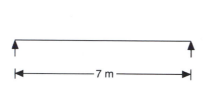

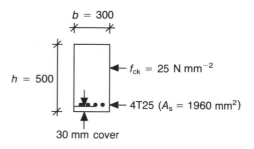

$b = 300$

$h = 500$

$f_{ck} = 25$ N mm^{-2}

4T25 ($A_s = 1960$ mm^2)

30 mm cover

7 m

MOMENT CAPACITY OF SECTION

Effective depth, d, is

$$d = h - \text{cover} - \phi/2$$

$$= 500 - 30 - 25/2 = 457 \text{ mm}$$

$$K_o = \frac{M}{f_{ck}bd^2} = \frac{M}{25 \times 300 \times 457^2}$$

$$z = d\left[0.5 + \sqrt{(0.25 - 3K_o/3.4)}\right]$$

$$= 457\left[0.5 + \sqrt{(0.25 - 3M/3.4 \times 25 \times 300 \times 457^2)}\right] \qquad (1)$$

$$A_{s1} = 1960 \text{ mm}^2 = \frac{M}{0.87f_{yk}z} = \frac{M}{0.87 \times 460z} \qquad (2)$$

Solving equations (1) and (2) simultaneously gives

$$M = 286.1 \text{ kN m} \qquad z = 364.7 \text{ mm}$$

Maximum uniformly distributed load, q_k

Permanent load, g_k $\quad = 0.5 \times 0.3 \times 24 = 3.6$ kN m^{-1}

Total ultimate load $(W) = (1.35g_k + 1.5q_k)\text{span}$

$$= (1.35 \times 3.6 + 1.5q_k)7$$

Substituting into

$$M = \frac{Wl}{8}$$

$$286.1 = \frac{(1.35 \times 3.6 + 1.5q_k)7^2}{8}$$

Hence

$$q_k = 27.9 \text{ kN m}^{-1}$$

Maximum point load, Q_k

Factored permanent load $(W_D) = (1.35g_k)$ span

$$= 1.35 \times 3.6 \times 7 = 34.02 \text{ kN}$$

Factored variable load (W_I) $= 1.5Q_k$

Substituting into

$$M = \frac{W_D l}{8} + \frac{W_I l}{4}$$

$$286.1 = \frac{34.02 \times 7}{8} + \frac{1.5Q_k 7}{4}$$

Hence

$$Q_k = 97.6 \text{ kN}$$

8.9 Design of one-way spanning solid slabs

8.9.1 DEPTH, BENDING, SHEAR

EC2 requires a slightly different approach to BS 8110 for designing one-way spanning solid slabs. To calculate the depth of the slab, the designer must first estimate the percentage of steel required in the slab for bending. Generally, most slabs will be lightly reinforced, i.e. $\rho < 0.5\%$. The percentage of reinforcement can be used, together with the support conditions, to select an appropriate span/depth ratio from *Table 8.10*. The effective depth of the slab can then be determined by multiplying the span/depth ratio by the span of the slab (*Example 8.7*). The actual area of steel required in the slab for bending can be calculated using the equations developed in *section 8.8.1*. Provided this agrees with the assumed value, the calculated depth of the slab is acceptable.

The designer will also need to check that the slab will not fail in shear. The shear resistance of the slab can be calculated as for beams (*section 8.8.2*). According to clause 4.3.2.1(2), where the design shear force (V_{Sd}) is less than the design shear resistance of the concrete alone (V_{Rd1}) no shear reinforcement need be provided. The same requirement appears in BS 8110. Where $V_{Sd} > V_{Rd1}$, shear reinforcement should be provided such that $V_{Sd} \leq V_{Rd3}$ where V_{Rd3} is the design shear force due to the concrete and shear reinforcement.

The actual area of shear reinforcement in slabs can be calculated as for beams and should not be less than the values in *Table 8.8* for beams (see clause 5.4.3.3(2) as amended by NAD). The maximum longitudinal spacing of successive series of links can be determined using the equations in *Table 8.9*. However, the limits in millimetres should be ignored and the maximum spacing of links should not exceed $0.75d$. Clause 5.4.3.3 also notes that shear reinforcement should not be provided in slabs less than 200 mm deep.

8.9.2 REINFORCEMENT DETAILS FOR SOLID SLABS

This section outlines EC2 requirements regarding the detailing of slabs with respect to:

1. reinforcement percentages
2. spacing of reinforcement
3. anchorage and curtailment of reinforcement
4. crack control.

8.9.2.1 Reinforcement areas (clause 5.4.2.1.1, EC2)

In EC2, the maximum and minimum percentages of longitudinal steel permitted in beams and slabs are the same, namely

$$\frac{0.6bd}{f_{yk}} \nleq 0.0015bd \leq A_s \leq 0.04 A_c$$

$$\text{(other than at laps)}$$

where b is the breadth of section, d the effective depth, A_c the cross-sectional area of concrete (bh) and f_{yk} the characteristic yield stress of reinforcement. The area of transverse or secondary reinforcement should not generally be less than 20% of the principal reinforcement, i.e.

$$A_s(\text{trans}) \not< 0.2 A_s(\text{main})$$

8.9.2.2 Spacing of reinforcement
The clear distance between reinforcing bars should not be less than the following:

1. maximum bar diameter
2. 20 mm
3. $d_g + 5$ mm if d_g (maximum aggregate size) > 32 mm.

The maximum bar spacing in slabs, s_{max}, for the main and secondary reinforcement should not exceed the following:

$$s_{max} < 3h \not> 500 \text{ mm}$$

where h denotes the overall depth of the slab.

8.9.2.3 Anchorage and curtailment
For detailing the main reinforcement in slabs, the same provision outlined earlier for beams will apply with $a_\ell = d$. Clause 5.4.3.2.2 of EC2 requires that in slabs near end supports, half the calculated span reinforcement should continue up to the support and be anchored in accordance with *Fig. 8.9*.

For continuous slabs, the *Concise Eurocode for the Design of Concrete Buildings* recommends that the simplified curtailment rules for continuous slabs, given in BS 8110, may still be suitable (*Fig. 8.11*).

8.9.2.4 Crack widths (clause 4.4.2.3, EC2)
According to EC2, where the overall depth of the slab does not exceed 200 mm and the code provisions with regard to reinforcement areas, spacing of reinforcement, anchorage and curtailment of bars, etc. discussed above have been applied, no further measures specifically to control cracking are necessary. For slab depths greater than 200 mm, where at least the minimum reinforcement area has been provided, the limitation of crack width to less than 0.3 mm will generally be achieved by limiting bar spacing (or bar size) in accordance with the provisions in sizes *Table 8.11*.

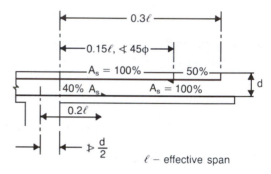

Fig. 8.11 *Curtailment of reinforcement in continuous slabs.*

Example 8.7 Design of a one-way spanning floor (EC2)

Design the floor shown below for an imposed load of 3.5 kN m^{-2}, assuming the following material strengths: $f_{ck} = 30$ N mm^{-2}, $f_{yk} = 460$ N mm^{-2}. The environmental conditions fall within exposure class 1.

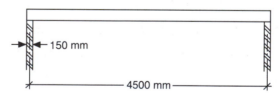

DETERMINE EFFECTIVE DEPTH OF SLAB AND AREA OF MAIN STEEL

Estimate overall depth of slab
Assume reinforcement ratio, ρ, is 0.35%. From *Table 8.10*, basic span/depth ratios for lightly and nominally reinforced, simply supported slabs are 25 and 34 respectively. Hence by interpolation, basic span/depth ratio for slab with $\rho = 0.35\%$ is approximately 28.8.

$$\text{Minimum effective depth } (d) = \frac{\text{span}}{\text{basic ratio}}$$

$$= \frac{4500}{28.8} \approx 156 \text{ mm}$$

Take $d = 160$ mm and assume diameter of main steel (Φ) $= 10$ mm and cover to reinforcement (c) $= 20$ mm. Overall depth of slab, h, is

$$h = d + \Phi/2 + c = 160 + 10/2 + 20 = 185 \text{ mm}$$

Loading

Permanent
Self-weight of slab (g_k) $= 0.185 \times 24$ (kN m^{-3}) $= 4.44$ kN m^{-2}

Variable
Total variable load (q_k) $= 3.5$ kN m^{-2}

Ultimate load. For 1 m width of slab, total ultimate load is

$$(1.35g_k + 1.5q_k)\text{span} = (1.35 \times 4.44 + 1.5 \times 3.5)4.5 = 50.6 \text{ kN}$$

Design moment

Maximum design moment (M) $= \dfrac{Wl}{8} = \dfrac{50.6 \times 4.5}{8} = 28.5$ kN m

Ultimate moment
Ultimate moment of resistance, M_u, is

$$M_u = 0.167 f_{ck} b d^2$$

$$= 0.167 \times 30 \times 1000 \times 160^2 \times 10^{-6} = 128 \text{ kN m}$$

Since $M_u > M$ no compression reinforcement is required.

Main reinforcement (A_{s1})

$$K_o = \frac{M}{f_{ck}bd^2} = \frac{28.5 \times 10^6}{30 \times 1000 \times 160^2} = 0.0371$$

$$z = d\left[0.5 + \sqrt{(0.25 - 3K_o/3.4)}\right]$$

$$= 160\left[0.5 + \sqrt{(0.25 - 3 \times 0.0371/3.4)}\right] = 160 \times 0.966 = 154 \text{ mm}$$

$$A_{s1} = \frac{M}{0.87 f_{yk} z} = \frac{28.5 \times 10^6}{0.87 \times 460 \times 154} = 462 \text{ mm}^2$$

Hence from *Table 3.17*, T10 at 150 mm centres ($A_{s1} = 523$ mm^2 m^{-1}) would be suitable.

Maximum bar spacing $< 3h = 3 \times 185 = 555$ mm $\not> 500$ mm   OK

Minimum reinforcement area $= 0.0015bd = 0.0015 \times 10^3 \times 160$

$$= 240 \text{ mm}^2 \text{ m}^{-1} \quad \text{OK}$$

The reinforcement ratio, ρ, in this case is

$$\rho = \frac{A_s}{bd} = \frac{523}{1000 \times 160} = 0.00327 = 0.327\%$$

Check estimated effective depth of slab

Design service stress, $\sigma_s = \dfrac{5 f_{yk} A_{req}}{8 A_{prov}} = \dfrac{5 \times 460 \times 462}{8 \times 523}$

$$= 254 \text{ N mm}^{-2}$$

Modification factor $= \dfrac{250}{\sigma_s} = \dfrac{250}{254} = 0.984$

Hence modified span/depth ratios for $\rho = 0.5$ and 0.15% are 24.6 ($= 25 \times 0.984$) and 33.46 ($= 34 \times 0.984$) respectively. Interpolating between these values gives a span/depth ratio of 28.9 for $\rho = 0.327\%$. Hence minimum effective depth, d is

$$d = \frac{\text{span}}{\text{span/depth ratio}}$$

$$= \frac{4500}{28.9} = 156 \text{ mm} < \text{assumed} \quad (= 160 \text{ mm})$$

Therefore, use a slab with an effective depth $= 160$ mm, overall depth $= 185$ mm and main steel $=$ T10 at 150 mm centres.

SECONDARY REINFORCEMENT

Provide T8 at 300 mm centres ($A_{s \text{ (trans)}} = 168 \text{ mm}^2 \text{ m}^{-1}$):

$$A_{s \text{ (trans)}} \nless 0.2 A_{s \text{ (main)}} = 0.2 \times 523 = 105 \text{ mm}^2 \text{ m}^{-1} \quad \text{OK}$$

$$\text{Maximum spacing} < 3h = 3 \times 185 = 555 \text{ mm} \ngtr 500 \text{ mm} \quad \text{OK}$$

SHEAR REINFORCEMENT

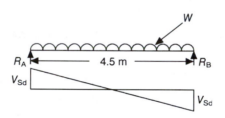

Ultimate load (W) = 50.6 kN

Shear force (V_{Sd})

$$V_{Sd} = \frac{W}{2} = \frac{50\,600}{2} = 25\,300 \text{ N}$$

Shear resistance of concrete alone, V_{Rd1}

$$f_{ck} = 30 \text{ N mm}^{-2} \quad \tau_{Rd} = 0.34 \text{ N mm}^{-2}$$

$$k = 1.6 - d = 1.6 - 0.160 = 1.440$$

Assuming that 50% of midspan steel does not extend to the supports $A_{s1} = 260 \text{ mm}^2/\text{m}$.

$$\rho_1 = \frac{A_{s1}}{b_w d} = \frac{260}{1000 \times 160} = 0.0016 \quad \sigma_{cp} = 0$$

$$V_{Rd1} = [\tau_{Rd} k (1.2 + 40\rho_1) + 0.15\sigma_{cp}] b_w d$$

$$= [0.34 \times 1.440(1.2 + 40 \times 0.0016)]1000 \times 160$$

$$= 99\,017 \text{ N}$$

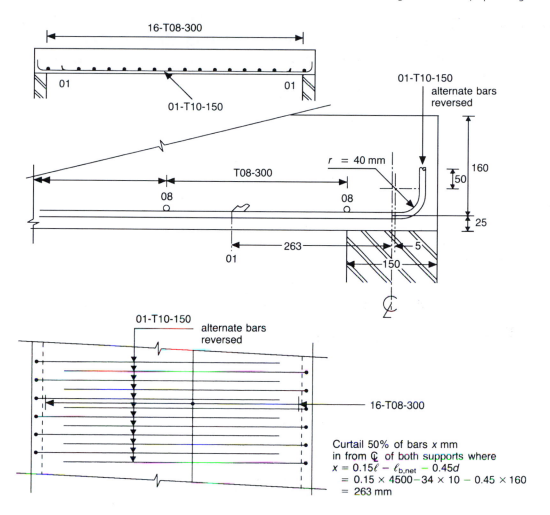

Since $V_{Rd1} > V_{Sd}$, no shear reinforcement is required.

REINFORCEMENT DETAILS
The sketches above show the main reinforcement requirements for the slab.

Reinforcement areas and bar spacing
Minimum reinforcement areas and bar spacing rules were checked above and found to be satisfactory.

Crack width
Since the overall depth of the slab does not exceed 200 mm and the rest of the code provisions have been met, no further measures specifically to control cracking are necessary.

Example 8.8 Analysis of a one–way spanning floor (EC2)

Calculate the maximum uniformly distributed variable load that the floor shown below can carry.

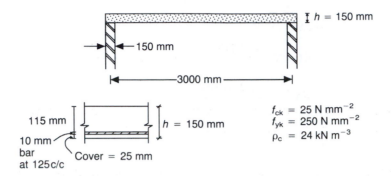

$f_{ck} = 25 \text{ N mm}^{-2}$
$f_{yk} = 250 \text{ N mm}^{-2}$
$\rho_c = 24 \text{ kN m}^{-3}$

EFFECTIVE SPAN

From clause 2.5.2.2.2 of EC2, effective span of slab, l_{eff}, is

$$l_{eff} = l_n + a_1 + a_2$$

$$= 2850 + \frac{150}{3} + \frac{150}{3} = 2950 \text{ mm}$$

MOMENT CAPACITY

Effective depth of slab $(d) = h - \text{cover} - \Phi/2$

$$= 150 - 25 - 10/2 = 120 \text{ mm}$$

Assume $z = 0.96d = 0.96 \times 120 = 115.2 \text{ mm}$:

$$A_{s1} = 628 \text{ mm}^2 \text{ m}^{-1} = \frac{M}{0.87 f_{yk} z} = \frac{M}{0.87 \times 250 \times 115.2}$$

Hence moment capacity $M = 15\,735\,000 \text{ N mm} = 15.735 \text{ kN m}$.

LEVER ARM (Z)

Check the assumed value of z:

$$K_o = \frac{M}{f_{ck} b d^2} = \frac{15.735 \times 10^6}{25 \times 10^3 \times 120^2} = 0.0437$$

$$z = d\left[0.5 + \sqrt{(0.25 - 3K_o/3.4)}\right]$$

$$= d\left[0.5 + \sqrt{(0.25 - 3 \times 0.0437/3.4)}\right]$$

$$= 0.96d \quad \text{(as assumed)}$$

MAXIMUM UNIFORMLY DISTRIBUTED VARIABLE LOAD (q_k)

Loading

Permanent

Self-weight of slab $(g_k) = 0.15 \times 24 \text{ kN m}^{-3} = 3.6 \text{ kN m}^{-2}$

Ultimate load

$$\text{Total ultimate load } (W) = (1.35g_k + 1.5q_k)\text{span}$$
$$= (1.35 \times 3.6 + 1.5q_k)2.95$$

Variable load
Maximum design moment M, is

$$M = 15.7 = \frac{Wl}{8} = (4.86 + 1.5q_k)\frac{2.95^2}{8}$$

$$q_k = 6.4 \text{ kN m}^{-2}$$

Hence the maximum uniformly distributed variable load the slab can support is 6.4 kN m^{-2}.

8.10 Design of pad foundations

The design of pad foundations was discussed in *Chapter 3* and found to be similar to that for slabs in respect of bending. However, the designer must also check that the pad will not fail due to face, transverse or punching shear. EC2 requirements in respect of these three modes of failure are discussed next.

8.10.1 FACE SHEAR
According to clause 6.4(d) of the NAD, the shear stress at the perimeter of the column should not exceed the following:

$$v_{Rd2} \not> 0.9\sqrt{f_{ck}} \qquad (8.30)$$

8.10.2 TRANSVERSE SHEAR (CLAUSE 4.3.4.5, EC2)
The critical section for transverse shear failure normally occurs at distance d from the face of the column. No shear reinforcement will be required provided $v_{Sd} < v_{Rd1}$, where v_{Sd} is the applied shear per unit length and is given by

$$v_{Sd} = \frac{\text{total design force (*Fig. 3.58*)}}{\text{length of critical section}} \qquad (8.31)$$

and where v_{Rd1} is the shear resistance per unit length and is given by

$$v_{Rd1} = \tau_{Rd}k(1.2 + 40\rho_1)d \qquad (8.32)$$

where τ_{Rd} is given in *Table 8.7*,

$$k = (1.6 - d) \geqslant 1.0 \quad (d \text{ in metres})$$

$$\rho_1 = \sqrt{\rho_{1x}\rho_{1y}} \not> 0.015$$

where ρ_{1x} and ρ_{1y} relate to the tension steel in the x and y dirctions respectively.

8.10.3 PUNCHING SHEAR (CLAUSE 4.3.4, EC2)
No shear reinforcement is required if $v_{Sd} < v_{Rd1}$, where v_{Rd1} is the design shear resistance per unit length of the critical perimeter (see above) and v_{Sd} is the applied shear per unit length and is given by

$$v_{Sd} = \frac{V_{Sd}\beta}{u} \qquad (8.33)$$

where V_{Sd} is total design shear force. For a foundation this is calculated along the perimeter of the base of the truncated punching shear cone, assumed to form at 33.7°, provided this falls within the foundation (*Fig. 8.12*). Here u is the perimeter of the

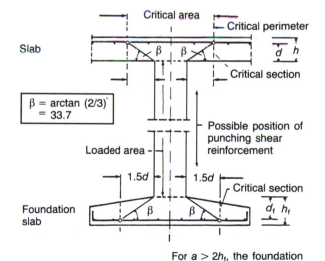

Fig. 8.12 *Design model for punching shear at ULS (Fig. 4.16, EC2).*

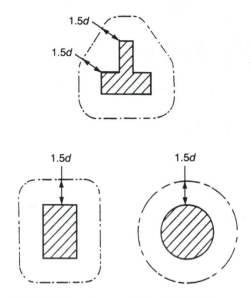

Fig. 8.13 *Critical perimeters (Fig. 4.18, EC2).*

critical section (*Fig. 8.13*) and β is a coefficient which takes account of the effects of eccentricity of loading. In cases where no eccentricity of loading is possible, β may be taken as 1.0. If v_{Sd} exceeds v_{Rd1}, shear reinforcement will need to be provided such that $v_{Sd} \leqslant v_{Rd3}$, where v_{Rd3} is the design shear resistance per unit length of the critical perimeter, for a slab with shear reinforcement calculated in accordance with clause 4.3.4.5.2 of EC2 and clause 6.4(d) of the NAD.

Example 8.9 Analysis of a pad foundation (EC2)

The pad footing shown below supports a column which is subject to axial characteristic permanent and variable actions of 900 and 300 kN respectively. Assuming the following material strengths and the cover to the main steel is 40 mm, check the suitability of the design in shear:

$$f_{ck} = 30 \text{ N mm}^{-2}$$
$$f_{yk} = 460 \text{ N mm}^{-2}$$

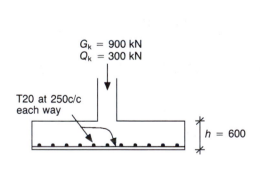

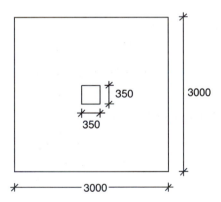

Shear failure could arise:

1. at the face of the column;
2. at a distance d from the face of the column;
3. punching failure of the slab.

FACE SHEAR
Load on footing due to column is

$$1.35 \times 900 + 1.5 \times 300 = 1665 \text{ kN}$$

Design shear stress at perimeter of column, v_{Sd}, is

$$v_{Sd} = \frac{V_{Sd}}{\text{perimeter} \times d} = \frac{1665 \times 10^3}{4 \times 350 \times 540} = 2.2 \text{ N mm}^{-2}$$

where $d = h - \text{cover} - \text{diameter of bar} = 600 - 40 - 20 = 540$ mm. Maximum shear stress at perimeter of column, v_{Rd2}, is

$$v_{Rd2} = 0.9\sqrt{f_{ck}} = 0.9\sqrt{30} = 4.93 \text{ N mm}^{-2} > v_{Sd} \quad \text{OK}$$

TRANSVERSE SHEAR

Design shear force on footing (V_{Sd}) is

$$V_{Sd} = \text{load from column}$$

$$= (1.35 \times 900 + 1.5 \times 300) = 1665 \text{ kN}$$

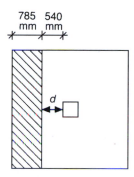

785 540
mm mm

d

$$\text{Earth pressure } (p_E) = \frac{\text{ultimate load}}{\text{area}} = \frac{1665}{9}$$

$$= 185 \text{ kN m}^{-2}$$

Ultimate load on shaded area is

$$p_E \times \text{area} = 185(0.785 \times 3) = 435.7 \text{ kN}$$

Design shear force per unit length, v_{Sd}, is

$$v_{Sd} = 435.7/3 = 145.2 \text{ kN m}^{-1}$$

Shear resistance of concrete alone, v_{Rd1}, is

$$v_{Rd1} = \tau_{Rd}k(1.2 + 40\rho_1)d$$

$$= [0.34 \times 1.06(1.2 + 40 \times 0.0023)]540$$

$$= 251 \text{ N mm}^{-1}$$

$$= 251.7 \text{ kN m}^{-1} > v_{Sd} \ (= 145.2 \text{ kN m}^{-1}) \quad \text{OK}$$

where $\tau_{Rd} = 0.34 \text{ N mm}^{-2}$ since $f_{ck} = 30 \text{ N mm}^{-2}$ from *Table 8.7* and where

$$k = 1.6 - d = 1.6 - 0.54 = 1.06$$

$$\rho_1 = \rho_{1y} = \rho_{1x} = \frac{A_{s1}}{b_w d}$$

$$= \frac{1260}{1000 \times 540} = 0.0023 \quad \sigma_{cp} = 0$$

PUNCHING SHEAR

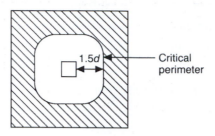

Perimeter of critical section, u, is

$$u = \text{column perimeter} + 2\pi(1.5d) = 4 \times 350 + 2\pi(1.5 \times 540) = 6489 \text{ mm}$$

Area within critical perimeter is

$$4 \times 350(1.5 \times 540) + 350^2 + \pi(1.5 \times 540)^2 = 3.32 \times 10^6 \text{ mm}^2$$

Design shear force, V_{Sd}, is

$$V_{Sd} = p_E A = 185(9 - 3.32) = 1050.8 \text{ kN}$$

Applied shear per unit length, v_{Sd}, is

$$v_{Sd} = \frac{V_{Sd}\beta}{u} = \frac{1050.8 \times 1}{6.489} = 161.9 \text{ kN m}^{-1} < v_{Rd1} = 251.7 \text{ kN m}^{-1} \quad \text{OK}$$

Hence no shear reinforcement is required.

8.11 Design of columns

Column design is covered in EC2 mostly within the chapter on buckling. The design procedure is slightly more complex than that used in BS 8110, although the final result is similar in both codes. The increased complexity is partly due to the style of presentation and partly the differences in design approach in the two codes. In EC2, for example, the slenderness ratio above which columns are considered as slender (λ_{min}) has to be evaluated while in BS 8110 it is taken as 15 for braced columns and 10 for unbraced columns. In addition, EC2 does not contain any formulae equivalent to equations 38 and 39 in BS 8110 (*section 3.13.5*) which relate to the design of short-braced columns and are so useful for initial sizing of members in compression.

In this chapter only the design of the most common types of columns found in building structures, i.e. those in non-sway structures subject to an axial load and bending will be described. Their design involves consideration of the following aspects which are discussed individually below:

1. braced columns
2. sway structures
3. slenderness ratio
4. slender columns
5. eccentricities.

8.11.1 BRACED COLUMNS (CLAUSE 4.3.5.3.2, EC2)

A column may be considered to be braced in a given plane if the bracing element or system (e.g. core or shear walls) is sufficiently stiff to resist at least 90% of all lateral forces in that plane. Otherwise it should be considered as unbraced. Note that this condition is more realistically stated in EC2 than in BS 8110 which requires that the bracing element/structure should be capable of resisting all lateral forces on the structure.

8.11.2 SWAY STRUCTURES (CLAUSE 4.3.5.3.3, EC2)

Unlike BS 8110, EC2 recognizes the fact that some braced structures may experience significant displacements at connections due to the lateral loading. These displacements will increase the design

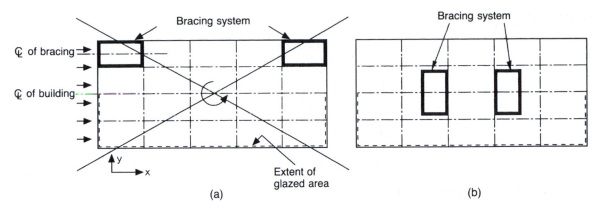

Fig. 8.14 *Braced sway and braced, non-sway structures.*

moment in the member and must therefore be taken into account. For example, the columns shown in both the floor plans in *Fig. 8.14* may well be braced. However, because the shear centre of the bracing system in plan (a) is not concentric with the axis of the building, the structure is liable to rotate when the direction of the wind loads are parallel to the x–x axis, thereby generating additional forces in the columns. Such a layout may be necessary because of architectural/client requirements for an open area which is glazed as shown in *Fig. 8.14*, for instance. For engineering purposes it would be better to adopt the arrangement of shear walls shown in plan (b).

Where there is doubt, Appendix 3 of EC2 should be used which gives a criterion for classifying braced structures as non-sway, based on the height of the structure, number of storeys, flexural stiffnesses of the bracing elements and service loads.

8.11.3 SLENDERNESS RATIO (CLAUSE 4.3.5.3.5, EC2)

In EC2 slenderness ratio (λ) is given by

$$\lambda = \frac{l_o}{i} \qquad (8.34)$$

where l_o is the effective height of the column and i the radius of gyration. Note that in BS 8110 the slenderness ratio is based on the lateral dimen-

sions of the column (b or h) and not the radius of gyration.

8.11.3.1 Effective height
The effective height of a column (l_o) is given by

$$l_O = \beta l_{col} \qquad (8.35)$$

where l_{col} is the height of column measured between centres of restraint and β is a coefficient. Values of β can be determined from the nomograph shown in *Fig. 8.15* via the coefficients k_A and k_B which denote the rigidity of restraint at the column ends and are given by

$$k_A(\text{or } k_B) = \frac{\Sigma E_{cm} I_{col}/l_{col}}{\Sigma E_{cm} \alpha I_b/l_{eff}} \not< 0.4 \qquad (8.36)$$

where E_{cm} is the modulus of elasticity of the concrete (*Table 8.16*), I_{col}, and I_b are the second moments of area of the column and beam respectively, l_{col} the height of column between centres of restraint, l_{eff} the effective span of beam and α the factor taking into account the condition of restraint of the beam at the opposite end to the joint:

$\alpha = 1.0$ opposite end elastically or rigidly restrained, i.e. a continuous end

$\alpha = 0.5$ opposite end free to rotate, i.e. a simply supported end

$\alpha = 0$ for a cantilever beam

Table 8.16 Values of the secant modulus of elasticity E_{cm} (kN mm^{-2}) (Table 3.2, EC2)

Strength class C	C12/15	C16/20	C20/25	C25/30	C30/37	C35/45	C40/50	C45/55	C50/60
E_{cm}	26	27.5	29	30.5	32	33.5	35	36	37

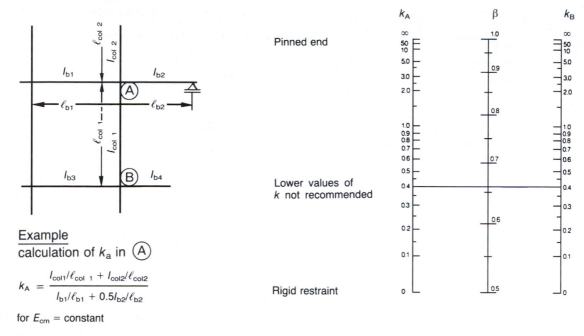

Example
calculation of k_a in (A)

$$k_A = \frac{I_{col1}/\ell_{col\ 1} + I_{col2}/\ell_{col2}}{I_{b1}/\ell_{b1} + 0.5I_{b2}/\ell_{b2}}$$

for E_{cm} = constant

Fig. 8.15 *Nomograph for evaluating effective lengths of columns in non-sway frames (based on Fig. 4.27, EC2).*

8.11.3.2 Radius of gyration
The radius of gyration (i) is given by

$$i = \sqrt{(I/A)} \qquad (8.37)$$

where I is the second moment of area of the column and A the gross cross-sectional area of the column.

8.11.4 SLENDER AND NON-SLENDER COLUMNS (CLAUSE 4.3.5.5.3, EC2)
If the slenderness ratio of an isolated column, λ, exceeds a critical value, λ_{min}, then it may be assumed to be slender. λ_{min} is taken as the greater of

$$\lambda_{min} = 25 \text{ and } \frac{15}{\sqrt{\upsilon_u}} \qquad (8.38)$$

where υ_u is the longitudinal force coefficient for an element and is given by

$$\upsilon_u = \frac{N_{Sd}}{A_c f_{cd}} \qquad (8.39)$$

where N_{Sd} is the design axial load, A_c the cross-sectional area of concrete and f_{cd} the design concrete strength $= \dfrac{f_{ck}}{1.5}$.

EC2 only gives simplified rules for the design of columns in non-sway structures. Since sway strac-

tures are fairly uncommon in practice, however, this is unlikely to present any major drawbacks to designers. The following discusses EC2 rules for the design of columns in non-sway structures.

8.11.5 DESIGN OF COLUMNS IN NON-SWAY STRUCTURES
A column in a non-sway structure is classified as slender if the slenderness ratio exceeds the greater of 25 or $\dfrac{15}{\sqrt{\upsilon_u}}$.

According to Flow chart 3 in Appendix A of EC2, where the column is non-slender, i.e. $\lambda < \lambda_{min}$, it should be designed simply for first order internal moments and forces, using the design charts shown in *Fig. 8.16*.

Where the column is slender, however, the design procedure depends upon whether the slenderness ratio lies above or below λ_{crit}, given by,

$$\lambda_{crit} = 25[2 - (e_{01}/e_{02})] \qquad (8.40)$$

where e_{01} and e_{02} the first-order eccentricities are defined as

$$e_{01} = M_{Sd1}/N_{Sd} \quad e_{02} = M_{Sd2}/N_{Sd} \qquad (8.41)$$

where M_{Sd1} and M_{Sd2} are the first-order design moments and N_{Sd} the design axial load. It is also assumed that $|e_{01}| \leqslant |e_{02}|$. The following subsections

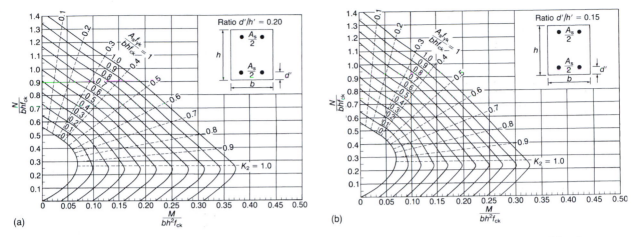

Fig. 8.16 *Typical column design charts for use with EC2 (Concise Eurocode for the Design of Concrete Buildings).*

describe the design of slender columns where (a) $\lambda_{min} < \lambda < \lambda_{crit}$ and (b) $\lambda > \lambda_{crit}$.

8.11.5.1 Slender columns where $\lambda_{min} < \lambda < \lambda_{crit}$ (clause 4.3.5.5.3, EC2)

If $\lambda_{min} < \lambda < \lambda_{crit}$ the column is designed for the design axial load (N_{Sd}) and the first-order design moment (M_{Sd1}) obtained by analysis. The latter is compared with the minimum design moment (M_{Rd}) and the larger value taken. Here M_{Rd} is given by

$$M_{Rd} = N_{Sd}\frac{h}{20} \tag{8.42}$$

Note that this is similar to BS 8110.

Once N_{Sd} and M_{Sd1} have been determined, the area of longitudinal steel can be calculated by strain compatibility using an iterative procedure. However, this is not practical for everyday design and therefore the British Cement Association have produced a series of design charts, similar to those contained in BS 8110: Part 3, which can be used to evaluate the required steel areas. Typical charts for the design of columns to EC2 are shown in *Fig. 8.16*. *Examples 8.10 and 8.13* show the procedure involved.

8.11.5.2 Slender columns where $\lambda > \lambda_{crit}$ (clause 4.3.5.6, EC2)

If the slenderness of a column exceeds λ_{crit} allowance has to be made for the additional moments caused by the deformations. This is achieved by firstly evaluating the various eccentricities and secondly calculating the resulting design moments. Critical conditions may occur at the top, middle or

bottom of the column. The total design eccentricity (e_{tot}) at the **top** and **bottom** of the column will be given by

$$e_{tot} = e_0 + e_a \tag{8.43}$$

where e_0 is the first-order eccentricity (equation 8.41) and e_a the allowance for imperfections given by:

$$e_a = \upsilon\frac{l_0}{2} \tag{8.44}$$

where l_0 is the effective height of the column (*section 8.11.3*) and υ is the angle of inclination of the structure and is equal to the greater of

$$\upsilon = \frac{1}{100\sqrt{l}} \quad \text{and} \quad \frac{1}{200} \tag{8.45}$$

where l is the total height of the structure in metres. This eccentricity will give rise to a design bending moment, M_{Sd}, given by

$$M_{Sd} = N_{Sd1}e_{tot} \tag{8.46}$$

The total design eccentricity at the **middle** of the column will depend on the relative positions of the first-order eccentricities. EC2 considers two cases: (i) $e_{01} = e_{02}$ (*Fig. 8.17(a)*) and (ii) $e_{01} \neq e_{02}$ (*Fig. 8.17(b), (c)*). In case (i)

$$e_{tot} = e_0 + e_a + e_2 \tag{8.47}$$

where e_0 is the first-order eccentricity (equation 8.41), e_a the allowance for imperfections (equation 8.44) and e_2 the second-order eccentricity given by

$$e_2 = K_1\frac{l_0^2}{10}(1/r) \tag{8.48}$$

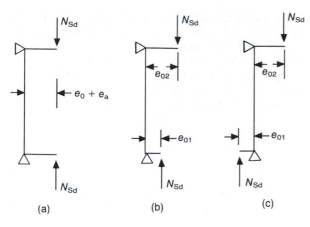

Fig. 8.17 Design model for the calculation of total eccentricity (Fig. 4.29, EC2).

where

$$K_1 = \frac{\lambda}{20} - 0.75 \quad \text{for } 15 \leqslant \lambda \leqslant 35 \quad (8.49)$$

and where $K_1 = 1$ for $\lambda > 35$ and l_0 the effective length of the column (*section 8.11.3*).

In cases where great accuracy is not required, the curvature $(1/r)$ may be derived from

$$1/r = 2K_2 \frac{\varepsilon_{yd}}{0.9d} \quad (8.50)$$

where ε_{yd} is the design yield strain of reinforcement:

$$\varepsilon_{yd} = f_{yd}/E_s \quad (8.51)$$

and where d is the effective depth of the section and K_2 the curvature factor:

$$K_2 = \frac{N_{ud} - N_{Sd}}{N_{ud} - N_{bal}} \leqslant 1 \quad (8.52)$$

where N_{ud} is the design ultimate capacity of the section:

$$N_{ud} = 0.85f_{cd}A_c + f_{yd}A_s \quad (8.53)$$

and where N_{Sd} is the actual design axial force and N_{bal} the design load capacity of the balanced section:

$$N_{bal} = 0.4f_{cd}A_c \quad (8.54)$$

Note that it is always conservative to assume $K_2 = 1$.

For case (ii), i.e. $e_{01} \neq e_{02}$, the total design eccentricity is given by

$$e_{tot} = e_e + e_a + e_2 \quad (8.55)$$

where e_a and e_2 are as previously defined in equations 8.44 and 8.48 respectively and e_e is the equivalent eccentricity taken as the greater of

$$0.6e_{02} + 0.4e_{01} \quad (8.56)$$

and

$$0.4e_{02} \quad (8.57)$$

Here e_{01} and e_{02} are the first-order eccentricities at the two ends, and $|e_{02}| > |e_{01}|$ *Fig. 8.17(b), (c)*). The design moment can then be evaluated using

$$M_{Sd} = N_{Sd1}e_{tot} \quad (8.58)$$

Summarizing, the design axial load (N_{Sd1}) is determined by analysing the loads acting on the structure. The design moment (M_{Sd}) is determined from equation 8.58 and e_{tot} is taken as the greater of equations 8.43 and 8.47 if $e_{01} = e_{02}$ or equations 8.43 and 8.55 if $e_{01} \neq e_{02}$. Once N_{Sd1} and M_{Sd} are known, the area of longitudinal steel can be evaluated using design charts similar to those in Fig. 8.16 (*Example 8.13*).

8.11.6 REINFORCEMENT DETAILS FOR COLUMNS

8.11.6.1 Longitudinal reinforcement (clause 5.4.1.2.1, EC2)

Number of bars. Columns with rectangular cross-sections should be reinforced with a minimum of four longitudinal bars; columns with circular cross-sections should be reinforced with a minimum of six. Each bar should have a diameter of not less than 12 mm.

Reinforcement percentages. The area of longitudinal reinforcement, A_s, should lie with the following limits:

$$\frac{0.15N_{Sd}}{f_{yd}} \not< 0.003A_c \leqslant A_s \leqslant 0.08A_c$$

where f_{yd} is the design yield strength of the reinforcement, N_{Sd} the design axial compression force and A_c the cross-section of the concrete. Note that the upper limit should not be exceeded even where lapped joints occur.

8.11.6.2 Transverse reinforcement (link) (clause 5.4.1.2.2, EC2)

Size and spacing of links. The diameter of the links should not be less than 6 mm or one-quarter of the maximum diameter of the longitudinal bar, whichever is the greater. However, as noted in *Chapter 3*, 6 mm bars may not be freely available and a minimum bar size of 8 mm is preferable. The spacing of links along the column should not exceed the smallest of the following three dimensions:

1. 12 times the minimum diameter of the longitudinal bars;
2. the smallest lateral dimension of the column;
3. 300 mm.

Arrangement of links. With regard to the arrangement of links around the longitudinal reinforcement, EC2 stipulates that (a) every longitudinal bar placed in a corner should be supported by a link passing around the bar and (b) a maximum of five bars in or close to each corner can be secured against buckling by any set of links.

Example 8.10 Design of a slender column (EC2)

A slender column for a non-sway structure is required to resist an ultimate axial load (N_{Sd}) of 2000 kN and bending moment (M_{Sd1}) of 60 kN m. Design the column using class C30/37 concrete and grade 460 reinforcement assuming that the slenderness ratio is $< \lambda_{crit}$.

CROSS-SECTION
Since the design bending moment is relatively small, equation 8.52 may be used to size the column:

$$N_{ud} = 0.85 f_{cd} A_c + f_{yd} A_{sc}$$

Clause 5.4.1.2.1 of EC2 stipulates that the percentage of longitudinal reinforcement, A_s, should generally lie within the following limits:

$$0.3\% A_c < A_s < 8\% A_c$$

Assuming that the percentage of reinforcement is equal to 3% (say) gives

$$A_s = 0.03 A_c$$

Substituting this into the above equation gives

$$2 \times 10^6 = \frac{0.85 \times 30 \times A_c}{1.5} + \frac{460 \times 0.03 A_c}{1.15}$$

therefore $A_c = 68\,966$ mm². For a square column $b = h = \sqrt{68\,966} = 263$ mm. Therefore a 300 mm square column is suitable.

LONGITUDINAL STEEL

Check minimum moment
Minimum eccentricity,

$$e_{min} = \frac{h}{20} = \frac{300}{20} = 15 \text{ mm}$$

Minimum design moment, M_{Rd}, is

$$M_{Rd} = N_{Sd} e_{min} = 2 \times 10^3 \times 15 \times 10^{-3} = 30 \text{ kN m} < M_{Sd1} = 60 \text{ kN m}$$

Therefore design moment is 60 kN m.

Design chart

Cover to links for exposure class 1	= 20 mm
Assume diameter of longitudinal bars (Φ)	= 32 mm
Assume diameter of links	= 8 mm

Therefore

$$d' = 20 + 32/2 + 8 = 44 \text{ mm}$$

$$\frac{d'}{h} = \frac{44}{300} = 0.147$$

Round up to 0.15 and use chart no. 1 (*Fig. 8.16*).

Longitudinal steel area

$$\frac{N}{bhf_{ck}} = \frac{2 \times 10^6}{300 \times 300 \times 30} = 0.74$$

$$\frac{M}{bh^2 f_{ck}} = \frac{60 \times 10^6}{300^3 \times 30} = 0.074$$

$$\frac{A_s f_{yk}}{bhf_{ck}} = \frac{A_s \times 460}{300^2 \times 30} = 0.43 \quad (\textit{Fig. 8.16})$$

$$A_s = 2524 \text{ mm}^2$$

Use 4T32 (3220 mm²).

$$\%A_s/A_c = 3220/300 \times 300 = 3.5\% \quad \text{(acceptable)}$$

LINKS

Diameter of links is the greater of (i) 8 mm and (ii) $^1/_4\Phi = {}^1/_4 \times 32 = 8$ mm. Spacing of links should not exceed the least of (i) $12\Phi = 12 \times 32 = 384$ mm, (ii) least dimension of column = 300 mm and (iii) 300 mm. Therefore, provide R8 links at 300 mm centres.

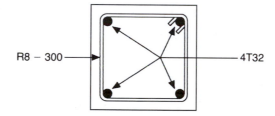

R8 – 300   4T32

Example 8.11 Classification of a column section (EC2)

Determine if column GH shown in *Fig. 8.18(a)* is slender assuming that it is designed to resist the design loads and moments in *Fig. 8.18(b)*. Assume that the structure is non-sway and $f_{ck} = 25$ N/mm².

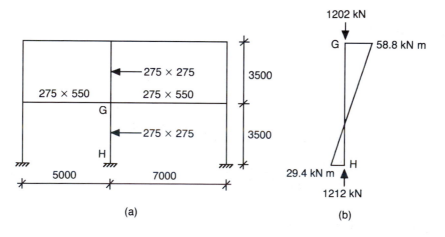

(a)   (b)

Fig. 8.18

SLENDERNESS RATIO OF COLUMN GH

Effective height

Moment of inertia of column, I_{col}, is given by

$$I_{col} = \frac{bd^3}{12} = \frac{275 \times 275^3 \times 10^{-12}}{12} = 4.766 \times 10^{-4} \text{ m}^4$$

Moment of inertia of beam, I_{beam}, is given by

$$I_{beam} = \frac{bd^3}{12} = \frac{275 \times 550^3 \times 10^{-12}}{12} = 3.813 \times 10^{-3} \text{ m}^4$$

From equation 8.36

$$k_G = \frac{E_{cm}(I_{col\ upper} / l_{col\ upper} + I_{col\ lower} / l_{col\ lower})}{E_{cm}(\alpha I_{beam\ 1} / l_{eff\ 1} + \alpha I_{beam\ 2} / l_{eff\ 2})} \not< 0.04$$

$$= \frac{(4.766 \times 10^{-4} / 3.5 + 4.766 \times 10^{-4} / 3.5)}{(1.0 \times 3.813 \times 10^{-3} / 5 + 1.0 \times 3.813 \times 10^{-3} / 7)}$$

$$= 0.208 \not< 0.4$$

To calculate the slenderness ratio it is conservative to assume that the column is pinned at H. Hence $\alpha = 0$ and $k_H = \infty$.
From nomograph in *Fig. 8.15*, $\beta = 0.8$. Effective height of column GH, l^O, is given by

$$l^O = \beta l_{col} = 0.8 \times 3500 = 2800 \text{ mm}$$

Radius of gyration

Radius of gyration, i, is given by

$$i = \sqrt{(I/A)} = \sqrt{[(bd^3/12)/bd]}$$

$$= d/\sqrt{12} = 275/\sqrt{12} = 79.4 \text{ mm}$$

Slenderness ratio

Slenderness ratio, λ, is given by

$$\lambda = \frac{l_o}{i} = \frac{2800}{79.4} = 35.3$$

Classification of column

Minimum slenderness ratio, λ_{min}, is

$$\lambda_{min} = 25 \text{ or } \frac{15}{\sqrt{\upsilon_u}} \text{ whichever is the greater}$$

$$\upsilon_u = \frac{N_{Sd}}{A_c f_{cd}} = \frac{1202 \times 10^3}{275^2 \times 25/1.5} = 0.954$$

$$\frac{15}{\sqrt{\upsilon_u}} = \frac{15}{\sqrt{0.954}} = 15.3$$

Hence $\lambda_{min} = 25$
Since $\lambda > \lambda_{min}$ column GH is slender

Example 8.12 Classification of a column (EC2)

Determine if column PQ is slender assuming that it is designed to resist the design loads and moments shown in *Fig. 8.19(b)*. Assume that the structure is non-sway and $f_{ck} = 25$ N mm^{-2}.

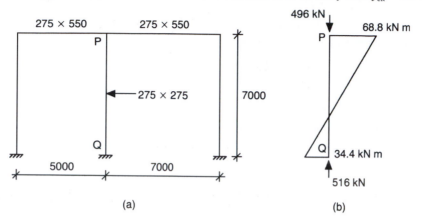

(a)

(b)

Fig. 8.19

SLENDERNESS RATIO OF COLUMN PQ

Effective height

Moment of inertia of column, I_{col}, is given by

$$I_{col} = \frac{bd^3}{12} = \frac{275 \times 275^3 \times 10^{-12}}{12} = 4.766 \times 10^{-4} \text{ m}^4$$

Moment of inertia of beam, I_{beam}, is given by

$$I_{beam} = \frac{bd^3}{12} = \frac{275 \times 550^3 \times 10^{-12}}{12} = 3.813 \times 10^{-3} \text{ m}^4$$

From equation 8.36

$$k_P = \frac{E_{cm}(I_{col\ upper}\ /\ l_{col\ upper} + I_{col\ lower}\ /\ l_{col\ lower})}{E_{cm}(\alpha I_{beam\ 1}\ /\ l_{eff\ 1} + \alpha I_{beam\ 2}\ /\ l_{eff\ 2})} \not< 0.4$$

$$= \frac{(4.766 \times 10^{-4}\ /\ 7)}{E_{cm}(1.0 \times 3.813 \times 10^{-3}\ /\ 5 + 1.0 \times 3.813 \times 10^{-3}\ /\ 7)} = 0.052 \not< 0.4$$

To calculate the slenderness ratio it is conservative to assume that the column is pinned at Q. Hence $\alpha = 0$ and $k_Q = \infty$.

From nomograph in *Fig. 8.15*, $\beta = 0.8$. Effective height of column PQ, l_O, is given by

$$l_O = \beta l_{col} = 0.8 \times 7000 = 5600 \text{ mm}$$

Radius of gyration

Radius of gyration, i, is given by

$$i = \sqrt{(I/A)} = \sqrt{[(bd^3/12)/bd]} = d/\sqrt{12} = 275/\sqrt{12} = 79.4 \text{ mm}$$

Slenderness ratio

Slenderness ratio, λ, is given by

$$\lambda = \frac{l_o}{i} = \frac{5600}{79.4} = 70.5$$

Classification of column

Critical slenderness ratio, λ_{min}, is

λ_{min} is the greater of 25 or $\dfrac{15}{\sqrt{v_u}}$

$$v_u = \frac{N_{Sd}}{A_c f_{cd}} = \frac{496 \times 10^3}{275^2 \times 25/1.5} = 0.394$$

$$\frac{15}{\sqrt{v_u}} = 23.9$$

$$\therefore \lambda_{min} = 25$$

Since $\lambda > \lambda_{min}$ column is slender

Example 8.13 Column design: (i) $\lambda_{min} < \lambda < \lambda_{crit}$; (ii) $\lambda > \lambda_{crit}$ (EC2)

Design the columns in *Examples 8.11* and *8.12*.

COLUMN GH

CRITICAL SLENDERNESS RATIO, λ_{crit}

Eccentricity at end G, $e_{02} = \dfrac{M_{Sd2}}{N_{Sd}} = \dfrac{58.8 \times 10^6}{1202 \times 10^3} = 48.9$ mm

Eccentricity at end H, $e_{01} = \dfrac{M_{Sd1}}{N_{Sd}} = \dfrac{-29.4 \times 10^6}{1212 \times 10^3} = -24.5$ mm

Critical slenderness ratio, λ_{crit}, is given by

$$\lambda_{crit} = 25(2 - e_{01}/e_{02}) = 25(2 - -24.5/48.9) = 62.5$$

Since $\lambda = 35.3$ (*Example 8.11*), and $\lambda < \lambda_{crit}$ column GH does not need to be designed for second order effects.

Longitudinal steel

$$N_{Sd} = 1202 \times 10^3 \text{ N}$$

$$M_{Sd} = 58.8 \times 10^6 \text{ N mm} > M_{Rd} = N_{Sd}\frac{h}{20}$$

$$= (1202 - 10^3)\frac{275}{20} = 16.5 \times 10^6 \text{ N mm} \quad \text{OK}$$

Assume:

Diameter of longitudinal steel (Φ) = 32 mm
Diameter of links (Φ') = 8 mm
Cover to all reinforcement (c) = 30 mm

Therefore

$$d' = \Phi/2 + \Phi' + c = 32/2 + 8 + 30 = 54 \text{ mm}$$

$$\frac{d'}{h} = \frac{54}{275} = 0.196$$

Use graph with $d'/h = 0.20$:

$$\frac{N}{bhf_{ck}} = \frac{1202 \times 10^3}{275^2 \times 25} = 0.635$$

$$\frac{M}{bh^2 f_{ck}} = \frac{58.8 \times 10^6}{275^3 \times 25} = 0.113$$

$$\frac{A_s f_{yk}}{bhf_{ck}} = \frac{A_s \times 460}{275^2 \times 25} = 0.5 \quad (\textit{Fig. 8.16})$$

$$A_s = 2055 \text{ mm}^2$$

Provide 4T32 ($A_s = 3220$ mm^2).

Links

Diameter of links is the greater of (i) 8 mm and (ii) $^1/_4\Phi = ^1/_4 \times 32 = 8$ mm. Spacing of links should not exceed the least of (i) $12\Phi = 12 \times 32 = 384$ mm, (ii) least dimension of column = 275 mm and (iii) 300 mm. Therefore, provide R8 at 275 mm centres.

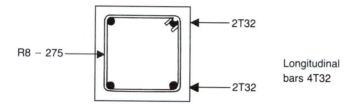

COLUMN PQ

Critical slenderness ratio, λ_{crit}

Eccentricity at end P, $e_{02} = \dfrac{M_{Sd2}}{N_{Sd}} = \dfrac{68.8 \times 10^6}{496 \times 10^3} = 139$ mm

Eccentricity at end Q, $e_{01} = \dfrac{M_{Sd1}}{N_{Sd}} = \dfrac{-34.4 \times 10^6}{516 \times 10^3} = -66.7$ mm

Critical slenderness ratio, λ_{crit}, is given by

$$\lambda_{crit} = 25(2 - e_{01}/e_{02})$$

$$= 25(2 - -66.7/139) = 62$$

Since $\lambda = 70.5$ (*Example 8.12*) and $\lambda > \lambda_{crit}$ column PQ needs to be designed for second order effects.

Eccentricities

Calculate e_e

e_e is the greater of

$$0.6e_{02} + 0.4e_{01} = 0.6 \times 139 + 0.4 \times -66.7$$

$$= 57 \text{ mm}$$

$$0.4e_{02} = 0.4 \times 139 = 56 \text{ mm}$$

Therefore take $e_e = 56$ mm.

Calculate e_a

$$l = 7000 \text{ mm}$$

$$\upsilon = \frac{1}{100\sqrt{l}} = \frac{1}{100\sqrt{7}} = \frac{1}{264.5} \not< \frac{1}{200}$$

$$l_O = 5600 \text{ mm} \quad (Example\ 8.12)$$

Hence

$$e_a = \frac{\upsilon l_o}{2} = \frac{(1/200)5600}{2} = 14 \text{ mm}$$

Calculate e_2**. Assume:**

Diameter of longitudinal steel (Φ) = 20 mm
Diameter of links (Φ') = 8 mm
Cover to all reinforcement (c) = 35 mm

Thus

$$d = h - (\Phi/2 + \Phi' + c)$$

$$= 275 - (20/2 + 8 + 35) = 222 \text{ mm}$$

$$K_1 = 1 \quad (\text{since } \lambda > 35)$$

Assume $K_2 = 1$:

$$\varepsilon_{yd} = \frac{f_{yk}}{E_s} = \frac{460}{200 \times 10^3} = 2.3 \times 10^{-3}$$

$$1/r = \frac{2K_2\varepsilon_{yd}}{0.9d} = \frac{2 \times 1 \times 2.3 \times 10^{-3}}{0.9 \times 222} = 2.3 \times 10^{-5}$$

$$e_2 = \frac{K_1 l_o^2 (1/r)}{10} = \frac{1 \times 5600^2 \times (2.3 \times 10^{-5})}{10} = 72 \text{ mm}$$

Calculate e_{tot}**.** Here e_{tot} is taken as the greater of:

$$e_{tot} = e_0 + e_a = 139 + 14 = 153 \text{ mm}$$

and

$$e_{tot} = e_e + e_a + e_2 = 57 + 14 + 72 = 143 \text{ mm}$$

Hence $e_{tot} = 153$ mm.

Design moment

$$M_{Sd} = N_{Sd}e_{tot} = 496 \times 0.153 = 75.9 \text{ kN m}$$

Longitudinal steel area
Assuming that the diameter of longitudinal steel (Φ) is 20 mm, the diameter of links (Φ') 8 mm and the cover to all reinforcement (c) 35 mm

$$d' = \Phi/2 + \Phi' + c$$

$$= 20/2 + 8 + 35 = 53 \text{ mm}$$

$$\frac{d'}{h} = \frac{53}{275} = 0.193$$

Use graph with $d'/h = 0.20$:

$$\frac{N}{bhf_{ck}} = \frac{496 \times 10^3}{275^2 \times 25} = 0.262$$

$$\frac{M}{bh^2 f_{ck}} = \frac{75.9 \times 10^6}{275^3 \times 25} = 0.146$$

$$\frac{A_s f_{yk}}{bhf_{ck}} = \frac{A_s \times 460}{275^2 \times 25} = 0.3$$

$$A_s = 1230 \text{ mm}^2$$

Provide 4T20 ($A_s = 1260 \text{ mm}^2$).

Check assumed value of K_2.

$$N_{ud} = 0.85 f_{cd} A_c + f_{yd} A_s$$

$$= \frac{0.85 \times 25 \times 275^2}{1.5} + \frac{460 \times 1260}{1.15} = 1575 \times 10^3 \text{ N}$$

$$N_{Sd} = 496 \times 10^3 \text{ N}$$

$$N_{bal} = 0.4 f_{cd} A_c = \frac{0.4 \times 25 \times 275^2}{1.5} = 504 \times 10^3 \text{ N}$$

$$K_2 = \frac{N_{ud} - N_{Sd}}{N_{ud} - N_{bal}} = \frac{(1575 - 496) \, 10^3}{(1575 - 504) \, 10^3} = 1.0$$

Therefore assumed value of K_2 is correct.

Links

Diameter of links is the greater of (i) 8 mm and (ii) $^1/_4 \Phi = ^1/_4 \times 20 = 5$ mm. Spacing of links should not exceed the least of (i) $12\Phi = 12 \times 20 = 240$ mm, (ii) least dimension of column = 275 mm and (iii) 300 mm. Therefore, provide R8 links at 240 mm centres.

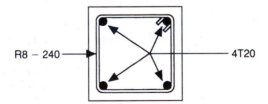

Chapter 9

Eurocode 3: Design of steel structures

This chapter describes the contents of Part 1.1 of Eurocode 3, the new European standard for the design of buildings in steel, which is expected to replace BS 5950 by about 2008. The chapter highlights the differences between ENV Eurocode 3: Part 1.1 and BS 5950 and illustrates the new design procedures by means of a number of worked examples on beams, columns and connections. To help comparison but primarily to ease understanding of the Eurocode, the material has here been presented in a similar order to that in Chapter 4 of this book on BS 5950, rather than strictly adhering to the sequence of chapters and clauses adopted in the Eurocode.

9.1 Introduction

Eurocode 3 applies to the design of buildings and civil engineering works in steel. It is based on limit state principles and comes in several parts as shown in *Table 9.1*.

Part 1.1 of Eurocode 3, which is the subject of this discussion, gives a general basis for the design of buildings and civil engineering works in steel. It is largely similar in scope to Part 1 of BS 5950, which was discussed in *Chapter 4*. Part 1.1 of

Table 9.1 Overall scope of Eurocode 3

Part	Subject
1.1	General rules and rules for buildings
1.2	Fire resistance
1.3	Cold formed thin gauge members and sheeting
2	Bridges and plated structures
3	Towers, masts and chimneys
4	Tanks, silos and pipelines
5	Piling
6	Crane structures
7	Marine and maritime structures
8	Agriculture

Eurocode 3, hereafter referred to as EC3, was published in draft form in 1984 and then a European pre-standard, reference no. DD ENV 1993–1–1: 1992, in September 1992. It is due to be issued as a European standard in 2003 and should replace BS 5950 by about 2008.

EC3 is published in two volumes and deals with the following subjects:

Volume 1
Chapter 1: Introduction (covers layout of code, important conventions and assumptions, symbols)
Chapter 2: Basis of design
Chapter 3: Materials
Chapter 4: Serviceability limit states
Chapter 5: Ultimate limit states
Chapter 6: Connections subject to static loading
Chapter 7: Fabrication and erection
Chapter 8: Design assisted by testing
Chapter 9: Fatigue

Volume 2
Annex B: Reference standards
Annex C: Design against brittle fracture
Annex E: Buckling length of a compression member
Annex F: Lateral – torsional buckling
Annex J: Beam-to-column connections
Annex K: Hollow section lattice girder connections
Annex L: Column bases
Annex M: Alternative method for fillet welds
Annex Y: Guidelines for loading tests

The purpose of the following discussion is to describe the contents of EC3 and to highlight the main differences between it and Part 1 of BS 5950. A number of design examples on beams, columns and connections are also included in *sections 9.11, 9.12* and *9.13* respectively to illustrate the new design procedures.

9.2 Structure of EC3

As can be appreciated from the above contents list, the organization of material in EC3 is quite different from that in BS 5950. BS 5950 contains separate sections on the design of individual elements, e.g. beams, columns and connections. EC3, however, divides the material on the basis of design criteria, e.g. deflection, tension, compression, bending, shear and buckling, which may apply to any element.

9.3 Principles and application rules (clause 1.2, EC3)

For the reasons discussed in *Chapter 7*, the clauses in EC3 have been divided into principles and application rules. Principles encompass statements, definitions, requirements and analytical methods for which there is no permitted alternative. Application rules are generally recognized rules which follow the principles, but for which EC3 permits the use of alternative techniques. The principles are printed in roman type. The application rules, on the other hand, are printed in italics (*Fig. 9.1*).

9.4 Boxed values (clause 1.3, EC3)

As pointed out in *Chapter 7*, a number of the safety elements given in EC3 appear in a box as shown below. This signifies that these values are meant to be for guidance only and that other values may be used by individual member states, for the time being.

$$\boxed{1.5}$$

6.5.2.2.   Design shear rupture resistance

(1) 'Block shear' failure at a group of fastener holes near the end of a beam web or bracket, see figure 6.5.5. shall be prevented by using appropriate hole spacing. This mode of failure generally consists of tensile rupture along the line of fastener holes on the tension face of the hole group, accompanied by gross section yielding in shear at the row of fastener holes along the shear face of the hole group, see figure 6.5.5.

(2) The design value of the effective resistance to block shear $V_{eff.Rd}$ should be determined from

$$V_{eff.Rd} = (f_y/\sqrt{3})A_{v.eff}/\gamma_{M0} \qquad (6.1)$$

where $A_{v.eff}$ is the effective shear area

Fig. 9.1 *Example of principle and application rules (page 145, EC3).*

The actual values to be used in the UK can be found in the *UK National Application Document* (NAD) for EC3.

9.5 Symbols (clause 1.6, EC3)

The symbols used in EC3 are extremely schematic but a little cumbersome. For example, $M_{pl.y.Rd}$ denotes the design plastic moment of resistance about the y–y (major) axis. Symbols such as this makes many expressions and formulae seem much longer than they actually are, but on the other hand the traditional list of definitions of symbols is no longer required. A number of symbols which are used to identify particular dimensions of universal sections have changed in EC3 as discussed in *section 9.6*. Furthermore, the elastic and plastic section moduli are denoted in EC3 by the symbols W_{el} and W_{pl} respectively.

9.6 Member axes (clause 1.6.7, EC3)

Member axes for commonly used steel members are shown in *Fig. 9.2*. This system is different from that used in BS 5950. Thus, x–x is the axis along the member, while axis y–y is what is termed in BS 5950 as axis x–x, the major axis. The minor axis of bending is taken as axis z–z in EC3. Considering steel design in isolation, this change need not have happened. It was insisted on for consistency with the structural Eurocodes for other materials, e.g. concrete and timber.

9.7 Basis of design

Like BS 5950, EC3 uses limit state principles and for design purposes considers the two principal categories of limit sates: ultimate and serviceability. The terms 'ultimate limit state' (ULS) and 'serviceability limit state' (SLS) apply in the same way as we understand them in BS 5950. Thus, ultimate limit states are 'those associated with collapse, or with other forms of structural failure which may endanger the safety of people' while serviceability limit states 'correspond to states beyond which specified service criteria are no longer met'.

The serviceability limit states requirements for steel structures are discussed in Chapters 2 and 4 of EC3. The principal serviceability limit states in EC3 are listed below. These are broadly similar to those specified in BS 5950 (*Table 4.1*):

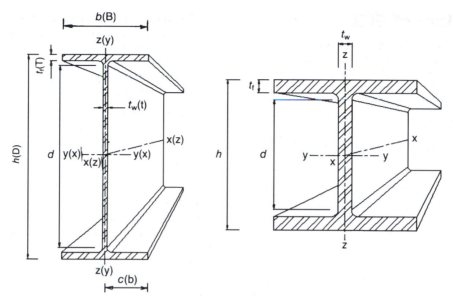

Fig. 9.2 *Member definitions and axes used in EC3 (N.B. symbols in parentheses denote symbols used in BS 5950).*

1. deformations or deflections which affect the appearance or effective use of the structure;
2. vibration, oscillation or sway (i.e. dynamic effects) which causes discomfort to the occupants of a building or damage to its contents;
3. damage to finishes or non-structural elements due to deformations, deflections or dynamic effect.

According to clause 2.3.2 of EC3, the ultimate limit states for steel structures and components are:

1. static equilibrium of the structure;
2. rupture or excessive deformation of a member;
3. transformation of the structure into a mechanism;
4. instability induced by second-order effects, e.g. lack of fit, thermal effects, sway;
5. fatigue;
6. accidental damage (may include fire resistance).

In addition, there is a separate (third) limit state of durability. As can be appreciated, these are roughly similar to the limit states used in BS 5950 (*Table 4.1*) although there are slight changes in the terminology.

In the context of elemental design, the designer is principally concerned with the ultimate limit states which affect the strength of the member, e.g. yield, buckling and rupture and the serviceability limit state of deflection. EC3 requirements in respect of these limit states will be discussed later for individual element types.

In a similar way to BS 5950, clause 5.2 of EC3 gives three possible sets of design assumptions for analysing structural frames:

1. Simple framing – this assumes that all joints are pinned and the structure is fully braced.
2. Continuous framing – this assumes that all joints are rigid (for elastic design) or full strength (for plastic analyses) and can transmit all moments and forces.
3. Semi-continuous framing – these methods take the real stiffness and strength of the joints into account.

9.8 Actions (clause 2.2.2, EC3)

'Actions' is Eurocode terminology for loads and imposed deformations. Dead and imposed loads are generally referred to in EC3 as permanent and variable actions respectively. The characteristic values of permanent and variable actions will be specified in Eurocode 1: *Actions on Structures*, which is at an advanced state of development. In the mean time, the UK NAD requires designers to continue using BS 648, BS 6399: Part 1 and CP 3: Chapter 5: Part 2 for characteristic values of actions. Indeed, the NAD requires these UK standards to be used throughout the life of the ENV, even after Eurocode 1 has been published.

The design values of actions (F_d) are obtained by multiplying the characteristic actions (F_k) by

Table 9.2 Partial safety factors for actions on building structures for persistent and transient design situations (Table 2.2, EC3)

	Permanent actions (γ_G)	Variable actions (γ_Q)	
		One with its characteristic value	Others with their combination values
Favourable effect $\gamma_{F,inf}$	1.0	0	0
Unfavourable effect $\gamma_{F,sup}$	1.35	1.5	1.5

Table 9.3 Nominal values of yield strength, f_y, and ultimate tensile strength, f_u, for structural steel to EN 10025 (Table 3.1, EC3)

Nominal steel grade	Thickness t (mm)			
	$t \leqslant 40$ mm		40 mm $< t \leqslant 100$ mm	
	f_y (N mm^{-2})	f_u (N mm^{-2})	f_y (N mm^{-2})	f_u (N mm^{-2})
Fe 360	235	360	215	340
Fe 430	275	430	255	410
Fe 510	355	510	335	490

Note. t is the nominal thickness of the element.

the appropriate partial safety factor for actions (γ_F) obtained from *Table 9.2*:

$$F_d = \gamma_F \times F_k \qquad (9.1)$$

For buildings, the partial safety factor for permanent actions (γ_G) is 1.35 and variable actions (γ_Q) is 1.5. The corresponding values in BS 5950 are 1.4 and 1.6 respectively.

9.9 Materials

9.9.1 DESIGN STRENGTHS (CLAUSE 2.2.3.2, EC3)
EC3 reflects the model Eurocode clause that design strengths, X_d, are obtained by dividing the characteristic strengths, X_k, by the partial safety factor for materials, γ_m, i.e.

$$X_d = X_k/\gamma_m \qquad (9.2)$$

But this is not in fact used in EC3. Instead, EC3 divides the cross-section resistance, R_k, by a partial safety factor to give the design resistance, R_d for the member, i.e.

$$R_d = R_k/\gamma_M \qquad (9.3)$$

where γ_M is the partial safety factor for the resistance. Thus, γ_M in EC3 is not the same as γ_m in BS 5950. Here γ_m is applied to strength; γ_M is applied to structures.

9.9.2 CHARACTERISTIC STRENGTHS (CLAUSE 3.2.2, EC3)
Table 9.3 shows the characteristic yield and ultimate strength of structural steelwork recommended for use by EC3. Note that the equivalents of grade S275 and S355 steel, which are referred to in BS 5950 are given, but not grade S460. Instead a lower-strength steel (Fe 360) is given which is the normal steel grade used on the Continent. This lower-grade steel has a higher fracture toughness, which enables the use of thicker ply in colder temperatures. Grade S460 steel can still be specified, however, since a grade Fe E 460 steel ($f_y = 460$ N mm^{-2}) similar to grade S460 is included in Annex D in Part 1A of Eurocode 3.

9.9.3 PARTIAL SAFETY FACTORS (CLAUSE 5.1.1, EC3)
The partial safety factor for materials is taken as 1 in BS 5950 and consequently there is no difference

between characteristic and design strength. However, as noted above, in EC3 the partial safety factor, γ_M, is applied to structures and components, rather than strengths and it is therefore not easy to compare the changes in the two standards.

According to clauses 5.1.1 and 3 of EC3 and the UK NAD respectively, the values for partial safety factors should be those given below:

Resistance of class 1, 2 or 3 cross-section γ_{M0}
$= 1.05$
Resistance of class 4 cross-section γ_{M1} $= 1.05$
Resistance of member to buckling γ_{M1} $= 1.05$
Resistance of net section at bolt holes γ_{M2} $= 1.25$

9.9.4 MATERIAL COEFFICIENTS (CLAUSE 3.2.5, EC3)

The following coefficients are specified in EC3:

Modulus of elasticity $E = 210\,000$ N mm^{-2}

Shear modulus G $= E/2(1+\nu)$ N mm^{-2} (9.4)

Poisson's ratio ν $= 0.3$

Coefficient of linear thermal expansion α $= 12 \times 10^6$ per °C

Density ρ $= 7850$ kg m^{-3}

Note that the value of E is slightly higher than that specified in BS 5950, being equal to 205 kN mm^{-2}.

9.10 Classification of cross-sections (clause 5.3, EC3)

Classification has the same purpose as in BS 5950, and the four classifications are identical:

Class 1 cross-sections: 'plastic' in BS 5950
Class 2 cross-sections: 'compact' in BS 5950
Class 3 cross-sections: 'semi-compact' in BS 5950
Class 4 cross-sections: 'slender' in BS 5950

Classification of a cross-section depends upon the proportions of each of its compression elements. The highest (least favourable) class number is generally quoted for a particular section.

Appropriate sections of EC3's classification are given in *Table 9.4*. Briefly, for rolled sections, d/t_w (d/t in BS 5950) ratios are slightly more onerous for webs, and c/t_f (b/T in BS 5950) ratios are slightly

Table 9.4 Maximum width-to-thickness ratios for compression elements (Table 5.3.1, EC3)

Type of element	Class 1	Class 2	Class 3
Outstand flange for rolled section	$\dfrac{c}{t_f} \leqslant 10\varepsilon$	$\dfrac{c}{t_f} \leqslant 11\varepsilon$	$\dfrac{c}{t_f} \leqslant 15\varepsilon$
Web with neutral axis at mid depth, rolled section	$\dfrac{d}{t_w} \leqslant 72\varepsilon$	$\dfrac{d}{t_w} \leqslant 83\varepsilon$	$\dfrac{d}{t_w} \leqslant 124\varepsilon$
Web subject to compression, rolled sections	$\dfrac{d}{t_w} \leqslant 33\varepsilon$	$\dfrac{d}{t_w} \leqslant 38\varepsilon$	$\dfrac{d}{t_w} \leqslant 42\varepsilon$
f_y	235	275	355
ε	1	0.92	0.81

less onerous for outstand flanges. It should be noted that the factor ε is given by

$$\varepsilon = (235/f_y)^{0.5} \qquad (9.5)$$

and not $(275/p_y)^{0.5}$ as in BS 5950. For class 4 sections effective cross-sectional properties can be calculated using effective widths of plate which in some cases put the member back into class 3.

9.11 Design of beams

The procedure for the design of beams is given in clause 5.1.5 of EC3. However, to ease understanding of this part of the code, as in *Chapter 4* of this book, we will consider the design of fully laterally restrained and unrestrained beams separately. Thus, the following section (*9.11.1*) will consider only the design of beams which are fully laterally restrained. *Section 9.11.2* will then look at EC3's rules for designing beams which are not laterally torsionally restrained.

9.11.1 FULLY LATERALLY RESTRAINED BEAMS

Generally, such members should be checked for:

1. resistance of cross-section to bending ULS;
2. resistance to shear buckling ULS;
3. resistance to flange-induced buckling ULS;
4. resistance of the web to transverse forces ULS;
5. deflection SLS.

9.11.1.1 Resistance of cross-sections – bending moment (clause 5.4.5, EC3)

When shear force is absent or of a low value, the design value of the bending moment, M_{Sd}, should at no point exceed the moment of resistance of the section, $M_{c.Rd}$, i.e.

$$M_{Sd} \leqslant M_{c.Rd} \qquad (9.6)$$

$M_{c.Rd}$ may be taken as follows:

1. The design plastic resistance moment of the gross section

$$M_{pl.Rd} = W_{pl} f_y / \gamma_{M0} \qquad (9.7)$$

where W_{pl} is the plastic section modulus, for class 1 and 2 sections only.

2. The design elastic resistance moment of the gross section

$$M_{el.Rd} = W_{el} f_y / \gamma_{M0} \qquad (9.8)$$

where W_{el} is the elastic section modulus for class 3 sections.

3. The design local buckling resistance moment of the gross section

$$M_{o.Rd} = W_{eff} f_y / \gamma_{M1} \qquad (9.9)$$

where W_{eff} is the effective section modulus, for class 4 cross-sections only.

4. The design ultimate resistance moment of the net section at bolt holes $M_{u.Rd}$, if this is less than the appropriate values above. In calculating this value, fastener holes in the compression zone do not need to be considered unless they are oversize or slotted. In the tension zone holes do not need to be considered provided that

$$0.9(A_{f.net}/A_f) \geqslant (f_y/f_u)(\gamma_{M2}/\gamma_{M0}) \qquad (9.10)$$

When this is not the case a reduced flange area may be assumed which satisfies the above. Consideration of fastener holes in bending is not clearly covered in BS 5950.

9.11.1.2 Resistance of cross-sections – shear (clause 5.4.6, EC3)

The design value of the shear force, V_{Sd}, at each cross-section should satisfy

$$V_{Sd} \leqslant V_{pl.Rd} \qquad (9.11)$$

where $V_{pl.Rd}$ is the design plastic shear resistance, given by

$$V_{pl.Rd} = A_v \left(f_y / \sqrt{3} \right) / \gamma_{M0} \qquad (9.12)$$

where A_v is the shear area, which for rolled I- and H-sections, loaded by gravity is

$$A_v = A - 2bt_f + (t_w + 2r)t_f \qquad (9.13)$$

where A is the cross-sectional area, b the overall breadth, r the root radius, t_f the flange thickness and t_w the web thickness.

As this will generally give a slightly higher shear area than in BS 5950, for simplicity A_v may be taken as

$$A_v = 1.04 h t_w \qquad (9.14)$$

where h is the overall depth of the section.

The shear resistance calculated as above will give almost identical answers to that calculated by BS 5950, if it is assumed that γ_{M0} is cancelled out by the lower loading partial safety factors γ_F.

Fastener holes in the web do not have to be considered provided that

$$A_{v.net}/A_v \geqslant f_y/f_u \qquad (9.15)$$

When this is not the case an effective shear area of $A_{v.net} f_u/f_y$ may be used.

The shear buckling resistance of unstiffened webs must additionally be considered when

$$d/t_w > 69\varepsilon \qquad (9.16)$$

When the changed definition of ε is taken into account, this value is slightly more onerous than the corresponding value in BS 5950.

For a stiffened web shear buckling resistance will need to be considered when

$$d/t_w > 30\varepsilon\sqrt{k_\tau} \qquad (9.17)$$

where k_τ is the buckling factor for shear and is given by

$k_\tau = 5.34$ (for webs with transverse stiffeners at supports only)

$k_\tau = 4 + \dfrac{5.34}{(a/d)^2}$ (for webs with transverse stiffeners at supports and intermediate transverse stiffeners with $a/d < 1$)

$k_\tau = 5.34 + \dfrac{4}{(a/d)^2}$ (for webs with transverse stiffeners at supports and intermediate transverse stiffeners with $a/d \geqslant 1$).

9.11.1.3 Resistance of cross-sections – bending and shear (clause 5.4.7, EC3)

The plastic resistance moment of the section is reduced by the presence of shear. When the design value of the shear force exceeds 50% of the design plastic shear resistance, the design resistance moment of the section should be reduced to $M_{V.Rd}$, obtained as follows for equal flanged sections:

$$M_{V.Rd} = f_y(W_{pl} - \rho A_v^2/4t_w)/\gamma_{M0} \leq M_{c.Rd} \quad (9.18)$$

where $\rho = (2V_{Sd}/V_{pl.Rd} - 1)^2$. In other cases the design strength should be reduced to $(1-\rho)f_y$ for the shear area only, when calculating the design resistance moment.

9.11.1.4 Shear buckling resistance (clause 5.6, EC3)

The shear buckling resistance of a web has to be checked when $d/t_w > 69\varepsilon$. For standard rolled beams and columns this check is rarely necessary since d/t_w is always less than 69ε, except for just one universal beam (UB) section, which has a d/t_w ratio which is marginally over this limit. See also *section 9.11.1.2.*

9.11.1.5 Flange–induced buckling (clause 5.7.7, EC3)

To prevent the possibility of the compression flange buckling in the plane of the web, EC3 requires that the ratio d/t_w of the web should satisfy the following criterion:

$$d/t_w \leq k(E/f_{yf})[A_w/A_{fc}]^{0.5} \quad (9.19)$$

where A_w is the area of the web, A_{fc} the area of the compression flange and f_{yf} is the yield strength of the compression flange. The factor k assumes the following values: class 1 flanges: 0.3; class 2 flanges: 0.4; class 3 or 4 flanges: 0.55.

9.11.1.6 Resistance of the web to transverse forces (clause 5.7, EC3)

EC3 identifies not two, as in BS 5950, but three possible modes of failure due to loads applied to the web through a flange (*Fig. 9.3*):

1. crushing of the web close to the flange, accompanied by plastic deformation of the flange;
2. crippling of the web in the form of localized buckling and crushing of the web close to the flange, accompanied by plastic deformation of the flange;
3. buckling of the web over most of the depth of the member.

A distinction is also made between two types of load application: (a) forces applied through a flange and resisted by shear in the web (*Fig. 9.4(a)*); (b) forces applied through one flange and transferred through the web directly to the other flange (*Fig. 9.4(b)*). For loading type (a) the web resistance should be taken as the smaller of (1) and (2) above, i.e. (1) the crushing resistance and (2) the crippling resistance. For loading type (b) the web resistance should be taken as the smaller of (1) and (3), i.e. (1) the crushing resistance and (3) the buckling resistance.

Crushing resistance. For an I- or H-section, the design crushing resistance is

$$R_{y.Rd} = (s_s + s_y)t_w f_{yw}/\gamma_{M1} \quad (9.20)$$

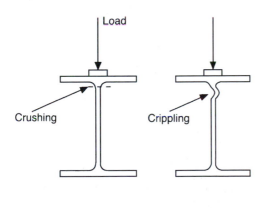

Loading type (a) (*Fig. 9.4*)

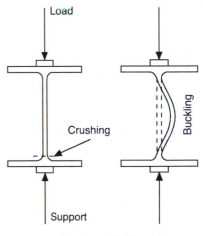

Loading type (b) (*Fig. 9.4*)

Fig. 9.3 *Web failure.*

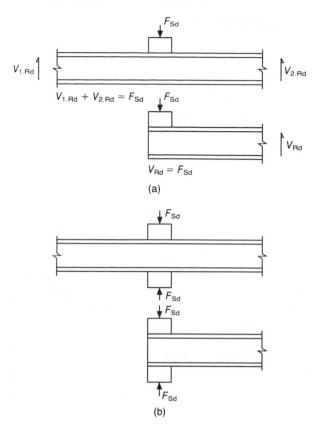

$V_{1.Rd} + V_{2.Rd} = F_{Sd}$

$V_{Rd} = F_{Sd}$

(a)

(b)

Fig. 9.4 *Forces applied through a flange: (a) forces resisted by shear in the web; (b) forces transmitted directly through the web (Fig. 5.7.1, EC3).*

where

$$s_y = 2t_f(b_f/t_w)^{1/2}(f_{yf}/f_{yw})^{1/2}(1 - (\sigma_{f.Ed}/f_{yf})^2)^{1/2} \quad (9.21)$$

in which b_f should not be more than $25t_f$, s_s is the length of stiff bearing and $\sigma_{f.Ed}$ is the longitudinal stress in the flange.

For a rolled I-, H- or U-section, s_y may alternatively be obtained from

$$s_y = \frac{2.5(h - d)(1 - (\sigma_{f.Ed}/f_{yf})^2)^{1/2}}{(1 + 0.8s_s/(h - d))} \quad (9.22)$$

At the end of a member s_y should be halved.

Crippling resistance. For an I- or H-section, the design crippling resistance is

$$R_{a.Rd} = 0.5t_w^2(Ef_{yw})^{1/2}((t_f/t_w)^{1/2} + 3(t_w/t_f)(s_s/d))/\gamma_{M1} \quad (9.23)$$

in which s_s/d should not be more than 0.2. Where the member is also subject to bending moments the following relationship should be satisfied:

Table 9.5 Recommended limiting values for vertical deflections (Table 4.1, EC3)

Conditions	Limits (see Fig. 9.5)	
	δ_{max}	δ_2
Roofs generally	$L/200$	$L/250$
Roofs frequently carrying personnel other than for maintenance	$L/250$	$L/300$
Floors generally	$L/250$	$L/300$
Floors and roofs supporting plaster or other brittle finish or non-flexible partitions	$L/250$	$L/350$
Floors supporting columns (unless the deflection has been included in the global analysis for ultimate limit state)	$L/400$	$L/500$
Where δ_{max} can impair the appearance of the building	$L/250$	–

$$\frac{F_{Sd}}{R_{a.Rd}} + \frac{M_{Sd}}{M_{c.Rd}} \leqslant 1.5 \quad (9.24)$$

Buckling resistance. For the web of I- or H-sections the design buckling resistance $R_{b.Rd}$ should be obtained by considering the web as a virtual compression member with an effective breadth

$$b_{eff} = (h^2 + s_s^2)^{1/2} \quad (9.25)$$

although b_{eff} must be reduced at the ends of a member. The buckling resistance can then be obtained as discussed in *section 9.12.1.2*, using buckling curve c and $\beta_A = 1$.

9.11.1.7 Deflections (clause 4.2, EC3)

Unlike BS 5950, EC3 recommends not one but two limiting values for vertical deflections, δ_2 and δ_{max}, as shown in *Table 9.5* and *Fig. 9.5*. The deflection in BS 5950 is comparable to the deflection δ_2, due to unfactored variable loads, in EC3. However, there are slight variations between the limiting values in the two codes, e.g. the recommended limiting deflection for beams carrying plaster or other brittle finishes is $L/360$ in BS 5950 and $L/350$ in EC3. The other deflection, δ_{max}, is not mentioned in BS 5950 and is the deflection due to the total permanent and variable loading. (*Note*: For cantilever beams the length L should be considered as twice the projecting length of the cantilever.)

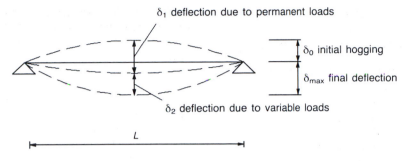

δ₁ deflection due to permanent loads

δ₀ initial hogging

δ_max final deflection

δ₂ deflection due to variable loads

L

Fig. 9.5 *Vertical deflections.*

Example 9.1 Analysis of a laterally restrained beam (EC3)

Check the suitability of $356 \times 171 \times 51$ kg m^{-1} UB section in grade Fe 430 steel loaded by uniformly distributed loading $g_k = 8$ kN m^{-1} and $q_k = 6$ kN m^{-1} as shown below. Assume that the beam is fully laterally restrained and that the beam sits on 100 mm bearings at each end. Ignore self-weight of beam.

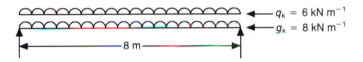

$q_k = 6$ kN m^{-1}

$g_k = 8$ kN m^{-1}

8 m

DESIGN BENDING MOMENT
Design action $(F_d) = (\gamma_G g_k + \gamma_Q q_k) \times$ span

$$= (1.35 \times 8 + 1.5 \times 6)\, 8 = 158.4 \text{ kN}$$

Design bending moment $(M_{Sd}) = \dfrac{F_d l}{8} = \dfrac{158.4 \times 8}{8}$

$$= 158.4 \text{ kN m}$$

STRENGTH CLASSIFICATION
Flange thickness $(t_f) = 11.5$ mm, steel grade Fe 430. Hence from *Table 9.3*, $f_y = 275$ N mm^{-2}.

SECTION CLASSIFICATION

$$\varepsilon = (235/f_y)^{0.5} = (235/275)^{0.5} = 0.924$$

$$\frac{c}{t_f} = 7.46 < 10\varepsilon = 10 \times 0.924 = 9.24$$

$$\frac{d}{t_w} = 42.8 < 72\varepsilon = 72 \times 0.924 = 66.5$$

Hence from *Table 9.4*, $356 \times 171 \times 51$ UB section belongs to class 1.

RESISTANCE OF CROSS-SECTION

Bending moment
Since the beam section belongs to class 1, design moment of resistance is equal to the plastic moment of resistance, $M_{pl.Rd}$, which is given by

$$M_{pl.Rd} = \frac{W_{pl}f_y}{\gamma_{M0}} = \frac{895 \times 10^3 \times 275}{1.05}$$

$$= 234.4 \times 10^6 \text{ N mm}$$

$$= 234.4 \text{ kN m} > M_{Sd} \ (= 158.4 \text{ kN m}) \quad \text{OK}$$

Shear

Design shear force, V_{Sd}, is

$$V_{Sd} = \frac{F_d}{2} = \frac{158.4}{2} = 79.2 \text{ kN}$$

For class 1 section design plastic shear resistance, $V_{pl.Rd}$, is given by

$$V_{pl.Rd} = A_v(f_y/\sqrt{3})/\gamma_{m0}$$

where

$$A_v = 1.04ht_w = 1.04 \times 355.6 \times 7.3 = 2699.7 \text{ mm}^2$$

Hence

$$V_{pl.Rd} = 2699.7\,(275/\sqrt{3})/1.05$$

$$= 408\,224 \text{ N} = 408 \text{ kN} > V_{Sd} \quad \text{OK}$$

Bending and shear

According to EC3 the theoretical plastic resistance moment of the section, i.e. $M_{pl.Rd}$, is reduced if

$$V_{Sd} > 0.5V_{pl.Rd}$$

But in this case

$$V_{Sd} = 79.2 \text{ kN} \ngtr 0.5V_{pl.Rd}$$

$$= 0.5 \times 408 = 204 \text{ kN}$$

Hence no check is required.

SHEAR BUCKLING RESISTANCE

As $d/t_w = 42.8 < 69\varepsilon = 69 \times 0.924 = 64$, no check on shear buckling is required.

FLANGE-INDUCED BUCKLING

Area of web is

$$A_w = (h - 2t_f)t_w = (355.6 - 2 \times 11.5)\,7.3 = 2428 \text{ mm}^2$$

Area of compression flange is

$$A_{fc} = bt_f = 171.5 \times 11.5 = 1972.2 \text{ mm}^2$$

For class 1 section, $k = 0.3$

$$k(E/f_{yf})[A_w/A_{fc}]^{1/2} = 0.3(210 \times 10^3/275)[2428/1972.2]^{1/2}$$

$$= 254.2 > d/t_w = 42.8 \quad \text{OK}$$

RESISTANCE OF THE WEB TO TRANSVERSE FORCES

According to EC3 when loads are resisted by shear in the web, the web should be checked for (a) crushing resistance and (b) crippling resistance.

Crushing resistance

For a rolled I-section

$$s_y = \frac{2.5(h - d)(1 - (\sigma_{f.Ed}/f_{yf})^2)^{1/2}}{(1 + 0.8s_s/(h - d))}$$

$$= \frac{2.5(355.6 - 312.3)(1 - (0)^2)^{1/2}}{(1 + 0.8 \times 100/(355.6 - 312.3))} = 38 \text{ mm}$$

Clause 5.7.3(3) of EC3 recommends that s_y should be halved at the ends of members. Hence, the design crushing resistance of web, $R_{y.Rd}$, is given by

$$R_{y.Rd} = (s_s + s_y)t_w f_{yw}/\gamma_{M1}$$

$$= (100 + 19)7.3 \times 275/1.05$$

$$= 227\,516 \text{ N} = 227.5 \text{ kN} > V_{Sd}$$

$$= 79.2 \text{ kN} \quad \text{OK}$$

Crippling resistance

Design crippling resistance of web, $R_{a.Rd}$, is given by

$$R_{a.Rd} = 0.5t_w^2 (Ef_{yw})^{1/2}[(t_f/t_w)^{1/2} + 3(t_w/t_f)(s_s/d)]/\gamma_{M1}$$

$$= 0.5 \times 7.3^2 (210 \times 10^3 \times 275)^{1/2}[(11.5/7.3)^{1/2} + 3(7.3/11.5)(100/312.3)]/1.05$$

$$= 359\,633 \text{ N} = 360 \text{ kN} > V_{Sd} \ (= 79.2 \text{ kN}) \quad \text{OK}$$

Hence section satisfies web crippling requirements.

DEFLECTION

The maximum bending moment due to working load is

$$M_{max} = 1/8(g_k + q_k)L^2$$

$$= 1/8(8 + 6)8^2 = 112 \text{ kN m}$$

Elastic resistance is

$$(M_{c.Rd})_{el} = \frac{W_{el}f_y}{\gamma_{M0}} = \frac{796 \times 10^3 \times 275}{1.05} = 208 \times 10^6 \text{ N mm}$$

$$= 208 \text{ kN m} > M_{max}$$

Hence deflection can be calculated elastically.
Deflection due to permanent and variable loading (w), i.e.

$$w = g_k + q_k = 8 + 6 = 14 \text{ kN m}^{-1} = 14 \text{ N mm}^{-1}$$

is

$$\delta_{max} = \frac{5}{384} \frac{wL^4}{EI} = \frac{5}{384} \times \frac{14 \times (8 \times 10^3)^4}{210 \times 10^3 \times 14\,200 \times 10^4}$$

$$= 25 \text{ mm} < \frac{\text{span}}{250} = \frac{8 \times 10^3}{250} = 32 \text{ mm} \quad \text{OK}$$

Deflection due to variable loading (q_k) = 6 kN m^{-1} = 6 N mm^{-1}.
Hence

$$\delta_2 = \frac{5wL^4}{384EI} = \frac{5 \times 6 \times (8 \times 10^3)^4}{384 \times 210 \times 10^3 \times 14\,200 \times 10^4}$$

$$= 11 \text{ mm} < \frac{\text{span}}{350} = \frac{8 \times 10^3}{350} = 22 \text{ mm} \quad \text{OK}$$

Example 9.2 Design of a laterally restrained beam (EC3)

Select and check a suitable beam section using grade Fe 360 steel to support the loads shown below. Assume beam is fully laterally restrained and that it sits on 125 mm bearings at each end. Ignore self-weight of beam.

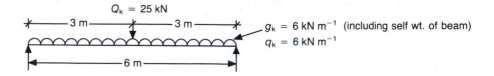

DESIGN BENDING MOMENT

Design action $(F_d) = (F_d)_{udl} + (F_d)_{pl}$

$$= (\gamma_G g_k + \gamma_Q q_k) \times span + \gamma_Q Q_k$$

$$= (1.35 \times 6 + 1.5 \times 6)6 + 1.5 \times 25$$

$$= 102.6 + 37.5$$

$$= 140.1 \text{ kN}$$

Design bending moment, M_{Sd}, is

$$M_{Sd} = \frac{(F_d)_{udl}\,l}{8} + \frac{(F_d)_{pl}\,l}{4} = \frac{102.6 \times 6}{8} + \frac{37.5 \times 6}{4}$$

$$= 133.2 \text{ kN m}$$

SECTION SELECTION

(Refer to 'Resistance of cross-sections – bending moments'). Assume suitable section belongs to class 1. Hence the minimum required plastic moment of resistance, W_{pl}, is given by

$$W_{pl} = \frac{M_{pl.Rd}\gamma_{M0}}{f_y}$$

Putting $M_{pl.Rd} = M_{Sd}$ and $f_y = 235 \text{ N mm}^{-2}$ (*Table 9.3*) gives

$$W_{pl} \geqslant \frac{133.2 \times 10^6 \times 1.05}{235} = 595.1 \times 10^3 \text{ mm}^3 = 595.1 \text{ cm}^3$$

From the steel table, try $356 \times 171 \times 45 \text{ kg m}^{-1}$ UB ($W_{pl} = 774 \text{ cm}^3$).

CHECK STRENGTH CLASSIFICATION

Flange thickness $(t_f) = 9.7$ mm. Hence from *Table 9.3*, $f_y = 235 \text{ N mm}^{-2}$ as assumed.

CHECK SECTION CLASSIFICATION

$$\varepsilon = (235/f_y)^{0.5} = (235/235)^{0.5} = 1$$

$$\frac{c}{t_f} = 8.81 < 10\varepsilon = 10 \quad \text{and}$$

$$\frac{d}{t_w} = 45.3 < 72\varepsilon = 72$$

Hence from *Table 9.4*, section belongs to class 1 as assumed.

RESISTANCE OF CROSS-SECTION

Bending moment

Plastic moment of resistance of $356 \times 171 \times 45$ UB is given by

$$M_{\text{pl.Rd}} = \frac{W_{\text{pl}} f_y}{\gamma_{\text{M0}}} = \frac{774 \times 10^3 \times 235}{1.05}$$

$$= 173 \times 10^6 \text{ N mm} = 173 \text{ kN m} > M_{\text{Sd}} \ (= 133.2 \text{ kN m}) \quad \text{OK}$$

Shear

Design shear force, $V_{\text{Sd}} = \dfrac{F_d}{2} = \dfrac{140.1}{2} = 70$ kN

For class 1 section, design plastic shear resistance, $V_{\text{pl.Rd}}$, is given by

$$V_{\text{pl.Rd}} = A_v(f_y/\sqrt{3})/\gamma_{\text{M0}}$$

where

$$A_v = 1.04 h t_w = 1.04 \times 352 \times 6.9 = 2526 \text{ mm}^2$$

Hence,

$$V_{\text{pl.Rd}} = 2526\,(235/\sqrt{3})/1.05$$

$$= 326.4 \times 10^3 \text{ N} = 326 \text{ kN} > V_{\text{Sd}} \ (= 70 \text{ kN}) \quad \text{OK}$$

Bending and shear

$$V_{\text{Sd}} \ (= 70 \text{ kN}) \not> 0.5 V_{\text{pl.Rd}} = 0.5 \times 326 = 163 \text{ kN}$$

Hence no check is required.

SHEAR BUCKLING RESISTANCE

As $d/t_w = 45.3 < 69\varepsilon = 69$, no check on shear buckling is required.

FLANGE-INDUCED BUCKLING

Area of web is

$$A_w = (h - 2t_f)t_w = (352 - 2 \times 9.7)6.9 = 2294.9 \text{ mm}^2$$

Area of compression flange is

$$A_{fc} = b t_f = 171 \times 9.7 = 1658.7 \text{ mm}^2$$

$$k(E/f_{yf})[A_w/A_{fc}]^{1/2} = 0.3(210 \times 10^3/235)[2294.9/1658.7]^{1/2}$$

$$= 315.3 > d/t_w = 45.3 \quad \text{OK}$$

RESISTANCE OF THE WEB TO TRANSVERSE FORCES

Since loads are resisted by shear in the web, check web for (a) crushing resistance and (b) crippling resistance.

Crushing resistance

For a rolled I-section

$$s_y = \frac{2.5(h - d)(1 - (\sigma_{f.Ed}/f_{yf})^2)^{1/2}}{(1 + 0.8 s_s/(h - d))}$$

$$= \frac{2.5\,(352 - 312.3)\,(1 - 0)^{1/2}}{1 + 0.8 \times 125/(352 - 312.3)} = 28 \text{ mm}$$

Clause 5.7.3(3) of EC3 recommends that s_y should be halved at the ends of members. Hence, design crushing resistance of web, $R_{y.Rd}$, is given by

$$R_{y.Rd} = (s_s + s_y) t_w f_{yw}/\gamma_{M1}$$
$$= (125 + 14)6.9 \times 235/1.05$$
$$= 214.6 \times 10^3 \text{ N}$$
$$= 214.6 \text{ kN} > V_{Sd} \text{ (= 70 kN)} \quad \text{OK}$$

Crippling resistance

Design crippling resistance of web, $R_{a.Rd}$, is given by

$$R_{a.Rd} = 0.5 t_w^2 (E f_{yw})^{1/2} [(t_f/t_w)^{1/2} + 3(t_w/t_f)(s_s/d)]/\gamma_{M1}$$
$$= 0.5 \times 6.9^2 (210 \times 10^3 \times 235)^{1/2} [(9.7/6.9)^{1/2} + 3(6.9/9.7)(125/312.3)]/1.05$$
$$= 324\,873 \text{ N}$$
$$= 324.8 \text{ kN} > V_{Sd} \text{ (= 70 kN)} \quad \text{OK}$$

Hence section satisfies web crippling requirements.

DEFLECTION

Deflection due to variable uniformly distributed loading $q_k = 4 \text{ kN m}^{-1} = 4 \text{ N mm}^{-1}$ and variable point load, $Q_k = 25\,000 \text{ N}$ is given by

$$\delta_2 = \frac{5 q_k L^4}{384 EI} + \frac{Q_k L^3}{48 EI}$$

Hence

$$\delta_2 = \frac{5 \times 4 \times (6 \times 10^3)^4}{384 \times 210 \times 10^3 \times 12\,100 \times 10^4} + \frac{25 \times 10^3 \times (6 \times 10^3)^3}{48 \times 210 \times 10^3 \times 12\,100 \times 10^4}$$
$$= 2.6 + 4.4 \text{ mm}$$

$$= 7 \text{ mm} < \text{allowable} \left(\frac{\text{span}}{350} = \frac{6 \times 10^3}{350} = 17 \text{ mm} \right) \quad \text{OK}$$

Deflection due to permanent and variable loading, w and W, where

$$w = g_k + q_k = 4 + 4 = 8 \text{ kN m}^{-1} = 8 \text{ N mm}^{-1}$$
$$W = Q_k = 25 \text{ kN} = 25\,000 \text{ N}$$

is

$$\delta_{max} = \frac{5 \times 8 (6 \times 10^3)^4}{384 \times 210 \times 10^3 \times 12\,100 \times 10^4} + \frac{25 \times 10^3 (6 \times 10^3)^3}{48 \times 210 \times 10^3 \times 12\,100 \times 10^4}$$
$$= 5.3 + 4.4 \text{ mm}$$

$$= 9.7 \text{ mm} < \text{allowable} \left(\frac{\text{span}}{250} = \frac{6 \times 10^3}{350} = 24 \text{ mm} \right)$$

Example 9.3 Design of a cantilever beam (EC3)

A cantilever beam is needed to resist the loading shown below. Select a suitable UB section in grade Fe 430 steel to satisfy bending and shear criteria only.

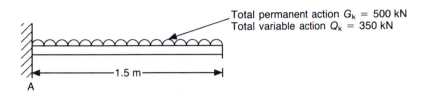

Total permanent action G_k = 500 kN
Total variable action Q_k = 350 kN

1.5 m

A

DESIGN BENDING MOMENT AND SHEAR FORCE

Design action $(F_d) = (\gamma_G G_k + \gamma_Q Q_k)$
$= (1.35 \times 500 + 1.5 \times 350)$
$= 1200$ kN

Design bending moment at A (M_{Sd}) is

$$M_{Sd} = \frac{F_d l}{2} = 1200 \times \frac{1.5}{2} = 900 \text{ kN m}$$

Design shear force at A $(V_{Sd}) = F_d = 1200$ kN

SECTION SELECTION

Since the cantilever beam is relatively short and subject to fairly high shear forces, the bending capacity or the shear strength of the section may be critical and both factors will need to be considered in order to select an appropriate section. Hence, plastic moment of section (assuming beam belongs to class 1), W_{pl}, must exceed the following to satisfy bending:

$$W_{pl} = \frac{M_{pl.Rd} \gamma_{M0}}{f_y} = \frac{900 \times 10^6 \times 1.05}{275}$$

$$= 3436 \times 10^3 \text{ mm}^2 = 3436 \text{ cm}^3$$

The shear area of the section, A_v, must exceed the following to satisfy shear:

$$A_v = \frac{V_{pl.Rd} \gamma_{M0}}{(f_y/\sqrt{3})} = \frac{1200 \times 10^3 \times 1.05}{(275/\sqrt{3})} = 7936 \text{ mm}^2$$

where $A_v = 1.04 h t_w$. Hence try $686 \times 254 \times 152$ kg m^{-1} UB section.

CHECK STRENGTH CLASSIFICATION

Flange thickness (t_f) = 21 mm, steel grade Fe 430. Hence from *Table 9.3*, f_y = 275 N mm^{-2} as assumed.

CHECK SECTION CLASSIFICATION

$$\varepsilon = (235/f_y)^{0.5} = (235/275)^{0.5} = 0.924$$

$$\frac{c}{t_f} = 6.06 < 10\varepsilon = 9.24$$

$$\frac{d}{t_w} = 46.6 < 72\varepsilon = 72 \times 0.924 = 66.5$$

Hence from *Table 9.4*, section belongs to class 1 as assumed.

RESISTANCE OF CROSS-SECTION

Bending moment

Plastic moment of resistance of $686 \times 254 \times 152$ UB is given by

$$M_{\text{pl.Rd}} = \frac{W_{\text{pl}} f_y}{\gamma_{\text{M0}}} = \frac{5000 \times 10^3 \times 275}{1.05}$$

$$= 1309 \times 10^6 \text{ N mm} > M_{\text{Sd}} \ (= 900 \text{ kN m}) \quad \text{OK}$$

Shear

For class 1 section, design plastic shear resistance is given by

$$V_{\text{pl.Rd}} = A_v (f_y / \sqrt{3}) / \gamma_{\text{M0}}$$

where

$$A_v = 1.04 h t_w = 1.04 \times 687.6 \times 13.2 = 9439 \text{ mm}^2$$

Hence

$$V_{\text{pl.Rd}} = 9439 (275/\sqrt{3})/1.05 = 1427.3 \times 10^3 \text{ N} = 1427.3 \text{ kN} > V_{\text{Sd}} \ (= 1200 \text{ kN}) \quad \text{OK}$$

Bending and shear

Since

$$V_{\text{Sd}} = 1200 \text{ kN} > 0.5 V_{\text{pl.Rd}} = 0.5 \times 1427.3 = 713.7 \text{ kN}$$

the section is subject to a 'high shear load' and the design moment of resistance of the section should be reduced to $M_{\text{V.Rd}}$ which is given by

$$M_{\text{V.Rd}} = f_y (W_{\text{pl}} - \rho A_v^2 / 4 t_w) / \gamma_{\text{M0}}$$

Where

$$\rho = (2V_{\text{Sd}} / V_{\text{pl.Rd}} - 1)^2 = (2 \times 1200/1427.3 - 1)^2 = 0.464$$

Hence

$$M_{\text{V.Rd}} = 275 \ (5000 \times 10^3 - 0.464 \times 9439^2 / 4 \times 13.2)/1.05$$

$$= 1104 \times 10^6 = 1104 \text{ kN m} > M_{\text{Sd}} \ (= 900 \text{ kN m}) \quad \text{OK}$$

Selected UB section, $686 \times 254 \times 152$ kg m^{-1}, is satisfactory in bending and shear.

Example 9.4 Design of a fully laterally restrained beam with overhang (EC3)

The figure shows a simply supported beam and cantilever with uniformly distributed loads applied to it. Using grade Fe 430 steel and assuming full lateral restraint, select and check a suitable beam section.

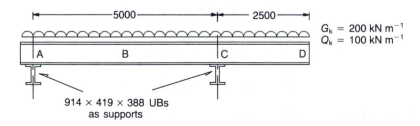

DESIGN BENDING MOMENT AND SHEAR FORCE

Load cases

(a) $1.35G_k + 1.5Q_k$ on span AC, $1.35G_k + 1.5Q_k$ on span CD;
(b) $1.35G_k + 1.5Q_k$ on span AC, $1.0G_k + 0.0Q_k$ on span CD;
(c) $1.0G_k + 0.0Q_k$ on span AC, $1.35G_k + 1.5Q_k$ on span CD.

Shear and bending

Shear force and bending moment diagrams for the three load cases are shown below. Hence $V_{Sd} = 1313$ kN and $M_{Sd} = 1313$ kN m.

STATIC EQUILIBRIUM

Overturning moment = $(1.35 \times 200 + 1.5 \times 100)\ 2.5 \times 1.25 = 1312.5$ kN m
Restorative moment = $(1.1 \times 200)\ 5 \times 2.5 = 2750$ kN m > 1312.5 kN m   OK

SECTION SELECTION

$$W_{pl} = \frac{M_{pl.Rd}\gamma_{M0}}{f_y} = \frac{1313 \times 10^6 \times 1.05}{275} = 5.01 \times 10^6 \text{ mm}^3$$

Try $762 \times 267 \times 173$ UB section.

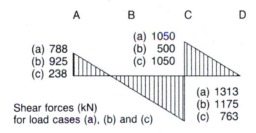

	A	B	C	D

(a) 788   (a) 1050
(b) 925   (b) 500
(c) 238   (c) 1050

Shear forces (kN)
for load cases (a), (b) and (c)

(a) 1313
(b) 1175
(c) 763

(a) 1313   (a) 1313
(b) 1313   (b) 625
(c) 625   (c) 1313

Bending moments (kN m)
for load cases (a), (b) and (c)

STRENGTH CLASSIFICATION

Flange thickness = 21.6 mm, steel grade Fe 430. Hence from *Table 9.3*, $f_y = 275$ N mm^{-2}.

SECTION CLASSIFICATION

$$\varepsilon = (235/f_y)^{1/2} = 0.924$$

$$\frac{c}{t_f} = 6.17 < 10\varepsilon = 9.24$$

$$\frac{d}{t_w} = 48 < 72\varepsilon = 66.5$$

Hence from *Table 9.4*, section belongs to class 1.

RESISTANCE OF CROSS-SECTIONS
Shear resistance

$$V_{pl.Rd} = A_v(f_y/\sqrt{3})/\gamma_{M0}$$

where

$$A_v = 1.04ht_w = 1.04 \times 762 \times 14.3 = 11\ 332\ \text{mm}^2$$

$$V_{pl.Rd} = \frac{11\ 332 \times 275}{\sqrt{3} \times 1.05}$$

$$= 1713 \times 10^3\ \text{N}$$

$$= 1713\ \text{kN} > V_{Sd}\ (= 1313\ \text{kN})$$

Bending and shear

$$V_{Sd} = 1313\ \text{kN} = 0.76V_{pl.Rd} > 0.5V_{pl.Rd}$$

Hence beam is subject to high shear load.

$$M_{V.Rd} = f_y(W_{pl} - \rho A_v^2/4t_w)/\gamma_{M0}$$

where

$$\rho = (2V_{Sd}/V_{pl.Rd} - 1)^2 = (2 \times 1313/1713 - 1)^2 = 0.284$$

$$M_{V.Rd} = 275(6200 \times 10^3 - 0.284 \times 11\ 332^2/4 \times 14.3)/1.05$$

$$= 1456 \times 10^6\ \text{N mm}$$

$$= 1456\ \text{kN m} > M_{Sd}\ (= 1313\ \text{kN m})\quad \text{OK}$$

SHEAR BUCKLING RESISTANCE
As will be seen under 'Resistance of web to transverse forces', the web needs stiffeners. Hence, provided d/t_w for the web is less than $30\varepsilon\sqrt{k_\tau}$ no check is required for resistance to shear buckling.

$$k_\tau = 5.34 \quad \text{(assuming stiffeners needed only at supports)}$$

$$\varepsilon = 0.924$$

Hence

$$30\varepsilon\sqrt{k_\tau} = 30 \times 0.924 \times \sqrt{5.34} = 64 > d/t_w\ (= 48.0)\quad \text{OK}$$

FLANGE-INDUCED BUCKLING
Area of web is

$$A_w = (h - 2t_f)t_w = (762 - 2 \times 21.6)14.3 = 10\ 279\ \text{mm}^2$$

Area of compression flange is

$$A_{fc} = bt_f = 266.7 \times 21.6 = 5761\ \text{mm}^2$$

$$k(E/f_{yf})[A_w/A_{fc}]^{1/2} = 0.3 \times (210 \times 10^3/275)[10\ 279/5761]^{1/2}$$

$$= 306 > d/t_w = 48.0$$

Hence no check is required.

RESISTANCE OF WEB TO TRANSVERSE FORCES
No check on web buckling is required when concentrated loads are resisted by shear in the web. We must, however, check the crippling and crushing resistance.

Web crippling

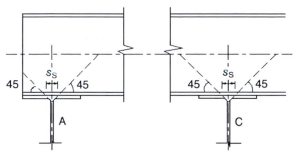

Stiff bearing length, $s_s = 2t_f + t_w + r = 2 \times 36.6 + 21.5 + 24.1 = 118.8$ mm (based on properties of $914 \times 419 \times 388$ UBs as supports)

Thus

$$R_{a.Rd} = 0.5 t_w^2 (E f_{yw})^{1/2} [(t_f/t_w)^{1/2} + 3(t_w/t_f)(s_s/d)]/\gamma_{M1}$$

$$= 0.5 \times 14.3^2 (210 \times 10^3 \times 275)^{1/2} [(21.6/14.3)^{1/2} + 3(14.3/21.6)(118.8/685.8)]/1.05$$

$$= 776\,994.7[1.573]/1.05$$

$$= 1164 \times 10^3 \text{ N} = 1164 \text{ kN} < F_{Sd} \quad (= 2363 \text{ kN at C, load case (a))} \quad \text{Not OK}$$

Where the member is also subject to bending moments, e.g. support C, clause 5.7.4(2) of EC3 recommends the following additional check:

$$\frac{F_{Sd}}{R_{a.Rd}} + \frac{M_{Sd}}{M_{c.Rd}} \leqslant 1.5 \qquad \frac{2363}{1164} + \frac{1313}{1456} = 3 > 1.5$$

Hence this is also not satisfied.

Web crushing

Support A

$$\sigma_{f.Ed} = 0$$

$$s_y = \frac{2.5\,(h-d)[1 - (\sigma_{f.Ed}/f_{yf})^2]^{1/2}}{(1 + 0.8 s_s/(h-d))}$$

$$= \frac{2.5\,(762 - 685.8)[1 - 0]^{1/2}}{(1 + 0.8 \times 118.8/76.2)} = 84.8 \text{ mm}$$

Hence

$$R_{y.Rd} = (s_s + s_y) t_w f_{yw}/\gamma_{M1}$$

$$= (118.8 + 84.8/2)14.3 \times 275/1.05$$

$$= 603.7 \times 10^3 \text{ N}$$

$$= 603 \text{ kN} \,(< 925 \text{ kN}) \quad \text{Not OK}$$

Support C

$$\sigma_{f.Ed} = \frac{M_{Sd}(h/2)}{I} = \frac{1313 \times 10^6 \times (762/2)}{2.05 \times 10^9} = 244 \text{ N mm}^{-2}$$

$$s_y = \frac{2.5\,(762 - 685.8)[1 - (244/275)^2]^{1/2}}{(1 + 0.8 \times 118.8/76.2)} = 39.1 \text{ mm}$$

Hence

$$R_{y.Rd} = (118.8 + 39.1)14.3 \times 275/1.05$$

$$= 591.3 \times 10^3 \text{ N} = 591 \text{ kN } (< 2363 \text{ kN}) \quad \text{Not OK}$$

Stiffeners are therefore required at A and C with the stiffener at C being the more critical.

Stiffener design at C

(Refer to clause 5.7.6 of EC3 and *section 9.12.1*)
Where loadbearing stiffeners are provided, e.g. at A and C, the effective cross-section of stiffener/web needs to be checked for buckling and crushing resistance.

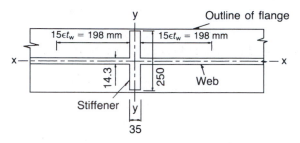

Buckling resistance. When checking the buckling resistance, the effective cross-section of a stiffener should be taken as including a width of web plate equal to $30\varepsilon t_w$, arranged with

$$15\varepsilon t_w = 15 \times (235/275)^{0.5}14.3 = 198 \text{ mm each side of the stiffener.}$$

$$\text{Buckling length, } l \geqslant 0.75d = 0.75 \times 685.8 = 514 \text{ mm}$$

Radius of gyration of stiffened section, i_x, is

$$i_x = (I/A)^{1/2} = [(1/12)35 \times 250^3/(2 \times 198 \times 14.3 + 250 \times 35)]^{1/2}$$

$$= (45\,572\,916.6/14\,412.8)^{1/2} = 56.2 \text{ mm}$$

$$\lambda = l/i = 514/56.2 = 9.15$$

$$\lambda_1 = 93.9\varepsilon = 93.9 \times 0.924 = 86.8$$

$$\bar{\lambda} = \lambda/\lambda_1[\beta_A]^{1/2} = 9.15/86.8 = 0.105$$

where $\beta_A = 1$ (for class 1 section). By substituting these values into equations 9.38 and 9.39 (*section 9.12.1*) and putting $\alpha = 0.49$ gives $\chi_c = 1.0$. Hence, design buckling resistance, $N_{b.Rd}$, is given by (equation 9.37, *section 9.12.1*)

$$N_{b.Rd} = \chi_c\beta_A A_s f_y/\gamma_{M1} = 1 \times 1 \times 14\,412.8 \times 275/1.05$$

$$= 3.77 \times 10^6 \text{ N} = 3774 \text{ kN } (> 2363 \text{ kN}) \quad \text{OK}$$

Crushing resistance

$$\sigma_{f.Ed} = \frac{M_{Sd}(h/2)}{I} = \frac{1313 \times 10^6 \times (762/2)}{2.05 \times 10^9} = 244 \text{ N mm}^{-2}$$

$$s_y = \frac{2.5\,(762 - 685.8)[1 - (244/275)^2]^{1/2}}{(1 + 0.8 \times 118.8/76.2)} = 39.1 \text{ mm}$$

Area of stiffener cross-section, A, is given by

$$A = (118.8 + 39.1)14.3 + 35(250 - 14.3) = 10\,507 \text{ mm}^2$$

Stiffener resistance, $N_{c.Rd}$, is given by

$$N_{c.Rd} = A f_y/\gamma_{M0} = 10\,507 \times 275/1.05 = 2751 \times 10^3 > 2363 \text{ kN} \quad \text{OK}$$

Hence double-sided stiffener made of two plates 118×35 is suitable at support C. A similar stiffener should be provided at support A.

DEFLECTION
For a simply supported span

$$\delta_2 = \frac{5wL^4}{384EI}$$

where w is the unfactored variable load = 100 kN m^{-1} = 100 N mm^{-1},

$$\delta_2 = \frac{5 \times 100 \times 5000^4}{384 \times 210 \times 10^3 \times 2.05 \times 10^9}$$

$$= 2 \text{ mm} = L/2580 \quad (< L/350) \quad \text{OK}$$

$$\delta_{max} = \frac{5w'L^4}{384EI}$$

where w', the total unfactored permanent and variable load, is $100 + 200 = 300$ kN m^{-1} = 300 N mm^{-1}. Hence

$$\delta_{max} = 6 \text{ mm} = L/833 \quad (< L/250) \quad \text{OK}$$

For a cantilever

$$\delta_2 = \frac{wL^4}{8EI} = \frac{100 \times 2500^4}{8 \times 210 \times 10^3 \times 2.05 \times 10^9}$$

$$= 1.1 \text{ mm} = L/2204 \quad (< L/175) \quad \text{OK}$$

$$\delta_{max} = 3.4 \text{ mm} = L/737 \quad (< L/125) \quad \text{OK}$$

9.11.2 LATERAL TORSIONAL BUCKLING OF BEAMS (CLAUSE 5.5.2, EC3)

In order to prevent the possibility of a beam failure due to lateral torsional buckling, the designer needs to ensure that the buckling resistance, $M_{b.Rd}$ exceeds the design moment, M_{Sd}, i.e.

$$M_{b.Rd} \geq M_{Sd} \tag{9.26}$$

This requires the designer to calculate the values of the following additional parameters:

1. geometrical slenderness ratio λ_{LT}
2. slenderness ratio $\bar{\lambda}_{LT}$
3. buckling factor χ_{LT}
4. buckling resistance $M_{b.Rd}$

9.11.2.1 Geometrical slenderness ratio λ_{LT} (Annex F, clause F.2.2, EC3)

The geometrical slenderness ratio is given by

$$\lambda_{LT} = \frac{L(W^2_{pl.y}/I_z I_y)^{1/4}}{C_1^{1/2}[1 + (L^2 GI_t/\pi^2 EI_w)]^{1/4}} \tag{9.27}$$

where L is the length of beam between points which have lateral restraint, $W_{pl.y}$ the plastic modulus about the major axis, I_z the second moment of area about the minor axis, I_y the second moment of area about the major axis, I_t the torsional constant, I_w

the warping constant, E the modulus of elasticity (210 000 N mm^{-2}), G the shear modulus equal to

$$\frac{E}{2(1+\upsilon)} = \frac{210\ 000}{2(1+0.3)} = 80\ 769 \text{ N mm}^{-2}$$

and C_1 the factor depending on the loading and end restraint conditions as indicated by Ψ and k (Table 9.6). Note that Ψ is the ratio of end moments over the length L between lateral restraints, and the effective length factor k varies from 0.5 for full fixity to 1.0 for no fixity, with 0.7 for one end fixed and one end free.

9.11.2.2 Slenderness ratio $\bar{\lambda}_{LT}$
The value of $\bar{\lambda}_{LT}$ may be determined from

$$\bar{\lambda}_{LT} = (\lambda_{LT}/\lambda_1)(\beta_W)^{1/2} \tag{9.28}$$

where

$$\lambda_1 = \pi(E/f_y)^{1/2} = 93.9\varepsilon \tag{9.29}$$

where $\varepsilon = (235/f_y)^{1/2}$, $\beta_W = 1$ for class 1 and 2 sections, $\beta_W = W_{el.y}/W_{pl.y}$ for class 3 cross-sections and $\beta_W = W_{eff.y}/W_{pl.y}$ for class 4 cross-sections.

9.11.2.3 Buckling factor χ_{LT} (clause 5.5.2, EC3)
Here χ_{LT} is the reduction factor for lateral torsional buckling which is given by

Table 9.6 Values of factors C_1, C_2, and C_3 corresponding to various k factors and end moment or transverse loading combinations (based on Tables F.1.1 & F.1.2, EC3)

Loading and support conditions	Bending moment diagram	Value of k	Values of factors		
			C_1	C_2	C_3
(a) $\psi = +1$		1.0	1.000		1.000
		0.7	1.000	–	1.113
		0.5	1.000		1.144
(b) $\psi = 0$		1.0	1.879		0.939
		0.7	2.092	–	1.473
		0.5	2.150		2.150
(c) $\psi = -1$		1.0	2.752		0.000
		0.7	3.063	–	0.000
		0.5	3.149		0.000
(d)		1.0	1.132	0.459	0.525
		0.5	0.972	0.304	0.980
(e)		1.0	1.285	1.562	0.753
		0.5	0.712	0.652	1.070
(f)		1.0	1.365	0.553	1.730
		0.5	1.070	0.432	3.050
(g)		1.0	1.565	1.267	2.640
		0.5	0.938	0.715	4.800
(h)		1.0	1.046	0.430	1.120
		0.5	1.010	0.410	1.890

$$\chi_{LT} = \frac{1}{\phi_{LT} + (\phi_{LT}^2 - \bar{\lambda}_{LT}^2)^{1/2}} \leqslant 1 \quad (9.30)$$

in which

$$\phi_{LT} = 0.5(1 + \alpha_{LT}(\bar{\lambda}_{LT} - 0.2) + \bar{\lambda}_{LT}^2) \quad (9.31)$$

The imperfection factor, α_{LT}, for lateral torsional buckling should be taken as 0.21 for rolled sections, more for welded sections.

9.11.2.4 Buckling resistance $M_{b.Rd}$ (clause 5.5.2, EC3)

The design buckling resistance moment of a laterally unrestrained beam is given by

$$M_{b.Rd} = \chi_{LT} \beta_w W_{pl.y} f_y / \gamma_{M1} \quad (9.32)$$

where f_y is the yield strength (*Table 9.4*) and γ_{M1} the partial safety factor for buckling = 1.05 (*section 9.9.3*).

Example 9.5 Analysis of a beam restrained at the supports (EC3)

Assuming that the beam in *Example 9.1* is only laterally and torsionally restrained at the supports, determine whether a $356 \times 171 \times 51$ kg m^{-1} UB section in grade Fe 430 steel is still suitable.

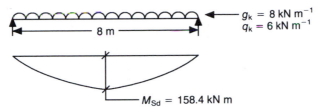

$g_k = 8$ kN m^{-1}
$q_k = 6$ kN m^{-1}

8 m

$M_{Sd} = 158.4$ kN m

SECTION PROPERTIES

From steel tables (*Appendix B*)

Depth of section, h	$= 355.6$ mm
Thickness of flange, t_f	$= 11.5$ mm
Second moment of area about the minor axis, I_z	$= 968 \times 10^4$ mm^4
Radius of gyration about z–z axis, i_z	$= 38.7$ mm
Elastic modulus about the major axis, $W_{el.y}$	$= 796 \times 10^3$ mm^3
Plastic modulus about the major axis, $W_{pl.y}$	$= 895 \times 10^3$ mm^3
Warping constant, I_w	$= 286 \times 10^9$ mm^6
Torsional constant, I_t	$= 236 \times 10^3$ mm^4
Yield strength of Fe 430 steel, f_y	$= 275$ N mm^{-2}

$$\text{Shear modulus, } G = \frac{E}{2(1+\upsilon)} = \frac{210 \times 10^3}{2(1+0.3)} \quad = 80\ 769 \text{ N mm}^{-2}$$

LATERAL BUCKLING RESISTANCE

Geometrical slenderness ratio, λ_{LT}, for lateral torsional buckling

The effective length factor $k = 1.0$ since compression flange is laterally unrestrained. From *Table 9.6*, factor $C_1 = 1.132$. Length of beam between points which are laterally restrained, $L = 8000$ mm. Buckling slenderness, λ_{LT}, is

$$\lambda_{LT} = \frac{L\left[\dfrac{W_{pl.y}^2}{I_z I_w}\right]^{1/4}}{C_1^{1/2}\left[1 + \dfrac{L^2 G I_t}{\pi^2 E I_w}\right]^{1/4}}$$

$$= \frac{8000\left[\dfrac{(895 \times 10^3)^2}{968 \times 10^4 \times 286 \times 10^9}\right]^{1/4}}{1.132^{1/2}\left[1 + \dfrac{8000^2 \times 80\ 769 \times 236 \times 10^3}{3.142^2 \times 210 \times 10^3 \times 286 \times 10^9}\right]^{1/4}} = 131.87$$

Slenderness ratio $\bar{\lambda}_{LT}$ for lateral torsional buckling

$$\beta_W = 1 \quad \text{(for class 1 cross-section)}$$

$$\varepsilon = (235/f_y)^{1/2} = (235/275)^{1/2} = 0.924$$

$$\lambda_1 = 93.9\varepsilon = 93.9 \times 0.924 = 86.8$$

$$\bar{\lambda}_{LT} = (\lambda_{LT}/\lambda_1)\beta_W^{1/2} = (131.87/86.8)1^{1/2} = 1.52$$

Buckling factor χ_{LT}

Imperfection factor $\alpha_{LT} = 0.21$ for rolled sections (clause 5.5.2(3) of EC3). Buckling factor, ϕ_{LT}, is

$$\phi_{LT} = 0.5[1 + \alpha_{LT}(\bar{\lambda}_{LT} - 0.2) + \bar{\lambda}_{LT}^2]$$
$$= 0.5[1 + 0.21(1.52 - 0.2) + 1.52^2] = 1.794$$

$$\chi_{LT} = \frac{1}{\phi_{LT} + [\phi_{LT}^2 - \bar{\lambda}_{LT}^2]^{1/2}} \leqslant 1$$

$$= \frac{1}{1.794 + [1.794^2 - 1.52^2]^{1/2}} = 0.364 < 1 \quad \text{OK}$$

Buckling resistance

Buckling resistance of beam, $M_{b.Rd}$ is

$$M_{b.Rd} = \chi_{LT}\beta_w W_{pl.y} f_y/\gamma_{M1}$$
$$= 0.364 \times 1 \times 895 \times 10^3 \times 275/1.05$$
$$= 85.3 \times 10^6 \text{ N mm} = 85.3 \text{ kN m} < M_{Sd} = 158.4 \text{ kN m} \quad (\textit{Example 9.1})$$

Since buckling resistance of beam (85.3 kN m) < design moment (158.4 kN m), the beam section is unsuitable.

Example 9.6 Analysis of a beam restrained at mid-span and supports (EC3)

Repeat *Example 9.5*, but this time assume that the beam is laterally and torsionally restrained at mid-span and at the supports.

SECTION PROPERTIES
See *Example 9.5*.

LATERAL BUCKLING RESISTANCE
$k = 1.0$ (since compression flange unrestrained between supports and mid-span)

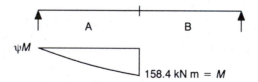

Ratio of end moments for spans A and B is Ψ, where $\Psi = 0/158.4 = 0$. Hence from *Table 9.6(b)*, i.e. $\Psi = 0$ and $k = 1$, $C_1 = 1.879$. Length of beam between points which are laterally restrained, $L = 4000$ mm.

$$\lambda_{LT} = \frac{L\left[\dfrac{W_{pl.y}^2}{I_z I_w}\right]^{1/4}}{C_1^{1/2}\left[1 + \dfrac{L^2 GI_t}{\pi^2 EI_w}\right]^{1/4}} = \frac{4000\left[\dfrac{(895 \times 10^3)^2}{968 \times 10^4 \times 286 \times 10^9}\right]^{1/4}}{1.879^{1/2}\left[1 + \dfrac{(4 \times 10^3)^2 \times 80\,769 \times 236 \times 10^3}{\pi^2 \times 210 \times 10^3 \times 286 \times 10^9}\right]^{1/4}} = 61.0$$

Note that clause F. 2.2.(4) of EC3 permits the following conservative approximation to be used to determine λ_{LT} for rolled I- or H-sections.

$$\lambda_{LT} = \frac{0.9L/i_z}{C_1^{1/2}\left[1 + \frac{1}{20}\left[\frac{L/i_z}{h/t_f}\right]^2\right]^{1/4}} \qquad (9.33)$$

$$= \frac{0.9 \times 4 \times 10^3/38.7}{1.879^{1/2}\left[1 + \frac{1}{20}\left[\frac{4 \times 10^3/38.7}{355.6/11.5}\right]^2\right]^{1/4}}$$

$$= 60.7$$

$$\bar{\lambda}_{LT} = (\lambda_{LT}/\lambda_1)\beta_W^{1/2} = (61/93.9 \times 0.924)1^{1/2} = 0.70$$

$$\alpha_{LT} = 0.21$$

$$\phi_{LT} = 0.5[1 + \alpha_{LT}(\bar{\lambda}_{LT} - 0.2) + \bar{\lambda}_{LT}^2]$$

$$= 0.5[1 + 0.21(0.70 - 0.2) + 0.70^2] = 0.80$$

Buckling factor

$$\chi_{LT} = \frac{1}{\phi_{LT} + [\phi_{LT}^2 - \bar{\lambda}_{LT}^2]^{1/2}} \leq 1$$

$$= \frac{1}{0.80 + [0.80^2 - 0.70^2]^{1/2}} = 0.84$$

Buckling resistance of beam, $M_{b.Rd}$, is

$$M_{b.Rd} = \chi_{LT}\beta_W W_{pl.y}f_y/\gamma_{M1} = 0.84 \times 1 \times 895 \times 10^3 \times 275/1.05$$

$$= 196.9 \times 10^6 \text{ N mm} = 196.9 \text{ kN m} > M_{Sd} = 158.4 \text{ kN m}$$

Hence the beam section is now suitable.

9.12 Design of columns

The design of columns is largely covered in Chapter 5 of EC3. However, unlike BS 5950, EC3 does not include the design of cased columns, which is left to Part 1.1 of Eurocode 4: *Design of Composite Steel and Concrete Structures*. EC4 is due to be published as an EN in 2004–2005 and it is reported that the design procedures for cased columns in EC4 and BS 5950 will be similar.

The following subsections will consider EC3 requirements in respect of the design of:

1. compression members;
2. members resisting combined axial force and moment;
3. simple column baseplates.

9.12.1 COMPRESSION MEMBERS

According to clause 5.1.4 of EC3, compression members (i.e. struts) should be checked for (1) resistance to compression failure and (2) resistance to buckling.

9.12.1.1 Resistance of cross-sections – compression (clause 5.4.4, EC3)

For members in axial compression, the design value of compressive force N_{Sd} at each cross-section should satisfy

$$N_{Sd} \leq N_{c.Rd} \qquad (9.34)$$

where $N_{c.Rd}$ is the design compressive resistance of cross-section, taken as the smaller of:

1. the design plastic compressive resistance of the gross section

$$N_{pl.Rd} = Af_y/\gamma_{M0} \qquad (9.35)$$

(for classes 1–3 cross-sections);
2. the design local buckling resistance of the gross section

Table 9.7 Imperfection factors (Table 5.5.1, EC3)

Buckling curve	a	b	c	d
Imperfection factor α	0.21	0.34	0.49	0.76

$$N_{o.Rd} = A_{eff}f_y/\gamma_{M1} \qquad (9.36)$$

(for class 4 cross-sections), where A_{eff} is the effective area of section.

9.12.1.2 Buckling resistance of members (clause 5.5.1, EC3)

The design buckling resistance of a compression member should be taken as

$$N_{b.Rd} = \chi\beta_A Af_y/\gamma_{M1} \qquad (9.37)$$

where $\beta_A = 1$ for classes 1–3 cross-sections, $\beta_A = A_{eff}/A$ for class 4 cross-sections and χ is the reduction factor for the relevant buckling mode, determined as follows or from EC3, Table 5.5.2 (not reproduced here).

For uniform members subjected to axial compression

$$\chi = \frac{1}{\phi + \left[\phi^2 - \bar{\lambda}^2\right]^{1/2}} \leq 1 \qquad (9.38)$$

where

$$\phi = 0.5(1 + \alpha(\bar{\lambda} - 0.2) + \bar{\lambda}^2) \qquad (9.39)$$

and where α is an imperfection factor from *Table 9.7*.

$$\bar{\lambda} = (\beta_A Af_y/N_{cr})^{1/2} = (\lambda/\lambda_1)(\beta_A)^{1/2} \qquad (9.40)$$

where λ is the slenderness for the relevant buckling mode (see below).

$$\lambda_1 = \pi(E/f_y)^{1/2} = 93.9\varepsilon \qquad (9.41)$$

where $\varepsilon = (235/f_y)^{1/2}$

Table 9.8 indicates which of the buckling curves is to be used.

The slenderness of a member, λ, is given by

$$\lambda = l/i \qquad (9.42)$$

where l is the buckling length of the member which can be obtained using the assumption in clause 5.5.1.5 of EC3 or from Annex E of EC3, and i the radius of gyration about the relevant axis. Note that buckling length in EC3 is the same as effective length in BS 5950. Clause 5.5.1.5 of EC3 states that the buckling length of a member may conservatively be taken as equal to its system (actual) length provided that both ends of the member are effectively held in position laterally.

Annex E – 'Buckling length of a compression member' goes into considerably more detail than BS 5950 on this subject. Essentially the buckling length is obtained from charts after calculation of distribution factors η_1 and η_2 based on member stiffness coefficients for appropriate subframe models. The buckling lengths are, however, identical to BS 5950 effective lengths, and could therefore be obtained from BS 5950, Table 22 (*Table 4.15*).

Table 9.8 Selection of buckling curve for a cross section (Table 5.5.3, EC3)

Cross-section	Limits		Buckling about axis	Buckling curve
Rolled I-sections	$h/b > 1.2$			
		$t_f \leq 40$ mm	y–y	a
			z–z	b
		40 mm $< t_f \leq 100$ mm	y–y	b
			z–z	c
	$h/b \leq 1.2$			
		$t_f \leq 100$ mm	y–y	b
			z–z	c
		$t_f > 100$ mm	y–y	d
			z–z	d

9.12.2 MEMBERS SUBJECT TO COMBINED AXIAL FORCE AND MOMENT

Clause 5.1.6 of EC3 states that members subject to combined axial (compression) force and moment should be checked for the following ultimate limit states: (1) resistance of cross-sections to the combined effects and (2) buckling resistance of member to the combined effects.

9.12.2.1 Resistance of cross-sections – bending and axial force (clause 5.4.8, EC3)

Classes 1 and 2 cross-sections. For classes 1 and 2 cross-sections the criterion to be satisfied in the absence of shear force is

$$M_{Sd} \leq M_{N.Rd} \qquad (9.43)$$

where $M_{N.Rd}$ is the reduced design plastic resistance moment allowing for the axial force. No reduction is necessary provided that the axial force does not exceed half the plastic tension resistance of the web, or a quarter of the plastic tension resistance of the whole cross-section, whichever is the smaller.

For larger axial loads the following approximations can be used for standard rolled I- and H-sections:

$$M_{Ny} = M_{pl.y}(1-n)/(1-0.5a) \leq M_{pl.y} \qquad (9.44)$$

$$M_{Nz} = M_{pl.z}\left[1-\left[\frac{n-a}{1-a}r\right]^2\right] \leq M_{pl.z} \qquad (9.45)$$

where

$$n = N_{Sd}/N_{pl.Rd} \qquad (9.46)$$

and

$$a = (A - 2bt_f)/A \leq 0.5 \qquad (9.47)$$

The above approximations may be further simplified for rolled I- and H-sections only to

$$M_{Ny} = 1.11 M_{pl.y}(1-n) \leq M_{pl.y} \qquad (9.48)$$

$$M_{Nz} = 1.56 M_{pl.z}(1-n)(n+0.6) \leq M_{pl.z} \qquad (9.49)$$

For biaxial bending, the following approximate criterion can be used:

$$\left[\frac{M_{y.Sd}}{M_{Ny.Rd}}\right]^\alpha + \left[\frac{M_{z.Sd}}{M_{Nz.Rd}}\right]^\beta \leq 1 \qquad (9.50)$$

in which $\alpha = 2$ and $\beta = 5n$ but ≥ 1 for I- and H-sections. As a further conservative approximation the following may be used:

$$\frac{N_{Sd}}{Af_y/\gamma_{M0}} + \frac{M_{y.Sd}}{W_{pl.y}f_y/\gamma_{M0}} + \frac{M_{z.Sd}}{W_{pl.z}f_y/\gamma_{M0}} \leq 1 \qquad (9.51)$$

Class 3 cross-sections. In the absence of a shear force, class 3 cross-sections will be satisfactory if the maximum longitudinal stress does not exceed the design yield strength, i.e.

$$\sigma_{x.Ed} \leq f_{yd} \qquad (9.52)$$

where $f_{yd} = f_y/\gamma_{M0}$. For cross-sections without fastener holes, this becomes

$$\frac{N_{Sd}}{Af_{yd}} + \frac{M_{y.Sd}}{W_y f_{yd}} + \frac{M_{z.Sd}}{W_z f_{yd}} \leq 1 \qquad (9.53)$$

Class 4 cross-sections. For class 4 cross-sections the above approach should also be used, but calculated using effective, rather than actual, widths of compression elements (clause 5.4.8.3 of EC3).

9.12.2.2 Buckling resistance of members – combined bending and axial compression (clause 5.5.4, EC3)

Again this is presented in a rather cumbersome manner in EC3. We will confine ourselves to the interaction formulae for classes 1 and 2 members:

$$\frac{N_{Sd}}{\chi_{min}Af_y/\gamma_{M1}} + \frac{k_y M_{y.Sd}}{W_{pl.y}f_y/\gamma_{M1}} + \frac{k_z M_{z.Sd}}{W_{pl.z}f_y/\gamma_{M1}} \leq 1 \qquad (9.54)$$

in which

$$k_y = 1 - \frac{\mu_y N_{Sd}}{\chi_y Af_y} \leq 1.5 \qquad (9.55)$$

$$\mu_y = \bar{\lambda}_y(2\beta_{My} - 4) + \left[\frac{W_{pl.y} - W_{el.y}}{W_{el.y}}\right] \leq 0.9 \qquad (9.56)$$

$$k_z = 1 - \frac{\mu_z N_{Sd}}{\chi_z Af_y} \leq 1.5 \qquad (9.57)$$

$$\mu_z = \bar{\lambda}_z(2\beta_{Mz} - 4) + \left[\frac{W_{pl.z} - W_{el.z}}{W_{el.z}}\right] \leq 0.9 \qquad (9.58)$$

Here χ_{min} is the lesser of χ_y and χ_z, which are the reduction factors for the y–y and z–z axes respectively; β_{My} and β_{Mz} are equivalent uniform moment factors for flexural buckling.

Members for which lateral torsional buckling is a potential failure mode should also satisfy

$$\frac{N_{Sd}}{\chi_z Af_y/\gamma_{M1}} + \frac{k_{LT} M_{y.Sd}}{\chi_{LT} W_{pl.y}f_y/\gamma_{M1}} + \frac{k_z M_{z.Sd}}{W_{pl.z}f_y/\gamma_{M1}} \leq 1 \qquad (9.59)$$

Moment diagram	Equivalent uniform moment factor β_M
End moments  M_1 ⟋⟍⟍ ψM_1 $-1 \leqslant \psi \leqslant 1$	$\beta_{M,\psi} = 1.8 - 0.7\psi$
Moments due to in-plane lateral loads 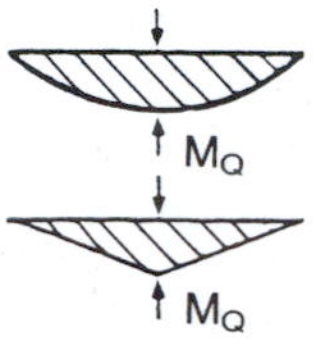 $\downarrow$ ⟍⟋ $\uparrow M_Q$ $\downarrow$ ⟍⟋ $\uparrow M_Q$	$\beta_{M,Q} = 1.3$ $\beta_{M,Q} = 1.4$

Fig. 9.6 *Equivalent uniform moment factors (based on Fig. 5.5.3 of EC3).*

in which

$$k_{LT} = 1 - \frac{\mu_{LT} N_{Sd}}{\chi_z A f_y} \leqslant 1 \qquad (9.60)$$

$$\mu_{LT} = 0.15 \bar{\lambda}_z \beta_{M.LT} - 0.15 \leqslant 0.90 \quad (9.61)$$

where $\beta_{M.LT}$ is an equivalent uniform moment factor for lateral torsional buckling and is obtained from *Fig. 9.6* according to the shape of the bending moment diagram between braced points.

9.12.3 SIMPLE COLUMN BASEPLATES (ANNEX L, CLAUSE L.1, EC3)

Generally, column baseplates should be checked to ensure (1) the bearing pressure does not exceed the design bearing strength of the foundations and (2) the bending moment in the compression or tension region of the baseplate does not exceed the resistance moment.

9.12.3.1 Bearing pressure and strength

The aim of design is to ensure that the bearing pressure does not exceed the bearing strength of the concrete, f_j, i.e.

$$\text{bearing pressure} = \frac{N_{Sd}}{A_{eff}} \leqslant f_j \qquad (9.62)$$

where N_{Sd} is the design axial force on column and A_{eff} the area in compression under the baseplate. The bearing strength of concrete foundations can be determined using

$$f_j = \beta_j k_j f_{cd} \qquad (9.63)$$

where f_{cd} is the design value of the concrete cylinder compressive strength of the foundation and is given by

$$f_{cd} = f_{ck}/\gamma_m \qquad (9.64)$$

and where β_j is the joint (concrete) coefficient which may generally be taken as two-thirds and k_j, the concentration factor, which may generally be taken as 1.0.

Figure 9.7 shows the effective bearing areas under axially loaded column baseplates. The additional bearing width x is given by

$$x = t[f_y/3f_j\gamma_{M0}]^{1/2} \qquad (9.65)$$

where t is the thickness of the baseplate, f_y the yield strength of baseplate material and f_j the bearing strength of the foundations.

9.12.3.2 Resistance moment

To prevent bending failure, the design bending moment in the baseplate, m_{Sd}, must not exceed the resistance moment, m_{Rd}:

$$m_{Sd} < m_{Rd} \qquad (9.66)$$

The bending moment in the baseplate is given by

$$m_{Sd} = (x^2/2)N_{Sd}/A_{eff} \qquad (9.67)$$

The resistance moment is given by

$$m_{Rd} = \frac{t^2 f_y}{6\gamma_{M0}} \qquad (9.68)$$

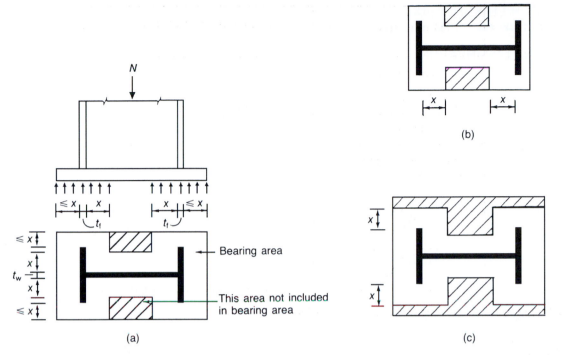

Fig. 9.7 *Area in compression under base plate: (a) general case; (b) short projection; (c) large projetion (Fig. L. 1, EC3).*

Example 9.7 Analysis of a column resisting an axial load (EC3)

Check the suitability of the $203 \times 203 \times 60$ kg m^{-1} UC section in grade Fe 430 steel to resist a design axial compression force of 1400 kN. Assume the column is pinned at both ends and that its height is 6 m.

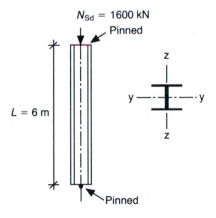

$N_{Sd} = 1600$ kN
Pinned

$L = 6$ m

z

y — $\cdots$ — y

z

Pinned

SECTION PROPERTIES
From steel tables (*Appendix B*)

Area of section (A)	$= 7580$ mm^2
Thickness of flange (t_f)	$= 14.2$ mm
Radius of gyration about the y–y axis (i_y)	$= 89.6$ mm
Radius of gyration about the z–z axis (i_z)	$= 51.9$ mm

STRENGTH CLASSIFICATION

Flange thickness = 14.2 mm, steel grade Fe 430. Hence from *Table 9.3*, $f_y = 275$ N mm^{-2}.

SECTION CLASSIFICATION

$$\varepsilon = (235/f_y)^{0.5} = (235/275)^{0.5} = 0.924$$

$$c/t_f = 7.23 < 10\varepsilon = 9.24$$

Also

$$d/t_w = 17.3 < 33\varepsilon = 33 \times 0.924 = 30.5$$

Hence from *Table 9.4*, section belongs to class 1.

RESISTANCE OF CROSS-SECTION – COMPRESSION

Plastic compressive resistance of section, $N_{pl.Rd}$, for class 1 section is given by

$$N_{pl.Rd} = \frac{Af_y}{\gamma_{M0}} = 7580 \times 275/1.05$$

$$= 1985 \times 10^3 \text{ N} = 1985 \text{ kN} > N_{Sd} = 1600 \text{ kN}$$

BUCKLING RESISTANCE OF MEMBER

Effective length of column about both axes is given by $L_{eff} = L_{eff\,y} = L_{eff\,z} = 1.0L = 1.0 \times 6000 = 6000$ mm

The column will buckle about the weak (z–z) axis. Slenderness ratio about z–z axis (λ_z) is

$$\lambda_z = \frac{L_{eff\,z}}{i_z} = \frac{6000}{51.9} = 115.6$$

Here $\beta_A = 1$ for class 1 section, $\lambda_1 = 93.9\varepsilon = 93.9 \times 0.924 = 86.76$.

$$\bar{\lambda}_z = (\lambda_z/\lambda_1)(\beta_A)^{1/2} = (115.6/86.76)(1)^{1/2} = 1.332$$

$$\frac{h}{b} = \frac{209.6}{205.2} = 1.02 < 1.2 \quad \text{and} \quad t_f = 14.2 \text{ mm}$$

Hence from *Table 9.8*, for buckling about z–z axis buckling curve c is appropriate and from *Table 9.7*, $\alpha = 0.49$.

$$\phi = 0.5[1 + \alpha(\bar{\lambda}_z - 0.2) + \bar{\lambda}_z^2]$$

$$= 0.5[1 + 0.49(1.332 - 0.2) + 1.332^2] = 1.665$$

$$\chi_z = \frac{1}{\phi + [\phi^2 - \bar{\lambda}_z^2]^{1/2}}$$

$$= \frac{1}{1.665 + [1.665^2 - 1.332^2]^{1/2}} = 0.375$$

Hence design buckling resistance, $N_{b.Rd}$, is given by

$$N_{b.Rd} = \chi_z \beta_A Af_y/\gamma_{M1} = 0.375 \times 1 \times 7580 \times 275/1.05$$

$$= 744 \times 10^3 \text{ N}$$

$$= 744 \text{ kN} < 1400 \text{ kN}$$

The section is therefore unsuitable to resist the design force.

Example 9.8 Analysis of a column with a tie-beam at mid-height (EC3)

Recalculate the axial compression resistance of the column in *Example 9.7* if a tie-beam is introduced at mid-height such that in-plane buckling about the z–z axis is prevented (see below).

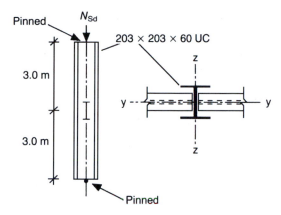

RESISTANCE OF CROSS-SECTION – COMPRESSION

Design plastic compressive resistance of section, $N_{pl.Rd} = 1985$ kN, as above.

BUCKLING RESISTANCE OF MEMBER

Buckling about y–y axis

Effective length of column about y–y axis is given by

$$L_{eff\,y} = 1.0L = 1.0 \times 6000 = 6000 \text{ mm}$$

Slenderness ratio about y–y axis (λ_y) is

$$\lambda_y = \frac{L_{eff\,y}}{i_y} = \frac{6000}{89.6} = 67$$

$$\beta_A = 1, \quad \lambda_1 = 86.76 \quad \text{(see above)}$$

$$\bar{\lambda}_y = (\lambda_y/\lambda_1)(\beta_A)^{1/2} = (67/86.76)(1)^{1/2} = 0.77$$

$$h/b = 1.02 \quad \text{and} \quad t_f = 14.2 \text{ mm}$$

Hence from *Table 9.8*, buckling curve b is appropriate, and from *Table 9.7* imperfection factor $\alpha = 0.34$.

$$\phi = 0.5[1 + \alpha(\bar{\lambda}_y - 0.2) + \bar{\lambda}_y^2]$$

$$= 0.5[1 + 0.34(0.77 - 0.2) + 0.77^2] = 0.893$$

$$\chi_y = \frac{1}{\phi + [\phi^2 - \bar{\lambda}_y^2]^{1/2}} = \frac{1}{0.893 + [0.893^2 - 0.77^2]^{1/2}} = 0.743$$

Hence design buckling resistance about the y–y axis is given by

$$N_{b.Rd} = \chi_y \beta_A A f_y/\gamma_{M1} = 0.743 \times 1 \times 7580 \times 275/1.05$$

$$= 1475 \times 10^3 \text{ N} = 1475 \text{ kN}$$

Buckling about z–z axis

Effective length of column about z–z axis, $L_{eff\,z}$, is equal to $L_{eff\,z} = 3000$ mm. Slenderness ratio about z–z axis (λ_z) is

$$\lambda_z = \frac{L_{eff\,z}}{i_z} = \frac{3000}{51.9} = 57.8$$

$$\beta_A = 1 \quad \lambda_1 = 86.76 \quad \text{(see above)}$$

$$\bar{\lambda}_z = (\lambda_z/\lambda_1)(\beta_A)^{1/2} = (57.8/86.76)(1)^{1/2} = 0.666$$

$$h/b = 1.02 \quad \text{and} \quad t_f = 14.2 \text{ mm}$$

Hence from *Table 9.8*, buckling curve c is appropriate and, from *Table 9.7*, imperfection factor $\alpha = 0.49$.

$$\phi = 0.5[1 + \alpha(\bar{\lambda}_z - 0.2) + \bar{\lambda}_z^2]$$

$$= 0.5[1 + 0.49\,(0.666 - 0.2) + 0.666^2] = 0.836$$

$$\chi_z = \frac{1}{\phi + [\phi^2 - \bar{\lambda}_z^2]^{1/2}} = \frac{1}{0.836 + [0.836^2 - 0.666^2]^{1/2}} = 0.746$$

$$N_{b.Rd} = \chi_z \beta_A A f_y/\gamma_{M1} = 0.746 \times 1 \times 7580 \times 275/1.05$$

$$= 1481 \times 10^3 \text{ N} = 1481 \text{ kN}$$

Hence compressive resistance of column is 1475 kN > 1400 kN   OK

Example 9.9 Analysis of a column resisting an axial load and moment (EC3)

A $305 \times 305 \times 137$ kg m^{-1} UC section extends through a height of 3.5 m and is pinned at both ends. Check whether this member is suitable to support a design axial permanent load of 600 kN together with a major axis variable bending moment of 300 kN m applied at the top of the element. Assume grade Fe 430 steel is to be used and that all effective length factors are unity.

ACTIONS
Factored axial loading is

$$N_{Sd} = 600 \times 1.35 = 810 \text{ kN}$$

Factored bending moment at top, middle and bottom of column is

$$M_{Sd.t} = 300 \times 1.5 = 450 \text{ kN m}$$

$$M_{Sd.m} = 225 \text{ kN m} \quad M_{Sd.b} = 0 \text{ kN m}$$

STRENGTH CLASSIFICATION
Flange thickness $t_f = 21.7$ mm, steel grade = Fe 430. Hence from *Table 9.3*, $f_y = 275$ N mm^{-2}.

SECTION CLASSIFICATION

$$c/t_f = 7.11 < 10\varepsilon = 9.24 \quad d/t_w = 17.9 < 33\varepsilon = 30.5$$

Hence from *Table 9.4*, section belongs to class 1.

RESISTANCE OF CROSS-SECTIONS: BENDING AND AXIAL FORCE
Squash load, $N_{pl.Rd}$, is

$$N_{pl.Rd} = A f_y/\gamma_{M0} = \frac{17\,500 \times 275}{1.05}$$

$$= 4\,583\,333 \text{ N} = 4583 \text{ kN}$$

Full plastic moment of resistance of section, $M_{pl.Rd}$, is

$$M_{pl.Rd} = W_{pl}f_y/\gamma_{M0} = \frac{2300 \times 10^3 \times 275}{1.05}$$

$$= 602 \times 10^6 \text{ N mm} = 602 \text{ kN m}$$

$$n = N_{Sd}/N_{pl.Rd} = 810/4583 = 0.177$$

$$a = (A-2bt_f)/A = \frac{17\,500 - 2(308.7 \times 21.7)}{17\,500} = 0.234$$

$$M_{Ny} = M_{pl.y}(1-n)/(1-0.5a)$$

$$= 602 \, (1 - 0.177)/(1 - 0.5 \times 0.234)$$

$$= 561 \text{ kN m} > M_{Sd} \, (= 450 \text{ kN m}) \quad \text{OK}$$

RESISTANCE OF MEMBER: COMBINED BENDING AND AXIAL COMPRESSION
Effective length of column about both axes is given by

$$L_{eff} = L_{eff\,y} = L_{eff\,z} = 1.0L = 1.0 \times 3500 = 3500 \text{ mm}$$

The column will buckle about the weak (z–z) axis. Slenderness ratio about z–z axis, λ_z is

$$\lambda_z = \frac{L_{eff\,z}}{i_z} = \frac{3500}{78.2} = 44.7$$

Here $\beta_A = 1$ for class 1 section and

$$\lambda_1 = 93.9\varepsilon = 93.9 \times 0.924 = 86.76$$

$$\bar{\lambda}_z = (\lambda_z/\lambda_1)(\beta_A)^{1/2} = (44.7/86.76)(1)^{1/2} = 0.515$$

$$\frac{h}{b} = 1.04 < 1.2 \quad \text{and} \quad t_f = 21.7 \text{ mm}$$

From *Table 9.8*, for buckling about z–z axis use buckling curve c. Hence from *Table 9.7*, $\alpha = 0.49$.

$$\phi = 0.5[1 + \alpha(\bar{\lambda}_z - 0.2) + \bar{\lambda}_z^2]$$

$$= 0.5[1 + 0.49(0.515 - 0.2) + 0.515^2] = 0.71$$

$$\chi_z = \frac{1}{\phi + [\phi^2 - \bar{\lambda}_z^2]^{1/2}} = \frac{1}{0.71 + [0.71^2 - 0.515^2]^{1/2}} = 0.834$$

Slenderness ratio about y–y axis, λ_y, is

$$\lambda_y = \frac{L_{eff\,y}}{i_y} = \frac{3500}{137} = 25.6$$

$$\bar{\lambda}_y = (\lambda_y/\lambda_1)(\beta_A)^{1/2} = (25.6/86.76)(1)^{1/2} = 0.294$$

$$\mu_y = \bar{\lambda}_y(2\beta_{My} - 4) + \left(\frac{W_{pl.y} - W_{el.y}}{W_{el.y}} \right)$$

$$= 0.294(2 \times 1.8 - 4) + \frac{(2300 - 2050)}{2050} = 0 \text{ approx.}$$

where $\beta_{My} = 1.8$ from *Fig. 9.6*. Thus

$$k_y = 1 - \frac{\mu_y N_{Sd}}{\chi_y A f_y} = 1$$

From *Table 9.8*, since $h/b < 1.2$ and $t_f < 100$ mm, for buckling about y–y axis use buckling curve b. Hence from *Table 9.7*, $\alpha = 0.34$.

$$\phi = 0.5[1 + \alpha(\bar{\lambda}_y - 0.2) + \bar{\lambda}_y^2]$$

$$= 0.5[1 + 0.34(0.294 - 0.2) + 0.294^2] = 0.56$$

$$\chi_y = \frac{1}{\phi + [\phi^2 - \bar{\lambda}_y^2]^{1/2}}$$

$$= \frac{1}{0.56 + [0.56^2 - 0.294^2]^{1/2}} = 0.964$$

χ_{min} is the smaller of χ_x (= 0.834) and χ_y (= 0.964).
Furthermore,

$$N_{c.Rd} = N_{pl.Rd} = 4583 \text{ kN} \quad M_{pl.y.Rd} = M_{pl.Rd} = 602 \text{ kN m}$$

Check using equation 9.54, i.e.

$$\frac{N_{Sd}}{\chi_{min} N_{c.Rd}} + \frac{k_y M_{y.Sd}}{M_{pl.y.Rd}} \leqslant 1$$

$$\frac{810}{0.834 \times 4583} + \frac{1 \times 450}{602} = 0.96$$

Hence the selected section is suitable.

Example 9.10 Analysis of a column baseplate (EC3)

Check that the column baseplate shown below is suitable to resist an axial design load, N_{Sd}, of 2200 kN. Assume that the foundations are of concrete of compressive cylinder strength, f_{ck}, of 30 N mm^{-2} and that the baseplate is made of steel grade Fe 430.

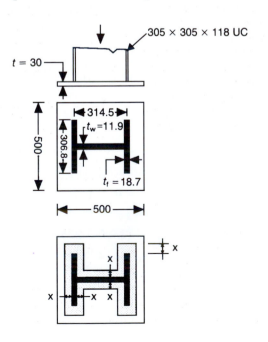

EFFECTIVE AREA

Additional bearing width, x, is

$$x = t(f_y/3f_j\gamma_{M0})^{1/2} = 30(275/3 \times 13.3 \times 1.05)^{1/2}$$

$$= 77 \text{ mm} < 0.5(500 - 314.5) = 92 \text{ mm} \quad \text{OK}$$

where

$$f_j = \beta_j k_j f_{cd} = 2/3 \times 1 \times 30/1.5 = 13.3 \text{ N mm}^{-2}$$

$$A_{eff} = (2x + h)(2x + b) - 2(c - t_w/2)(h - 2t_f - 2x)$$

$$= (2 \times 77 + 314.5)(2 \times 77 + 306.8) - 2(153.4 - 11.9/2)(314.5 - 2 \times 18.7 - 2 \times 77)$$

$$= 215\,885 - 36\,302 = 179\,583 \text{ mm}^2$$

AXIAL LOAD CAPACITY

Axial load capacity of baseplate is equal to

$$A_{eff}f_j = 179\,583 \times 13.3 = 2388 \times 10^3 \text{ N} > N_{Sd} (= 2200 \text{ kN}) \quad \text{OK}$$

BENDING IN BASEPLATE

Bending moment per unit length in baseplate, m_{Sd}, is

$$m_{Sd} = (x^2/2)N_{Sd}/A_{eff} = (77^2/2)2200/179\,583 = 36.3 \text{ kN mm mm}^{-1}$$

Moment of resistance, m_{Rd}, is

$$m_{Rd} = t^2 f_y/6\gamma_{M0} = 30^2 \times 275/6 \times 1.05 = 39.3 \times 10^3 \text{ N mm mm}^{-1}$$

$$= 39.3 \text{ kN mm mm}^{-1} > m_{Sd} \quad \text{OK}$$

9.13 Connections

In EC3, Chapter 6 on connections appears more comprehensive than BS 5950, but the principles are essentially the same. In general the results for bolting and welding seem slightly more conservative than BS 5950. This is largely because of the larger material safety factors for connections $\gamma_M = 1.25$.

To help comparison of the design methods in BS 5950 and EC3 with regards to connections, the material in this section is presented under the following headings:

1. material properties
2. clearances in holes for fasteners
3. positioning of holes for bolts
4. bolted connections
5. high-strength bolts in slip-resistant connections
6. welded connections
7. design of connections.

9.13.1 MATERIAL PROPERTIES

9.13.1.1 Nominal bolt strengths (clause 3.3, EC3)

The nominal values of the yield strength, f_{yb}, and the ultimate tensile strength, f_{ub}, of bolts are shown in Table 9.9. According to clause 3.3.2.2 of EC3, high-strength bolts may be used as preloaded bolts

Table 9.9 Nominal values of f_{yb} and f_{ub} for bolts (Table 3.3, EC3)

Bolt grade	4.6	4.8	5.6	5.8	6.8	8.8	10.9
f_{yb} (N mm^{-2})	240	320	300	400	480	640	900
f_{ub} (N mm^{-2})	400	400	500	500	600	800	1000

with controlled tightening provided that they conform with the requirements in reference 3 of Annex B. With regard to welded connections, EC3 requires that the specified yield strength, ultimate tensile strength, etc. should be equal to or greater than the values specified for the steel grade being welded.

9.13.1.2 Partial safety factors

Partial safety factors, γ_M, are taken as follows:

$$\text{Resistance of bolts, } \gamma_{Mb} = 1.25$$

$$\text{Resistance of welds, } \gamma_{Mw} = 1.25$$

9.13.2 CLEARANCES IN HOLES FOR FASTENERS (CLAUSE 7.5.2, EC3)

The nominal clearance in standard holes for bolted connections should be as follows:

1. 1 mm for M12 and M14 bolts
2. 2 mm for M16 and M24 bolts
3. 3 mm for M27 and larger bolts.

The nominal clearance in oversize holes for slip-resistant connections should be:

1. 3 mm for M12 bolts
2. 4 mm for M14 to M22 bolts
3. 6 mm for M24 bolts
4. 8 mm for M27 and larger bolts.

9.13.3 POSITIONING OF HOLES FOR BOLTS (CLAUSE 6.5.1, EC3)

9.13.3.1 Minimum end distance

The end distance e_1 from the centre of a fastener hole to the adjacent end of any part, measured in the direction of load transfer (*Fig. 9.8*), should be not less than $1.2d_0$, where d_0 is the hole diameter.

9.13.3.2 Minimum edge distance

The edge distance e_2 from the centre of a fastener hole to the adjacent edge of any part, measured at right angles to the direction of load transfer (*Fig. 9.8*), should generally be not less than $1.5d_0$.

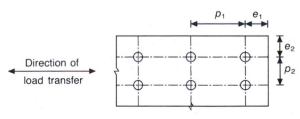

Fig. 9.8 Spacing of fasteners.

9.13.3.3 Maximum end and edge distances

Under normal conditions, the end and edge distance should not exceed $12t$ or 150 mm, whichever is the larger, where t is the thickness of the thinner outer connected part.

9.13.3.4 Minimum spacing

The spacing p_1 between centres of fasteners in the direction of load transfer (*Fig. 9.8*), should be not less than $2.2d_0$. The spacing p_2 between rows of fasteners, measured perpendicular to the direction of load transfer (*Fig. 9.8*), should normally be not less than $3d_0$.

9.13.4 BOLTED CONNECTIONS (CLAUSE 6.5.5, EC3)

9.13.4.1 Design shear resistance per shear plane

If the shear plane passes through the threaded portion of the bolt, the design shear resistance per shear plane, $F_{v.Rd}$, for strength grades 4.6, 5.6 and 8.8 bolts, is given by

$$F_{v.Rd} = \frac{0.6f_{ub}A_s}{\gamma_{Mb}} \qquad (9.69)$$

and for strength grades 4.8, 5.8 and 6.8 and 10.9 bolts, is given by

$$F_{v.Rd} = \frac{0.5f_{ub}A_s}{\gamma_{Mb}} \qquad (9.70)$$

If the shear plane passes through the unthreaded portion of the bolt, the design shear resistance is given by

$$F_{v.Rd} = \frac{0.6f_{ub}A}{\gamma_{Mb}} \qquad (9.71)$$

where A is the gross cross-section area of the bolt, A_s the tensile stress area of the bolt, d the bolt diameter and d_0 the hole diameter. Note that these values for design shear resistance apply only where the bolts are used in holes with nominal clearances specified in *section 9.13.2*.

9.13.4.2 Bearing resistance

The design bearing resistance, $F_{b.Rd}$, is given by

$$F_{b.Rd} = \frac{2.5\alpha f_{ub}dt}{\gamma_{Mb}} \qquad (9.72)$$

where α is the smallest of:

$$\frac{e_1}{3d_0}; \qquad \frac{p_1}{3d_0} - \frac{1}{4}; \qquad \frac{f_{ub}}{f_u} \quad \text{or} \quad 1.0$$

Note that the values of the design bearing resistance only apply where the edge distance e_2 is not less than $1.5d_0$ and the spacing p_2 is not less than $3.0d_0$.

9.13.5 HIGH-STRENGTH BOLTS IN SLIP-RESISTANT CONNECTIONS (CLAUSE 6.5.8, EC3)

9.13.5.1 Slip resistance
The design slip resistance of a preloaded high-strength bolt, $F_{s.Rd}$, is given by

$$F_{s.Rd} = \frac{k_s n \mu F_{p.Cd}}{\gamma_{Ms}} \qquad (9.73)$$

where $k_s = 1.0$ where the holes in all the plies have standard nominal clearances as outlined in *section 9.13.2* above, n is the number of friction interfaces, μ the slip factor (see below) and γ_{Ms} the partial safety factor. For bolt holes in standard nominal clearance holes, $\gamma_{Ms} = 1.25$ and 1.10 for the ultimate and serviceability limit states respectively. Finally, $F_{p.Cd}$ is the design preloading force. It is given by

$$F_{p.Cd} = 0.7 f_{ub} A_s \qquad (9.74)$$

9.13.5.2 Slip factor
The value of the slip factor μ is dependent on the class of surface treatment. The value of μ should be taken as follows:

$\mu = 0.5$ for class A surfaces

$\mu = 0.4$ for class B surfaces

$\mu = 0.3$ for class C surfaces

$\mu = 0.2$ for class D surfaces

The surface descriptions are as follows:

Class A: surfaces blasted with shot or grit, with any loose rust removed, no pitting; surfaces blasted with shot or grit, and spray-metallized with aluminium; surface blasted with shot or grit, and spray-metallized with a zinc-based coating certified to provide a slip factor of not less than 0.5.

Class B: surfaces blasted with shot or grit, and painted with an alkali–zinc silicate paint to produce a coating thickness of 50–80 μm.

Class C: surfaces cleaned by wire brushing or flame cleaning, with any loose rust removed;

Class D: surfaces not treated.

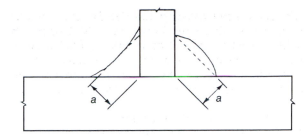

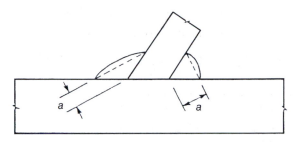

Fig. 9.9 *Throat thickness of a fillet weld.*

9.13.6 WELDED CONNECTIONS (CLAUSE 6.6, EC3)

9.13.6.1 Design resistance of a fillet weld
The design resistance per unit length of a fillet weld, $F_{w.Rd}$, is given by

$$F_{w.Rd} = f_{vw} a \qquad (9.75)$$

where a is the throat thickness of the weld and is taken as the height of the largest triangle which can be inscribed within the fusion faces and weld surface, measured perpendicular to the outer side of this triangle (*Fig. 9.9*). Note that a should not be less than 3 mm. Here f_{vw} is the design shear strength of the weld and is given by

$$f_{vw} = \frac{f_u / \sqrt{3}}{\beta_w \gamma_{Mw}} \qquad (9.76)$$

where f_u is the nominal ultimate tensile strength of the weaker part joined and β_w is a correlation factor whose value should be taken as follows. Linear interpolation for intermediate values of f_u is allowed.

EN10025 steel grade	Ultimate tensile strength f_u (N mm^{-2})	Correlation factor β_w
Fe 360	360	0.8
Fe 430	430	0.85
Fe 510	510	0.9

9.13.7 DESIGN OF CONNECTIONS

The use of the above equations is illustrated by means of the following design examples:

1. tension splice connection;
2. welded end plate to beam connection;
3. bolted beam-to-column connection using end plates;
4. bolted beam-to-column connection using web cleats.

9.13.7.1 Splice connections

The design of splice connections (*Fig. 9.10*) in EC3 is essentially the same as that used in BS 5950 and involves determining the design values of the following parameters:

1. design shear resistance of fasteners (*section 9.13.4* or *9.13.5*);
2. critical bearing resistance (*section 9.13.4*);
3. critical tensile resistance (next section).

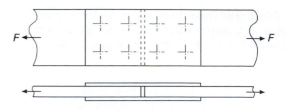

Fig. 9.10 *Splice connection.*

Tension resistance of cross-sections (clause 5.4.3, EC3). For members in axial tension, the design value of the tensile force, N_{Sd}, at each cross-section should satisfy the following:

$$N_{Sd} \leq N_{t.Rd} \qquad (9.77)$$

where $N_{t.Rd}$ is the design tension resistance of the cross-section taken as the design ultimate resistance of the net section at holes for fasteners, $N_{u.Rd}$, which is given by

$$N_{u.Rd} = 0.9 A_{net} f_u / \gamma_{M2} \qquad (9.78)$$

Example 9.11 Analysis of tension splice connections (EC3)

Calculate the design resistance of the splice connection shown below. The cover plates are made of grade Fe 430 steel and connected with either (a) non-preloaded bolts of diameter 20 mm and grade 4.6 or (b) prestressed bolts also of diameter 20 mm and grade 4.6. Assume that in both cases, the shear plane passes through the unthreaded portions of the bolts.

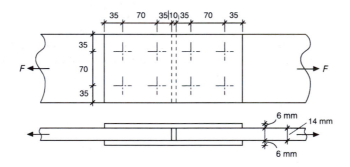

NON-PRELOADED BOLTS

Design shear resistance

Design shear resistance per shear plane, $F_{v.Rd}$, is given by

$$F_{v.Rd} = \frac{0.6 f_{ub} A}{\gamma_{Mb}} = \frac{0.6 \times 400 \times 314.16}{1.25}$$

$$= 60\,318 \text{ N} = 60 \text{ kN}$$

where $f_{ub} = 400$ N mm^{-2} (*Table 9.9*), $\gamma_{Mb} = 1.25$ and $A = \pi 20^2 / 4 = 314.16$ mm^2. All four bolts are in double shear. Hence, shear resistance, F_{sd}, of connection is

$$F_{sd} = 4 \times (2 \times 60) = 480 \text{ kN}$$

Bearing resistance

Bearing failure will tend to take place in the cover plates since they are thinner. According to clause 6.1.1 (2) of EC3, α is the smallest of

$$e_1/3d_0 = 35/3 \times 22 = 0.530$$

$$p_1/3d_0 - 1/4 = 70/3 \times 22 - 1/4 = 0.810$$

$$f_{ub}/f_u = 400/430 = 0.930 \quad (\textit{Tables 9.3 and 9.9})$$

$$\alpha = 1.000$$

Hence, $\alpha = 0.530$ and the bearing resistance of one shear plane, $F_{b.Rd}$ is given by

$$F_{b.Rd} = \frac{2.5\alpha f_u dt}{\gamma_{Mb}} = \frac{2.5 \times 0.530 \times 430 \times 20 \times 6}{1.25}$$

$$= 54\,696 \text{ N} = 54.7 \text{ kN}$$

Bearing resistance of double shear plane is

$$2 \times 54.7 = 109.4 \text{ kN}$$

Bearing resistance of bolt group is

$$4 \times 109.4 = 437.6 \text{ kN}$$

Tensile resistance of cover plates

Net area of cover plate, A_{net}, is

$$A_{net} = 6 \times 140 - 2 \times 6 \times 22 = 576 \text{ mm}^2$$

Design ultimate resistance, $N_{u.Rd}$, is given by

$$N_{u.Rd} = \frac{0.9 A_{net} f_u}{\gamma_{M2}} = \frac{0.9 \times 576 \times 430}{1.25}$$

$$= 178\,330 \text{ N} = 178 \text{ kN}$$

Total ultimate resistance of connection, i.e. two cover plates = $(2 \times 178) = 356$ kN. Hence design resistance of the connection with grade 4.6, 20 mm diameter non-preloaded bolts is 356 kN.

PRESTRESSED BOLTS

Slip resistance

Preloading force, $F_{p.Cd}$, is given by

$$F_{p.Cd} = 0.7 f_{ub} A_s = 0.7 \times 400 \times 245$$

$$= 68\,600 \text{ N} = 68.6 \text{ kN}$$

Assuming the surfaces have been shot blasted, i.e. class A, take $\mu = 0.5$. For two surfaces $n = 2$, standard clearances $k_s = 1.0$ and $\gamma_{Ms} = 1.25$. Hence slip resistance for each bolt, $F_{s.Rd}$, is given by

$$F_{s.Rd} = \frac{k_s n \mu}{\gamma_{Ms}} F_{p.Cd} = \frac{1 \times 2 \times 0.5}{1.25} \times 68.6 = 54.9 \text{ kN}$$

Hence design resistance = $4 \times 54.9 = 219.6$ kN.

Tensile resistance of cover plate

Tensile resistance of cover plates = 356 kN (from above). Hence design resistance of connection with grade 4.6, 20 mm diameter prestressed bolts = 219 kN.

Example 9.12 Welded end plate to beam connection (EC3)

Calculate the shear resistance of the welded end plate to beam connection shown below. Assume the throat thickness of the fillet weld is 4 mm and the steel grade is Fe 430.

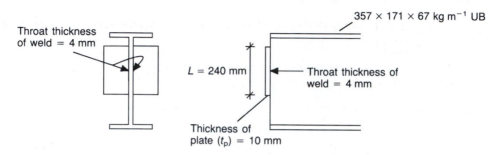

Ultimate tensile strength of Fe 430 steel (f_u) = 430 N mm^{-2} (*Table 9.3*); correlation factor (β_w) = 0.85; throat thickness of weld (a) = 4 mm; partial safety factor for welds (γ_{Mw}) = 1.25. Design shear strength of weld, f_{vw}, is given by

$$f_{VW} = \frac{f_u/\sqrt{3}}{\beta_w \gamma_{Mw}} = \frac{430/\sqrt{3}}{0.85 \times 1.25} = 233.6 \text{ N mm}^{-2}$$

Design resistance of weld per unit length, $F_{w.Rd}$, is given by

$$F_{w.Rd} = f_{vw}a = 233.6 \times 4 = 934.4 \text{ N mm}^{-1}$$

Hence for weld length of 2 × 240 mm the shear resistance of the weld, $V_{w.Rd}$, is

$$V_{w.Rd} = 2 \times (F_{w.Rd}L) = 2 \times (934.4 \times 240) = 448\,512 \text{ N} = 448 \text{ kN}$$

9.13.7.2 Welded end plate to beam connection

The relevant equations for the design of such connections were discussed in *section 9.13.6*.

9.13.7.3 Bolted beam-to-column connection using end plate

The design of such connections (*Fig. 9.11*) involves determining the design values of the following parameters:

1. design shear resistance of fasteners (*section 9.13.4*);
2. bearing resistance of fasteners (*section 9.13.4*);
3. resistance of welded connection (*section 9.13.6*);
4. shear resistance of end plate (section below);
5. local shear resistance of beam web (*section 9.11.1.2*).

Shear resistance of end plate (clause 5.4.6, EC3). The design value of the shear force, V_{Sd}, at each cross-section should satisfy the following:

$$V_{Sd} \leq V_{pl.Rd} \qquad (9.79)$$

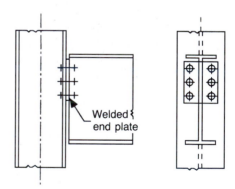

Fig. 9.11 *Typical bolted beam-to-column connection.*

where $V_{pl.Rd}$ is the design plastic shear resistance given by

$$V_{pl.Rd} = A_v(f_y/\sqrt{3})/\gamma_{M0} \qquad (9.80)$$

where A_v is the shear area and may be taken as being equal to the cross-sectional area, A, for plate sections.

Clause 5.4.6(8) of EC3 also states that fastener holes in webs need not be allowed for provided that

$$A_{v.net}/A_v \geqslant f_y/f_u \qquad (9.81)$$

However, when $A_{v.net}$ is less than this limit, the following value of the effective shear area may be used:

$$\text{Effective shear area} = A_{v.net}f_u/f_y \qquad (9.82)$$

Example 9.13 Bolted beam-to-column connection using end plate (EC3)

If the beam in *Example 9.12* is to be connected to a column using 8 no. grade 4.6, M16 bolts as shown below, calculate the maximum shear resistance of the connection.

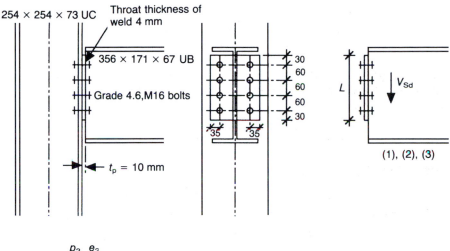

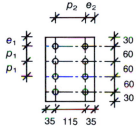

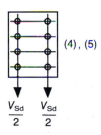

CHECK POSITIONING OF HOLES FOR BOLTS

Diameter of bolt, $d = 16$ mm

Diameter of bolt hole, $d_0 = 18$ mm

End distance, $e_1 = 30$ mm

Edge distance, $e_2 = 35$ mm

Spacing between centres of bolts in the direction of load transfer, $p_1 = 60$ mm

Spacing between rows of bolts, $p_2 = 115$ mm

Thickness of end plate, $t_p = 10$ mm

The following conditions need to be met:

End distance, $e_1 \nless 1.2d_0 = 1.2 \times 18 = 21.6$ mm < 30 mm   OK

Edge distance, $e_2 \nless 1.5d_0 = 1.5 \times 18 = 27$ mm < 35 mm   OK

$$e_1 \text{ and } e_2 \not> \text{ larger of } 12t \ (= 12 \times 10 = 120 \text{ mm}) \text{ or } 150 \text{ mm} > 30, 35 \quad \text{OK}$$
$$\text{Spacing, } p_1 \geqslant 2.2d_0 = 2.2 \times 18 = 39.6 < 60 \quad \text{OK}$$
$$\text{Spacing, } p_2 \geqslant 3d_0 = 3 \times 18 = 54 < 115 \quad \text{OK}$$
$$p_1 \text{ and } p_2 \geqslant \text{ lesser of } 14t \ (= 14 \times 10 = 140 \text{ mm}) \text{ or } 200 \text{ mm} > 60, 115 \quad \text{OK}$$

SHEAR RESISTANCE OF BOLT GROUP

Cross-sectional area of bolt $(A) = \dfrac{\pi 16^2}{4} = 201 \text{ mm}^2$

Shear resistance per shear plane $(F_{v.Rd})$ is given by

$$F_{v.Rd} = \frac{0.6 f_{ub} A}{\gamma_{Mb}} = \frac{0.6 \times 400 \times 201}{1.25}$$
$$= 38\,592 \text{ N}$$

Hence, shear resistance of bolt group is

$$V_{v.Rd} = 8 \times F_{v.Rd} = 8 \times 38.6 = 308.7 \text{ kN}$$

BEARING RESISTANCE OF BOLT GROUP

Bolt diameter $(d) = 16$ mm, hole diameter $(d_0) = 18$ mm and α is the smallest of the following:

$$e_1/3d_0 = 30/3 \times 18 = 0.555$$
$$p_1/3d_0 - 1/4 = 60/3 \times 18 - 1/4 = 0.861$$
$$f_{ub}/f_u = 400/430 = 0.930 \quad (\textit{Tables 9.3 and 9.9}) \quad \text{or} \quad 1.000$$

Hence $\alpha = 0.555$.
   Bearing resistance of one bolt

$$F_{b.Rd} = \frac{2.5 \alpha f_u d t_p}{\gamma_{Mb}} = \frac{2.5 \times 0.555 \times 430 \times 16 \times 10}{1.25}$$
$$= 76\,368 \text{ N} = 76.3 \text{ kN}$$

Bearing resistance of bolt group is

$$= 8 F_{b.Rd} = 8 \times 76.3 = 610.4 \text{ kN}$$

RESISTANCE OF WELDED CONNECTION BETWEEN BEAM AND END PLATE

$$V_{w.Rd} = 448 \text{ kN} \quad (\textit{Example 9.12})$$

SHEAR RESISTANCE OF END PLATE

Bolt holes in plate do not need to be taken into account provided that

$$A_{v.net}/A_v \geqslant f_y/f_u \quad \text{where} \quad A_v = 10 \times 240 = 2400 \text{ mm}^2$$

and

$$A_{v.net} = 2400 - 4 \times 18 \times 10 = 1680 \text{ mm}^2$$

Substituting above gives

$$1680/2400 = 0.7 > f_y/f_u = \frac{275}{430} = 0.64$$

Hence shear resistance of end plate per section is given by

$$V_{pl.Rd} = A_v (f_y/\sqrt{3})/\gamma_{M0}$$
$$= 2400 \, (275/\sqrt{3})/1.05$$
$$= 362\,906 \text{ N} = 362.9 \text{ kN}$$

For failure, two planes have to shear. Hence shear resistance of end plate is

$$2 \times V_{pl.Rd} = 2 \times 362.9 = 725.8 \text{ kN}$$

LOCAL SHEAR RESISTANCE OF BEAM WEB
Shear area of web is

$$(A_v)_{web} = Lt_{web} = 240 \times 9.1 = 2184 \text{ mm}^2$$

where t_{web} = beam web thickness = 9.1 mm. Hence local shear resistance of web, $V_{pl.Rd}$, is given by

$$V_{pl.Rd} = (A_v)_{web}(f_y/\sqrt{3})/1.05$$
$$= 2184 \times (275/\sqrt{3})/1.05 = 330\ 244 \text{ N} = 330 \text{ kN}$$

Hence the maximum shear resistance of the connection is controlled by the shear resistance of the bolt group and is equal to 308.7 kN.

9.13.7.4 Bolted beam–to–column connection using web cleats

The design of such connections (*Fig. 9.12*) involves determining the design values of the following parameters:

1. design shear resistance of fasteners (*section 9.13.4*);
2. bearing resistance of fasteners (*section 9.13.4*);
3. shear resistance of cleats (see previous example);
4. distribution of shear forces between fasteners (see below);
5. bearing resistance of cleats (based on (2) and (4));
6. bearing resistance of beam web (*section 9.13.4*);
7. shear rupture strength (see below).

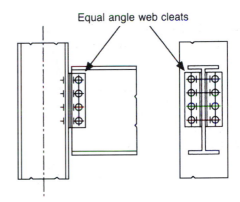

Fig. 9.12 Typical bolted beam-to-column connection using wet cleat.

Distribution of forces between fasteners (clause 6.5.4, EC3). The distribution of internal forces between fasteners at the ultimate limit state can be assumed to be proportional to the distance from the centre of rotation (*Fig. 9.13*) where the design shear resistance of a fastener, $F_{v.Rd}$, is less than the design bearing resistance $F_{b.Rd}$, i.e.

$$F_{v.Rd} < F_{b.Rd} \qquad (9.83)$$

Thus, for the connection detail shown in *Fig. 9.13*, the maximum horizontal shear force on the bolts, $F_{h.Sd}$, is given by

$$F_{h.Sd} = M_{Sd}/5p \qquad (9.84)$$

and the vertical shear force per bolt is = $V_{Sd}/5$. The design shear force, $F_{v.Sd}$, is given by

$$F_{v.Sd} = \left[\left(\frac{M_{Sd}}{5p}\right)^2 + \left(\frac{V_{Sd}}{5}\right)^2\right]^{1/2} \qquad (9.85)$$

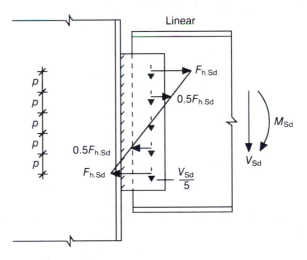

Fig. 9.13 Distribution of loads between fasteners (Fig. 6.5.7, EC3).

where M_{Sd} is the design bending moment, V_{Sd} the design shear force and p the spacing between fasteners.

Design shear rupture resistance (clause 6.5.2.2, EC3). 'Block shear' failure at a group of fastener holes near the end of a beam web may occur as shown in *Fig. 9.14*. The design value of the effective resistance to block shear, $V_{eff.Rd}$, is given by

$$V_{eff.Rd} = (f_y/\sqrt{3})A_{v.eff}/\gamma_{M0}$$

where $A_{v.eff}$ is the effective shear area and is given by

$$A_{v.eff} = tL_{v.eff} \quad \text{where}$$
$$L_{v.eff} = L_v + L_1 + L_2 \leq L_3$$

in which

$$L_1 = a_1 \leq 5d \quad L_2 = (a_2 - kd_{0.t})(f_u/f_y)$$

and

$$L_3 = L_v + a_1 + a_3 \leq (L_v + a_1 + a_3 - nd_{0.v})(f_u/f_y)$$

where a_1, a_2, a_3 and L_v are as indicated in *Fig. 9.14*, d is the nominal diameter of fasteners, $d_{0.t}$, $d_{0.v}$ is generally the hole diameter, n the number of fastener holes on the shear face, t is the thickness of the web or bracket and k is a coefficient with values as follows:

For a single row of bolts $k = 0.5$

For two rows of bolts $k = 2.5$.

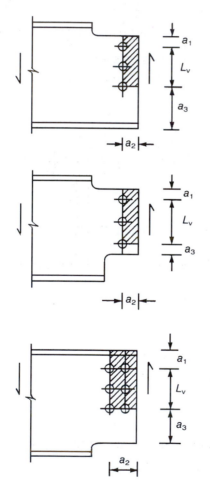 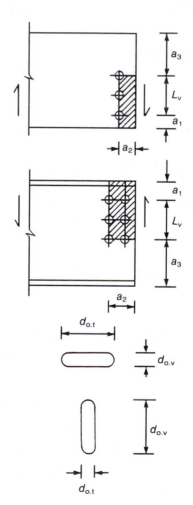

Fig. 9.14 *Block shear–effective shear areas (Fig. 6.5.5, EC3).*

Example 9.14 Bolted beam-to-column connection using web cleats (EC3)

Show that the double angle web cleat beam-to-column connection detail below is suitable to resist the design shear force, V_{Sd}, of 225 kN. Assume the steel grade is Fe 430 and the bolts are of grade 5.8 and diameter 16 mm.

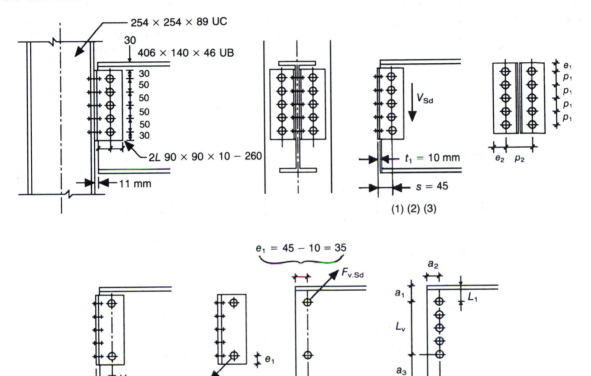

(1) (2) (3)

(4) (5) (6) (7)

CHECK POSITIONING OF HOLES FOR BOLTS

Diameter of bolt, $d = 16$ mm

Diameter of bolt hole, $d_0 = 18$ mm

End distance, $e_1 = 30$ mm

Edge distance, $e_2 = 45$ mm

Spacing between centres of bolts in the direction of load transfer, $p_1 = 50$ mm

Thickness of angle cleat, $t_p = 10$ mm

The following conditions need to be met:

End distance, $e_1 \not< 1.2d_0 = 1.2 \times 18 = 21.6$ mm < 30 mm   OK

Edge distance, $e_2 \not< 1.5d_0 = 1.5 \times 18 = 27$ mm < 45 mm   OK

e_1 and $e_2 \not> $ larger of $12t$ ($= 12 \times 10 = 120$ mm) or 150 mm $> 30, 45$ OK

$$\text{Spacing, } p_1 \geqslant 2.2d_0 = 2.2 \times 18 = 39.6 < 50 \quad \text{OK}$$

$$\text{Spacing, } p_1 \geqslant \text{lesser of } 14t \ (= 14 \times 10 = 140 \text{ mm}) \text{ or } 200 \text{ mm} > 50 \quad \text{OK}$$

SHEAR RESISTANCE OF BOLT GROUP

Assume that the shear plane passes through threaded portion of the bolt. Hence tensile stress area of bolt $A_s = 157$ mm^2. Shear resistance per bolt, $F_{v.Rd}$, is

$$F_{v.Rd} = \frac{0.5 f_{ub} A_s}{\gamma_{Mb}} = \frac{0.5 \times 500 \times 157}{1.25} = 31\,400 \text{ N} = 31.4 \text{ kN}$$

where $f_{ub} = 500$ N mm^{-2} (*Table 9.9*) and $\gamma_{Mb} = 1.25$. Shear resistance for bolt group is

$$10 \times F_{v.Rd} = 10 \times 31.4 \text{ kN} > V_{Sd} = 225 \text{ kN} \quad \text{OK}$$

BEARING RESISTANCE OF BOLT GROUP

Diameter of bolts $(d) = 16$ mm; hole diameter $(d_0) = 18$ mm; end distance in the direction of load transfer $(e_1) = 30$; spacing between bolts in the direction of load transfer $(p_1) = 50$; ultimate tensile strength of grade 5.8 bolts $(f_{ub}) = 500$ N mm^{-2} (*Table 9.9*); ultimate tensile strength of grade Fe 430 steel $(f_u) = 430$ N mm^{-2} (*Table 9.3*).

Here α is the smallest of

$$\frac{e_1}{3d_0} = \frac{30}{3 \times 18} = 0.555$$

$$\frac{p_1}{3d_0} - \frac{1}{4} = \frac{50}{3 \times 18} - \frac{1}{4} = 0.676$$

$$\frac{f_{ub}}{f_u} = \frac{500}{430} = 1.163 \quad \text{or} \quad 1.000$$

Take $\alpha = 0.555$:

$$F_{b.Rd} = 2.5 \alpha f_u d t_c / \gamma_{Mb}$$

$$= \frac{2.5 \times 0.555 \times 430 \times 16 \times 10}{1.25} = 76\,368 \text{ N} = 76.3 \text{ kN}$$

Bearing resistance of bolt group is

$$10 \times F_{b.Rd} = 10 \times 76.3 = 763 \text{ kN} > V_{Sd} = 225 \text{ kN} \quad \text{OK}$$

SHEAR RESISTANCE OF LEGS OF CLEATS

Fastener holes need not be allowed for provided that

$$\frac{A_{v.net}}{A_v} \geqslant \frac{f_y}{f_u} = \frac{1700}{2600} = 0.65 > \frac{275}{430} = 0.64$$

where

$$A_{v.net} = 10(260 - 5 \times 18) = 1700 \text{ mm}^2$$

$$A_v = 10 \times 260 = 2600 \text{ mm}^2$$

and for steel Fe 430 $f_u = 430$ N mm^{-2} and $f_y = 275$ N mm^{-2} (*Table 9.3*). Hence design shear resistance of each leg of cleat, $V_{pl.Rd}$, is

$$V_{pl.Rd} = A_v (f_y/\sqrt{3}) \gamma_{M0} = 2600 (275/\sqrt{3})/1.05 = 393\,148 \text{ N} = 393.1 \text{ kN}$$

where $\gamma_{M0} = 1.05$. Total shear resistance of both legs of cleats is

$$2 \times V_{pl.Rd} = 2 \times 393.1 = 786.2 \text{ kN} > V_{Sd} = 225 \text{ kN} \quad \text{OK}$$

SHEAR ACTION ON BOLTS DUE TO SHEAR FORCE AND RESULTANT BENDING MOMENT

Shear force per bolt in vertical direction, $F_{v.Sd}$, is

$$F_{v.Sd} = \frac{V_{Sd}}{n} = \frac{225}{5} = 45 \text{ kN}$$

Maximum shear force on bolt assembly in horizontal direction, $F_{h.Sd}$, is

$$F_{h.Sd} = \frac{M_{Sd}}{5p_1} = \frac{V_{Sd}s}{5p_1} = \frac{225 \times 45}{5 \times 50} = 40.5 \text{ kN}$$

Resultant force, $F_{V.Sd}$, is

$$F_{V.Sd} = (F_{v.Sd}^2 + F_{h.Sd}^2)^{1/2} = (45^2 + 40.5^2)^{1/2} = 60.5 \text{ kN}$$

Shear resistance per bolt, $F_{v.Rd}$, is 31.4 kN (see 'Shear resistance of bolt group' above). Since the resultant shear force is resisted by bolts in double shear, the total shear resistance is

$$F_{v.Rd} = 2 \times 31.4 = 62.8 \text{ kN} > F_{v.Sd} = 60.5 \text{ kN} \quad \text{OK}$$

BEARING RESISTANCE OF LEGS OF CLEATS CONNECTED TO WEB OF BEAM

Here α is the smallest of

$$\frac{e_1}{3d_0} = \frac{30}{3 \times 18} = 0.555$$

$$\frac{p_1}{3d_0} - \frac{1}{4} = \frac{50}{3 \times 18} - \frac{1}{4} = 0.676$$

$$\frac{f_{ub}}{f_u} = \frac{500}{430} = 1.163 \quad \text{or} \quad 1.000$$

Take $\alpha = 0.555$:

$$F_{b.Rd} = 2.5\alpha f_u dt_c / \gamma_{Mb}$$

$$= \frac{2.5 \times 0.555 \times 430 \times 16 \times 10}{1.25} = 76\,368 \text{ N} = 76.4 \text{ kN}$$

$$F_{v.Sd} = 60.5 \text{ kN} < 2F_{b.Rd} = 2 \times 76.4 = 152.8 \text{ kN} \quad \text{OK}$$

BEARING RESISTANCE OF WEB OF BEAMS

Here α is the smallest of

$$\frac{e_1}{3d_0} = \frac{35}{3 \times 18} = 0.648$$

$$\frac{p_1}{3d_0} - \frac{1}{4} = \frac{50}{3 \times 18} - \frac{1}{4} = 0.676$$

$$\frac{f_{ub}}{f_u} = \frac{500}{430} = 1.163 \quad \text{or} \quad 1.000$$

Hence $\alpha = 0.648$. Bearing resistance per bolt, $F_{b.Rd}$, is

$$F_{b.Rd} = \frac{2.5\alpha f_u dt_w}{\gamma_{Mb}} = \frac{2.5 \times 0.648 \times 430 \times 16 \times 6.9}{1.25}$$

$$= 61\,523 \text{ N} = 61.5 \text{ kN} > F_{v.Sd}$$

$$= 60.5 \text{ kN}$$

BLOCK SHEAR RESISTANCE

$$t = \text{thickness of web of beam} = 6.9 \text{ mm}$$

$$L_1 = a_1 = 60 \text{ mm} < 5d = 5 \times 16 = 80 \text{ mm}$$

$$L_v = 200 \text{ mm and } a_3 = 95.6 \text{ mm}$$

$$L_2 = (a_2 - kd_{0.t})(f_u/f_y)$$

$$= (35 - 0.5 \times 18)(430/275) = 40.6$$

$$L_3 = L_v + a_1 + a_3 \leqslant (L_v + a_1 + a_3 - nd_{0.v})(f_u/f_y)$$

$$= 200 + 60 + 95.6 = 355.6$$

$$< (200 + 60 + 95.6 - 5 \times 18)(430/275) = 415.3$$

$$L_{v.eff} = L_v + L_1 + L_2 \leqslant L_3$$

$$= 200 + 60 + 40.6 = 300.6 \text{ mm} < L_3$$

$$A_{v.eff} = tL_{v.eff}$$

$$= 6.9 \times 300.6 = 2074 \text{ mm}^2$$

$$V_{eff.Rd} = (f_y/\sqrt{3})A_{v.eff}/\gamma_{M0}$$

$$= (275/\sqrt{3})2074/1.05$$

$$= 313.6 \times 10^3 \text{ N} = 313.6 \text{ kN} > V_{Sd} = 225 \text{ kN} \quad \text{OK}$$

Chapter 10

Eurocode 6: Design of masonry structures

This chapter briefly describes the purpose, scope and progress to date of Eurocode 6: Design of Masonry Structures. The chapter highlights some of the difficulties in drafting the document and summarizes the layout and contents of Part 1.1 of Eurocode 6, which was issued in DD ENV format together with the UK National Application Document by the British Standards Institution in 1996.

10.1 Introduction

Eurocode 6 is the new structural European standard for the design of masonry structures. Eurocode 6 is similar in scope to BS 5628 (Chapter 5), which it will eventually replace. Part 1.1 of Eurocode 6 was finally published in DD (Draft for Development) ENV format in 1996. This is some three to four years after its counterparts for concrete, steel and timber. The delay was partly attributable to problems with the Comité Européen de Normalisation (CEN) mandate and agreement of production timescales during the transition of Eurocode 6 from the Commission of European Community responsibility to CEN responsibility, and partly to financial difficulties. In common with the other structural Eurocodes, there have also been a number of technical difficulties in drafting the document, some of which are still ongoing and will be discussed more fully in section 10.2.

Like BS 5628, Eurocode 6 is based on limit state principles and will be published in a number of parts as follows:

Part 1.1: General Rules for buildings – Rules for reinforced and unreinforced masonry;
Part 1.2: Structural fire design;
Part 1.3: Detailed rules on lateral loading;
Part 1-X: Complex shape sections in masonry structures;
Part 2: Design, selection of materials and execution of masonry;

Part 3: Simplified and simple rules for masonry structures;
Part 4: Construction with lesser requirements for reliability and durability.

Part 1.1, hereafter referred to as EC6, shares many features with Parts 1 and 2 of BS 5628. However, only general design procedures for laterally loaded wall panels are given in EC6. For detailed guidance on this aspect of masonry design, designers must turn to Part 1.3, which was published in DD ENV form in 2001. Similarly, Parts 2 and 3 were only published in DD ENV form in 2001. Part 1.2 was published much earlier in 1997. Parts 1-X and 4 have never been worked on and there are no ENV's available, neither will there be EN's. They have effectively been deleted from the masonry work programme in CEN.

10.2 Technical difficulties

As mentioned above, apart from administrative problems, the drafting panel of EC6, CEN Technical Committee SC6, faced various technical difficulties in drafting the document. These principally related to:

(a) the diverse range of masonry units produced throughout Europe and
(b) national differences in design methodologies.

The way in which these difficulties were tackled is discussed next.

10.2.1 CEN SUPPORTING STANDARDS FOR PRODUCTS

One of the stated aims of the Single Act is that there should be no barriers to EU trade in construction products and related services (i.e. design), which has generally been interpreted as meaning that design rules must not disadvantage available products from EC countries. Yet there is a wide variety of masonry units used throughout Europe. The units differ in many respects including appearance, size, strength and percentage and orientation of voids or perforations. Thus, the EC6 drafting panel had to develop methods that would allow the majority of masonry products currently available throughout Europe to be used in design.

The first step towards solving this problem was to agree a European specification for masonry units, which is the responsibility of CEN Technical Committee 125. There are six proposed harmonised masonry unit standards for: clay, calcium silicate, aggregated concrete, autoclaved aerated concrete, manufactured stone and natural stone. All of these, apart from natural stone, which is delayed, are currently out for formal voting. The European standard for clay masonry units, reference no. prEN 777-1, uses a declaration system for specifying product characteristics. This is somewhat similar to that already adopted in BS 3921 (see Chapter 5) and covers a number of essential requirements, all of which can be assessed by means of standard tests. These include: compressive strength, water absorption, dimensional tolerance, frost resistance, soluble salt content and density. The latter is an additional parameter to those mentioned in BS 3921 and has been included because of its correlation with noise attenuation, thermal insulation and other characteristics. Note that appearance is not mentioned as it requires subjective judgement and cannot be supported with an appropriate test method. It is worth remembering that the masonry product standards have not been developed by the EC6 drafting panel and, therefore, they contain general details on unit specification in terms of individual product types and will be relevant to all uses, not just structural applications. The majority of the supporting standards (non-harmonised) have already been voted through and the British Standards Institution is in the process of or has already published them in the UK, but they cannot be used until the EN product standards becomes available and are fully implemented.

Masonry units which have been characterised according to the appropriate European standard could be used to size masonry elements. However, it was considered by the EC6 drafting panel that this would result in a large volume of design data covering the range of masonry units produced throughout Europe, which would be impractical to include in a working design standard. Therefore, to simplify the process, it was decided that masonry units with similar characteristics should be grouped together into unit groups, for which typical design parameters e.g. characteristic compressive strength, characteristic shear strength, etc., would be available. Clause 3.1.1(6) of EC6 requires that masonry units be classified for design purposes into one of four Groups: 1, 2a, 2b or 3, with the superior structural use masonry units being placed in Group 1. It would appear that all U.K. bricks currently manufactured to British Standards, including frogged and perforated bricks, and most concrete blocks to British Standards, with the exception of a few cellular and hollow blocks, fit into Group 1 unit specification. Furthermore, UK bricks compare very favourably from an appearance point of view with those available in many other EC countries, which should help to maintain present levels of UK brick production.

10.2.2 DESIGN PROCEDURE

Although there were differences over certain aspects of compressive design e.g. slender members, shear resistance and concentrated load design, EC6 committee members were able to reach a consensus on the design approach relatively easily. However, it was considerably more difficult to agree a methodology for flexural design (i.e. design of laterally loaded panel walls) for the reasons discussed below and this was responsible for the significant delay in publication of Part 1.3 of Eurocode 6.

In the U.K. code of practice, the design of panel walls is based on 'yield line' principles. This assumes that the wall behaves plastically when subject to lateral loading, an assumption which correlates well with the available test data based on the BS 5628 panel design range. These tests were used to establish the characteristic flexural strength values give in Table 3 of BS 5628 (see Table 5.12). Many continental standards, however, either do not contain a method for flexural design as their walls tend to be much thicker than those normally used in the UK, or they are based on alternative design principles. In both cases, the increased thickness of wall would lead to increased costs of masonry construction. It was therefore difficult to achieve agreement on a pan-European basis. Thankfully these difficulties have now been resolved and, fortunately for UK designers, the design method for laterally loaded panel walls given in Part 1.3 of Eurocode 6 is almost identical to that used in BS 5628.

10.3 Layout and contents

The material in EC6 is divided into six sections as follows

Section 1: General
Section 2: Basis of design
Section 3: Materials
Section 4: Design of masonry
Section 5: Structural detailing
Section 6: Construction

In addition, supplementary information is provided in a number of annexes, classified normative or informative. The material in the normative annexes has the same status as the rest of the code whereas the material in the informative annexes does not have the same text status and is included merely for information. The following subsections give brief summaries of the contents of each section of the code.

SECTION 1: GENERAL

Section 1 describes the scope of EC6, which is somewhat broader than that of BS5628. Thus, EC6 covers the design of elements, buildings and civil engineering works in unreinforced, reinforced, pre-stressed and confined masonry. However, beyond a general statement on the need to provide adequate lateral strength resistance in flexure, little guidance is given in Part 1.1 as to how this should be achieved in practice. As previously noted, this aspect is addressed in Part 1.3 of Eurocode 6.

As with the other Eurocodes, the material in EC6 is divided into principles and application rules. Principles are general statements and definitions for which no alternative is permitted. Application rules are generally recognised rules which follow the principles and satisfy their requirements. In theory it is permissible to use alternatives to the application rules, provided that they satisfy the principles and achieve the same safety levels.

A number of the safety elements in EC6 appear in boxes which signifies that these values can be varied by CEN Member Standards Bodies, such as the British Standards Institution on behalf of the UK building regulations. The actual values to be used by any Member State will be specified in the accompanying National Application Document (section 7.4.3). Section 1 also contains a list of definitions and symbols used in EC6.

SECTION 2: BASIS OF DESIGN

As previously noted, EC6 is a limit state code. Section 2 therefore outlines ultimate and serviceability limit states relevant to masonry design. In order to assess the effect of particular limit states on the structure, the designer will need to estimate both the design loading (termed design action in

Table 10.1 Partial safety factors for material properties, γ_M (based on Table 2.3 of EC6 and Table 1 of UK NAD)

		Category of execution	
		A	B
Category of manufacturing	I	2.0	2.5
control of masonry units	II	2.3	2.8

EC6) and the design strength resistance of the materials and elements.

The design value of action is obtained by multiplying the characteristic value of action by the partial safety factor for the action. Characteristic dead (termed characteristic permanent in EC6), imposed (termed variable in EC6) and wind actions should be obtained from BS 648, BS 6399: Parts 1–3 and CP3: Chapter V: Part 2. In general, the partial safety factor for permanent action is 1.35 and variable action 1.5, but when the load case considers wind action also, all actions are multiplied by 1.35.

The design value of strength is obtained by dividing the characteristic value of strength by a partial safety factor for the material. The characteristic compressive strength of masonry can be assessed via a number of equations in section 3 of EC6. Like BS 5628, the partial factor of safety for masonry, γ_M, is a function of the masonry unit manufacturing and of the relevant construction control (termed execution control in EC6). Two categories of manufacturing control are specified in EC6: category I and category II. They are defined in clause 3.1.1 of EC6. Category I corresponds to the 'special' category of BS5628 whereas category II corresponds to the 'normal' category.

Provision is made for three categories of execution control in EC6: A, B and C, but, as in BS 5628, only two categories – A and B – are used in the UK National Application Document (NAD). Table 10.1 shows the range of partial safety factors recommended for use in the NAD. Comparison with the corresponding table in BS 5628 (Table 5.9) shows that the values in EC6 are somewhat lower.

SECTION 3: MATERIALS

As noted above, Section 3 contains a number of equations for assessing the strength of unreinforced masonry. The characteristic compressive strength of unreinforced masonry, f_k, built with general-purpose mortar can be determined using the following expression

$$f_k = K f_b^{0.65} f_m^{0.25} \qquad (10.1)$$

where

f_m is the compressive strength of general-purpose mortar but not exceeding 20 Nmm^{-2} or $2f_b$, whichever is the smaller

f_b is the normalised compressive strength of the masonry units

K is a constant

The compressive strength of general-purpose mortar is obtained from Table 5 of the NAD. This assumes that mortar designations (i), (ii), (iii) and (iv) (as defined in Table 1 of BS 5628, reproduced here as Table 5.5) have compressive strengths of, respectively, 12 Nmm^{-2}, 6 Nmm^{-2}, 4 Nmm^{-2} and 2 Nmm^{-2}. These values are not the same as the site and laboratory mortar strengths indicated in BS 5628, which are intended for compliance testing and are not design values as such.

The normalised compressive strength of masonry, f_b, is equivalent to the compressive strength of an "air dry" 100 mm cube of material. The specified compressive strength of masonry units manufactured to British Standards are wet strengths. According to clause 6.3(a) of the NAD, the air dry strengths can be obtained by multiplying the wet strengths by 1.2. The normalised compressive strength (of a 100 mm cube of the material) is then determined by further multiplying by a shape factor δ. Values for δ are given in Table 3.2 of EC6, to allow for the height and width of the units. For 102.5 mm $\times$ 65 mm bricks, $\delta = 0.85$ and for 215 mm $\times$ 100 mm blocks, $\delta = 1.38$.

The constant K is a function of the group of masonry, type of mortar and type of bonding. In EC6, masonry units are grouped according to the percentage of voids present. Units with not more than 25 per cent voids are classified as Group 1 and between 25–70 per cent as Groups 2a, 2b or 3. All bricks currently manufactured to British Standards fall within Group 1. A few UK cellular and hollow units (Fig. 5.2) fall within Group 2. Walls may be constructed with or without longitudinal joints through the wall thickness (see Fig. 5.1 of EC6 for details). All other things being equal, the latter bond type normally results in higher compressive strengths. Table 3 of the UK NAD should be consulted in order to determine values of the constant K, which lie in the range 0.3–0.8. Walls made of clay units and general purpose mortar with no longitudinal joint through all or part of the wall have a K-value of 0.7.

The NAD recommends that the flexural strength of masonry should be obtained from Table 3 of BS 5628, reproduced here as Table 5.12. Note that flexural strength parallel to the bedding plane is denoted by the symbol f_{xk1} and flexural strength perpendicular to the bedding plane by the symbol f_{xk2}. As previously noted, the actual design methodology is covered in Part 1.3 of Eurocode 6 but is in fact very similar to that used in BS 5628 and will not be discussed further.

SECTION 4: DESIGN OF MASONRY

EC6 rules for the design of vertically loaded walls are outlined in section 4. The approach is similar to that in BS 5628 and principally involves calculating the slenderness ratio and design resistance of the wall.

The slenderness ratio of a wall, defined in clause 4.4.6 as the effective height, h_{ef}, divided by the effective thickness, t_{ef}, should not exceed 27. The effective thickness of a single leaf wall is equal to the actual thickness, but for cavity walls in which the leaves are connected by suitable wall ties it is generally given by

$$t_{ef} = \sqrt[3]{t_1^3 + t_2^3} \qquad (10.2)$$

where t_1 and t_2 are the thicknesses of the two leaves.

According to clause 6.4(c) of the NAD, the effective thickness of walls stiffened with piers should be taken from Fig. 3 (reproduced here as Fig. 5.13) and Table 5 (reproduced here as Table 5.11), both of BS 5628.

The effective wall height is a function of the actual wall height, h, and end/edge restraints. It can be taken as

$$h_{ef} = \rho_n h \qquad (10.3)$$

ρ_n is a reduction factor where $n = 2$, 3 or 4 depending on the number of restrained and stiffened edges. Thus, $n = 2$ for walls restrained at the top and bottom only, $n = 3$ for walls restrained top and bottom and stiffened on one vertical edge with the other vertical edge free and $n = 4$ for walls restrained top and bottom and stiffened on two vertical edges.

For walls with simple resistance (Fig. 5.10) at the top and bottom, $\rho_2 = 1$ as in BS 5628. For walls with enhanced resistance (Fig. 11) at the top and bottom, $\rho_2 = 0.75$ as in BS 5628. However, in the case of walls supported by timber floors EC6 recommends that ρ_2 be taken as 1.0 rather than 0.75 as in BS 5628. ρ_n assumes other values if the wall is additionally stiffened along one or more vertical edges. See clause 4.4.4.3 of EC6 for details.

According to clause 4.4.2 of EC6, the vertical design resistance of a single leaf unreinforced wall per unit length, N_{Rd}, is given by

$$N_{Rd} = \frac{\Phi_{i,m} t f_k}{\gamma_M} \qquad (10.4)$$

where

$\Phi_{i,m}$ is the capacity reduction factor, Φ_i or Φ_m, as appropriate

t is the thickness of the wall

f_k is the characteristic compressive strength of masonry obtained from equation 10.1

γ_M is the material partial safety factor for masonry determined from Table 10.1.

The factor Φ_i is given by

$$\Phi_i = 1 - 2e_i/t \qquad (10.5)$$

where e_i is the eccentricity at the top or bottom of the wall calculated from

$$e_i = M_i/N_i + e_{hi} + e_a \geqslant 0.05t \qquad (10.6)$$

where

M_i is the design bending moment at the top or bottom of the wall

N_i is the design vertical load at the top or bottom of the wall

e_{hi} is the eccentricity at the top or bottom of the wall, if any, resulting from horizontal loads e.g. wind

e_a is the accidental eccentricity resulting from construction inaccuracies and can be taken as

$$h_{ef}/450 \qquad (10.7)$$

The factor Φ_m is obtained using Fig. 4.2 of EC6 and involves determining the eccentricity within the middle one-fifth of the wall height, e_{mk}, given by

$$e_{mk} = M_m/N_m + e_{hm} \pm e_k \qquad (10.8)$$

where

M_m is the greatest bending moment within the middle one-fifth height of the wall

N_m is the greatest vertical load within the middle one-fifth height of the wall

e_{hm} is the greatest eccentricity within the middle one-fifth height of the wall, if any, due to lateral loads

e_k is the eccentricity due to creep and is equal to zero for walls built of clay and natural stone units.

A simplified sub-frame analysis procedure described in Annex C of EC6 can be used to calculate the values of the design bending moments and thence the corresponding values of e_i and e_{mk}. Finally the smaller of Φ_i and Φ_m is used to estimate the vertical design resistance of the wall.

Clause 4.4.2(3) notes that where the cross-sectional area of a wall is less than 0.1 m², the characteristic compressive strength of masonry should be multiplied by $(0.7 + 3A)$, where A is the loaded horizontal gross cross-sectional area of the member, expressed in square metres. The modification factor for narrow walls is accounted for in EC6 in the choice of the "K" factor (equation 10.1).

SECTION 5: STRUCTURAL DETAILING

Section 5 describes the detailing rules which must be followed if the design methods are to be valid. It covers aspects such as the selection of masonry materials for durability, minimum thickness of walls and bonding of masonry.

SECTION 6: CONSTRUCTION

Section 6 gives minimum requirements for the standards of workmanship relating to, amongst others, handling and storage of masonry units and other materials, tolerance limits on verticality and straightness of walls and the location and formation of movement joints. Execution control is described in Clause 6.9.

10.4 Conclusions and future direction

A review of ENV Eurocode 6: Part 1.1 rules for the design of vertically loaded walls has shown that they are not too dissimilar to current practice. The rules for the design of laterally loaded panels given in ENV Eurocode 6: Part 1.3 are almost identical to those in BS 5628.

Work on converting Parts 1.1, 1.2 and 1.3 of Eurocode 6 to full EN standard is at an advanced stage. The latest draft of EN Eurocode 6: Part 1.1 has been reorganised and it is more concise than the ENV. Boxed values have been removed and replaced by classes. For classes a range of useable values is given for those issues where single figure agreement cannot be reached (e.g. γ_M) and countries can choose one as their appropriate safety level. Classes have been adopted for many of the previous boxed values, but the number of these instances has been reduced by about half from the ENV (about 160 boxed values) to the EN (about 80 class situations). Another significant change is that

Part 1.3 has now been fully incorporated into Part 1.1 of the EN draft. Parts 1.1 and 1.2 are currently being converted to EN's and 1.2 is only a few months behind 1.1 (because it needs to draw on some 1.1 information which has not up to now been fully agreed – like the unit groupings). Part 1.1 should go for voting late summer 2003 with Part 1.2 following on. Parts 2 and 3 are being converted to EN's at the time of writing and fairly developed drafts are available to the project teams working on them. They will follow-on from 1.1 and 1.2. After publication of the EN's, a period of co-existence of around five years with BS 5628 will be permitted. It is highly likely therefore that full implementation of Eurocode 6 will take place by the end of this decade.

Chapter 11

Eurocode 5: Design of timber structures

This chapter briefly describes the content of Part 1.1 of Eurocode 5, the new European standard for the design of buildings in timber, which was published as an ENV in the late summer of 1993. The chapter highlights the principal differences between the standard and its British equivalent, BS 5268: Part 2. It also includes a number of worked examples to illustrate the new procedures for designing flexural and compression members.

11.1 Introduction

Eurocode 5 applies to the design of building and civil engineering structures in timber. It is based on limit state principles and comes in several parts as shown in *Table 11.1*.

Part 1–1 of Eurocode 5, which is the subject of this discussion, gives a general basis for the design of buildings and civil engineering works in timber. It is largely similar in scope to Part 2 of BS 5268, which was discussed in *Chapter 6*. Part 1–1 of Eurocode 5, hereafter referred to as EC5, was published as a preliminary standard, reference no: DD ENV 1995–1–1, in the late summer of 1993. Part 1–2 of Eurocode 5 on fire was published as a preliminary standard, reference no: DD ENV 1995–1–2, in 1994.

In contrast, work on Part 2 of Eurocode 5 on bridges only began in January 1993 and it was eventually published in 1996. This delay was partly attributable to problems in issuing the CEN mandate and partly the fact that there is not the same

Table 11.1 Overall scope of Eurocode 5

Part	Subject
1–1	General rules and rules for buildings
1–2	Structural fire design
2	Bridges

commercial pressure for timber bridges as those made from concrete and steel.

At the time of writing, EC5 was only available as a draft pre-standard, prENV 1995–1–1, and the comments that follow are largely based on the July 1992 document and incorporates the November amendments.

11.2 Layout

In common with the other structural Eurocodes, EC5 was drafted by a panel of experts drawn from the various EC member state countries. It is based on studies carried out by Working Commission W18: *Timber Structures of the CIB International Council for Building Research Studies and Documentation*, in particular on *CIB Structural Timber Design Code: Report No. 66*, published in 1983. The following subjects are covered in EC5:

Chapter 1: Introduction
Chapter 2: Basis of design
Chapter 3: Material properties
Chapter 4: Serviceability limit states
Chapter 5: Ultimate limit states
Chapter 6: Joints
Chapter 7: Structural detailing and control
Annex A: Determination of 5-percentile characteristic values from test results and acceptance criteria for a sample
Annex B: Mechanically jointed beams
Annex C: Built-up columns
Annex D: The design of trusses with punched metal plate fasteners

As can be appreciated from the above contents list, the organization of material is different from that used in BS 5268 but follows the layout adopted in the other structural Eurocodes. Generally, the design rules in EC5 are sequenced on the basis of action effects rather than on the type of member, as in BS 5268.

All the annexes in EC5 are labelled 'Normative'. As explained in *section 7.4*, this signifies that this material has the same status as the rest of the code but appears here rather than in the body of the code in order to make the document easier to use.

This chapter briefly describes the contents of EC5, in so far as it is relevant to the design of flexural and compression members in solid timber. The rules governing the design of joints are not discussed.

11.3 Principles/application rules, shaded values, symbols

As with the other structural Eurocodes and for the reasons discussed in *Chapter 7*, the clauses in EC5 have been divided into principles and application rules. Principles comprise general statements, definitions, requirements and models for which no alternative is permitted. Principles are preceded by the letter P. The application rules are generally recognized rules which follow the statements and satisfy the requirements given in the principles.

Some of the numerical values in EC5, particularly partial safety coefficients, are shaded. These values are meant to be for guidance only and the appropriate UK values will be specified in the UK *National Application Document* (NAD) which will be published in conjunction with the ENV.

Chapter 1 of EC5 lists the symbols used in EC5. Those relevant to this discussion are reproduced below.

GEOMETRICAL PROPERTIES

b	breadth of beam
h	depth of beam
A	area
i	radius of gyration
I	second moment of area
Z	section modulus

BENDING

l	span
M	bending moment
G	permanent action
Q	variable action
$\sigma_{m,d}$	design normal bending stress
$f_{m,k}$	characteristic bending strength
$f_{m,d}$	design bending strength
k_{mod}	modification factor for strength values
k_{ls}	load-sharing factor
k_{inst}	instability factor for lateral buckling

$E_{0,05}$	characteristic modulus of elasticity (parallel) to grain
$E_{0,mean}$	mean modulus of elasticity (parallel) to grain
G_{mean}	mean shear modulus = $E_{0,mean}/16$
γ_G	partial coefficient for permanent actions
γ_Q	partial coefficient for variable actions
γ_m	partial coefficient for material properties

DEFLECTION

$u_{2,inst}$	instantaneous deflection due to variable loads
$u_{2,fin}$	final deflection due to variable loads
$u_{net,fin}$	net, final deflection due to total load plus precamber (if applied)
k_{def}	deformation factor

VIBRATION

f_1	fundamental frequency of vibration
b	floor width
l	floor length
υ	unit impulse velocity
ζ	damping coefficient
n_{40}	number of first-order modes with natural frequencies below 40 Hz

SHEAR

V_d	design shear force
τ_d	design shear stress
$f_{v,k}$	characteristic shear strength
$f_{v,d}$	design shear strength

BEARING

$F_{90,d}$	design bearing force
l	length of bearing
$\sigma_{c,90,d}$	design compression stress perpendicular to grain
$f_{c,90,k}$	characteristic compression strength perpendicular to grain
$f_{c,90,d}$	design compression strength perpendicular to grain

COMPRESSION

l_{ef}	effective length of column
λ_y, λ_z	slenderness ratios about y–y and z–z axes
$\lambda_{rel,y}, \lambda_{rel,z}$	relative slenderness ratios about y–y and z–z axes
N	design axial force

$\sigma_{c,0,d}$ design compression stress parallel to grain

$f_{c,0,k}$ characteristic compression strength parallel to grain

$f_{c,0,d}$ design compression strength parallel to grain

$\sigma_{m,y,d}$,   design bending stresses parallel to
$\sigma_{m,z,d}$ grain

$f_{m,y,d}$, $f_{m,z,d}$ design bending strengths parallel to grain

k_c compression factor

11.4 Basis of design

As pointed out above, EC5, unlike BS 5268, is based on limit state principles. However, in common with other limit state codes, EC5 recommends that the two principal categories of limit states to be considered in design are the ultimate and serviceability limit states. The terms ultimate state and serviceability state apply in the same way as is understood in other limit state codes. Thus ultimate limit states are those associated with collapse or with other forms of structural failure which may endanger the safety of people, while serviceability limit states correspond to states beyond which specific service criteria are no longer met.

The serviceability limit states which must be checked in EC5 are deflection and vibration. The ultimate limit states, which must be checked singly or in combination, are bending, shear, compression and tension. The various design rules for checking these limit states are discussed later.

11.4.1 ACTIONS

Actions is the Eurocode terminology for loads and imposed deformations. Permanent actions, G, are all the dead loads acting on the structure, including the finishes, fixtures and self-weight of the structure. Variable actions, Q, include the imposed, wind and snow loads.

The characteristic permanent actions, G_k, and variable actions, Q_k, are specified in Eurocode 1: *Actions on Structures*, which is in an advanced state of development. For the time being designers should, therefore, continue using BS 648, BS 6399: Part 1 and CP 3: Chapter 5: Part 2 or BS 6399: Part 2 for characteristic values of actions.

The design values of actions, F_d, are obtained by multiplying the characteristic actions, F_k, by the appropriate partial safety factor for actions, γ_F:

$$F_d = \gamma_F F_k \qquad (11.1)$$

Table 11.2   Partial safety factors for actions in building structures for persistent and transient design situations (based on Table 2.3.3.1, EC5)

	Permanent actions (γ_G)	Variable actions (γ_Q)	
		One with its characteristic value	Others with their combination values
Favourable effect	1.0	0	0
Unfavourable effect	1.35	1.5	1.5

According to EC5, the partial safety factors for permanent actions, γ_G, and variable actions, γ_Q, should normally be taken as 1.35 and 1.5 respectively (*Table 11.2*).

11.4.2 MATERIAL PROPERTIES

EC5, unlike BS 5268, does not contain the material properties, e.g. bending and shear strengths, necessary for sizing members. This information is to be found in a CEN supporting standard for timber products, namely BS EN 338: 1995. Like BS 5268, EN 338 provides for a number of strength classes and gives typical strength and stiffness values and densities for each (*Table 11.3*). However, the European standard specifies 15 strength classes, rather than the sixteen identified in BS 5268 (*Table 6.3*). Grade TR 26 timber is not mentioned in EN 338. More significant, perhaps, is the fact that in EN 338 the bending strengths correspond to the strength class designations but in BS 5268 this is not the case. This is because EN 338 uses characteristic strengths which are generally fifth percentile values derived directly from laboratory tests of five minutes' duration whereas BS 5268 uses grade stress which has been reduced for long-term duration and already includes a safety factor.

One benefit of using characteristic values of material properties rather than grade stresses is that it will make it easier to sanction the use of new materials and components for structural purposes, since such values can be utilized immediately, without first having to determine what reduction factors are needed to convert them to permissible or working values.

The characteristic strength values given in *Table 11.3* are related to a depth in bending and width

Table 11.3 Characteristic values for structural timber strength classes (Table 1, BS EN 338)

Species type		Poplar and conifer species									Deciduous species					
Strength class		C14	C16	C18	C22	C24	C27	C30	C35	C40	D30	D35	D40	D50	D60	D70
Strength properties (N mm^{-2})																
Bending	$f_{m,k}$	14	16	18	22	24	27	30	35	40	30	35	40	50	60	70
Tension parallel	$f_{t,0,k}$	8	10	11	13	14	16	18	21	24	18	21	24	30	36	42
Tension perpendicular	$f_{t,90,k}$	0.3	0.3	0.3	0.3	0.4	0.4	0.4	0.4	0.4	0.6	0.6	0.6	0.6	0.7	0.9
Compression parallel	$f_{c,0,k}$	16	17	18	20	21	22	23	25	26	23	25	26	29	32	34
Compression perpendicular	$f_{c,90,k}$	4.3	4.6	4.8	5.1	5.3	5.6	5.7	6.0	6.3	8.0	8.4	8.8	9.7	10.5	13.5
Shear	$f_{v,k}$	1.7	1.8	2.0	2.4	2.5	2.8	3.0	3.4	3.8	3.0	3.4	3.8	4.6	5.3	6.0
Stiffness properties (kN mm^{-2})																
Mean modulus of elasticity parallel	$E_{0,mean}$	7	8	9	10	11	12	12	13	14	10	10	11	14	17	20
5% modulus of elasticity parallel	$E_{0,05}$	4.7	5.4	6.0	6.7	7.4	8.0	8.0	8.7	9.4	8.0	8.7	9.4	11.8	14.3	16.8
Mean modulus of elasticity perpendicular	$E_{90,mean}$	0.23	0.27	0.30	0.33	0.37	0.40	0.40	0.43	0.47	0.64	0.69	0.75	0.93	1.13	1.33
Mean shear modulus	G_{mean}	0.44	0.50	0.56	0.63	0.69	0.75	0.75	0.81	0.88	0.60	0.65	0.70	0.88	1.06	1.25
Density (kg m^{-2})																
Characteristic density	ρ_k	290	310	320	340	350	370	380	400	420	530	560	590	650	700	900
Average density	ρ_{mean}	350	370	380	410	420	450	460	480	500	640	670	700	780	840	1080

Table 11.4 Partial coefficients for material properties, γ_m (based on Table 2.3.3.2, EC5)

Limit states	γ_m
Ultimate limit states	
Fundamental combinations	
Timber- and wood-based materials	1.3
Steel used in joints	1.1
Accidental combinations	1.0
Serviceability limit states	1.0

Table 11.5 Values of k_{mod} for solid and glued laminated timber plywood (based on Table 3.1.7, EC5)

Load duration class	Service class		
	1	*2*	*3*
Permanent	0.60	0.60	0.50
Long term	0.70	0.70	0.55
Medium term	0.80	0.80	0.65
Short term	0.90	0.90	0.70
Instantaneous	1.10	1.10	0.90

Table 11.6 Service classes

Service class	Moisture content	Typical service conditions
1	$\leq 12\%$	20°C, 65% RH
2	$\leq 20\%$	20°C, 85% RH
3	$> 20\%$	Climatic conditions leading to a higher moisture content than in service class 2

Table 11.7 Load duration classes (Table 3.1.6, EC5)

Load duration class	Order of duration	Examples of loading
Permanent	> 10 years	Self-weight
Long term	6 months–10 years	Imposed storage load
Medium term	1 week–6 months	Imposed occupational loads
Short term	< 1 week	Snow[a] and wind
Instantaneous		Accidental impact

Note: [a] Depending on local conditions – in parts of Scotland snow may be medium term.

in tension of solid timber of 150 mm. For depths in bending or widths in tension of solid members, h, less than 150 mm the characteristic bending or tension strengths may be increased by the factor, k_h, which is given by

$$k_h = (150/h)^{0.2} \qquad (11.2)$$

The characteristic strengths, X_k, are converted to design values, X_d, by dividing them by a partial coefficient for material properties, γ_m, generally taken from *Table 11.4*, and multiplying by a factor k_{mod}, taken from *Table 11.5*.

$$X_d = k_{mod} X_k / \gamma_m \qquad (11.3)$$

Here k_{mod} is a modification factor which takes into account the effect on the strength parameters of the duration of loading and the climatic conditions that the structure will experience in service. EC5 defines three service classes and five load duration classes as summarized in *Tables 11.6* and *11.7* respectively.

Where a load combination consists of actions belonging to different load duration classes the value of k_{mod} should correspond to the action with the shortest duration. For example, for members sub-ject to permanent and variable (imposed) loading, a value of k_{mod} corresponding to the long-term load duration class should be used.

EC5, like BS 5268, allows the design strengths calculated using equation 11.3 to be multiplied by a load-sharing factor, k_{ls}, where several equally spaced similar members are able to resist a common load. Typical members which fall into this category may include joists in flat roofs or floors with a maximum span of 6 m and wall studs with a maximum height of 4 m. According to clause 5.4.6 of EC5, a value of $k_{ls} = 1.1$ may generally be assumed.

Having discussed these more general aspects it is now possible to describe in detail EC5 rules governing the design of flexural and compression members.

11.5 Design of flexural members

The design of flexural members to EC5 principally involves checking the effect of the following actions as discussed next:

Fig. 11.1 Beam axes.

1. bending
2. deflection
3. vibration
4. lateral buckling
5. shear
6. bearing.

11.5.1 BENDING (CLAUSE 5.1.6, EC5)

If members are not to fail in bending, the following conditions should be satisfied:

$$k_m \frac{\sigma_{m,y,d}}{f_{m,y,d}} + \frac{\sigma_{m,z,d}}{f_{m,z,d}} \leq 1 \qquad (11.4)$$

$$\frac{\sigma_{m,y,d}}{f_{m,y,d}} + k_m \frac{\sigma_{m,z,d}}{f_{m,z,d}} \leq 1 \qquad (11.5)$$

where $\sigma_{m,y,d}$ and $\sigma_{m,z,d}$ are the design bending stresses about axes y–y and z–z as shown in *Fig. 11.1*, $f_{m,y,d}$ and $f_{m,z,d}$ the design bending strengths from equation 11.3 and k_m the bending factor. It should be noted that in EC5 the x–x axis is the axis along the member and that axes y–y and z–z are the major and minor axes respectively. These definitions are consistent with the other structural Eurocodes.

For a beam with rectangular cross-sections

$$\sigma_{m,y,d} = \frac{M_y}{Z_y} = \frac{M_y}{bh^2/6} \qquad (11.6)$$

$$\sigma_{m,z,d} = \frac{M_z}{Z_z} = \frac{M_z}{hb^2/6} \qquad (11.7)$$

where M_y and M_z are the design bending moments about axes y–y and z–z, Z_y and Z_z the moduli of elasticity about axes y–y and z–z, b the breadth of beam and h the depth of beam. The value of the factor k_m assumes the following values: for rectangular sections $k_m = 0.7$; for other cross-sections $k_m = 1.0$.

11.5.2 DEFLECTION (CLAUSE 4.3, EC5)

To prevent the possibility of damage to surfacing materials, ceilings, partitions and finishes, and to the functional needs as well as aesthetic requirements, EC5 recommends various limiting values of deflection for beams. The components of deflection are shown in *Fig. 11.2*, where the symbols are defined as

u_0 = precamber (if applied)
u_1 = deflection due to permanent loads
u_2 = deflection due to variable loads
u_{net} = net deflection = $u_1 + u_2 - u_0$

EC5 recommends that the following limiting values will normally need to be observed:

1. Instantaneous deflection due to variable load, $u_{2,inst}$, should not exceed
   $u_{2,inst} \leq 1/300 \times$ span
   $u_{2,inst} \leq 1/150 \times$ span   (for cantilever)
2. Final deflection due to variable load only, $u_{2,fin}$, should not exceed
   $u_{2,fin} \leq 1/200 \times$ span
   $u_{2,fin} \leq 1/100 \times$ span   (for cantilever)
3. Final deflection due to all the loads and any precamber, $u_{net,fin}$, should not exceed
   $u_{net,fin} \leq 1/200 \times$ span
   $u_{net,fin} \leq 1/100 \times$ span   (for cantilever)

The instantaneous deflection of a member in a low-sharing system due to variable loads, $u_{2,inst}$, and the final deflection due to the total load, $u_{net,fin}$, can be calculated using the formulae given in *Table 6.9* and should be based on $E_{0,mean}$ or $E_{90,mean}$. The final deflection due to variable loading, $u_{2,fin}$, is derived from the instantaneous deflection using the following expression:

$$u_{fin} = u_{inst}(1 + k_{def}) \qquad (11.8)$$

where k_{def} is the deformation factor which takes into account the increase in deformation with time due to the combined effect of creep and moisture. Values of k_{def} are given in *Table 11.8*.

11.5.3 VIBRATION (CLAUSE 4.4, EC5)

EC5, unlike BS 5268, gives a procedure for calculating the vibrational characteristics of residential

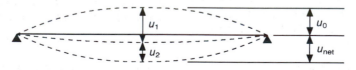

Fig. 11.2 Components of deflection.

Table 11.8 Values of k_{def} for solid and glue-laminated timber (based on Table 4.1, EC5)

Load duration class	Service class		
	1	*2*	*3*
Permanent	0.60	0.80	2.00
Long term	0.50	0.50	1.50
Medium term	0.25	0.25	0.75
Short term	0.00	0.00	0.30

floors. These must satisfy certain requirements otherwise the vibrations may impair the proper functioning of the structure or cause unacceptable discomfort to the users. The method given in EC5 assumes that the floor is supported on four edges which is rarely achieved in UK design; there is some doubt as to the relevance of the EC5 design method to UK floors.

The fundamental frequency of vibration of a rectangular residential floor supported on four edges, f_1, can be estimated using

$$f_1 = \frac{\pi}{2l^2} \sqrt{\frac{(EI)_l}{m}} \qquad (11.9)$$

where m is the mass equal to the self-weight of the floor and other permanent actions per unit area (kg m^{-2}), l the floor span (m) and $(EI)_l$ the equivalent bending stiffness in the beam direction (N m^2 m^{-1}).

For residential floors with a fundamental frequency greater than 8 Hz the following conditions should be satisfied:

$$u/F \le 1.5 \text{ mm kN}^{-1} \qquad (11.10)$$

and

$$v \le 100^{(f_1\zeta - 1)} \qquad (11.11)$$

where ζ is the damping coefficient, normally taken as 0.01, u the maximum vertical deflection caused by a concentrated static force $F = 1$ kN and v the unit impulse velocity.

In carrying out the deflection check in equation 11.10, the transverse distribution of load can be taken as 50%, i.e. 0.5 kN on the loaded joist and 25% on the adjacent ones. The value of v may be estimated from

$$v = 4(0.4 + 0.6n_{40})/(mbl + 200) \text{ m N}^{-1} \text{ s}^{-2} \qquad (11.12)$$

where b is the floor width (m) and n_{40} the number of first-order modes with natural frequencies below 40 Hz given by

$$n_{40} = \left\{ \left[\left(\frac{40}{f_1} \right)^2 - 1 \right] \left(\frac{b}{l} \right)^4 \frac{(EI)_l}{(EI)_b} \right\}^{0.25} \qquad (11.13)$$

where $(EI)_b$ is the equivalent plate bending stiffness parallel to the beams.

11.5.4 LATERAL BUCKLING (CLAUSE 5.2.2, EC5)
Clause 5.1.6(2) of EC5 also requires that beams should be checked for lateral instability. Generally, it will be necessary to show that the following condition is satisfied:

$$\sigma_{m,d} \le k_{inst} f_{m,d} \qquad (11.14)$$

where $\sigma_{m,d}$ is the design bending stress, $f_{m,d}$ the design bending strength and k_{inst} the instability factor. Here k_{inst} is given by

$$k_{inst} = 1 \quad \text{for} \quad \lambda_{rel,m} \le 0.75 \qquad (11.15)$$

$$k_{inst} = 1.56 - 0.75\lambda_{rel,m} \quad \text{for} \quad 0.75 < \lambda_{rel,m} \le 1.4 \qquad (11.16)$$

$$k_{inst} = 1/\lambda_{rel,m}^2 \quad \text{for} \quad 1.4 < \lambda_{rel,m} \qquad (11.17)$$

where $\lambda_{rel,m}$ is the relative slenderness ratio for bending. For beams with rectangular cross-section, $\lambda_{rel,m}$ can be calculated from the following expression:

$$\lambda_{rel,m} = \sqrt{\left[\frac{l_{ef} h f_{m,k}}{\pi b^2 E_{0,k05}} \sqrt{\left(\frac{E_{0,mean}}{G_{mean}} \right)} \right]} \qquad (11.18)$$

where l_{ef} is the effective length of the beam and is obtained from *Fig. 11.3*, b the width of beam, h the depth of beam, $f_{m,k}$ the characteristic bending strength (*Table 11.3*), $E_{0,k05}$ the characteristic modulus of elasticity parallel to grain (*Table 11.3*), $E_{0,mean}$ the mean modulus of elasticity parallel to grain (*Table 11.3*) and G_{mean} the mean shear modulus = $E_{0,mean}/16$.

11.5.5 SHEAR (CLAUSE 5.1.7, EC5)
If flexural members are not to fail in shear, the following condition should be satisfied:

$$\tau_d \le f_{v,d} \qquad (11.19)$$

where τ_d is the design shear stress and $f_{v,d}$ the design shear strength.

	The load is acting at the		
	top	mid-depth	bottom
	0.95	0.9	0.85
	0.8α	0.75α	0.7α
		$\alpha = 1.35 - 1.4\dfrac{x}{\ell}\dfrac{\ell-x}{\ell}$	
		1	
		0.6	
			0.85

Fig. 11.3 *Ratios of l_{ef}/l.*

For beams with a rectangular cross-section, the design shear stress occurs at the neutral axis and is given by

$$\tau_d = \frac{3V_d}{2A} \qquad (11.20)$$

where V_d is the design shear force and A the cross-sectional area. The design shear strength, $f_{v,d}$, is given by

$$f_{v,d} = \frac{k_{mod}f_{v,k}}{\gamma_m} \qquad (11.21)$$

where $f_{v,k}$ is the characteristic shear strength (*Table 11.3*).

For beams notched at the ends as shown in *Fig. 11.4*, the following condition should be checked:

$$\tau_d \leq k_v f_{v,d} \qquad (11.22)$$

where k_v is the shear factor which may attain the following values: (a) for beams notched on the

unloaded side $k_v = 1$ and (b) for beams of solid timber notched on the loaded side k_v is taken as the lesser of $k_v = 1$ and

$$k_v = \frac{5\left(1 + \dfrac{1.1i^{1.5}}{\sqrt{h}}\right)}{\sqrt{h}\left[\sqrt{\alpha(1-\alpha)} + 0.8\dfrac{x}{h}\sqrt{\left(\dfrac{1}{\alpha} - \alpha^2\right)}\right]} \qquad (11.23)$$

where h is the beam depth (mm), x the distance from line of action to the corner, $\alpha = h_e/h$ (*Fig. 11.4*) and i is defined in *Fig. 11.4*.

11.5.6 COMPRESSION PERPENDICULAR TO GRAIN (CLAUSE 5.1.4, EC5)

For compression perpendicular to the grain the following condition should be satisfied:

$$\sigma_{c,90,d} \leq k_{c,90}f_{c,90,d} \qquad (11.24)$$

where $\sigma_{c,90,d}$ is the design compressive stress perpendicular to grain, $f_{c,90,d}$ the design compressive strength perpendicular to grain from equation 11.3 and $k_{c,90}$ the compressive strength factor. Here, $k_{c,90}$ takes into account that the load can be increased if the loaded length, l in *Fig. 11.5*, is short. *Table 11.9* shows values for $k_{c,90}$ for various combinations of a, l and l_1.

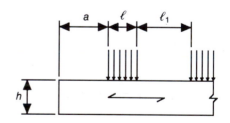

Fig. 11.5 *Compression perpendicular to grain (Fig. 5.1.5(a), EC5).*

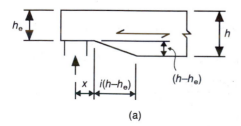

(a)

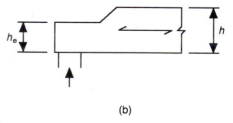

(b)

Fig. 11.4 *End notched beams: (a) notch on loaded side; (b) notch on unloaded side (Fig. 5.1.7.2, EC5).*

Table 11.9   Values of $k_{c,90}$ (Table 5.1.5, EC5)

	$l_1 \leqslant 150$ mm	$l_1 > 150$ mm	
		$a \geqslant 100$ mm	$a < 100$ mm
$l \geqslant 150$ mm	1	1	1
150 mm $> l \geqslant 15$ mm	1	$1 + \dfrac{150 - l}{170}$	$1 + \dfrac{a(150 - l)}{17\,000}$
15 mm $> l$	1	1.8	$1 + a/125$

Example 11.1   Design of timber floor joists (EC5)

Design the timber floor joists for a domestic dwelling using timber of strength class C22 given that the:

1. Floor width, b, is 3.6 m and floor span, l, is 3.4 m.
2. Joists are spaced at 600 mm centres.
3. Flooring is tongue and groove boarding of thickness 21 mm and has a self-weight of 0.1 kN m^{-2}.
4. Ceiling is of plasterboard with a self-weight of 0.2 kN m^{-2}.
5. The bearing length is 100 mm.

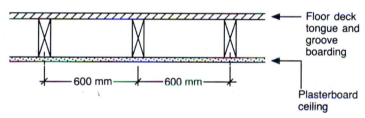

DESIGN LOADING

Permanent loading, G_k

Tongue and groove boarding	$= 0.10$ kN m^{-2}
Ceiling	$= 0.20$ kN m^{-2}
Joists (say)	$= 0.10$ kN m^{-2}
Total characteristic permanent load	$= 0.40$ kN m^{-2}

Variable load, Q_k

Imposed floor load for domestic dwelling (BS 6399: Part 1) is 1.50 kN m^{-2}.

Design load

Total design load is

$$\gamma_G G_k + \gamma_Q Q_k = 1.35 \times 0.40 + 1.5 \times 1.5 = 2.79 \text{ kN m}^{-2}$$

Design load/joist, F_d, is

$$F_d = \text{joist spacing} \times \text{effective span} \times \text{load} = 0.6 \times 3.4 \times 2.79 = 5.7 \text{ kN}$$

CHARACTERISTIC STRENGTHS AND MODULUS OF ELASTICITY FOR TIMBER OF STRENGTH CLASS C22

Values in N mm^{-2} are given as follows:

Bending strength ($f_{m,k}$)	Compression perpendicular to grain ($f_{c,90,k}$)	Shear parallel to grain ($f_{v,k}$)	Modulus of elasticity ($E_{0,mean}$)
22.0	5.1	2.4	10 000

BENDING

Bending moment

$$(M) = \frac{Wl}{8} = \frac{5.7 \times 3.4}{8} = 2.42 \text{ kN m}$$

Assuming that the average moisture content of the timber joists does not exceed 20% during the life of the structure, the design should be based on service class 2 (*Table 11.6*). Further, since the joists are required to carry permanent and variable (imposed) loads, the critical load duration class is 'medium term' (*Table 11.7*). Hence from *Table 11.5*, $k_{mod} = 0.8$. From *Table 11.4*, γ_m (for ultimate limit state) = 1.3. Since the joists satisfy the conditions for a load-sharing system outlined in clause 5.4.6 of EC5, the design strengths can be multiplied by a load-sharing factor, $k_{ls} = 1.1$. Assuming $k_h = 1$, design bending strength about the y–y axis is

$$f_{m,y,d} = k_h k_{ls} k_{mod} f_{m,k}/\gamma_m = 1.0 \times 1.1 \times 0.8 \times 22/1.3 = 14.9 \text{ N mm}^{-2}$$

The design bending stress is obtained by substituting into equation 11.5. Hence

$$\frac{\sigma_{m,y,d}}{f_{m,y,d}} + k_m \frac{\sigma_{m,z,d}}{f_{m,z,d}} \leqslant 1$$

$$\frac{\sigma_{m,y,d}}{14.9} + 0.7 \frac{0}{14.9} \leqslant 1$$

$$\sigma_{m,y,d} = 14.9 \text{ N mm}^{-2}$$

$$Z_y \text{req} \geqslant \frac{M_y}{\sigma_{m,y,d}} = \frac{2.42 \times 10^6}{14.9} = 162 \times 10^3 \text{ mm}^3$$

From *Table 6.8* a 50 × 200 mm joist would be suitable ($Z_y = 333 \times 10^3 \text{ mm}^3$, $I_y = 33.3 \times 10^6 \text{ mm}^4$, $A = 10 \times 10^3 \text{ mm}^2$).

Since $h > 150$ mm, $k_h = 1$ (as assumed).

DEFLECTION

Check $u_{2,inst}$

From *Table 11.4*, γ_m (for serviceability limit state) is 1.0 and factored variable load, Q, is

$$Q = \gamma_m Q_k = 1.0 \times 1.5 = 1.5 \text{ kN m}^{-2}$$

Factored variable load per joist is

$$\text{Total load} \times \text{joist spacing} \times \text{span length} = 1.5 \times 0.6 \times 3.4 = 3.06 \text{ kN}$$

From *Table 6.9*, instantaneous deflection due to variable load, $u_{2,inst}$, is given by

$u_{2,\text{inst}}$ = bending deflection + shear deflection

$$= \frac{5}{384} \times \frac{Wl^3}{EI} + \frac{12}{5} \times \frac{Wl}{EA}$$

$$= \frac{5}{384} \times \frac{3.06 \times 10^3 \times (3.4 \times 10^3)^3}{10 \times 10^3 \times 33.3 \times 10^6} + \frac{12}{5} \times \frac{3.06 \times 10^3 \times 3.4 \times 10^3}{10 \times 10^3 \times 10 \times 10^3}$$

$$= 4.7 + 0.2 = 4.9 \text{ mm}$$

Permissible instantaneous deflection is $1/300 \times \text{span} = 3.4 \times 10^3/300 = 11.3 \text{ mm} > 4.9 \text{ mm}$ OK

Check $u_{2,\text{fin}}$

From *Table 11.8*, for solid timber members subject to service class 2 and medium-term loading, $k_{\text{def}} = 0.25$. Final deflection due to variable load only is given by

$$u_{2,\text{fin}} = u_{\text{inst}}(1 + k_{\text{def}}) = 4.9(1 + 0.25) = 6.1 \text{ mm}$$

Permissible final deflection is $1/200 \times \text{span} = 1/200 \times 3.4 \times 10^3 = 17 \text{ mm} > 6.1 \text{ mm}$ OK

Check $u_{\text{net,fin}}$

Factored permanent load, $G = \gamma_m G_k = 1.0 \times 0.40 = 0.40 \text{ kN m}^{-2}$ and factored permanent load per joist is

Total load $\times$ joist spacing $\times$ span length $= 0.4 \times 0.6 \times 3.4 = 0.82 \text{ kN}$

Instantaneous deflection due to permanent load is given by

$u_{1,\text{inst}}$ = bending deflection + shear deflection

$$= \frac{5}{384} \times \frac{Wl^3}{EI} + \frac{12}{5} \times \frac{Wl}{EA}$$

$$= \frac{5}{384} \times \frac{0.82 \times 10^3 \times (3.4 \times 10^3)^3}{10 \times 10^3 \times 33.3 \times 10^6} + \frac{12}{5} \times \frac{0.82 \times 10^3 \times 3.4 \times 10^3}{10 \times 10^3 \times 10 \times 10^3}$$

$$= 1.26 + 0.07 = 1.3 \text{ mm}$$

From *Table 11.8*, for solid timber members subject to service class 2 and permanent loading, $k_{\text{def}} = 0.8$. Hence final deflection due to permanent loading, $u_{1,\text{fin}}$, is given by

$$u_{1,\text{fin}} = u_{1,\text{inst}}(1 + k_{\text{def}}) = 1.3(1 + 0.8) = 2.3 \text{ mm}$$

and final deflection due to permanent and variable loading is given by

$$u_{\text{net,fin}} = u_{1,\text{fin}} + u_{2,\text{fin}} = 2.3 + 6.1 = 8.4 \text{ mm}$$

Permissible deflection is

$$1/200 \times \text{span} = 1/200 \times 3.4 \times 10^3 = 17 \text{ mm} > 8.4 \text{ mm}$$ OK

Therefore a 50 $\times$ 200 mm joist is adequate in deflection.

VIBRATION

Assuming that $f_1 > 8$ Hz, check that $u/F \leqslant 1.5 \text{ mm kN}^{-1}$ and $v \leqslant 100^{(f_1\zeta-1)}$

Check u/F ratio

Assuming that due to transverse distribution, 50% of the concentrated load occurs on the joist, i.e. $W = F/2 = 0.5$ kN, from *Table 6.9* the maximum deflection, u, is given by

$$u = \frac{1}{48} \times \frac{Wl^3}{EI} + \frac{24}{5} \times \frac{Wl}{EA}$$

$$= \frac{1}{48} \times \frac{0.50 \times 10^3 \times (3.4 \times 10^3)^3}{10 \times 10^3 \times 33.3 \times 10^6} + \frac{24}{5} \times \frac{0.50 \times 10^3 \times 3.4 \times 10^3}{10 \times 10^3 \times 10 \times 10^3}$$

$$= 1.23 + 0.08 = 1.3 \text{ mm}$$

Hence $u/F = 1.3/1 = 1.3$ mm kN^{-1} < permissible = 1.5 mm kN^{-1} OK

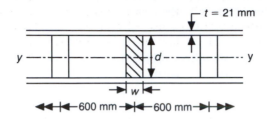

Check impulse velocity

Floor width, $b = 3.6$ m and floor span, $l = 3.4$ m. I_y = second moment of area of joist (ignore tongue and groove boarding unless a specific shear calculation at the interface of joist and board is made):

$$I_y = 33.3 \times 10^6 \text{ mm}^4 = 33.3 \times 10^{-6} \text{ m}^4$$

$$E_{0,\text{mean}} = 10\,000 \text{ N mm}^{-2} = 10 \times 10^9 \text{ N m}^{-2} \quad (\textit{Table 11.3})$$

$$(EI)_1 = E_{0,\text{mean}} I_y / \text{joist spacing}$$

$$= 10 \times 10^9 \times 33.3 \times 10^{-6}/0.6 = 555 \times 10^3 \text{ Nm}^2 \text{ m}^{-1}$$

Mass due to permanent actions per unit area, m, is

$$m = \text{permanent action/gravitational constant} = 0.40 \times 10^3/9.81 = 40.8 \text{ kg m}^{-2}$$

$$f_1 = \frac{\pi}{2l^2} \sqrt{\frac{(EI)_1}{m}}$$

$$= \frac{\pi}{2 \times 3.4^2} \sqrt{\frac{555 \times 10^3}{40.8}}$$

$$= 15.8 \text{ Hz} > 8 \text{ Hz as assumed}$$

I (parallel to beam) is tongue and groove boarding and is

$$I = bt^3/12 = 1000 \times 21^3/12$$

$$= 0.772 \times 10^6 \text{ mm}^4 = 0.772 \times 10^{-6} \text{ m}^4$$

$$(EI)_b = (E_{0,\text{mean}} I) = 10 \times 10^9 \times 0.772 \times 10^{-6} = 7.72 \times 10^3 \text{ Nm}^2 \text{ m}^{-1}$$

$$n_{40} = \left\{ \left[\left(\frac{40}{f_1} \right)^2 - 1 \right] \left(\frac{b}{l} \right)^4 \frac{(EI)_l}{(EI)_b} \right\}^{0.25}$$

$$= \left\{ \left[\left(\frac{40}{15.8} \right)^2 - 1 \right] \left(\frac{3.6}{3.4} \right)^4 \frac{555 \times 10^3}{7.72 \times 10^3} \right\}^{0.25} = 4.7$$

$$v = 4(0.4 + 0.6n_{40})/(mbl + 200) \text{ m N}^{-1}\text{ s}^{-2}$$

$$= 4(0.4 + 0.6 \times 4.7)/(40.8 \times 3.6 \times 3.4 + 200)$$

$$= 0.0184 \text{ m N}^{-1}\text{ s}^{-2}$$

Assume damping coefficient $\zeta = 0.01$. Permissible floor velocity is

$$100^{(f_1\zeta-1)} = 100^{(15.8 \times 0.01 - 1)} = 0.02 \text{ m N}^{-1}\text{ s}^{-2} > v = 0.0184 \text{ m N}^{-1}\text{ s}^{-2} \quad \text{OK}$$

As discussed in section 11.5.3, there is some doubt as to the relevance of this method for checking the vibrational characteristics of UK floors. The UK NAD therefore suggests an alternative method of establishing compliance which simply involves checking that the total instantaneous deflection, u_{inst}, of the floor joist under load does not exceed 14 mm or span/333, whichever is the lesser. In this example, $u_{inst} = u_{1,inst} + u_{2,inst} = 1.3 + 4.9 = 6.2$ mm < 14 mm and span/333 = 3400/333 = 10.2 mm. Hence the joist is satisfactory.

LATERAL BUCKLING

From *Fig. 11.3*, $l_{ef}/l = 0.95$. Hence

$$l_{ef} = 0.95l = 0.95 \times 3.4 \times 10^3 = 3230 \text{ mm}$$

$$\lambda_{rel,m} = \sqrt{\left[\frac{l_{ef}h}{\pi b^2} \times \frac{f_{m,k}}{E_{0,05}} \sqrt{\left(\frac{E_{0,mean}}{G_{mean}}\right)}\right]}$$

$$= \sqrt{\left[\frac{3230 \times 200}{\pi \times 50^2} \times \frac{22}{6700} \sqrt{\left(\frac{E_{0,mean}}{E_{0,mean}/16}\right)}\right]}$$

$$= 1.04$$

For $0.75 < \lambda_{rel,m} < 1.4$

$$k_{inst} = 1.56 - 0.75\lambda_{rel,m} = 1.56 - 0.75 \times 1.04 = 0.78$$

Buckling strength is

$$k_{inst}f_{m,d} = 0.78 \times 14.9 = 11.6 \text{ N mm}^{-2}$$

Buckling stress is

$$\sigma_{m.d} = \frac{M}{Z} = \frac{2.42 \times 10^6}{333 \times 10^3} = 7.3 \text{ N mm}^{-2} < 11.6 \text{ N mm}^{-2} \quad \text{OK}$$

SHEAR

Design shear strength is

$$f_{v,d} = k_{ls}k_{mod}f_{v,k}/\gamma_m = 1.1 \times 0.8 \times 2.4/1.3 = 1.62 \text{ N mm}^{-2}$$

Maximum shear force is

$$F_v = W/2 = \frac{5.7 \times 10^3}{2} = 2.85 \times 10^3 \text{ N}$$

Design shear stress at neutral axis is

$$\tau_d = \frac{3}{2} \times \frac{F_v}{A} = \frac{3}{2} \times \frac{2.85 \times 10^3}{10 \times 10^3}$$

$$= 0.43 \text{ N mm}^{-2} < \text{permissible}$$

Therefore joist is adequate in shear.

BEARING

Design compressive stress
Design bearing force is

$$F_{90,d} = W/2 = 5.7 \times 10^3/2 = 2.85 \times 10^3 \text{ N}$$

Assuming that the floor joists span on to 100 mm wide walls as shown above, the bearing stress is given by

$$\sigma_{c,90,d} = \frac{F_{90,d}}{bl} = \frac{2.85 \times 10^3}{50 \times 100} = 0.57 \text{ N mm}^{-2}$$

Design compressive strength
Design compressive strength perpendicular to grain, $f_{c,90,d}$, is given by

$$f_{c,90,d} = k_{ls} k_{mod} f_{c,90,k}/\gamma_m$$
$$= 1.1 \times 0.8 \times 5.1/1.3 = 3.4 \text{ N mm}^{-2}$$

Factor $k_{c,90}$
By comparing the above diagram with *Fig. 11.5*, it can be seen that $a = 0$, $l = 100$ mm and $l_1 > 150$ mm. From *Table 11.9*

$$k_{c,90} = 1 + \frac{a(150 - l)}{17\,000} = 1$$

From above,

$$\sigma_{c,90,d} (= 0.57 \text{ N mm}^{-2}) < k_{c,90} f_{c,90,d} = 1.0 \times 3.4 = 3.4 \text{ N mm}^{-2}$$

Hence the joist is adequate in bearing.

CHECK ASSUMED SELF-WEIGHT OF JOISTS
From *Table 11.3*, density of timber of strength class C22 is 410 kg m^{-3}. Hence self-weight of the joists is

$$\frac{50 \times 200 \times 10^{-6} \times 410 \text{ kg m}^{-3} \times 9.81 \times 10^{-3}}{0.6} = 0.07 \text{ kN m}^{-2} < \text{assumed} \quad \text{OK}$$

Example 11.2 Design of a notched floor joist (EC5)

The joists in *Example 11.1* are to be notched at the bearings with a 75 mm deep notch as shown below. Check the notched section is still adequate.

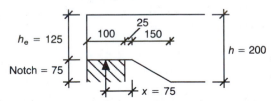

The presence of the notch only affects the shear stresses in the joists.

FACTOR k_v

For beams notched on the loaded side, k_v is taken as the lesser of 1 and the value calculated using equation 11.23. Comparing the above diagram with *Fig. 11.4(a)* gives

$$i = 2 \quad x = 775 \text{ mm}$$

$$\alpha = h_e/h = 125/200 = 0.625$$

$$k_v = \frac{5\left(1 + \dfrac{1.1 i^{1.5}}{\sqrt{h}}\right)}{\sqrt{h}\left(\sqrt{\alpha}\,(1 - \alpha) + 0.8\dfrac{x}{h}\sqrt{\left(\dfrac{1}{\alpha} - \alpha^2\right)}\right)}$$

$$k_v = \frac{5\left(1 + \dfrac{1.1 \times 2^{1.5}}{\sqrt{200}}\right)}{\sqrt{200}\left(\sqrt{0.625}(1 - 0.625) + 0.8\dfrac{75}{200}\sqrt{\left(\dfrac{1}{0.625} - 0.625^2\right)}\right)} = 0.53$$

The shear strength, $f_{v,d}$, is

$$f_{v,d} = k_{ls}k_{mod}f_{v,k}/\gamma_m = 1.1 \times 0.8 \times 2.4/1.3 = 1.62 \text{ N mm}^{-2}$$

For a notched member, the design shear strength is given by $k_v f_{v,d} = 0.53 \times 1.62 = 0.86 \text{ N mm}^{-2}$

Design shear stress is $\tau_d = 1.5 V_d/bh_e = 1.5 \times 2.85 \times 10^3/50 \times 125$

$$= 0.68 \text{ N mm}^{-2} < 0.86 \text{ N mm}^{-2}$$

Therefore the section is also adequate when notched with a 75 mm deep bottom edge notch at the bearing.

11.6 Design of compression members

11.6.1 MEMBERS SUBJECT TO AXIAL COMPRESSION ONLY (CLAUSE 5.1.4, EC5)

Members subject to axial compression only should be designed according to the following expression provided there is no tendency for buckling to occur:

$$\sigma_{c,0,d} \leq f_{c,0,d} \qquad (11.25)$$

where $f_{c,0,d}$ is the design compressive strength parallel to the grain obtained from equation 11.3 and $\sigma_{c,0,d}$ the design compressive stress parallel to the grain given by

$$\sigma_{c,0,d} = \frac{N}{A} \qquad (11.26)$$

in which N is the axial load and A is the cross-sectional area.

11.6.2 COLUMNS SUBJECT TO BENDING AND AXIAL COMPRESSION

EC5 gives two sets of conditions for designing columns resisting combined bending and axial compression. Provided that the relative slenderness ratios about both the y–y and z–z axes of the column, $\lambda_{rel,y}$ and $\lambda_{rel,z}$ respectively, are not greater than 0.5, i.e.

$$\lambda_{rel,y} \leq 0.5 \quad \text{and} \quad \lambda_{rel,z} \leq 0.5$$

the suitability of the design can be assessed using the more stringent of the following conditions:

$$\left(\frac{\sigma_{c,0,d}}{f_{c,0,d}}\right)^2 + \frac{\sigma_{m,y,d}}{f_{m,y,d}} + k_m\frac{\sigma_{m,z,d}}{f_{m,z,d}} \leq 1 \quad (11.27)$$

$$\left(\frac{\sigma_{c,0,d}}{f_{c,0,d}}\right)^2 + k_m\frac{\sigma_{m,y,d}}{f_{m,y,d}} + \frac{\sigma_{m,z,d}}{f_{m,z,d}} \leq 1 \quad (11.28)$$

where $\sigma_{c,0,d}$ is the design compressive stress from equation 11.26, $f_{c,0,d}$ the design compressive strength

Eurocode 5: Design of timber structures

from equation 11.3 and $k_m = 0.7$ for rectangular sections and 1.0 for other cross-sections. Note that

$$\lambda_{\mathrm{rel},y} = \sqrt{\left(\frac{f_{c,0,k}}{\sigma_{c,\mathrm{crit},y}}\right)} \qquad (11.29)$$

and

$$\lambda_{\mathrm{rel},z} = \sqrt{\left(\frac{f_{c,0,k}}{\sigma_{c,\mathrm{crit},z}}\right)} \qquad (11.30)$$

where

$$\sigma_{c,\mathrm{crit},y} = \frac{\pi^2 E_{0,05}}{\lambda_y^2} \qquad (11.31)$$

$$\sigma_{c,\mathrm{crit},z} = \frac{\pi^2 E_{0,05}}{\lambda_z^2} \qquad (11.32)$$

in which λ is the slenderness ratio given by

$$\lambda = \frac{l_{ef}}{i} \qquad (11.33)$$

where l_{ef} is the effective length and i the radius of gyration. EC5 does not include a method for determining the effective length of a column. Therefore, designers should follow the recommendation contained in BS 5268: Part 2, as discussed in *section 6.7.1*.

In all other cases the stresses should satisfy the more stringent of the following conditions:

$$\frac{\sigma_{c,0,d}}{k_{c,y}f_{c,0,d}} + \frac{\sigma_{m,y,d}}{f_{m,y,d}} + k_m\frac{\sigma_{m,z,d}}{f_{m,z,d}} \leq 1 \quad (11.34)$$

$$\frac{\sigma_{c,0,d}}{k_{c,z}f_{c,0,d}} + k_m\frac{\sigma_{m,y,d}}{f_{m,y,d}} + \frac{\sigma_{m,z,d}}{f_{m,z,d}} \leq 1 \quad (11.35)$$

where σ_m is the bending stress due to any lateral or eccentric loads

$$k_c = \frac{1}{k + \sqrt{\left(k^2 - \lambda_{\mathrm{rel}}^2\right)}}$$

where $k = 0.5(1 + \beta_c(\lambda_{\mathrm{rel}} - 0.5) + \lambda_{\mathrm{rel}}^2)$ and $\beta_c = 0.2$ (for solid timber).

Example 11.3 Analysis of a column resisting an axial load (EC5)

A machine graded timber column of strength class C16 consists of a 100 mm square section which is restrained at both ends in position but not in direction. Assuming that the service conditions comply with service class 2 and the actual height of the column is 3.75 m, calculate the design axial medium-term load that the column can support.

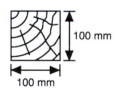

SLENDERNESS RATIO

$$\lambda_y = \lambda_z = \frac{l_{ef}}{i}$$

$$l_{ef} = 1.0 \times h = 1.0 \times 3750 = 3750 \text{ mm} \quad (Table\ 6.11)$$

$$i = \sqrt{(I/A)} = \sqrt{\left(\frac{db^3/12}{db}\right)} = \sqrt{\frac{b^2}{12}} = \frac{100}{\sqrt{12}} = 28.867$$

$$\lambda_y = \lambda_z = \frac{3750}{28.87} = 129.9$$

CHARACTERISTIC STRENGTH AND STIFFNESSES FOR GRADE C16 TIMBER
Values in N mm^{-2} are as follows:

Compressive strength parallel to grain $f_{c,0,k}$	Modulus of elasticity (5-percentile) $E_{0,05}$
17	5400

(EULER) CRITICAL STRESS

$$\sigma_{c,crit,y} = \sigma_{c,crit,z} = \frac{\pi^2 E_{0,05}}{\lambda_y^2} = \frac{\pi^2 5400}{129.9^2} = 3.16$$

$$\lambda_{rel,y} = \lambda_{rel,z} = \sqrt{\left(\frac{f_{c,0,k}}{\sigma_{c,crit,y}}\right)} = \sqrt{\left(\frac{17}{3.16}\right)} = 2.32$$

Since $\lambda_{rel,y}$ and $\lambda_{rel,z} > 0.5$, use equations 11.34 or 11.35.

AXIAL LOAD CAPACITY

$$k = 0.5(1 + \beta_c(\lambda_{rel} - 0.5) + \lambda_{rel}^2)$$
$$= 0.5(1 + 0.2(2.32 - 0.5) + 2.32^2) = 3.37$$

$$k_{c,y} = k_{c,z} = \frac{1}{k + \sqrt{(k^2 - \lambda_{rel}^2)}} = \frac{1}{3.37 + \sqrt{(3.37^2 - 2.32^2)}}$$
$$= 0.17$$

Design compressive strength parallel to grain is given by

$$f_{c,0,d} = k_{mod}f_{c,0,k}/\gamma_m = 0.8 \times 17/1.3 = 10.46 \text{ N mm}^{-2}$$

where $\gamma_m = 1.3$ (*Table 11.4*) and $k_{mod} = 0.8$ (service class 2 and medium-term loading). Since column is axially loaded $\sigma_{m,y,d} = \sigma_{m,z,d} = 0$.
Substituting into equation 11.34 or 11.35 gives $\sigma_{c,0,d} = k_{c,y}f_{c,0,d} = 0.17 \times 10.46 = 1.77$ N mm^{-2}
Hence axial load capacity of column, N, is given by $N = \sigma_{c,0,d}A = 1.77 \times 10^4 = 17.7 \times 10^3$ N = 17.7 kN.

Example 11.4 Analysis of an eccentrically loaded column (EC5)
Check the adequacy of the column in *Example 11.3* to resist a medium-term design (ultimate) axial load of 10 kN applied 35 mm eccentric to its y–y axis.

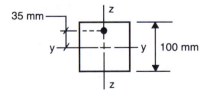

Plan showing column loading

SLENDERNESS RATIO

$$\lambda_y = \lambda_z = 129.9 \quad (\text{Example 11.3})$$

CHARACTERISTIC STRENGTHS AND MODULUS OF ELASTICITY
Values in N mm^{-2} for machine-graded timber of strength class C16:

Bending parallel to grain $f_{m,k}$	Compression parallel to grain $f_{c,0,k}$	Modulus of elasticity 5-percentile value $E_{0,05}$
16	17	5400

COMPRESSION AND BENDING STRESSES AND STRENGTHS
Design compression stress is

$$\sigma_{c,0,d} = \frac{\text{design axial load}}{A} = \frac{10 \times 10^3}{10^4} = 1 \text{ N mm}^{-2}$$

Design compression strength is

$$f_{c,0,d} = k_{mod} f_{c,0,k}/\gamma_m = 0.8 \times 17/1.3 = 10.46 \text{ N mm}^{-2}$$

Design bending moment about y–y axis, M_y, is

$$M_y = \text{Axial load} \times \text{eccentricity} = 10 \times 35 = 350 \text{ kN mm}$$

Design bending stress about y–y axis, $\sigma_{m,y,d}$,

$$\sigma_{m,y,d} = \frac{M_y}{Z_y} = \frac{350 \times 10^3}{167 \times 10^3} = 2.10 \text{ N mm}^{-2}$$

Design bending strength about y–y axis is $f_{m,d}$ and

$$f_{m,d} = k_h k_{mod} f_{m,k}/\gamma_m = 1.08 \times 0.8 \times 16/1.3 = 10.63 \text{ N mm}^{-2}$$

where $k_h = (150/h)^{0.2} = (150/100)^{0.2} = 1.08$. Design bending stress about z–z axis is $\sigma_{m,y,d} = 0$.
  Compression factor is $k_{c,y} = 0.17$ (*Example 11.3*). Check the suitability of the column by using equation 11.34:

$$\frac{\sigma_{c,0,d}}{k_{c,y} f_{c,0,d}} + \frac{\sigma_{m,y,d}}{f_{m,y,d}} + k_m \frac{\sigma_{m,z,d}}{f_{m,z,d}} \leqslant 1$$

$$\frac{1}{0.17 \times 10.46} + \frac{2.1}{10.63} + k_m \frac{0}{f_{m,z,d}} = 0.56 + 0.20 = 0.76 < 1$$

Therefore a 100×100 mm column is adequate.

Appendix A

Permissible stress and load factor design

The purpose of this Appendix is to illustrate the salient features and highlight essential differences between the following philosophies of structural design:

1. permissible stress approach, i.e. elastic design;
2. load factor approach, i.e. plastic design.

The reader is referred to *Chapter 2* for revision of some basic concepts of structural analysis.

Example A.1

Consider the case of a simply supported, solid rectangular beam (*Fig. A.1*), depth (*d*) 200 mm, span (*l*) 10 m and subject to a uniformly distributed load (*w*) of 12 kN m^{-1}. Calculate the minimum width of beam (*b*) using permissible stress and load factor approaches to design assuming the following:

$$\sigma_{\text{yield}} = 265 \text{ N mm}^{-2}$$

Factor of safety (f.o.s.) = 1.5

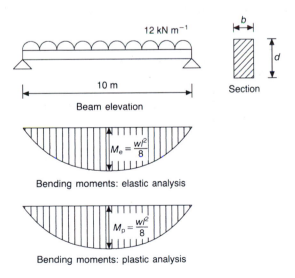

Fig. A.1 *Simply supported beam.*

PERMISSIBLE STRESS APPROACH

Elastic design moment, M_e

$$M_e = \frac{wl^2}{8} = \frac{12 \times 10^2}{8} \quad (Table\ 2.4)$$

$$= 150 \text{ kN m}$$

$$= 150 \times 10^6 \text{ N mm}$$

LOAD FACTOR APPROACH

Plastic design moment, M_p

Plastic section modulus, $S = \dfrac{bd^2}{4}$

Elastic section modulus, $Z = \dfrac{bd^2}{6}$ (*Table 2.4*)

Shape factor (s.f.) = S/Z
Hence

s.f. $= (bd^2/4)/(bd^2/6) = 1.5$

Load factor = s.f. × f.o.s.

$$= 1.5 \times 1.5 = 2.25$$

Actual load (w)   = 12 kN m^{-1}
Factored load (w') = $12 \times 2.25 = 27$ kN m^{-1}

$$M_p = \frac{w'l^2}{8} = \frac{27 \times 10^2}{8}$$

$$= 337.5 \text{ kN m} = 337.5 \times 10^6 \text{ N mm}$$

Moment of resistance, M_r

Permissible stress, σ_{perm}, is

$$\sigma_{perm} = \frac{\sigma_{yield}}{f.o.s.} = \frac{265}{1.5}$$

$$M_r = \sigma_{perm} Z$$

where Z is the elastic section modulus = $bd^2/6$
(*Table 2.4*). Hence

$$M_r = \frac{(265)}{1.5} \times \frac{200^2 b}{6}$$

$$= 1.178 \times 10^6 b \text{ N mm}^{-2}$$

Moment of resistance, M_r

$$M_r = \sigma_{yield} S = \frac{265 \times bd^2}{4}$$

$$= \frac{265 \times 200^2 b}{4} = 2.65 \times 10^6 b$$

Breadth of beam
At equilibrium, $M_e = M_r$

$$150 \times 10^6 = 1.178 \times 10^6 b$$

Hence breadth of beam, b, is

$$b = 150/1.178 = 127 \text{ mm}$$

Breadth of beam
At equilibrium, $M_p = M_r$

$$337.5 \times 10^6 = 2.65 \times 10^6 b$$

Hence breadth of beam, b, is

$$b = 337.5/2.65 = 127 \text{ mm}$$

It can therefore be seen that for the case of a simply supported beam, provided all the factors are taken into account, both approaches will give the same result and this will remain true irrespective of the shape of the section.

Example A.2

Repeat *Example A.1* but this time assume that the beam is built in at both ends as shown in *Fig. A.2*.

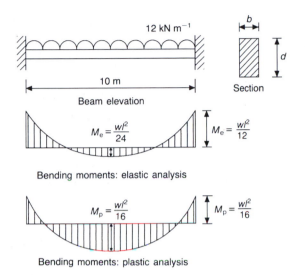

Fig. A.2 Beam with built-in supports.

PERMISSIBLE STRESS APPROACH

Elastic design moment, M_e

$$M_e = \frac{wl^2}{12} = \frac{12 \times 10^2}{12}$$

$$= 100 \text{ kN m}$$

$$= 100 \times 10^6 \text{ N mm}$$

Moment of resistance, M_r

$$M_r = \sigma_{perm}Z = \frac{(265)}{1.5} \times \frac{200^2 b}{6}$$

$$= 1.178 \times 10^6 b \text{ N mm}^{-2}$$

LOAD FACTOR APPROACH

Plastic design moment, M_p

Shape factor (s.f.) $= S/Z$

Hence
s.f. $= (bd^2/4)/(bd^2/6)$
$= 1.5$
Load factor $=$ s.f. $\times$ f.o.s.
$= 1.5 \times 1.5 = 2.25$
Actual load (w) $= 12 \text{ kN m}^{-1}$
Factored load $(w') = 12 \times 2.25$
$= 27 \text{ kN m}^{-1}$

$$M_p = \frac{w'l^2}{16} = \frac{27 \times 10^2}{16}$$

$$= 168.7 \text{ kN m} = 168.7 \times 10^6 \text{ N mm}$$

Moment of resistance, M_r

$$M_r = \sigma_{yield}S = 265 \times \frac{bd^2}{4}$$

$$= 265 \times \frac{200^2 b}{4} = 2.65 \times 10^6 b$$

Breadth of beam

At equilibrium, $M_e = M_r$

$100 \times 10^6 = 1.178 \times 10^6 b$

Hence breadth of beam, b, is

$b = 100/1.178 = 85$ mm

Breadth of beam

At equilibrium, $M_p = M_r$

$168.7 \times 10^6 = 2.65 \times 10^6 b$

Hence breadth of beam, b, is

$b = 168.7/2.65 = 75$ mm

Hence, it can be seen that the load factor approach gives a more conservative estimate for the breadth of the beam, a fact which will be generally found to hold for other sections and indeterminate structures.

The basic difference between these two approaches to design is that while the permissible stress method models behaviour of the structure under working loads, and realistic predictions of behaviour in service can be calculated, the load factor method only models failure, and no information on behaviour in service is obtained.

Dimensions and properties of steel universal beams and columns

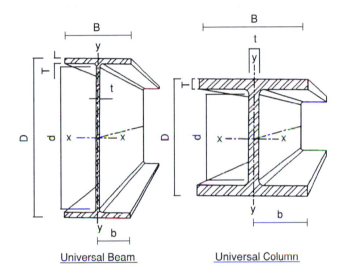

Universal Beam   Universal Column

Table B1 Dimensions and properties of steel universal beams (structural sections to BS 4: Part 1 and BS 4848: Part 4)

Serial size (mm)	Mass per metre (kg)	Depth of section D (mm)	Width of section B (mm)	Thickness Web t (mm)	Thickness Flange T (mm)	Root radius r (mm)	Depth between fillets d (mm)	Flange b/T	Web d/t	Second moment of area Axis x–x (cm^4)	Axis y–y (cm^4)	Radius of gyration Axis x–x (cm)	Axis y–y (cm)	Elastic modulus Axis x–x (cm^3)	Axis y–y (cm^3)	Plastic modulus Axis x–x (cm^3)	Axis y–y (cm^3)	Buckling parameter u	Torsional index x	Warping constant H (dm^6)	Torsional constant J (cm^4)	Area of section A (cm^2)
914×419	388	920.5	420.5	21.5	36.6	24.1	799.1	5.74	37.2	719 000	45 400	38.1	9.58	15 600	2 160	17 700	3 340	0.884	26.7	88.7	1 730	494
	343	911.4	418.5	19.4	32.0	24.1	799.1	6.54	41.2	625 000	39 200	37.8	9.46	13 700	1 870	15 500	2 890	0.883	30.1	75.7	1 190	437
914×305	289	926.6	307.8	19.6	32.0	19.1	824.5	4.81	42.1	505 000	15 600	37.0	6.51	10 900	1 010	12 600	1 600	0.867	31.9	31.2	929	369
	253	918.5	305.5	17.3	27.9	19.1	824.5	5.47	47.7	437 000	13 300	36.8	6.42	9 510	872	10 900	1 370	0.866	36.2	26.4	627	323
	224	910.3	304.1	15.9	23.9	19.1	824.5	6.36	51.9	376 000	11 200	36.3	6.27	8 260	738	9 520	1 160	0.861	41.3	22.0	421	285
	201	903.0	303.4	15.2	20.2	19.1	824.5	7.51	54.2	326 000	9 430	35.6	6.06	7 210	621	8 360	983	0.853	46.8	18.4	293	256
838×292	226	850.9	293.8	16.1	26.8	17.8	761.7	5.48	47.3	340 000	11 400	34.3	6.27	7 990	773	9 160	1 210	0.87	35.0	19.3	514	289
	194	840.7	292.4	14.7	21.7	17.8	761.7	6.74	51.8	279 000	9 070	33.6	6.06	6 650	620	7 650	974	0.862	41.6	15.2	307	247
	176	834.9	291.6	14.0	18.8	17.8	761.7	7.76	54.4	246 000	7 790	33.1	5.90	5 890	534	6 810	842	0.856	46.5	13.0	222	224
762×267	197	769.6	268.0	15.6	25.4	16.5	685.8	5.28	44.0	240 000	8 170	30.9	5.71	6 230	610	7 170	959	0.869	33.2	11.3	405	251
	173	762.0	266.7	14.3	21.6	16.5	685.8	6.17	48.0	205 000	6 850	30.5	5.57	5 390	513	6 200	807	0.864	38.1	9.38	267	220
	147	753.9	265.3	12.9	17.5	16.5	685.8	7.58	53.2	169 000	5 470	30.0	5.39	4 480	412	5 170	649	0.857	45.1	7.41	161	188
686×254	170	692.9	255.8	14.5	23.7	15.2	615.1	5.40	42.4	170 000	6 620	28.0	5.53	4 910	518	5 620	810	0.872	31.8	7.41	307	217
	152	687.6	254.5	13.2	21.0	15.2	615.1	6.06	46.6	150 000	5 780	27.8	5.46	4 370	454	5 000	710	0.871	35.5	6.42	219	194
	140	683.5	253.7	12.4	19.0	15.2	615.1	6.68	49.6	136 000	5 180	27.6	5.38	3 990	408	4 560	638	0.868	38.7	5.72	169	179
	125	677.9	253.0	11.7	16.2	15.2	615.1	7.81	52.6	118 000	4 380	27.2	5.24	3 480	346	4 000	542	0.862	43.9	4.79	116	160
610×305	238	633.0	311.5	18.6	31.4	16.5	537.2	4.96	28.9	208 000	15 800	26.1	7.22	6 560	1 020	7 460	1 570	0.886	21.1	14.3	788	304
	179	617.5	307.0	14.1	23.6	16.5	537.2	6.50	38.1	151 000	11 400	25.8	7.08	4 910	743	5 520	1 140	0.886	27.5	10.1	341	228
	149	609.6	304.8	11.9	19.7	16.5	537.2	7.74	45.1	125 000	9 300	25.6	6.99	4 090	610	4 570	937	0.886	32.5	8.09	200	190
610×229	140	617.0	230.1	13.1	22.1	12.7	547.3	5.21	41.8	112 000	4 510	25.0	5.03	3 630	392	4 150	612	0.875	30.5	3.99	217	178
	125	611.9	229.0	11.9	19.6	12.7	547.3	5.84	46.0	98 600	3 930	24.9	4.96	3 220	344	3 680	536	0.873	34.0	3.45	155	160
	113	607.3	228.2	11.2	17.3	12.7	547.3	6.60	48.9	87 400	3 440	24.6	4.88	2 880	301	3 290	469	0.87	37.9	2.99	112	144
	101	602.2	227.6	10.6	14.8	12.7	547.3	7.69	51.6	75 700	2 910	24.2	4.75	2 510	256	2 880	400	0.863	43.0	2.51	77.2	129
533×210	122	544.6	211.9	12.8	21.3	12.7	476.5	4.97	37.2	76 200	3 390	22.1	4.67	2 800	320	3 200	501	0.876	27.6	2.32	180	156
	109	539.5	210.7	11.6	18.8	12.7	476.5	5.60	41.1	66 700	2 940	21.9	4.60	2 470	279	2 820	435	0.875	30.9	1.99	126	139
	101	536.7	210.1	10.9	17.4	12.7	476.5	6.04	43.7	61 700	2 690	21.8	4.56	2 300	257	2 620	400	0.874	33.1	1.82	102	129
	92	533.1	209.3	10.2	15.6	12.7	476.5	6.71	46.7	55 400	2 390	21.7	4.51	2 080	229	2 370	356	0.872	36.4	1.60	76.2	118
	82	528.3	208.7	9.6	13.2	12.7	476.5	7.91	49.6	47 500	2 010	21.3	4.38	1 800	192	2 060	300	0.865	41.6	1.33	51.3	104
457×191	98	467.4	192.8	11.4	19.6	10.2	407.9	4.92	35.8	45 700	2 340	19.1	4.33	1 960	243	2 230	378	0.88	25.8	1.17	121	125
	89	463.6	192.0	10.6	17.7	10.2	407.9	5.42	38.5	41 000	2 090	19.0	4.28	1 770	217	2 010	318	0.879	28.3	1.04	90.5	114
	82	460.2	191.3	9.9	16.0	10.2	407.9	5.98	41.2	37 100	1 870	18.8	4.23	1 610	196	1 830	304	0.877	30.9	0.923	69.2	105
	74	457.2	190.5	9.1	14.5	10.2	407.9	6.57	44.8	33 400	1 670	18.7	4.19	1 460	175	1 660	272	0.876	33.9	0.819	52.0	95.0
	67	453.6	189.9	8.5	12.7	10.2	407.9	7.48	48.0	29 400	1 450	18.5	4.12	1 300	153	1 470	237	0.873	37.9	0.706	37.1	85.4

Dimensions and Properties

Designation Serial size (mm)	Mass per metre (kg)	Depth of section D (mm)	Width of section B (mm)	Thickness Web t (mm)	Thickness Flange T (mm)	Root radius r (mm)	Depth between fillets d (mm)	Ratios for local buckling Flange b/T	Ratios for local buckling Web d/t	Second moment of area Axis x–x (cm⁴)	Second moment of area Axis y–y (cm⁴)	Radius of gyration Axis x–x (cm)	Radius of gyration Axis y–y (cm)	Elastic modulus Axis x–x (cm³)	Elastic modulus Axis y–y (cm³)	Plastic modulus Axis x–x (cm³)	Plastic modulus Axis y–y (cm³)	Buckling parameter u	Torsional index x	Warping constant H (dm⁶)	Torsional constant J (cm⁴)	Area of section A (cm²)
457×152	82	465.1	153.5	10.7	18.9	10.2	407.0	4.06	38.0	36 200	1 140	18.6	3.31	1 560	149	1 800	235	0.872	27.3	0.569	89.3	104
	74	461.3	152.7	9.9	17.0	10.2	407.0	4.49	41.1	32 400	1 010	18.5	3.26	1 410	133	1 620	209	0.87	30.0	0.499	66.6	95.0
	67	457.2	151.9	9.1	15.0	10.2	407.0	5.06	44.7	28 600	878	18.3	3.21	1 250	116	1 440	182	0.867	33.6	0.429	47.5	85.4
	60	454.7	152.9	8.0	13.3	10.2	407.7	5.75	51.0	25 500	794	18.3	3.23	1 120	104	1 280	163	0.869	37.5	0.387	33.6	75.9
	52	449.8	152.4	7.6	10.9	10.2	407.7	6.99	53.6	21 300	645	17.9	3.11	949	84.6	1 090	133	0.859	43.9	0.311	21.3	66.5
406×178	74	412.8	179.7	9.7	16.0	10.2	360.5	5.62	37.2	27 300	1 540	17.0	4.03	1 320	172	1 500	267	0.881	27.6	0.608	63.0	95.0
	67	409.4	178.8	8.8	14.3	10.2	360.5	6.25	41.0	24 300	1 360	16.9	4.00	1 190	153	1 350	237	0.88	30.5	0.533	46.0	85.5
	60	406.4	177.8	7.8	12.8	10.2	360.5	6.95	46.2	21 500	1 200	16.8	4.00	1 060	135	1 190	208	0.88	33.9	0.464	32.9	76.0
	54	402.6	177.6	7.6	10.9	10.2	360.5	8.15	47.4	18 600	1 020	16.5	3.97	925	114	1 050	177	0.872	38.5	0.39	22.7	68.4
406×140	46	402.3	142.4	6.9	11.2	10.2	359.7	6.36	52.1	15 600	539	16.3	3.85	778	75.7	888	118	0.87	38.8	0.206	19.2	59.0
	39	397.3	141.8	6.3	8.6	10.2	359.7	8.24	57.1	12 500	411	15.9	3.02	627	58.0	721	91.1	0.859	47.4	0.155	10.6	49.4
356×171	67	364.0	173.2	9.1	15.7	10.2	312.3	5.52	34.3	19 500	1 360	15.1	3.99	1 070	157	1 210	243	0.887	24.4	0.413	55.5	85.4
	57	358.6	172.1	8.0	13.0	10.2	312.3	6.62	39.0	16 100	1 110	14.9	3.92	896	129	1 010	199	0.884	28.9	0.331	33.1	72.2
	51	355.5	171.5	7.3	11.5	10.2	312.3	7.46	42.8	14 200	968	14.8	3.87	796	113	895	174	0.882	32.2	0.286	23.6	64.6
	45	352.0	171.0	6.9	9.7	10.2	312.3	8.81	45.3	12 100	812	14.6	3.78	687	95.0	774	147	0.875	36.9	0.238	15.7	57.0
356×127	39	352.8	126.0	6.5	10.7	10.2	311.2	5.89	47.9	10 100	357	14.3	2.69	572	56.6	654	88.7	0.872	35.3	0.104	14.9	49.4
	33	348.5	125.4	5.9	8.5	10.2	311.2	7.38	52.7	8 200	280	14.0	2.59	471	44.7	540	70.2	0.864	42.2	0.081	8.68	41.8
305×165	54	310.9	166.8	7.7	13.7	8.9	265.7	6.09	34.5	11 700	1 060	13.1	3.94	753	127	845	195	0.89	23.7	0.234	34.5	68.4
	46	307.1	165.7	6.7	11.8	8.9	265.7	7.02	39.7	9 950	897	13.0	3.90	648	108	723	166	0.89	27.2	0.196	22.3	58.9
	40	303.8	165.1	6.1	10.2	8.9	265.7	8.09	43.6	8 520	763	12.9	3.85	561	92.4	624	141	0.888	31.1	0.164	14.7	51.5
305×127	48	310.4	125.2	8.9	14.0	8.9	264.6	4.47	29.7	9 500	460	12.5	2.75	612	73.5	706	116	0.874	23.3	0.101	31.4	60.8
	42	306.6	124.3	8.0	12.1	8.9	264.6	5.14	33.1	8 140	388	12.4	2.70	531	62.5	610	98.2	0.872	26.5	0.0842	21.0	53.2
	37	303.8	123.5	7.2	10.7	8.9	264.6	5.77	36.7	7 160	337	12.3	2.67	472	54.6	540	85.7	0.871	29.6	0.0724	14.9	47.5
305×102	33	312.7	102.4	6.6	10.8	7.6	275.9	4.74	41.8	6 490	193	12.5	2.15	415	37.8	480	59.8	0.866	31.7	0.0441	12.1	41.8
	28	308.9	101.9	6.1	8.9	7.6	275.9	5.72	45.2	5 420	157	12.2	2.08	351	30.8	407	48.9	0.858	37.0	0.0353	7.63	36.3
	25	304.8	101.6	5.8	6.8	7.6	275.9	7.47	47.6	4 390	120	11.8	1.96	288	23.6	338	38.0	0.844	43.8	0.0266	4.65	31.4
254×146	43	259.6	147.3	7.3	12.7	7.6	218.9	5.80	30.0	6 560	677	10.9	3.51	505	92.0	568	141	0.889	21.1	0.103	24.1	55.1
	37	256.0	146.4	6.4	10.9	7.6	218.9	6.72	34.2	5 560	571	10.8	3.47	434	78.1	485	120	0.889	24.3	0.0858	15.5	47.5
	31	251.5	146.1	6.1	8.6	7.6	218.9	8.49	35.9	4 440	449	10.5	3.35	353	61.5	396	94.5	0.879	29.4	0.0662	8.73	40.0
254×102	28	260.4	102.1	6.4	10.0	7.6	225.1	5.10	35.2	4 010	178	10.5	2.22	308	34.9	353	54.8	0.873	27.5	0.0279	9.64	36.2
	25	257.0	101.9	6.1	8.4	7.6	225.1	6.07	36.9	3 410	148	10.3	2.14	265	29.0	306	45.8	0.864	31.4	0.0228	6.45	32.2
	22	254.0	101.6	5.8	6.8	7.6	225.1	7.47	38.8	2 870	120	10.00	2.05	226	23.6	262	37.5	0.854	35.9	0.0183	4.31	28.4
203×133	30	206.8	133.8	6.3	9.6	7.6	172.3	6.97	27.3	2 890	384	8.72	3.18	279	57.4	313	88.1	0.882	21.5	0.0373	10.2	38.0
	25	203.2	133.4	5.8	7.8	7.6	172.3	8.55	29.7	2 360	310	8.54	3.10	232	46.4	260	71.4	0.876	25.4	0.0295	6.12	32.3
203×102	23	203.2	101.6	5.2	9.3	7.6	169.4	5.46	32.6	2 090	163	8.49	2.37	206	32.1	232	49.5	0.89	25.6	0.0153	6.87	29.0
178×102	19	177.8	101.6	4.7	7.9	7.6	146.8	6.43	31.2	1 360	138	7.49	2.39	153	27.2	171	41.9	0.889	22.6	0.00998	4.37	24.2
152×89	16	152.4	88.9	4.6	7.7	7.6	121.8	5.77	26.5	838	90.4	6.40	2.10	110	20.3	124	31.4	0.889	19.5	0.00473	3.61	20.5
127×76	13	127.0	76.2	4.2	7.6	7.6	96.6	5.01	23.0	477	56.2	5.33	1.83	75.1	14.7	85	22.7	0.893	16.2	0.002	2.92	16.8

429

Table B2 Dimensions and properties of steel universal columns (structural sections to BS 4: Part 1 and BS 4848: Part 4)

Serial size (mm)	Mass per metre (kg)	Depth of section D (mm)	Width of section B (mm)	Web t (mm)	Flange T (mm)	Root radius r (mm)	Depth between fillets d (mm)	Flange b/T	Web d/t	Second moment of area Axis x-x (cm⁴)	Axis y-y (cm⁴)	Radius of gyration Axis x-x (cm)	Axis y-y (cm)	Elastic modulus Axis x-x (cm³)	Axis y-y (cm³)	Plastic modulus Axis x-x (cm³)	Axis y-y (cm³)	Buckling parameter u	Torsional index x	Warping constant H (dm⁶)	Torsional constant J (cm⁴)	Area of section A (cm²)
356×406	634	474.7	424.1	47.6	77.0	15.2	290.2	2.75	6.10	275 000	98 200	18.5	11.0	11 600	4630	14 200	7110	0.843	5.46	38.8	13 700	808
	551	455.7	418.5	42.0	67.5	15.2	290.2	3.10	6.91	227 000	82 700	18.0	10.9	9960	3950	12 100	6060	0.841	6.05	31.1	9240	702
	467	436.6	412.4	35.9	58.0	15.2	290.2	3.56	8.08	183 000	67 900	17.5	10.7	8390	3290	10 000	5040	0.839	6.86	24.3	5820	595
	393	419.1	407.0	30.6	49.2	15.2	290.2	4.14	9.48	147 000	55 400	17.1	10.5	7000	2720	8230	4160	0.837	7.86	19.0	3550	501
	340	406.4	403.0	26.5	42.9	15.2	290.2	4.70	11.0	122 000	46 800	16.8	10.4	6030	2320	6990	3540	0.836	8.85	15.5	2340	433
	287	393.7	399.0	22.6	36.5	15.2	290.2	5.47	12.8	100 000	38 700	16.5	10.3	5080	1940	5820	2950	0.835	10.2	12.3	1440	366
	235	381.0	395.0	18.5	30.2	15.2	290.2	6.54	15.7	79 100	31 000	16.2	10.2	4150	1570	4690	2380	0.834	12.1	9.54	812	300
COLCORE	477	427.0	424.4	48.0	53.2	15.2	290.2	3.99	6.05	172 000	68 100	16.8	10.6	8080	3210	9700	4980	0.815	6.91	23.8	5700	607
356×368	202	374.7	374.4	16.8	27.0	15.2	290.2	6.93	17.3	66 300	23 600	16.0	9.57	3540	1260	3980	1920	0.844	13.3	7.14	560	258
	177	368.3	372.1	14.5	23.8	15.2	290.2	7.82	20.0	57 200	20 500	15.9	9.52	3100	1100	3460	1670	0.844	15.0	6.07	383	226
	153	362.0	370.2	12.6	20.7	15.2	290.2	8.94	23.0	48 500	17 500	15.8	9.46	2680	944	2960	1430	0.844	17.0	5.09	251	195
	129	355.6	368.3	10.7	17.5	15.2	290.2	10.5	27.1	40 200	14 600	15.6	9.39	2260	790	2480	1200	0.843	19.9	4.16	153	165
305×305	283	365.3	321.8	26.9	44.1	15.2	246.6	3.65	9.17	78 800	24 500	14.8	8.25	4310	1530	5100	2340	0.855	7.65	6.33	2030	360
	240	352.6	317.9	23.0	37.7	15.2	246.6	4.22	10.7	64 200	20 200	14.5	8.14	3640	1270	4250	1950	0.854	8.73	5.01	1270	306
	198	339.9	314.1	19.2	31.4	15.2	246.6	5.00	12.8	50 800	16 200	14.2	8.02	2990	1030	3440	1580	0.854	10.2	3.86	734	252
	158	327.2	310.6	15.7	25.0	15.2	246.6	6.21	15.7	38 700	12 500	13.9	7.89	2370	806	2680	1230	0.852	12.5	2.86	379	201
	137	320.5	308.7	13.8	21.7	15.2	246.6	7.11	17.9	32 800	10 700	13.7	7.82	2050	691	2300	1050	0.851	14.1	2.38	250	175
	118	314.5	306.8	11.9	18.7	15.2	246.6	8.20	20.7	27 600	9010	13.6	7.75	1760	587	1950	892	0.851	16.2	1.97	160	150
	97	307.8	304.8	9.9	15.4	15.2	246.6	9.90	24.9	22 200	7270	13.4	7.68	1440	477	1590	723	0.85	19.3	1.55	91.1	123
254×254	167	289.1	264.5	19.2	31.7	12.7	200.3	4.17	10.4	29 900	9800	11.9	6.79	2070	741	2420	1130	0.852	8.49	1.62	625	212
	132	276.4	261.0	15.6	25.3	12.7	200.3	5.16	12.8	22 600	7520	11.6	6.67	1630	576	1870	879	0.85	10.3	1.18	322	169
	107	266.7	258.3	13.0	20.5	12.7	200.3	6.30	15.4	17 500	5900	11.3	6.57	1310	457	1490	695	0.848	12.4	0.894	173	137
	89	260.4	255.9	10.5	17.3	12.7	200.3	7.40	19.1	14 300	4850	11.2	6.52	1100	379	1230	575	0.849	14.4	0.716	104	114
	73	254.0	254.0	8.6	14.2	12.7	200.3	8.94	23.3	11 400	3870	11.1	6.46	894	305	989	462	0.849	17.3	0.557	57.3	92.9
203×203	86	222.3	208.8	13.0	20.5	10.2	160.9	5.09	12.4	9460	3120	9.27	5.32	851	299	979	456	0.85	10.2	0.317	138	110
	71	215.9	206.2	10.3	17.3	10.2	160.9	5.96	15.6	7650	2540	9.16	5.28	708	246	802	374	0.852	11.9	0.25	81.5	91.1
	60	209.6	205.2	9.3	14.2	10.2	160.9	7.23	17.3	6090	2040	8.96	5.19	581	199	652	303	0.847	14.1	0.195	46.6	75.8
	52	206.2	203.9	8.0	12.5	10.2	160.9	8.16	20.1	5260	1770	8.90	5.16	510	174	568	264	0.848	15.8	0.166	32.0	66.4
	46	203.2	203.2	7.3	11.0	10.2	160.9	9.24	22.0	4560	1540	8.81	5.11	449	151	497	230	0.846	17.7	0.142	22.2	58.8
152×152	37	161.8	154.4	8.1	11.5	7.6	123.5	6.71	15.2	2220	709	6.84	3.87	274	91.8	310	140	0.848	13.3	0.04	19.5	47.4
	30	157.5	152.9	6.6	9.4	7.6	123.5	8.13	18.7	1740	558	6.75	3.82	221	73.1	247	111	0.848	16.0	0.0306	10.5	38.2
	23	152.4	152.4	6.1	6.8	7.6	123.5	11.2	20.2	1260	403	6.51	3.68	166	52.9	184	80.9	0.837	20.4	0.0214	4.87	29.8

Buckling resistance of unstiffened webs

The previous version of BS 5950–1 calculated the buckling resistance of unstiffened webs assuming the web behaved as a strut with a slenderness of $2.5d/t$. Comparison with test results suggested that this approach could, in a limited number of cases, lead to unconservative estimates of the web's buckling resistance. It was therefore decided in BS 5950–1:2000 to revise the design approach and base the web's buckling resistance on the well-known theory of plate buckling, which represents the behaviour more realistically compared to the previous assumption of the web acting as a strut.

The buckling resistance, P_X, of a plate is given by

$$P_X = \rho P_{bw}$$

where ρ is a reduction factor based on the effective width concept of representing the inelastic post buckling of plates, and P_{bw} is the bearing capacity. The reduction factor can, approximately, be given by[1]

$$\rho = \frac{0.65}{\lambda_p}$$

where the slenderness of the plate, λ_p, is given by

$$\lambda_p = \sqrt{\frac{P_{bw}}{P_{elastic}}}$$

Representing an unstiffened web as a plate, BS 5950–1:2000 defines the bearing capacity, P_{bw}, as

$$P_{bw} = (b_1 + nk)t p_{yw}$$

$P_{elastic}$ is the elastic buckling load of the web. Assuming that the web is restrained by the flanges of the section and the web behaves as a long plate, the elastic buckling load is given by[2]

$$P_{elastic} = \frac{2\pi E t^3}{3(1 - v^2)d}$$

Substitution gives a slenderness of

$$\lambda_p = 0.659 \sqrt{\frac{(b_1 + nk)d p_{yw}}{E t^2}}$$

The buckling resistance, P_X, can be written as

$$P_X = \frac{0.65}{0.659 \sqrt{\dfrac{(b_1 + nk)d p_{yw}}{E t^2}}} P_{bw}$$

Re-arranging gives

$$P_X = \frac{27.2 t \sqrt{\dfrac{275}{p_{yw}}}}{\sqrt{(b_1 + nk)d}} P_{bw}$$

Comparison of the above equation with available test data highlights the approximations made in the above formulation and led to a reduction of the factor 27.2 by 8%, to 25.0.

Letting

$$\varepsilon = \sqrt{\frac{275}{p_{yw}}}$$

results in

$$P_X = \frac{25\varepsilon t}{\sqrt{(b_1 + nk)d}} P_{bw}$$

which represents the equation given in BS 5950–1:2000 for an unstiffened web.

When the applied load or reaction is less than $0.7d$ from the end of the member, the buckling resistance of an unstiffened web is reduced by the factor

$$\frac{a_e + 0.7d}{1.4d}$$

where $a_e < 0.7d$ and is the distance from the load or reaction to the end of the member.

References

1. Bradford, M. A. *et al. Australian Limit State Design Rules for the Stability of Steel Structures*. First National Structural Engineering Conference. pp 209–216. Melbourne 1987.
2. Timoshenko, S. P. and Gere, J. M. *Theory of Elastic Stability*. McGraw-Hill 1961.

Second moment of area of a composite beam

Deflections of composite beams are normally calculated using the gross value of the second moment of area of the uncracked section, I_g. This appendix derives the formula for I_g given in section 4.10.3.6.

Consider the beam section shown in *Fig. D1* which consists of a concrete slab of effective width, B_e, and depth, D_s, acting compositely with a steel beam of cross-sectional area, A, and overall depth, D.

Assuming the modular ratio is α_e, the transformed area of concrete slab is $(B_e/\alpha_e)D_s$. Taking moments about x–x, the distance between the centroids of the concrete slab and the steel beam $\bar{y}$, is

$$\bar{y} = \frac{A\left(\dfrac{D}{2} + \dfrac{D_s}{2}\right)}{\left(A + \dfrac{B_e D_s}{\alpha_e}\right)} = \frac{\alpha_e A(D_s + D)}{2(\alpha_e A + B_e D_s)} \quad \text{(D1)}$$

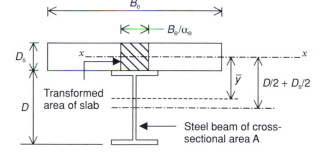

Fig. D1

The second moment of area of the composite section, I_g, is then

$$I_g = I_s + \frac{B_e D_s^3}{12\alpha_e} + A\left(\left(\frac{D_s}{2} + \frac{D}{2}\right) - \bar{y}\right)^2 + \frac{B_e D_s}{\alpha_e}\bar{y}^2 \quad \text{(D2)}$$

where I_s is the second moment of area of the steel section.

Making $(D_s + D)/2$ the subject of equation (D1) and substituting into (D2) gives

$$I_g = I_s + \frac{B_e D_s^3}{12\alpha_e} + A\left(\left(\frac{A + B_e D_s/\alpha_e}{A}\right)\bar{y} - \bar{y}\right)^2$$

$$+ \frac{B_e D_s}{\alpha_e}\bar{y}^2 \quad \text{(D3)}$$

Simplifying and substituting (D1) into (D3) gives

$$I_g = I_s + \frac{B_e D_s^3}{12\alpha_e} + \frac{AB_e^2 D_s^2}{A^2\alpha_e^2}\left(\frac{\alpha_e A(D_s - D)}{2(\alpha_e A + B_e D_s)}\right)^2$$

$$+ \frac{B_e D_s}{\alpha_e}\left(\frac{\alpha_e A(D_s + D)}{2(\alpha_e A + B_e D_s)}\right)^2 \quad \text{(D4)}$$

Collecting terms and simplifying obtains the equation given for the gross value of the second moment of area of the uncracked composite section quoted in *section 4.10.3.6*

$$I_g = I_s + \frac{B_e D_s^3}{12\alpha_e} + \frac{AB_e D_s(D_s + D)^2}{4(\alpha_e A + B_e D_s)} \quad \text{(D5)}$$

References and further reading

References

BRITISH STANDARDS

BS 4:   *Structural Steel Sections*; Part 1: *Specification for Hot-rolled Sections*

BS 18:   *Method of Tensile Testing of Metals Including Aerospace Materials* (now superseded by BS EN10002: Part 1)

BS 449:   *The Use of Structural Steel in Buildings*: Part 2: 1969

BS 639:   *Specification for Covered Carbon and Manganese Steel Electrodes for Manual Metal Arc Welding*

BS 648:   *Schedule of Weights of Building Materials*

BS 709:   *Methods of Destructive Testing Fusion Welded Joints and Weld Metal in Steel*

BS 1243:   *Specification for Metal Ties for Cavity Wall Construction*

BS 1881:   *Methods of Testing Concrete*

BS 3921:   *Specification for Clay Bricks*

BS 4360:   *Specification for Weldable Structural Steels*

BS 4848:   *Specification for Hot-rolled Structural Steel Sections*

BS 4978:   *Specification for Softwood Grades for Structural Use*

BS 5268:   *Structural Use of Timber*; Part 2: *1996 Code of Practice for Permissible Stress Design, Materials and Workmanship*

BS 5400:   *Code of Practice for the Design of Steel, Concrete and Composite Bridges*

BS 5628:   *Code of Practice for Use of Masonry*; Part 1: 1992: *Unreinforced Masonry*

BS 5950:   *Structural Use of Steelwork in Buildings*; Part 1: *2000 Code of Practice for Design in Simple and Continuous Construction: Hot Rolled Sections*

BS 6073:   *Precast Concrete Masonry Units.* Part 1: *Specification for Precast Concrete Masonry Units*

BS 6399:   *Design Loading for Buildings*; Part 1: *Code of Practice for Dead and Imposed Loads*; Part 2: *Code of Practice for Wind Loads*

BS 8007:   *Code of Practice for the Design of Concrete Structures for Retaining Aqueous Liquids*

BS 8110:   *Structural Use of Concrete*;

Part 1:   1997 *Code of Practice for Design and Construction*;

Part 2:   1985 *Code of Practice for Special Circumstances*;

Part 3:   1985 *Design Charts for Singly Reinforced Beams, Doubly Reinforced Beams and Rectangular Columns*

CP 3:   *Code of Basic Data for the Design of Buildings*; Chapter V: Part 2: *Wind Loads*

CP 114:   *Structural Use of Reinforced Concrete in Buildings* (now withdrawn)

EUROCODES AND EUROPEAN STANDARDS

ENV 1991:   Eurocode 1: *Basis of Design and Actions on Structures*

ENV 1992–1–1:   Eurocode 2: *Design of Concrete Structures*; Part 1–1: 1992 *General Rules and Rules for Buildings*

ENV 1993–1–1:   Eurocode 3: *Design of Steel Structures*; Part 1–1: 1992 *General Rules and Rules for Buildings*

ENV 1995–1–1:   Eurocode 5: *Design of Timber Structures*; Part 1–1: 1994 *Common Rules and Rules for Buildings*

ENV 1996–1–1:   Eurocode 6: *Design of Masonry Structures*; Part 1–1: 1995 *General Rules for Buildings. Rules for Reinforced and Unreinforced Masonry*

ENV 206:   *Concrete – Performance, Production, Placing and Compliance Criteria*: 1992

EN 338:   *Structural Timber; Strength Classes*: 1995

prEN 77–1:   *Specification for Masonry Units*; Part 1: 1992 *Clay Masonry Units*

BS EN 10002:   *Tensile Testing of Metallic Materials*; Part 1: *Method of Test at Ambient Temperature*

prEN 10080:   *Steel for the Reinforcement of Concrete Weldable Ribbed Reinforcing Steel B 500*; *Technical Delivery Condition for Bars, Coils and Weldable Fabrics*

GENERAL

British Cement Association *Concise Eurocode for the Design of Concrete Building Structures*

British Cement Association *Shearheads*

British Standards Institution *Extracts from British Standards for Students of Structural Design*

CIB (International Council for Building Research Studies and Documentation) *CIB Report 66*, CIB Structural Timber Design Code, Luxembourg

Concrete Society and Institution of Structural Engineers *Standard Methods of Detailing Structural Concrete*, London

Construction Industry and Research Information Association 'Composite beams and slabs with profiled steel sheating' Report 99

Corus Construction and Industrial *Structural Sections to BS 4: Part 1: 1993 and BS EN 10056*: 1999

Higgins, J. B. and Rogers, B. R. *Design and Detailing (BS 8110: 1997)*, British Cement Association, Crowthorne

Precision Metal Flooring 'Composite Floor Decking Systems' Cheltenham, 2002

Rowe, R. E. *et al. Handbook to BS 8110: 1985: Structural Use of Concrete*, Spon, London

Further Reading

Curtin, W. G., Shaw, G., Beck, J. K. and Bray, W. A. *Structural Masonry Designers' Manual*, BSP Professional Books, Oxford

Institution of Structural Engineers, Institution of Civil Engineers *Manual for the Design of Reinforced Concrete Building Structures*, London

Kermani, A. *Structural Timber Design*, Blackwell Science, London

Mosley, W. H., Bungey, J. H. and Hulse, R. *Reinforced Concrete Design*, Macmillan, London

Nethercot, D. A. Limit state design of structural steelwork, Spon, London

Pask, J. W. *Manual on Connections*, BCSA Ltd, London

Index